CELL BIOLOGY

The Jones and Bartlett Series in Biology

Aquatic Entomology
Patrick McCafferty, Arwin
Provonsha

Basic Genetics, Second Edition
Daniel L. Hartl

Biochemistry
Robert H. Abeles, Perry A. Frey,
William P. Jencks

*Biological Bases of Human Aging
and Disease*
Cary S. Kart, Eileen K. Metress,
Seamus P. Metress

Biology in the Laboratory
Jeffrey A. Hughes

*The Biology of AIDS, Second
Edition*
Hung Fan, Ross F. Conner, and
Luis P. Villarreal

*Cell Biology: Organelle Structure
and Function*
David E. Sadava

*Cells: Principles of Molecular
Structure and Function*
David M. Prescott

*Concepts and Problem Solving in
Basic Genetics: A Study Guide*
Rowland H. Davis, Stephen G.
Weller

*Cross Cultural Perspectives in
Medical Ethics: Readings*
Robert M. Veatch

Early Life
Lynn Margulis

Electron Microscopy
John J. Bozzola, Lonnie D. Russell

Elements of Human Cancer
Geoffrey M. Cooper

*Essentials of Molecular Biology,
Second Edition*
David Freifelder, George M.
Malacinski

Evolution
Monroe W. Strickberger

*Experimental Techniques in Bacterial
Genetics*
Stanley R. Maloy

*From Molecules to Cells: A Lab
Manual*
Nancy Guild, Karen Bever

*Functional Diversity of Plants in the
Sea and on Land*
A.R.O. Chapman

General Genetics
Leon A. Snyder, David Freifelder,
Daniel L. Hartl

Genetics of Populations
Philip W. Hedrick

The Global Environment
Penelope ReVelle, Charles ReVelle

Handbook of Protoctista
Lynn Margulis, John O. Corliss,
Michael Melkonian, and David J.
Chapman, Editors

*The Illustrated Glossary of
Protoctista*
Lynn Margulis, Heather I.
McKhann, and Lorraine
Olendzenski, Editors

*Human Anatomy and Physiology
Coloring Workbook and Study Guide*
Paul D. Anderson

Human Biology
Donald J. Farish

*Human Genetics: A Modern
Synthesis*
Gordon Edlin

*Human Genetics: The Molecular
Revolution*
Edwin H. McConkey

*Introduction to Human Disease,
Third Edition*
Leonard V. Crowley

Living Images
Gene Shih, Richard Kessel

Major Events in the History of Life
J. William Schopf, Editor

*Methods for Cloning and Analysis of
Eukaryotic Genes*
Al Bothwell, George D.
Yancopoulos, Fredrick W. Alt

Microbial Genetics
David Freifelder

Molecular Biology, Second Edition
David Freifelder

Molecular Evolution
E.A. Terzaghi, A.S. Wilkins,
D. Penny

*100 Years Exploring Life, 1888-
1988, The Marine Biological
Laboratory at Woods Hole*
Jane Maienschein

Oncogenes
Geoffrey M. Cooper

*Origins of Life: The Central
Concepts*
David W. Deamer, Gail Raney
Fleischaker

*Plant Nutrition: An Introduction to
Current Concepts*
A. D. M. Glass

Population Biology
Philip W. Hedrick

Protoctista Glossary
Lynn Margulis

Vertebrates: A Laboratory Text
Norman K. Wessels

*Writing a Successful Grant
Application, Second Edition*
Liane Reif-Lehrer

CELL BIOLOGY

Organelle Structure and Function

DAVID E. SADAVA

The Claremont Colleges   Claremont, California

Jones and Bartlett Publishers
Boston   London

Editorial, Sales, and Customer Service Offices

Jones and Bartlett Publishers
One Exeter Plaza
Boston, MA 02116

Jones and Bartlett Publishers International
P O Box 1498
London W6 7RS
England

Library of Congress Cataloging-in-Publication Data
Sadava, David E.
    Cell biology : organelle structure and function / David E. Sadava.
        p. cm.
    Includes bibliographical references and index.
        ISBN 0-86720-228-9
        1. Cell organelles.   2. Eukaryotic cells.   I. Title.
    QH581.2.S2   1992
    574.87′34 — dc20   92-27777
                                                                                                    CIP

Production service: The Book Company
Design: Wendy Calmenson
Typesetting: The Clarinda Company
Cover design: Susan Paradise
Printing and binding: Courier Westford

Cover photo courtesy of Mark Ladinsky, University of Colorado, Health Science Center. *Rat kidney cells stained with flourescent marker for lysosomes.*

Chapter opening photo credits: Chapter 1, T. Agre; Chapter 2, K. Burger; Chapter 3, M. A. Bonneville and K. R. Porter; Chapter 4, L. A. Staehelin; Chapter 5, L. P. Vernon; Chapter 6, T. Ogata; Chapter 7, J. J. Bozzola and L. D. Russell; Chapter 8, E. Flynn; Chapter 9, J. V. Small; Chapter 10, E. D. Allen; Chapter 11, A. K. Christensen; Chapter 12, W. Heneen; Chapter 13, A. Bairati; Chapter 14, J. P. Revel; Appendix 1, G. Edelman; Appendix 2, J. Gross; Appendix 3, K. Brasch

Printed in the United States of America
95 94 93 92   10 9 8 7 6 5 4 3 2 1

BRIEF CONTENTS

CONTENTS

Chapter 3   Mitochondria   91

Chapter 4   Plastids   127

This book is about the biochemistry and molecular biology of eukaryotic cells and their organelles. This has been, and continues to be, one of the most exciting fields of science. A century ago, cell biologists were limited to looking at cells through the microscope. Today they have a powerful array of experimental tools to probe into cell functions. The most remarkable of these tools are from the allied and overlapping fields of biochemistry, genetics, and molecular biology. The insights revealed by the use of these methods have provided important advances in both basic knowledge of how cells work and applications for human betterment.

Four major themes permeate this book. The first is the relationship between biological **structure** and **function.** The elegance and economy of cell components specifically arranged to perform their roles are shown by such examples as the locations of electron carriers in energy-transducing membranes and the microtubules and associated proteins in cilia and flagella. Common structure-function mechanisms, such as membrane transport, unite and relate the different parts of the cell to each other.

The second theme is the **unity** and yet **diversity** in nature, and both plant and animal cells are given equal treatment in this book to reflect this. Many cellular functions are well illustrated by both cell types, and the specialized functions in plants are generally integrated into the text.

The third theme is an emphasis not only on what we know but on how we know it. The **experimental approaches** of modern cell biology include the use of biochemical techniques for direct analysis of cell composition, genetic and inhibitor studies to show cause and effect, and the powerful methods of molecular biology to derive sequence information as well as to probe function through mutagenesis. An appendix describes these experimental approaches in more detail.

The fourth theme comes from my belief that far from being a purely academic pursuit, cell biology is an **applied science,** with its discoveries quickly used in the two major areas of applied biology—medicine and agriculture. Examples ranging from mitochondrial diseases to cell interactions in crop plant reproduction not only make the relevant basic science more interesting but also reinforce the idea that the study of cells provides limitless opportunities for human benefit.

This book is aimed at readers who have studied biology and chemistry at the introductory university level. It is not a molecular biology textbook. Absent are detailed discussions of DNA synthesis and gene regulation. Nor is it a biochemistry text. Absent are extensive discussions of molecular structure, enzymology, and intermediary metabolism. To do justice to these two huge areas of knowledge (which I also teach, in addition to cell biology) would require two additional books. However, the reader will perceive that these two areas permeate the text. The outlines of biochemistry and molecular biology necessary to an understanding of this book are included as an appendix.

To write this book, I have extensively reviewed literally thousands of papers from the original literature of many journals in basic and applied research. The book represents my synthesis of this information, and specific references are given only for cited data. The references at the ends of the chapters are mostly to the review articles and books that I think best give the reader an entrance to the literature of the topic. The study of cell organelles is an exciting and huge undertaking, with new discoveries being made and reported almost daily. The outstanding achievements that led to the knowledge outlined in this book are, as is the case in much of science, only the bases of future research. In some instances, I try to point out what I think these future directions may be.

The book begins with the cell membrane, since it is the boundary between a cell and its environment and it delineates many subcellular organelles. Subsequent chapters then discuss the major membrane-bound organelles. The two important energy-producing structures, mitochondria and plastids, are described in Chapters 3 and 4. These organelles have their own DNA and some autonomy, and these remarkable systems are discussed in Chapter 5. Discussion of the internal membrane system of the endoplasmic reticulum (Chapter 6) and Golgi (Chapter 7) is followed by a description of specialized hydrolytic compartments, lysosomes, vacuoles, and microbodies (Chapter 8).

Chapter 9 deals with the various filaments that make up the cytoskeleton and give cells structure as well as participate in internal and external cell movements. Discussion of the structure of the nucleus and its boundary envelope (Chapter 10) is followed by a description of the major intranuclear structure, the nucleolus, and the ribosomes whose assembly it directs (Chapter 11). Chapter 12 describes the roles of the nucleus and other organelles in the cell division cycle. Finally, two chapters examine the structure (Chapter 13) and function (Chapter 14) of the extracellular matrix that surrounds most cells.

I am grateful to many colleagues from around the world who shared insights from their work in discussions with me and supplied some of the illustrations in this book. Many scientific reviewers provided frank and helpful advice on the manuscript. Finally, my students used successive versions of this book in manuscript and gave me many useful critical comments on both writing and science.

On the publishing side, Art Bartlett had the confidence in this project to see it to fruition. His consistent enthusiasm for it kept things moving along during the inevitable ups and downs of the publishing process. While reading and synthesizing research papers was intellectually stimulating for me, I did not look forward to the detailed process of converting the manuscript into a book. George and Wendy Calmenson at The Book Company in Oregon, and Joni Hopkins McDonald at the main office in Boston, made this process not only painless for me, but interesting and even pleasurable. Finally, I thank Herb Brown, who gave me the initial push to write this book.

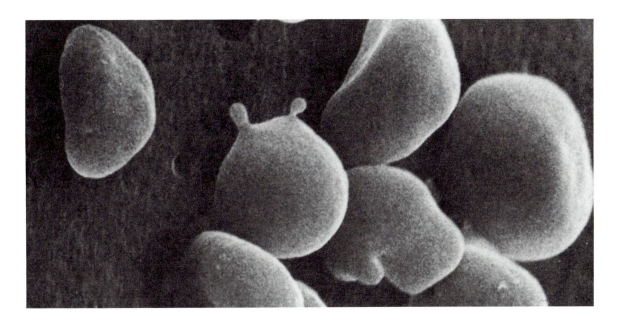

Cell Membrane Structure

Cell Boundary

The invention of the light microscope in the 17th century was the key event that led to the recognition of the cell and its internal structure. When R. Hooke looked at slices of cork tissue under microscope, he saw a honeycombed arrangement and termed the individual units cells. But it was not until the 1800s that cell biology emerged as a scientific discipline. M. J. Schleiden, a botanist, and T. Schwann, a zoologist, almost simultaneously concluded that all living tissues are composed of individual cellular units. In 1858, the same year that Darwin published On the Origin of Species, *R. Virchow formally stated the* **cell theory,** *that cells are the basic building blocks of life that make up the complex structures of organisms.*

If each cell had its own identity, then there must be a boundary separating one cell from its neighbors and from the environment. Although the nature of this boundary was unknown, it could be visualized when tissues were stained with colored dyes, and from the 1850s onward, microscopists included this barrier in their drawings.

Some History

The chemical nature of the cell boundary could be inferred from its properties. During the 1890s, E. Overton performed an elegant series of experiments, first on plant root-hair cells and later on red blood cells. He used these tissues as finely tuned osmometers, observing the rate at which a solute (followed by water) entered the cell. This permitted an estimation of the relative permeability of cells—and their boundaries—to over 500 molecules. Two trends were apparent: First, small molecules (e.g., alcohols) entered cells much more readily than large ones (e.g., proteins). Second, the rate of entry was directly proportional to the molecules' solubilities in organic (nonpolar) solvents and inversely proportional to their solubilities in water. From these experiments came the concept of a relatively solid, water-insoluble (nonpolar) barrier.

As biochemists began to characterize the molecules making up various organisms, it became apparent that lipids were by far the most common nonpolar components. Measurements of the physical properties of the cell boundary (e.g., electrical resistance and capacitance) were within the range of those of isolated lipids. I. Langmuir investigated the surface properties of lipid films and concluded that when a lipid containing polar groups was placed on water, it became oriented so that the polar "head" projected into the water and the nonpolar "tail" (hydrocarbon) projected away from the water. Careful application of force across the film's boundaries compressed it to a monomolecular, tightly packed structure whose area could be measured.

E. Gorter and F. Grendel then combined Langmuir's technique with microscopy. First, they measured the surface area of a red blood cell under the microscope. Then, they extracted the lipids from a defined number of cells and measured their surface area. Remarkably, the lipid surface area was twice that needed to cover the cell surface. Therefore, they proposed an idea that has dominated thinking about the lipid structure of membranes—a **biomolecular lipid layer.** It is ironic that their experimental data cannot be replicated with modern methods: Their lipid extraction was incomplete, and their crude microscopy underestimated the red blood cell surface area. Thus, two experimental errors that cancelled each other out led to a historic (and correct) conclusion.

The idea that a membrane is made up only of lipid had to be revised when surface tension measurements were made during the early 1930s. When a starfish egg was placed between two coverslips, the force needed to compress it was significantly less than that expected for a lipid-only membrane. Also, experiments with lipid films showed that if the film also contained protein, its surface tension lowered to the value seen in cells. Taking all of the experimental data into account, H. Davson and J. Danielli in 1935 proposed that the membrane is a bimolecular lipid layer, with the polar ends of the lipids facing outward and hydrophilic ("water-loving") proteins coating these polar ends.

This "sandwich model" was attractive not only because of its simplicity but also because it accounted for the physical properties of the membrane. Electron microscopy of plasma membranes revealed a three-layered structure (Figure 1–1)—two 2.5-nm thick dense lines separated by a clear 4-nm space. This trilaminate pattern was seen because the heavy metal salts used in staining bound to the outer, hydrophilic regions but not to the inner, hydrophobic ("water-fearing") regions. That the length of two hydrophobic lipid chains laid end to end is 4 nm did not escape the notice of microscopists.

Careful examination of the preservation of the outer and inner stained regions under different conditions of fixation led to the proposal that there can be differences in the chemical compositions of the outer and inner surfaces. This led J. D. Robertson to propose in 1957 that all membranes are based on a "unit" with a lipid bilayer sandwiched between proteins and different proteins on the exterior and interior surfaces (Figure 1–2).

Advances in electron microscopy and biophysical chemistry during the 1960s led to a revision of the unit membrane model. Freeze-

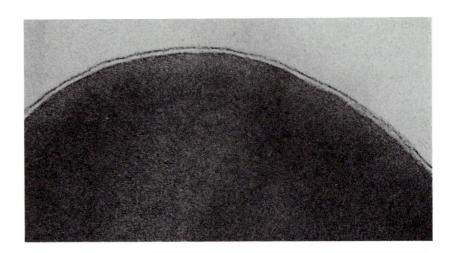

FIGURE 1–1

The trilaminate structure of the erythrocyte cell membrane. Note the "railroad track" appearance, with two densely staining lines. (Photograph courtesy of Dr. J. David Robertson.)

fracturing (see Chapter 4) allowed microscopists to visualize particles in the interior (hydrophobic) region of the membrane. These particles were probably proteins, an indication that these molecules were not only on the exterior and interior surfaces. Sophisticated physical chemical techniques, such as magnetic circular dichroism, showed that membrane proteins were globular, with both hydrophilic and hydrophobic regions, as opposed to the flat, hydrophilic proteins assumed in the model. Several physical chemical techniques applied to the membrane lipids showed them to be a fluid state. Finally, elegant studies demonstrated that proteins can move within the plane of the membrane. These four lines of evidence led J. Singer and G. Nicolson to propose, in 1972, the currently accepted **fluid mosaic** model for cell membrane

FIGURE 1–2

The unit membrane model in cross section.

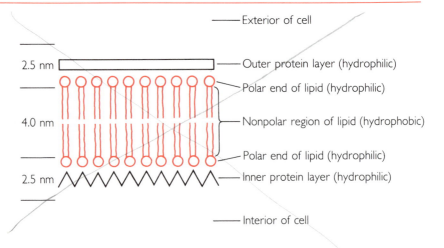

structure. This model has dominated scientific thinking about membranes.

Fluid Mosaic Model

The fluid mosaic model (Figure 1–3) builds on the asymmetric bilayer proposed in the unit membrane. There are two types of proteins: **extrinsic** (peripheral) proteins, which coat much of the outer and inner surfaces, and **intrinsic** (integral) proteins, many of which can insert into the hydrophobic part of the membrane. Some intrinsic proteins (termed transmembrane proteins) span the entire membrane, projecting out to the exterior and into the cell interior, whereas others are embedded entirely in the lipid. Proteins on the exterior surface, as well as certain lipids, have short carbohydrate chains (oligosaccharides). Both the lipids and proteins exist in the hydrophobic region in a fluid state and can move through the plane of the membrane as well as perform occasional "flip-flops" between the membrane surfaces. The protein and lipid are not covalently linked together; however, ionic charges probably hold extrinsic proteins to the hydrophilic lipid surface or to intrinsic proteins.

   Studies of the chemical composition and molecular organization of cellular membranes generally require their **isolation,** intact and in pure form, but this has been a difficult task. For example, until the recent advances in isolating cell-wall–free protoplasts, data on plant plasma membranes were unreliable because of contamination from the cell wall. Because membranes often are a minor proportion of the mass of a cell, substantial purification is necessary. This is usually achieved by the use of density gradient centrifugation (see Appendix I for descriptions of this

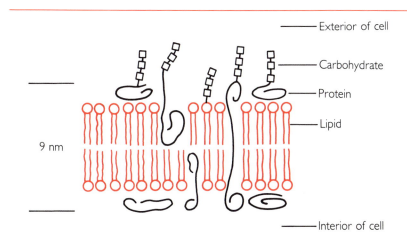

FIGURE 1–3

The fluid mosaic model of cell membrane structure in cross section. Compare the locations of the proteins with those in Figure 1–2.

TABLE I-I   **Enzyme Markers for Cell Components**

Cell Component	Marker
Plasma membrane	Na^+-K^+ ATPase
Nucleus	NAD pyrophosphorylase
Golgi	Glycoprotein glycosyl transferases
Lysosome	Acid phosphatase
Endoplasmic reticulum	Arylesterase
Mitochondria	Succinate dehydrogenase
Chloroplast	Ribulose-biphosphate carboxylase

and other methods cited in the text), since different cell components have different densities.

The purity of a membrane fraction can be monitored by the use of specific enzymatic markers unique for each cell constituent (Table 1–1). But even if a membrane fraction contains only its enzyme marker and appears homogeneous with the electron microscope, it is possible that extrinsic proteins have been lost, or intrinsic proteins have been rearranged, or lipids have been exchanged with others in the cell during the isolation procedures. It is not surprising, therefore, that a well-studied membrane (see Chapter 2) is the mammalian erythrocyte plasma membrane, since this cell has only the one membrane.

Recently, the characterization of membrane proteins has been greatly aided by **molecular biology** techniques. Instead of determining the amino acid sequence of a membrane protein directly, its gene (DNA) can be isolated and cloned in a suitable host cell. Sequencing of the gene leads to a proposed sequence for the protein. If the gene is expressed in a suitable host cell, the protein can be isolated in adequate amounts for study.

Whereas the unit membrane model stressed the unity of the asymmetric lipid bilayer, the fluid mosaic model has focused attention on the diversity of membranes. The overall compositions of different membranes (Table 1–2) reflect this diversity. Although all are composed of lipid and protein (the exterior surface carbohydrates do not exist as separate molecules but covalently attach to the much larger main components), the proportion varies.

Myelin is the membrane that has the highest lipid content. This membrane surrounds axons of nerve fibers in many highly ordered parallel layers. Its high lipid content probably allows it to function as an insulator for the electrical energy of the neuron, preventing its dissipation of the energy to the outside medium. At the other extreme is the overall composition of the inner mitochondrial membrane. Its high pro-

TABLE 1-2 Overall Compositions of Membranes

Source	Percentage Dry Weight	
	Lipid	Protein
Rat liver		
Plasma membrane	38	62
Nuclear	29	71
Golgi	60	40
Lysosomes	25	75
Endoplasmic reticulum (rough)	20	80
Endoplasmic reticulum (smooth)	60	40
Mitochondria (outer)	40	60
Mitochondria (inner)	20	80
Bovine myelin	78	22
Human erythrocyte	40	60
Tomato cultured cell	41	59
Spinach chloroplast	52	48

tein content reflects its function in electron transport and energy transduction, with the electron carriers and enzymes involved in these processes arranged as integral proteins in a highly organized fashion (see Figure 3–14).

Plasma membranes are typically 40% lipid and 60% protein by weight. Because proteins are much larger than lipids, this means that there are about 25 lipid molecules per protein molecule. Given the three-dimensional nature of these molecules, there are only a few molecules of lipid between the proteins in most cases.

The following discussion of these molecules focuses on the plasma membrane largely because it is the membrane that has been most intensively studied. Organelle membranes will be considered in later chapters.

Membrane Lipids

Structure

Most of the major lipids in cell membranes are built on the framework of the three-carbon sugar alcohol, glycerol (Figure 1–4). In addition, the sterol, cholesterol, is often present (Figure 1–5). Membrane lipid composition varies from cell to cell and from organelle to organelle (Table 1–3).

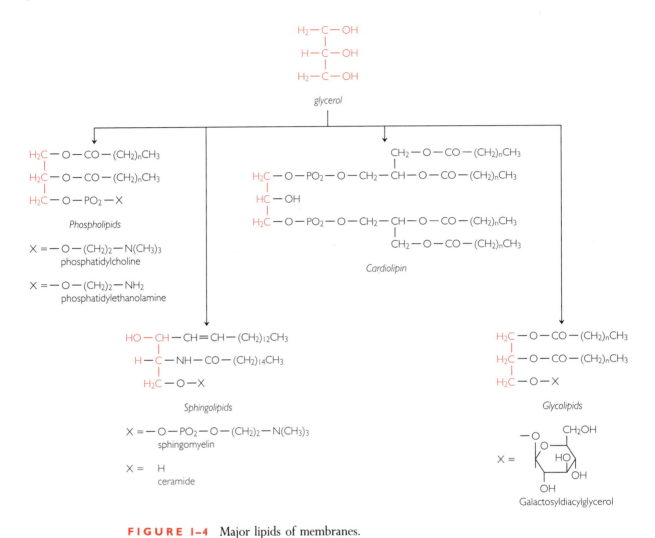

FIGURE I–4 Major lipids of membranes.

Fatty acids usually attach to two of the three carbon atoms of glyceride lipids. Only 2% of the fatty acids in mammalian cells have an odd number of carbon atoms; the majority have 16, 18, or 20 carbon atoms. Often, one or two chains in a lipid may have unsaturated bonds between carbon atoms, with the majority of these occurring in the 18- and 20-carbon–long chains. Thus, the major fatty acids in membranes are:

Palmitic	$C_{16:0}$	Linoleic	$C_{18:2}$
Stearic	$C_{18:0}$	Linolenic	$C_{18:3}$
Oleic	$C_{18:1}$	Arachidonic	$C_{20:4}$

(Note that the number after the colon shows the number of unsaturated carbon-carbon bonds in the molecule: For example, $C_{18:1}$ means there are 18 carbon atoms in the chain with one double bond in the molecule.)

Neutral lipids, or triglycerides, are a major form in which fatty acids are transported between tissues. These lipids contain three fatty acids, which acylate the three hydroxyl groups of glycerol. As their name indicates, they are relatively uncharged. Although triglycerides are rare in most plasma membranes, they do occur in some malignant tumor cell membranes where they appear to form small vesicles within the lipid domain of the membrane. When normal cells are induced to become malignant in culture, these vesicles are formed and can make up to 5% of the total membrane lipid. Their function is not known.

Phospholipids are by far the most abundant membrane lipids. Here, two of the carbon atoms of glycerol are acylated with bulky, long-chain fatty acids, while the third has a much smaller, polar, phosphate-containing moiety. The polarity of this group derives from the phosphate (negatively charged) and, often, the presence of an amine group (positively charged). Thus, phospholipids exhibit distinct hydrophilic (polar) and hydrophobic (nonpolar) regions.

Sphingolipids contain a long hydrocarbon chain attached to one glycerol carbon atom, a second hydrocarbon chain attached to the second glycerol carbon via an aminoacyl linkage, and a third, usually polar, small group on the remaining carbon. In sphingomyelin, the third group

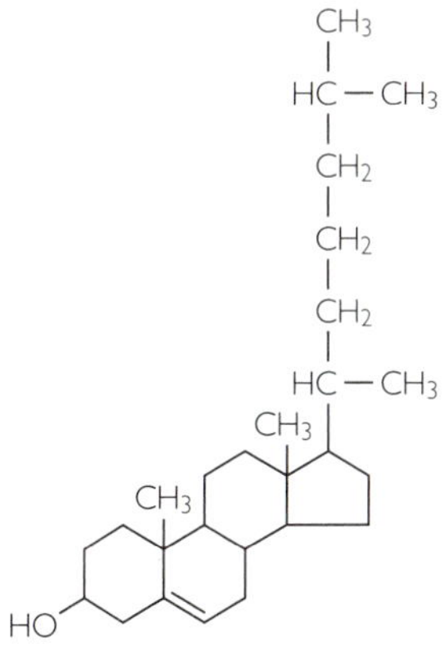

FIGURE 1–5
Cholesterol. The hydroxyl group can be acylated with a fatty acid to form a cholesterol ester.

TABLE 1–3 Lipid Compositions of Membranes

Source	Percentage of Lipid Dry Weight				
	Phospholipids	Sphingolipids	Cardiolipids	Glycolipids	Cholesterol
Rat liver					
Plasma membrane	70				21
Nuclear envelope	85	3			10
Golgi	69	10			8
Lysosomes	47	24	5		14
Endoplasmic reticulum (rough)	81	6			6
Endoplasmic reticulum (smooth)	76	10	2		10
Mitochondria (outer)	83	5	4		
Mitochondria (inner)	79	3	18		
Bovine myelin	42	7		22	17
Human erythrocyte	70	9			9
Spinach chloroplast	12			55	

is phosphorylcholine, making the lipid an analogue of the most abundant phospholipid.

Cardiolipids are essentially disphosphatidyl glycerols. They are especially abundant in plant membranes but much less so in animal tissues. Cardiolipin, originally isolated from heart mitochondria, has two of its glycerol carbons bonded to the diphosphatidyl group. But in plants, the group usually occurs only on one of the three carbons of the glycerol skeleton.

Glycolipids, as the name implies, contain sugar residues covalently attached to the lipid. These can derive from either glycerolipids (shown in Figure 1–4) or sphingolipids. The mono- and digalactose-substituted glycerolipids are the most abundant species in chloroplast membranes. In animals, glycosphingolipids account for a minor fraction of the membrane (except in myelin where galactosylceramide is abundant). The carbohydrate chains may have up to 60 sugar units, as in the blood group antigens.

Cholesterol occurs most abundantly in animal plasma membranes, and other, related, sterols occur in plants. It is incorporated into the membrane noncovalently, and its flat ring structure interdigitates between long hydrocarbon chains.

Fluidity

A lipid bilayer (or intact membrane) changes its physical structure at a certain temperature. When cooled below this temperature, it is in a highly ordered, gel phase. When heated above the temperature, it undergoes a transition to a more disordered liquid crystalline phase. These abrupt organizational changes as the temperature rises can be observed in several physical parameters: The volume expands, the area per molecule expands, molecular motions increase (see below), and the membrane becomes more soluble in nonpolar solvents. The fact that cell membranes exist at temperatures above that of their phase transition implies that this fluid state is essential for membrane function. Many studies have borne this out.

The **phase transition temperature** can be measured in a calorimeter (a closed vessel in which the heat absorbed by the lipids can be measured as the external temperature rises). For pure lipids, this transition typically occurs over a 1–3°C range. But membranes are made up of complex lipid mixtures, and so it is not surprising that their transitions typically occur over a 20–30°C span.

Three lipid compositional factors influence the phase transition temperature of a membrane. The first is fatty acid **chain length:** Longer-chain fatty acids have a higher transition temperature (Table 1–4). This factor is relatively unimportant in membranes, since most of the fatty acids are of similar length.

TABLE 1-4 Thermal Phase Transition
Temperatures for Diphosphatidylcholines

Fatty Acids in Phospholipid	Transition Temperature (°C)
Both $C_{14:0}$	24
Both $C_{16:0}$	41
Both $C_{18:0}$	58
Both $C_{22:0}$	75
One $C_{18:0}$ and one $C_{18:1}$	26
Both $C_{18:1}$	5

ADAPTED FROM: Jain, M., and Wagner, R. *Introduction to biological membranes.* New York: Wiley, 1988.

The second factor is the degree of **unsaturation** of the fatty acids: Lipids containing more unsaturated fatty acids have considerably lower transition temperatures than their saturated counterparts (Table 1–4). The introduction of double bonds changes the conformation of the fatty acid chains. Because unsaturated chains have kinks of 30° angles, they are less likely to be as tightly packed together than are the saturated straight chains, and thus it takes less energy to induce disorder in an unsaturated lipid.

The overall degree of unsaturation of membrane phospholipids can be a misleading indicator of fluidity. For example, an artificial membrane made up of pure phosphatidylcholine with one $C_{18:0}$ and one $C_{18:1}$ fatty acid has a phase transition temperature of 26°C. However, a membrane made up of half phosphatidylcholine with two $C_{18:0}$ fatty acids and half with two $C_{18:1}$ fatty acids has two transition temperatures—one at 35–55°C and the other at 5°C. Note that both of these membranes have overall unsaturated fatty acid contents of 50%.

Fatty acid saturation can be of key importance in the thermal tolerance of poikilotherms, organisms whose body temperature is that of the environment. These organisms have two methods to ensure a fluid membrane: Either live above the phase transition temperature of their lipids or change their lipid composition to, for example, greater unsaturation at lower temperatures. Both mechanisms occur in certain animals and microbes. For example, catfish, which typically live in water below 15°C, have membrane phase transition temperatures of less than 5°C. Certain plants and bacteria can survive the winter, in part, by varying their lipid phase transition temperatures by 20°C.

Hibernating mammals such as ground squirrels and bats face a similar challenge to the poikilotherms. During hibernation, the body temperatures of these normally homeothermic animals may be lowered to less than 10°C. Studies of these animals show that there are changes in their membrane lipid content that could result in a more fluid state at this lower temperature. Typically, the degree of lipid unsaturation is increased, and this effectively lowers the phase transition temperature by over 20°C.

The third factor influencing the phase transition temperature is **cholesterol** content: Higher contents of cholesterol can lead to higher or lower transition temperatures, depending on the ratio of saturated to unsaturated fatty acids. If the ratio is high, cholesterol lowers the temperature; if the ratio is low, cholesterol raises the temperature. This can be important in the plasma membranes of cells lining the blood vessels. If the cholesterol content of these cells is high because of hereditary or dietary influences (see Chapter 2), membranes may become less fluid. This change may be involved in the pathology of atherosclerosis, the disease involving hardening of the arteries. The reason for the stabilizing effect of cholesterol on fluidity is not clear. Perhaps the sterol prevents fatty acid chains from interacting with one another.

Above the transition temperature, lipids are in a fluid state. Kinetically, this means that they are in motion. Three types of lipid motion have been shown by physical-chemical techniques: rotation, diffusion, and transversion.

Rotational motion occurs when a lipid rotates along its longitudinal axis, perpendicular to the plane of the membrane. Both electron paramagnetic resonance (EPR) spectroscopy and fluorescence spectroscopy have been used to demonstrate this phenomenon. In the first technique, a "reporter" group with an unpaired electron (Figure 1–6) is covalently attached to a fatty acid, which is then incorporated into a membrane. The unpaired electron is a magnetic dipole, and its orientation relative to neighboring dipoles in the hydrophobic environment can be measured. Spectra of phospholipids with the label further from the polar end (more toward the interior of the membrane) show an increasingly changing orientation of the label, indicating constant motion. At the hydrophilic end, motion is restricted.

Similar results have been obtained using fluorescence spectroscopy, in which polarized light is shone on a reporter attached to a fatty acid (Figure 1–6). The group fluoresces in the plane of the incoming light if it is immobile. But rotational mobility, especially in the hydrophobic region of the membrane, reduces the signal considerably. Typically, lower signals are obtained when the reporter is attached at the distal, nonpolar end of the lipid. Once again, this indicates that rotation is more rapid in the hydrophobic interior of the membrane and less so near the membrane surface.

It is interesting that reporters for EPR or fluorescence that are attached to lipids adjoining intrinsic membrane proteins show little, if any, mobility. These "boundary lipids," probably a single layer, associate

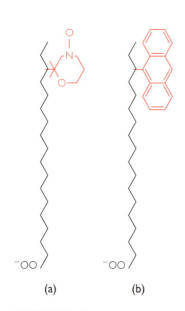

(a)　　　　　(b)

FIGURE 1–6

Reporter groups used for **(a)** electron paramagnetic resonance and **(b)** fluorescence spectroscopy of membrane lipids. The groups are shown attached to the hydrophobic end of the fatty acid, distal to the polar carboxyl group.

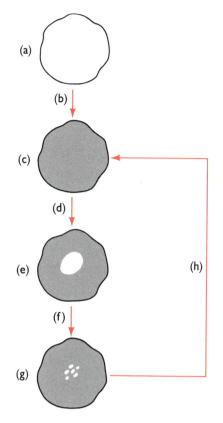

A cell (a) is incubated in fluorescently labeled lipids (b) to label its entire surface (c).

A laser microbeam is shone on a region of the cell (d) to bleach the lipids in this region (e).

Over time (f), labeled lipids from adjacent membrane regions diffuse into the bleached region (g), until a uniform fluorescence is restored (h).

FIGURE 1–7

Fluorescence microphotolysis to demonstrate membrane fluidity.

noncovalently but rigidly with the proteins and are the exceptions to the rule of rotational motion by lipids.

Diffusion occurs when lipids move in the plane of the membrane. Fluorescence microphotolysis (also called fluorescent recovery after photobleaching [FRAP]) has been the major technique used to show diffusion (Figure 1–7). Lipid is again labeled with a fluorescent probe and incorporated into a cell's membrane. A small area of the cell surface (several square micrometers) is irradiated with very high-intensity laser light, which bleaches the label on those lipids. When the light intensity is reduced, a fluorescence microscope can be used to follow the locations of the bleached and unbleached molecules on the cell surface, since all other regions have unbleached label and will fluoresce. If unbleached molecules diffuse into the formerly bleached region, fluorescence will increase, and the rate of this increase is proportional to the lipid diffusion rate. This rate is calculated to be $1-50$ μm/sec, indicating that a lipid molecule can diffuse completely around a cell in a second!

Transversion, more descriptively termed "flip-flop" motion, involves the movement of a lipid from one half of a bilayer to the other.

The familiar observation that oil (nonpolar) and water (polar) do not mix leads to the conclusion that transversion is thermodynamically difficult. Chemically, transversion involves interaction of the polar region on the membrane with its nonpolar interior and, probably, the interaction of the nonpolar lipid hydrocarbon tails with the hydrophilic exterior. Indeed, the free energy required for this process is greater than 20 Kcal/mole, making it an unlikely event. (This can be compared to the familiar reaction of ATP synthesis from ADP and inorganic phosphate, which requires about 8 Kcal/mole.)

Studies using phospholipid vesicles as artificial membranes have shown that spin-labeled phospholipids have half-times for transverse motion in the order of days to weeks. (Half-time is the time needed for half of a group of atoms or molecules to complete a specific process. In this case, half of the spin-labeled lipids would have undergone transversion in days to weeks.) Addition of certain membrane proteins somewhat accelerates this phenomenon, as in immature red blood cell membranes with half-lives for transversion of 8 to 27 hours. Cholesterol moves from one layer to another quite rapidly, with a half-time of under 1 minute in the red blood cell membrane.

Since Overton's observation that alcohols can penetrate the plasma membrane according to how hydrophobic they are, the idea that **local anesthetics** act by altering membrane properties has been investigated. Studies on chloroform, for example, show that it increases lipid fluidity in model and intact membranes, and this increase somehow leads to an alteration in permeability to a cation (potassium), which in turn affects cell electrical potentials. This would adversely affect the functioning of the nervous system. A number of anesthetics have been tested, first for their lipid solubility, then on model membranes, and finally on animals. The general relationship of hydrophobicity → cation permeability disruption → anesthesia holds true in many instances.

Another agent that tends to increase membrane fluidity is **ethanol.** This observation has led to the hypothesis that this change in membrane structure leads to changes in permeability that in turn produce the intoxicating effects of ethanol on the nervous system. Experiments on two genetically different lines of mice support this idea. One line falls asleep readily when exposed to a minimal dose of ethanol, and this is a dose that also makes the membranes more fluid. In contrast, another line of mice stays awake at the same ethanol dose and has membranes whose fluidity is apparently unchanged. The erythrocyte membranes of people with alcoholism also tend to be relatively unaffected by moderate doses of ethanol, as shown by physical measurements of fluidity. Although the exact connection between ethanol-induced membrane fluidity and intoxication is unclear, implications of this theory for public health warrant its intensive investigation.

Asymmetry

The asymmetric localization of lipids in the membrane does not necessarily follow from the fluid mosaic model. Glycolipids, certainly, are on
the outer layer so that their antigenic determinants (e.g., blood cell
groups) face the environment. But for the bulk of the lipids, there is
no reason to postulate asymmetry unless certain lipids are preferentially
associated with certain proteins. Nevertheless, M. Bretscher did
so in 1972, and the evidence gathered since then has confirmed his
hypothesis.

Experiments localizing specific lipids on one side of the membrane
or the other have depended on the use of impermeant probes (Figure
1–8). These probes can covalently bind to membrane lipids but do not
permeate the membrane. Thus, if the probe binds to a lipid species A,
that species is accessible to the outside of the membrane. Now, if the
membrane is made leaky (by bursting the cell and incompletely resealing
the membrane into a vesicle), the probe can now bind to lipid species on
both sides of the membrane. If no further species A is bound, this shows
that lipid A is entirely on the outer surface. Similar experiments have
been done using various lipases, which hydrolyze lipids. By analysis of
membrane lipids before and after lipase digestion of intact and leaky
membrane vesicles, lipid species' locations can be deduced.

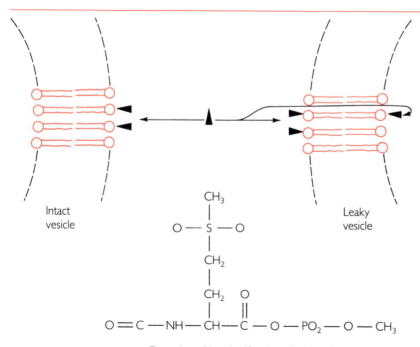

Intact
vesicle

Leaky
vesicle

Formyl-methionyl-sulfonyl-methylphosphate

FIGURE 1–8

Use of an impermeant probe to
label a lipid species in intact and
leaky membrane vesicles. In this
case, two-thirds of the species is
in the outer lipid leaflet. The
structure of an impermeant
probe is shown.

TABLE I-5　Asymmetric Distributions of Membrane Lipids

Membrane	Lipids Preferential to:	
	Outer Surface	**Inner Surface**
Human erythrocyte	Cholesterol	Phosphatidylserine
	Phosphatidylcholine	Phosphatidylserine
	Sphingomyelin	
Bovine rod	Phosphatidylserine	Phosphatidylcholine
Influenza virus in bovine kidney cells	Phosphatidylinositol	Phosphatidylserine
		Sphingomyelin
Inner mitochondria	Phosphatidylcholine	Cardiolipin
	Phosphatidylserine	
Endoplasmic reticulum	Phosphatidylcholine	Phosphatidylserine
	Sphingomyelin	
Intestinal epithelial plasma membrane	Glycosphingolipid	Phosphatidylcholine

Data from various sources.

The results of these experiments for several plasma and organelle membranes show considerable lipid asymmetry (Table 1–5). In many cases, the reasons for this asymmetry are not clear, but in at least one instance the asymmetry has definite biological implications. Platelets, like erythrocytes, have most of the phosphatidylserine (with its negatively charged phosphate group) in the inner plasma membrane leaflet. This has been shown by the use of impermeant probes. When platelets bind to collagen exposed on the wall of the blood vessel endothelium because of an injury, they adhere to the wall in clumps, temporarily stopping the flow of blood. At the same time, the platelets provide a catalytic surface on which the reactions of fibrin clot formation occur, and this more effectively stops blood flow. This is termed procoagulant activity (Figure 1–9).

Two key reactions in the cascade that forms fibrin require a membrane with negatively charged phospholipids on its surface. These reactions are the conversion of factor X to a protease, Xa, and the subsequent conversion of prothrombin to thrombin. When platelets bind to collagen, transversion of phosphatidylserine to the outer leaflet occurs, possibly mediated by detachment from the submembrane cytoskeleton (see Chapter 2). This provides the necessary surface for the two cascade reactions to take place. Then, the phosphatidylserine is returned to the inner leaflet of the bilayer via a translocase.

The same transversion phenomenon occurs in the hereditary blood disease sickle cell anemia. When an erythrocyte with this aberrant he-

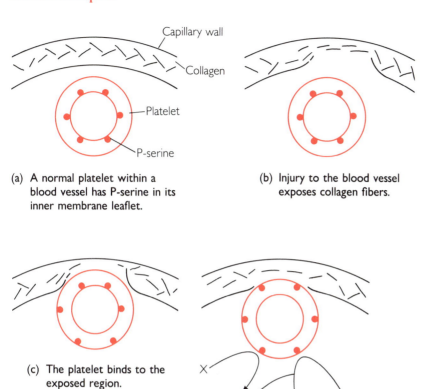

(a) A normal platelet within a blood vessel has P-serine in its inner membrane leaflet.

(b) Injury to the blood vessel exposes collagen fibers.

(c) The platelet binds to the exposed region.

(d) This causes P-serine to transverse to the outer membrane leaflet, where it is involved in the formation of the blood clot.

FIGURE 1–9

Transversion of phosphatidyl-serine during platelet activation. A normal platelet within a blood vessel has P-serine in its inner membrane leaflet **(a).** Injury to the blood vessel exposes collagen fibers **(b),** to which the platelet binds **(c).** This causes the P-serine to transverse to the outer membrane leaflet, where it is involved in the formation of the blood clot **(d).**

moglobin is repeatedly oxygenated and deoxygenated, its plasma membrane forms protrusions, which give the cell a sickled shape. Vesicles bud from these protrusions, and during the process their membranes change their asymmetry. Instead of phosphatidylserine in the inner leaflet, it is transversed to the outer leaflet. Thus the vesicles now have procoagulant activity as they travel through the bloodstream. When they adhere to capillary walls (in the kidney, for example), this activity leads to clot formation, and blood flow is blocked. This results in the clinical symptoms of a sickle cell crisis.

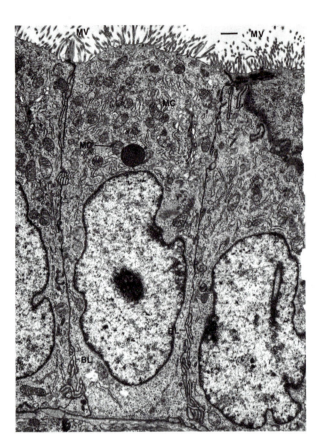

FIGURE 1–10

Polarized epithelial cells in the rat endometrium. There are three plasma membrane domains: The apical domain has microvilli (MV); the lateral domain has specialized intercellular junctions; the basal domain sits on a basal lamina. BL: basolateral; MC: mitochondrion. (×8280. Photograph courtesy of Dr. H. Bender. From: *J Ultrastr Res* 101 (1988):139. Reprinted by permission of Academic Press.)

Another example of genetic involvement in transversion is certain patients with a severe bleeding disorder. Examination of the patients' clotting cascade shows that the two membrane-catalyzed reactions mentioned above scarcely occur and that this deficiency is due to a reduction in phosphatidylserine transversion to the outer leaflet of the platelet plasma membrane.

Another kind of asymmetry is seen in cells where different regions in the plane of the membrane have different lipid compositions. The polarized mammalian epithelial cell has two extracellular faces with quite different environments. The apical part of the cell faces the lumen, for example, of the endometrium (Figure 1–10) or gut, and the basal part faces the underlying extracellular matrix (see Chapter 13) and blood supply. These two regions of the cell surface have very different membrane lipid (and protein—see below) compositions. The apical membrane is rich in glycosphingolipids and relatively poor in phospholipids, while this situation is reversed in the basal membrane, whose composition is typical of those of other cells in the gut. The two different regions are termed **membrane domains.** The glycolipid is mostly in the outer leaflet of the apical region, where its role is to protect the membrane from the harmful effects of the ions and other factors in the lumen. This is apparently achieved by the formation of hydrogen bonds between the fatty acids and sphingosine base, leading to a more stable structure.

Membrane Proteins

Structure

In the fluid mosaic model, there are two classes of membrane proteins: extrinsic proteins and intrinsic proteins.

Extrinsic (peripheral) proteins attach to the outer and inner membrane surfaces. These proteins can be detached from the surface by solutions with a high ionic strength or by chelation of divalent cations, rather mild chemical environments often encountered during membrane isolation. The possibility exists that true extrinsic proteins are lost and cytoplasmic contaminants adsorb during isolation.

After detachment from the membrane, these proteins are water-soluble and free of lipid. Because they contain a high (over 70%) proportion of hydrophilic amino acids, extrinsic proteins probably bind to the hydrophilic regions of the lipid bilayer ionically. Alternatively, they may bind to polar parts of intrinsic proteins. Three proteins, all lying on the cytoplasmic side of cellular membranes, are well-studied examples of extrinsic proteins: spectrin in erythrocytes, clathrin in coated vesicles, and the fatty acid desaturase complex containing cytochrome b_5 in endoplasmic reticulum. These will be discussed in later chapters.

Intrinsic (integral) proteins are embedded in the fatty acid hydrophobic core of the membrane. They can be removed from the membrane only by rather severe measures, such as detergents or organic solvents. Once removed, intrinsic proteins are water-insoluble and must be in a nonaqueous or detergent environment. The amino acid composition of these proteins is usually high (over 40%) in hydrophobic side chains (Table 1–6). This makes these proteins most stable in a nonaqueous environment.

Each intrinsic protein has at least one domain that is composed of hydrophobic amino acids and is embedded in the membrane lipid (Figure 1–11). This region adopts a typical alpha-helical conformation. In some cases, such as cytochrome b_5 of the endoplasmic reticulum (Chapter 6), the embedded sequence does not go through the membrane to the other surface. These are termed **monotopic** proteins. But in most intrinsic proteins, the hydrophobic region is connected to two more hy-

TABLE I-6 **Hydrophobicity of Amino Acid Side Chains**

Side chain	ΔG (Kcal/mole)*
Tryptophan (trp)	−3.4
Phenylalanine (phe)	−2.5
Tyrosine (tyr)	−2.3
Leucine (leu)	−1.8
Valine (val)	−1.5
Methionine (met)	−1.3
Alanine (ala)	−0.5
(−CO, NH−)	+4.1

*ΔG refers to the free energy change when the side chain is transferred from water to ethanol at 25°C.

FROM: Jain, M., and Wagner, R. *Introduction to biological membranes.* New York: Wiley, 1988.

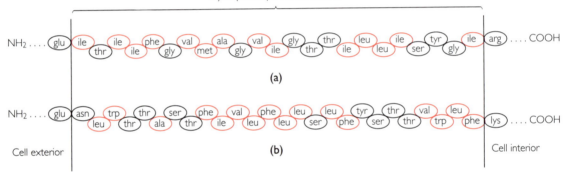

FIGURE I-11 The amino acid sequences of the part of glycophorin from erythrocytes **(a)** and immunoglobulin M **(b)** that spans the plasma membrane. Note the abundance of hydrophobic residues.

drophilic sequences that project out from the membrane surfaces (Figure 1–12). Such proteins span the lipid bilayer and can be classified on the basis of how many such spanning regions the protein contains.

Bitopic proteins span the membrane only once. Examples include some hydrolases in intestinal epithelial cells, the spike glycoproteins of enveloped viruses, and red blood cell glycophorin (see Chapter 2). In the hydrolases, most of the enzyme molecule extends out from the cell exterior face of the membrane, with a hydrophobic "tail" anchoring the protein to the bilayer. The extracellular portion of this molecule is fully catalytic, while the only role of the hydrophobic part is to insert into the

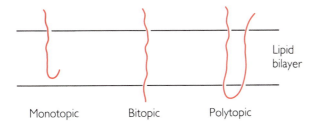

Monotopic   Bitopic   Polytopic

Lipid bilayer

FIGURE 1–12

Three types of intrinsic membrane protein topology are (1) monotopic—embedded in the lipid bilayer; (2) bitopic—traverses the bilayer once; (3) polytopic—traverses the bilayer more than once.

membrane. This arrangement suits the enzymes' function in breaking down proteins and carbohydrates in the intestine.

Certain animal viruses (e.g., Semliki Forest virus, human immunodeficiency virus [HIV]) contain an RNA-protein core surrounded by a membrane whose major proteins, the spike proteins, are coded for by the viral genome. These proteins extend out from the viral exterior surface and are anchored to the membrane by a hydrophobic sequence. Once again, the arrangement is functional, since the spike proteins allow the virus to bind to the host cell membrane during infection. For example, in HIV, the protruding spike protein gp160 is the key to recognition, and hence infection, of white blood cells.

Many intrinsic proteins are **polytopic,** with multiple membrane-spanning domains. These include many of the proteins that are involved in the transduction of extracellular signals across the membrane to the interior of the cell (see Chapter 2). A typical polytopic protein is the glucose transporter in the human erythrocyte membrane (Figure 1–13). Of the approximately 500 amino acids in the single polypeptide chain, about half comprise 12 membrane-spanning, hydrophobic regions, each 21 amino acids long.

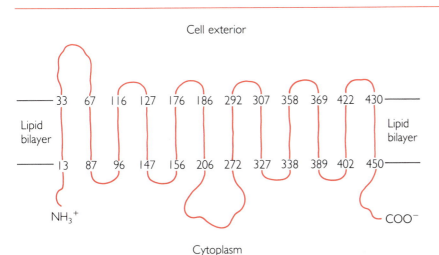

Cell exterior

Lipid bilayer   33   67   116   127   176   186   292   307   358   369   422   430   Lipid bilayer

13   87   96   147   156   206   272   327   338   389   402   450

NH_3^+ COO^-

Cytoplasm

FIGURE 1–13

Diagram of the membrane-spanning structure of the human erythrocyte glucose transporter. There are 12 domains that span the bilayer. Numbers refer to position of amino acid residues in the polypeptide chain.

The orientation of the hydrophobic region (that is, which end faces the cytoplasm and which end faces the exterior) is not random. The polar amino acids that surround it determine which way it faces. Specifically, the region that flanks the membrane-spanning domain on the cytoplasmic side usually has positively charged amino acids such as lysine and arginine, with few of these on the extracellular domain.

Although most intrinsic proteins are directly embedded in the hydrophobic lipid leaflet, some lack a hydrophobic insertion sequence and are attached instead to a lipid that is the membrane anchor. The carboxyl terminus (amino acid at the end of the protein that still has a free $-COO^-$ group) covalently attaches through ethanolamine to a complex phospholipid that contains the cyclic sugar inositol. The latter is connected to a glycerol backbone with two fatty acids, and these embed hydrophobically in the membrane (Figure 1–14). These proteins are called **glycolipoproteins.**

Attachment of the lipid anchor apparently occurs in the endoplasmic reticulum while the protein is being made on ribosomes bound to this organelle (see Chapter 6). Over 40 proteins having this anchor have been identified, including many surface hydrolases such as acetylcholinesterase in nervous tissues, lymphoid antigens such as the Thy-1 antigen of mice, intercellular adhesion molecules such as neural cell adhesion molecule (NCAM) (see Chapter 14), and several coat proteins from Protozoa. A particularly interesting case is carcinoembryonic antigen (CEA), which occurs normally in many types of epithelial cells. When these cells become cancerous, CEA appears in the blood serum, and its presence there is used as a marker for the presence of certain types of tumors. This serum CEA is correlated with the disappearance of CEA from the tumor cell membranes, probably by degradation of the glycosyl-phosphatidylinositol membrane anchor.

Glycoproteins, like glycolipids, occur only on the exterior-facing part of the lipid bilayer. Indeed, most cell surface proteins are glycoproteins. Whereas animal cell membranes are relatively lightly glycosylated (up to 3%), plant plasma membranes contain about 20% sugars by weight, most of this as glycoproteins. The reason for protein glycosylation is not clear; recognition of other cells or extracellular substances plays a prominent role in theories of its function.

The sugar contents of these proteins range from a single residue to oligosaccharide chains several hundred units long. Simple sugars (e.g., galactose, mannose, fucose) and amino sugars (N-acetylglucosamine, N-acetylgalactosamine, N-acetylmuramic acid) are linked to the protein by N-glycosidic (attachment of the sugar to the free amino group on asparagine of the protein) or O-glycosidic (attachment of the sugar to a hydroxyl group on serine or threonine) bonds. The hydrophilic, exposed

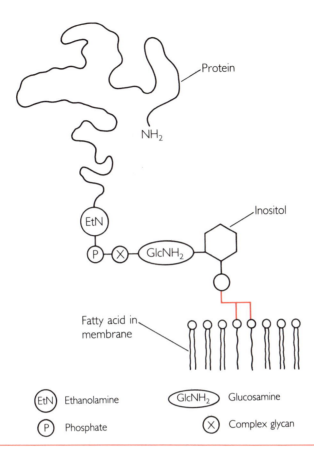

EtN Ethanolamine

P Phosphate

GlcNH₂ Glucosamine

X Complex glycan

FIGURE 1-14
A model for the glycophospho-lipid anchor for a membrane protein.

nature of the sugars allows them to perform important cell recognition functions (see Chapters 2 and 14).

Mobility

The fluid mosaic model predicts that, because the lipid bilayer is fluid and proteins are embedded in it through noncovalent interactions, the proteins should be free to move in the membrane. As is the case for lipids, both rotation and diffusion have been demonstrated.

Rotation of proteins embedded in the hydrophobic lipid is affected by the lipid's viscosity. Because the motions of the proteins in such a medium are quite slow, physical techniques used for proteins in aqueous solutions are inappropriate. For example, the fluorescence of a probe attached to a protein membrane decays long before any significant rotation of the molecule occurs. On the other hand, the histochemical dye eosin has a much longer decay time when excited with polarized light, and eosin-labeled proteins have been used to detect rotation by virtue of changes in the light emitted as the excited dye decays in the membrane. Proteins that are naturally colored with long-lived interme-diates on excitation (e.g., the visual pigment rhodopsin in the vertebrate rod) have also been used to detect rotation. Finally, EPR, using spin-labeled probes, has had some success in certain instances.

All of these studies show considerable rotation of intrinsic proteins, ranging from low rates (of chloroplast ferrodoxin) to higher rates (of the acetylcholine receptor). The rate of rotation is dependent on the state of the lipids: If they are fluid, there is more rotation, whereas below the phase transition temperature, rotation is greatly reduced.

Diffusion of membrane proteins occurs when they move laterally in the plane of the bilayer. This phenomenon has been demonstrated by both physical and biochemical techniques. Fluorescence microphotolysis, used to demonstrate lipid diffusion (see above), has also been used on proteins. Vertebrate rhodopsin was actually the first membrane compo-nent whose diffusion rate was measured in this way. Whereas this pro-tein absorbs excitation energy by itself, other membrane proteins require attachment to a fluorescent probe; in most cases, these have been fluorescein-labeled antibodies or lectins targeted at specific membrane proteins.

The diffusion rates of integral proteins (Table 1–7) are less than those of lipids, as would be expected, since the proteins are larger mol-ecules. But protein diffusion is still rather rapid and within a limited range. Some membrane proteins do not diffuse, indicating physical re-striction on their mobility, whereas others, such as those anchored via phospholipids, diffuse as rapidly as a typical lipid. Most proteins exhibit diffusion rates between the two extremes, indicating both mobility and limited restrictions.

The biochemical method of detecting membrane protein diffusion involves examining, microscopically, the global membrane redistribution

TABLE 1–7 **Diffusion Rates of Integral Membrane Proteins**

Protein: Cell Type	Approx. Diffusion Rate (μm/sec)
Surface proteins: fish fibroblasts	1.54
Rhodopsin: vertebrate rod	1.26
Nerve growth factor receptor: neurites	0.52
Insulin receptor: mouse fibroblast	0.40
Band 3 protein: human erythrocyte	0.35
Acetylcholine receptor: rat myoblasts	0.06

DATA CALCULATED FROM: Lenaz, G. Lipid fluidity and membrane protein dynamics. *Biosci Rep* 7 (1987):826 and other sources. Diffusion coefficients are usually represented in units of cm²/sec but have been linearized here.

of labeled proteins initially concentrated in a restricted part of the membrane. The most influential experiment along these lines was reported by L. Frye and M. Edidin, who used mouse and human tissue culture cells (Figure 1–15).

Mouse membrane surface proteins (histocompatibility antigens) and human membrane surface proteins (those antigenic to a rabbit) were

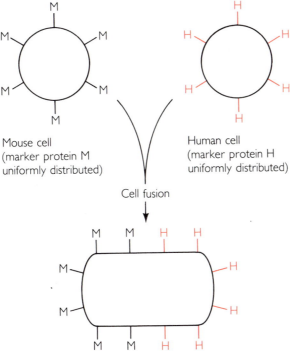

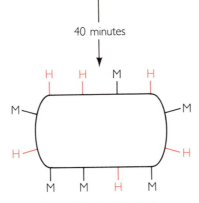

FIGURE 1–15

The Frye-Edidin experiment showing protein diffusion in membranes.

specifically labeled with antibodies that fluoresced green (mouse) or red (human). The two types of cells were fused together in culture and examined under a fluorescence microscope. Initially, green fluorescence was confined to half of the surface of the hybrid cell, and red was confined to the other half. This segregation of membrane proteins did not last long, and by 40 minutes after fusion complete intermixing had occurred. Careful control experiments ruled out all other explanations (e.g., breakdown and resynthesis of the proteins) except diffusion in the plane of the membrane. For example, the intermixing occurred in the presence of antibiotics, which block protein synthesis.

Experiments along similar lines have also been done using proteins whose distribution on a cell membrane is initially concentrated by experimental manipulation (cultured cells' lectin receptors) or tissue morphology prior to isolation (intestinal epithelial hydrolases). Movement of the protein toward a random distribution on the membrane can then be studied. Once again, diffusion occurs at rates similar to those obtained spectroscopically.

If lymphocytes, the white blood cells that produce antibodies, are stained with a fluorescently labeled antibody to their cell surface globulin, uniform staining is initially observed (Figure 1–16). After 15 minutes at room temperature, the stain exhibits a patched appearance, and by 30 minutes, it has migrated, along with its membrane protein target, to a localized cap at one end of the cell.

This aggregation of initially diffuse proteins has also been observed for receptors to lectins on the surfaces of nonimmune system cells and may be important in intercellular interactions (Chapter 14). Inhibitor experiments indicate that while patching is independent of cell metabolism, capping does depend on cellular energy. Technically, these movements are not diffusion, since they involve a movement to, rather than from, greater concentration. But they do show rapid movement of a macromolecular complex in the plane of the membrane. The mechanism of capping may involve an interaction of the membrane protein with cytoplasmic structures such as microtubules and microfilaments (see Chapter 9).

Cytoskeletal interactions with certain membrane proteins are probably important in restricting their movement. Membrane proteins that are normally restricted in their diffusion because of presumed interactions with the cytoskeleton can be made by molecular cloning so that they lack a cytoplasmic "tail." In these cases, their rate of diffusion increases dramatically. Thus cytoplasmic structure can create membrane protein domains, restricting the global distribution of certain proteins.

Epithelial cell hydrolases (see above) are usually confined to the apical region of the cell membrane, next to their targets in the lumen extracellular medium. Many explanations have been offered to account for the apparent lack of diffusion of these proteins through the plane of

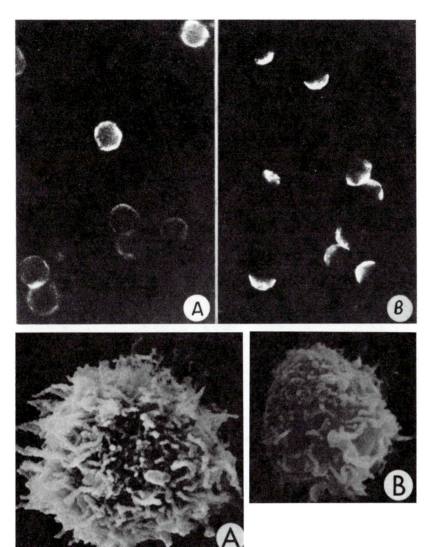

FIGURE I–16

"Cap" formation in white blood cells. *A:* Binding of antibody before capping shows random distribution of a protein antigen over the cell surface. *B:* After capping, the antigen is concentrated at one part of the membrane. *Top:* Fluorescent antibody stain. (Courtesy of Dr. G. Edelman.) *Bottom:* Scanning electron microscopy. (Courtesy of Dr. F. Naeim.)

the membrane. It has been attributed to the cytoskeleton, or to an asymmetric, directed insertion of the proteins into the membrane (see Chapter 6), or to the presence of tight junctions (see Chapter 14), or to membrane domains. A final restrictive factor may be molecular crowding. Studies of immune system cells indicate that the surface proteins are so crowded that they continually bump into each other.

Asymmetry

Protein asymmetry is a major feature of the unit membrane and fluid mosaic models. The methods to demonstrate it have been somewhat similar to those used to show lipid asymmetry. Impermeant labeling reagents that react with amino or sulfhydryl groups to form covalent linkages have been used in intact and leaky vesicles (see Figure 1–8). Subsequent isolation of membrane proteins gives an indication of their bilayer location. Proteolytic enzymes have also been used as impermeant probes: After an intact membrane is incubated with proteases, only those proteins facing the outside will be digested. The latter can be assayed by sizing the proteins on gel electrophoresis. Third, the ability of a membrane protein to bind to a specific marker (antibody, hormone, virus, or enzyme substrate) can be assayed. In this manner, receptors have been localized to the outer membrane surface and certain enzymes to both surfaces.

The picture that emerges from the topological analysis of membrane proteins is one of absolute specificity. Proteins are arranged in the bilayer in a highly invariant manner. Those that protrude from the exterior surface contain carbohydrate. Those on the inner (cytoplasmic) surface are often associated with cytoplasmic structures. Proteins that span the membrane may be ion pumps or transport molecules. Elegant arrangements of functional proteins in membranes of the erythrocyte, chloroplast, and mitochondrion will be described in later chapters.

Membrane Turnover

Membranes are not static structures. There are relatively slow but constant changes as individual components are broken down and replaced or exchanged. These new molecules may or may not be chemically identical to the old ones. In addition to these continuous changes, environmental fluctuations (e.g., temperature) can lead to changes in membrane lipids (e.g., more or less unsaturated fatty acids to maintain fluidity).

Estimates of the rates of **turnover** (loss from the membrane) of lipids and proteins have been made by radioactively labeling individual components and estimating the rate of disappearance of membrane-bound label. These rates (Table 1–8) show that a metabolically inert membrane, such as myelin, is more stable than one with a great deal of activity, such as the mitochondrial inner membrane.

Lipids may be lost from membranes through the action of phospholipid exchange proteins. As their name implies, these molecules can exchange specific phospholipids between membranes or between lipid-carrying proteins (lipoproteins) and membranes. Because they contain the hydrophobicity necessary to carry lipids and insert them into membranes, proteins may be responsible for moving lipids from one half of the bilayer to the other. A specific carrier protein that transfers phosphatidylserine and phosphatidylethanolamine from one leaflet of a lipid bilayer to the other has been characterized. This "flippase," responsible

TABLE 1-8 Half-Lives of Membrane Molecules

Component	Half-Life (days)	
	Myelin	**Mitochondria**
Phosphatidylcholine	41	12
Phosphatidylserine	120	17
Sphingomyelin	>200	33
Cholesterol	>200	39
Proteins	35	21

FROM: Jain, M., and Wagner, R. *Introduction to biological membranes.* New York: Wiley, 1988.

for some lipid asymmetry, requires the hydrolysis of ATP for energy for its activity.

Membrane lipids can be hydrolyzed by a host of lipases specific for each chemical species: phospholipases, sphingomyelinase, ceramidase, and so on. These breakdown enzymes act on lipids in intact membranes rather than on individual lipid molecules; the reactions occur in specialized hydrolytic organelles, the lysosomes. The metabolic importance of membrane turnover is shown by the lipidoses. In these hereditary diseases, patients lack one or other of the lipases, and the result is often severe mental retardation and even early death.

An important aspect of membrane turnover does not involve the loss of lipid components but rather a *reassembly* into different molecules. This "retailoring" of membrane lipids occurs in three steps:

1. A specific lipase (e.g., a phospholipase for a phospholipid) removes a fatty acid from the lipid.

2. The fatty acid is activated by the addition of Coenzyme A.

3. The activated fatty acid is transferred to another lipid (Figure 1–17).

The original lipid may obtain a different fatty acid from the one it lost.

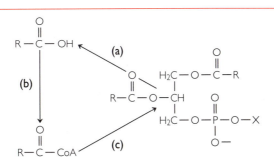

FIGURE 1-17

Three enzymes involved in membrane phospholipid turnover are **(a)** phospholipase A; **(b)** fatty acyl CoA synthetase; **(c)** fatty acyl CoA transferase.

Studies of the green alga *Dunaliella* and the ciliate *Tetrahymena* show that retailoring is a rapid response to environmental changes. When the temperature is reduced, $C_{18:1}$ fatty acids are transferred from phospholipids where they are present along with saturated fatty acids to produce lipids with two unsaturated fatty acids. The result is a more fluid membrane. The key enzyme regulating the net transfer appears to be the phospholipase. In addition, there is an increase in fatty acid desaturase, the enzyme responsible for the generation of double bonds in preexisting fatty acids. Because these enzymes are under hormonal control in vertebrates, rapid lipid retailoring may occur there as well in response to external signals.

Membrane Biogenesis

Membranes are self-assembling structures in the test tube. If a membrane is dissociated into individual lipids and proteins by the use of a detergent, removal of the latter can lead to the spontaneous reassembly of a lipid bilayer with embedded proteins. Although this is a satisfying experiment in terms of the fluid mosaic model, a major problem arises if it is hypothesized that membranes are made by insertion of whole, self-assembled pieces: The spontaneously assembled membrane is not asymmetric by design; its lipids and proteins are arranged in a compositionally random fashion. Because specific distribution of molecular species in the bilayer is an important feature of all membranes, cytoplasmic self-assembly is unlikely. Instead, components are inserted individually into preexisting membranes.

Membrane synthesis is needed to replace components that turn over as well as for the requirements of growth. For example, neurons sending out axons or plant cells expanding because of hormones grow at the rate of μm/min, and this implies considerable assembly of additional plasma membrane. The **lipids** come from two sources: diet (in animals) and synthesis (in all organisms). Feeding experiments using diets radically different in the proportions of saturated and unsaturated fatty acids show that their proportions in membranes remain relatively constant, even to the point of which fatty acids are at the 1- and 2- positions on glycerol of phospholipids. The mechanisms for maintaining lipid constancy are largely unknown.

The assembly of fatty acid chains onto glycerol phosphate is the key step in phospholipid synthesis. The enzyme catalyzing this assembly occurs in the smooth endoplasmic reticulum, generally on the cytoplasmic side of the membrane. In animal cells, cardiolipids are made in the mitochondria, where they are localized (Table 1–3). Cholesterol is synthesized in a multistep pathway in the endoplasmic reticulum. Some of the lipids are then transported to other membrane systems by bulk membrane flow and vesicles. Alternatively, exchange proteins can be used to shuttle the phospholipids to their ultimate destination.

A major problem in the incorporation of lipids into membranes is that of sorting. One example of how this might occur is that of sphingolipids, which are assembled in the Golgi, in the outer half of the membrane, and do not show tranversion. Thus a vesicle budding from the Golgi and fusing with a target membrane will result in sphingolipids in the outer membrane leaflet. For other membrane lipids, which do transverse and can exchange readily with a cytoplasmic pool, the explanation of how membrane lipids are assembled in a specific topographical and compositional fashion is much less clear.

The biosynthesis and assembly of membrane **proteins** have been intensively studied, and like the lipids, many of them are made on the endoplasmic reticulum. Proteins anchored to the inner (cytoplasmic) membrane surface present the least conceptual problems. Extrinsic proteins can be made in the cytoplasm and will bind ionically to the hydrophilic surface. An intrinsic protein has a hydrophobic region by which the molecule is inserted into the lipid and remains there. The proteins that appear on the outer (noncytoplasmic) surface present a conceptual challenge. Simply stated, the problem is: How does the hydrophilic portion of a protein made in the cytoplasm cross through the hydrophobic lipid barrier? This thermodynamic problem exists not only for integral membrane proteins but also for molecules that are destined for other cell organelles (e.g., mitochondria) or the exterior of the cell (e.g., secreted protein hormones).

Although the details of membrane traverse have not been elucidated, a useful hypothesis has emerged. This is the **signal peptide hypothesis** of C. Milstein and G. Blobel (Figure 1–18). Proteins destined for the noncytoplasmic surfaces of membrane or the interior membrane compartments are envisaged to be made initially on free ribosomes, unattached to the endoplasmic reticulum. Following the initial translation of some 20 to 40 amino acids, the synthesizing complex migrates to the surface of the endoplasmic reticulum, where translation of the protein resumes. The first amino acids (about 25) translated by the ribosome are not part of the finished protein but rather are a hydrophobic "signal" sequence.

Signal sequences have been described for a number of membrane proteins, and there are generally three regions, or domains. First, at the amino terminus (the first part of the protein that is made), there are positively charged amino acids. This is followed by a central core of 7 to 15 hydrophobic amino acids and then 3 to 7 more polar amino acids.

FIGURE 1–18

The biosynthesis and insertion of an integral membrane protein.

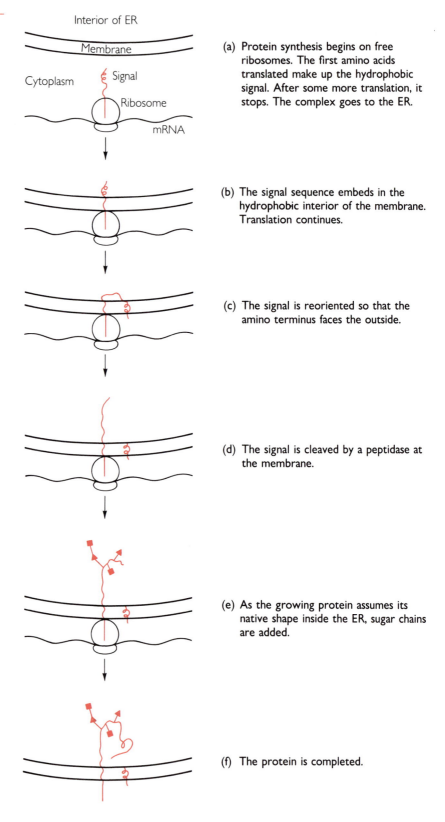

Interior of ER

Membrane

Cytoplasm   Signal

Ribosome

mRNA

(a) Protein synthesis begins on free ribosomes. The first amino acids translated make up the hydrophobic signal. After some more translation, it stops. The complex goes to the ER.

(b) The signal sequence embeds in the hydrophobic interior of the membrane. Translation continues.

(c) The signal is reoriented so that the amino terminus faces the outside.

(d) The signal is cleaved by a peptidase at the membrane.

(e) As the growing protein assumes its native shape inside the ER, sugar chains are added.

(f) The protein is completed.

As the protein proper is made, the signal region leads it into the membrane's hydrophobic region. A translocating machinery then mediates passage of the protein through the hydrophobic membrane bilayer. Once the signal crosses the membrane, it is reinserted so that it faces the outside again. This looping event exposes the carboxyl end of the signal to the active site of specific signal peptidase. This enzyme now cleaves off the signal sequence, as the newly made protein continues to "tunnel" through the membrane.

As the parts of the protein sequence leave the bilayer and arrive in the polar environment outside the membrane, two events occur to ensure that the protein will not reverse its travels and back out of the membrane. First, it folds into its native three-dimensional configuration, making it too wide to go through the narrow "tunnel." Second, it is glycosylated with polar sugars, increasing its charge to make interaction with nonpolar hydrocarbons less likely.

The translocation machinery must be able to distinguish those proteins that are destined for complete passage (e.g., the ones that end up inside the endoplasmic reticulum) from those that stop in the membrane and stay there as integral membrane proteins. This essentially means that it must be able to "see" the hydrophobic stretches that come after the signal and instead of letting them through to the other side of the membrane, transfer them to the lipid bilayer. J. Singer has proposed that the translocator is a transmembrane protein with a central core that is hydrophilic. Sequences destined to pass through and be protein domains facing the cytoplasm of extracellular environment would be allowed to traverse this core. But those sequences that become membrane-spanning regions would be transferred to the lipid environment of the membrane.

Experimental evidence for the signal hypothesis will be described later in the context of protein secretion (see Chapter 6). Suffice it to say that, for a number of membrane proteins, the overall scheme of biosynthesis confirms the hypothesis.

Two types of linkages occur between carbohydrates and proteins or lipids in membranes: N-glycosidic linkages to asparagine and O-glycosidic linkages to threonine and serine. Oligosaccharides are attached via these linkages in different locations and by different mechanisms. Chains with the **N-links** are first assembled and then attached as completed chains to the growing protein chain as it emerges from the membrane (Figure 1–19). The enzyme responsible for this attachment is in the interior of the endoplasmic reticulum, and the carrier for the oligosaccharide chains is dolichol phosphate (Figure 1–19). This molecule has a long hydrocarbon chain by which it can be embedded in the membrane, and its polar phosphate end at which the sugar chain is attached protrudes at the membrane surface.

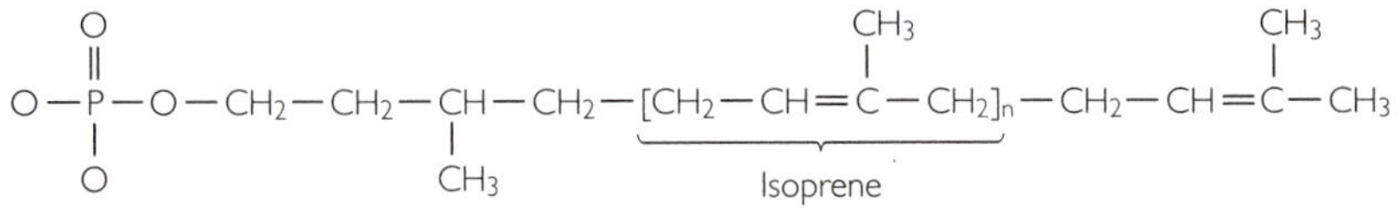

FIGURE I–19

Dolichol phosphate, the molecule that transfers N-linked oligosaccharide chains.

$$O=\overset{\displaystyle O}{\underset{\displaystyle O}{P}}-O-CH_2-CH_2-\underset{CH_3}{CH}-CH_2-\underbrace{[CH_2-CH=\underset{CH_3}{C}-CH_2]_n}_{\text{Isoprene}}-CH_2-CH=\underset{CH_3}{C}-CH_3$$

Chains with **O-links,** on the other hand, are synthesized on the protein one sugar at a time via glycosyl transferases located inside the Golgi complex. Thus the location of these enzymes within the organelles can lead to the asymmetry of membrane glycosylation.

Further Reading

Aloia, R. (Ed.). *Membrane fluidity in biology.* Multivolume series. New York: Academic Press, 1983–.

Aloia, R., and Raison, J. Membrane function in mammalian hibernation. *Biochim Biophys Acta* 988 (1989):123–146.

Bevers, E., Rosing, J., and Zwaal, R. F. A. Platelets and coagulation. In *Platelets in biology and pathology* (pp. 127–160). McIntyre, E., and Gordon, J. (Eds.). Amsterdam: Elsevier, 1987.

Boyd, D., and Beckwith, J. The role of charged amino acids in the localization of secreted and membrane proteins. *Cell* 62 (1990):1031–1033.

Datta, D. B. *Comprehensive introduction to membrane biochemistry.* Madison: Floral Press, 1987.

Dawidowicz, E. A. Dynamics of membrane lipid metabolism and turnover. *Ann Rev Biochem* 56 (1987):43–61.

Devaux, P. Phospholipid flippases. *FEBS Lett* 234 (1988):8–12.

Edidin, M. Rotational and lateral diffusion of membrane proteins and lipids. *Curr Top Membr Trans* 29 (1987):91–124.

Elson, E., Frazier, W., and Glaser, L. (Eds.). *Cell membranes: Methods and reviews.* Multivolume series. New York: Plenum, 1984–.

Ferguson, M. A., and Williams, A. F. Cell surface anchoring of proteins via glycosyl-phosphatidylinositol structures. *Ann Rev Biochem* 57 (1988):285–320.

Finean, J. B., Coleman, R., and Michell, R. *Membranes and their cellular functions.* London: Blackwell, 1984.

Fleischer, S., and Fleischer, B. (Eds.). *Biomembranes.* Multivolume series. New York: Academic Press, 1983–.

Frye, L., and Edidin, M. The rapid intermixing of cell surface antigens after formation of mouse-human heterokaryons. *J Cell Sci* 7 (1970):319–335.

Gennis, R. B. *Biomembranes.* New York: Springer-Verlag, 1990.

Gierasch, L. Signal sequences. *Biochemistry* 28 (1989):923–930.

Harwood, J., and Walton, T. (Eds.). *Plant membranes—Structure, assembly and function.* London: The Biochemical Society, 1988.

Hilderson, H. (Ed.). *Fluorescence studies of biological membranes.* New York: Plenum, 1988.

Hodges, T. K., and Mills, D. Isolation of the plasma membrane. *Meth Enzymol* 118 (1986):41–54.

Jacobson, K., Ishihara, A., and Inman, R. Lateral diffusion of proteins in membranes. *Ann Rev Physiol* 49 (1987):163–175.

Jain, M., and Wagner, R. *Introduction to biological membranes,* 2nd ed. New York: Wiley, 1988.

Jennings, M. Topography of membrane proteins. *Ann Rev Biochem* 58 (1989):999–1027.

Khorana, H. G. Bacteriorhodopsin: A membrane protein that uses light to translate photons. *J Biol Chem* 263 (1988):7439–7442.

Knowles, P., and Marsh, D. Magnetic resonance of membranes. *Biochem J* 274 (1991):625–641.

Lenaz, G. Lipid fluidity and membrane protein dynamics. *Biosci Rep* 7 (1987):823–839.

Lipowsky, R. The conformation of membranes. *Nature* 349 (1991):475–481.

Lisanti, M., Rogriguez-Boulan, E., and Saltiel, A. Emerging functional roles for the glycosyl-phosphatidylinositol membrane protein anchor. *J Membr Biol* 117 (1990):1–10.

Morre, D. J., Howell, K., Cook, G., and Evans, W. (Eds.). *Cell-free analysis of membrane traffic.* New York: Alan R. Liss, 1988.

Mountford, C., and Wright, L. Organization of lipids in plasma membranes of malignant and stimulated cells. *Trends Biochem Sci* 13 (1988):172–177.

Pattus, F. Membrane protein structure. *Curr Op Cell Biol* 2 (1990):681–685.

Pugelsy, A. P. Translocation of proteins with signal sequences across membranes. *Curr Op Cell Biol* 2 (1990):609–616.

Quinn, P. J., Joo, F., and Vigh, L. The role of unsaturated lipids in membrane structure and stability. *Prog Biophys Molec Biol* 53 (1989):71–103.

Robinson, D. G. *Plant membranes.* New York: Wiley, 1985.

Rueckert, D. G., and Schmidt, K. Lipid transfer proteins. *Chem Phys Lipids* 56 (1990):1–20.

Simons, K., and van Meer, G. Lipid sorting in epithelial cells. *Biochemistry* 27 (1988):6197–6202.

Singer, S. J. The structure and insertion of integral proteins in membranes. *Ann Rev Cell Biol* 6 (1990):247–296.

Singer, S. J., and Nicolson, G. The fluid mosaic model of the structure of cell membranes. *Science* 175 (1972):720–731.

Sussman, M., and Harper, J. Molecular biology of the plasma membrane of higher plants. *Plant Cell* 1 (1989):953–960.

Thompson, G. Membrane acclimation by unicellular organisms in response to temperature change. *J Bioenerg Biomembr* 21 (1989):43–60.

VanDeenen, L. Topology and dynamics of phospholipids in membranes. *FEBS Let* 123 (1981):3–13.

Van Meer, G. Lipid traffic in animal cells. *Ann Rev Cell Biol* 5 (1989):247–275.

Von Heijne, G. Transcending the impenetrable: How proteins come to terms with membranes. *Biochim Biophys Acta* 947 (1988):307–333.

Walko, R. M., and Nothnagel, E. A. Lateral diffusion of proteins and lipids in the plasma membrane of rose protoplast. *Protoplasma* 152 (1989):46–56.

Warren, R. C. *Physics and architecture of cell membranes.* Philadelphia: Taylor and Francis, 1987.

Watts, A. Membrane structure and dynamics. *Curr Opin Cell Biol* 1 (1989):691–700.

Wood, W. G., and Schroeder, F. Membrane effects of ethanol. *Life Sci* 43 (1988):467–475.

Yeagle, P. *The membranes of cells.* New York: Academic Press, 1988.

Zwaal, R., Bevers, E., Comfurius, P., Rosing, J., Tilly, R., and Verhallen, P. Loss of membrane phospholipid asymmetry during activation of blood platelets and sickled red cells: Mechanisms and physiological significance. *Mol Cell Biochem* 91 (1989):23–31.

CHAPTER 2

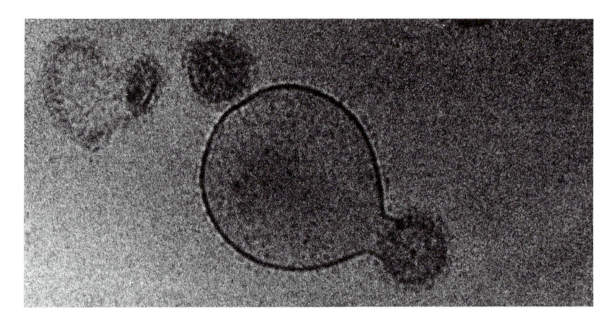

Cell Membrane Function

Cell and Environment

The cell does not exist in isolation. Just outside the plasma membrane is either another cell (for example, in an organ such as the liver) or an aqueous environment (for example, the blood serum surrounding a red blood cell). In both cases, the plasma membrane acts as the boundary that delineates a cell from its environment.

As a boundary, the cell membrane has two major roles. First, it regulates ionic and molecular traffic into and out of the cell; this helps keep the cytoplasm different in its chemical composition from the exterior of the cell. Second, the membrane recognizes important exterior substances and begins the process of transducing the signal of recognition into an appropriate response by the cell; thus, many hormones and similar messengers bind to and affect the cell first at its plasma membrane. As will be shown, the structure of the membrane is well suited to these two vital functions.

Diffusion and Facilitated Transport

Diffusion

To understand how the cell membrane regulates the transport of substances into and out of the cell, it is important to define some chemical terms.

The concentrations of molecules and ions inside a cell often differ significantly from their concentrations in the extracellular medium. These imbalances lead to a tendency of a given chemical species to move from the region where it is highly concentrated to where it is less concentrated, a process termed **diffusion.** In thermodynamic terms, the species tends to randomize its distribution:

$$\Delta G = RT \log (c_2/c_1)$$

where ΔG is the net free energy change, c_2/c_1 is the concentration gradient, R is the gas constant, and T is the absolute temperature.

Prediction of spontaneity of the chemical reaction can be made from the sign of the free energy change. If ΔG is positive (there is an increase in free energy going from reactants to products), there must be some energy input and the reaction by itself is not spontaneous. On the other hand, if ΔG is negative (there is a net decrease in free energy from reactants to products), there is no need for an energy input and the reaction by itself is spontaneous. Relating this to the transport of a substance into the cell and the equation above, if c_2 is greater than c_1 (movement to a region of higher concentration), ΔG will be positive and the movement of the species cannot occur without an input of energy. However, if c_2 is less than c_1 (movement to a region of lower concentration), ΔG is negative (free energy is decreasing) and movement will occur spontaneously.

Osmosis is the diffusion of water across a semipermeable membrane. This process occurs when the concentration of water differs between the cell and its surrounding medium. Rather than refer to the water concentration, it is more convenient to measure its figurative mirror image, the total solute concentration. Thus, water will move from a region of low solute concentration (high in water) to one of high solute concentration (low in water). The force generated by this flow of water is osmotic pressure (π, in atmospheres):

$$\pi = cRT$$

where c is the total solute concentration (molal).

Note that for osmotic pressure purposes, the nature of the chemical species making up c is irrelevant. All that matters for osmotic equilibrium of a cell is that the total concentration of osmotically active substances on both sides of the plasma membrane is the same. For instance, a blood cell is in osmotic equilibrium with its environment, the serum, although the chemical constituents of cell and serum are very different.

If the osmotic pressure of blood serum exceeds that of red cells (for example, due to defective renal clearance of a major solute), water leaves the blood cell and it shrinks. Red cell plasma membranes are somewhat flexible, giving the cell a "doughnut" (biconcave) shape (see Figure 2–29). This is important for the cell, since it swells when it becomes oxygenated and then shrinks when deoxygenated. If the membrane is not flexible, it is more readily damaged and the cell is removed from the circulation. On the other hand, if the osmotic pressure of the serum is less than that of the cell, water will enter the cell and it will balloon in size until it bursts. This occurs in certain hemolytic anemias.

In plant tissues, the situation is rather different (Figure 2–1). The plant cell is contained within a typical plasma membrane, but this membrane is surrounded by a rigid cell wall (see Chapter 13). In a medium of higher osmotic pressure, water moves out of the cell, and the membrane and cytoplasm shrink from the wall. This process of plasmolysis leads to a reduction in the strength of the structure of the plant. For instance, if plants are overfertilized, the osmotic pressure of their medium (the soil water) is much higher than that of the root cells. Water leaves the roots, additional water is transported there to replace the water that left, and the plants wilt.

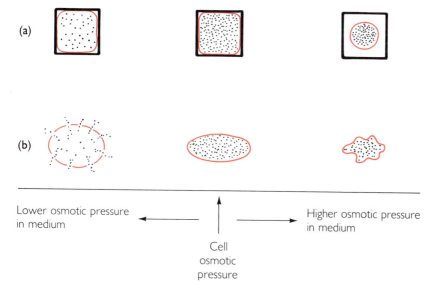

Lower osmotic pressure
in medium

Higher osmotic pressure
in medium

Cell
osmotic
pressure

FIGURE 2–1

Effect of changing medium osmotic pressures on plant root
cells **(a)** and animal erythrocytes
(b). At a lower osmotic pressure
in the medium than in the cells,
water tends to enter the cells.
The plant cell becomes turgid,
while the erythrocyte lyses. At a
higher medium osmotic pressure,
water leaves the cells and they
shrink.

More typically, the osmotic concentration of the soil water is less
than that inside the root cells. In this case, water moves into the cells,
and they tend to expand. This osmotic pressure is counteracted by the
cell wall's rigidity and resistance to expansion. Thus the plant cell does
not swell but becomes more rigid because of its "hydrostatic skeleton."
It must be emphasized that even when a cell is in osmotic equilibrium
with its environment and there is no net flow of water, it still passes
quite readily and rapidly through the plasma membrane. This passage
can be shown if a cell is placed in water whose hydrogen atoms are
radioactive (tritium, a hydrogen isotope). In a medium in osmotic equilibrium, the labeled water enters the cell.

The diffusion of uncharged substances into cells has been intensively studied since the early experiments of Overton. Both animal tissues (e.g., toad bladder, mammalian erythrocyte) and plant cells (e.g.,
large algal cells) have yielded similar results. Two consistent characteristics of this diffusion are shown in Figure 2–2. First, permeability is
directly proportional to a molecule's lipid solubility, usually expressed as
the partition coefficient between oil and water. Second, for molecules of
equivalent polarity, size inversely relates to ability to cross the membrane. This means that large polymers such as proteins and nucleic acids
do not pass directly through the membrane. Situations that reduce
membrane fluidity, such as a temperature below the phase transition or
increased cholesterol content, also reduce diffusion of molecules through
the membrane.

The diffusion of ions involves not only a concentration gradient but
also a gradient of electrical potential:

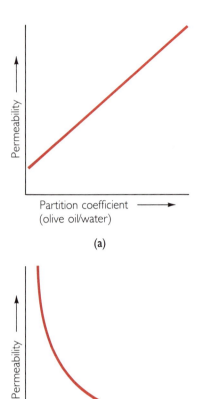

(a)

(b)

FIGURE 2–2

Effects of lipid solubility **(a)** and size **(b)** on molecular permeability through the plasma membrane.

$$\Delta G = RT \log_e (c_2/c_1) + ZF\Delta V$$

where Z is the charge of the ion, F is Faraday's constant, and ΔV is the potential difference across the membrane. If, as is usually the case, ΔV is negative (the cell has an interior net negative charge relative to the medium—see below), ion movements became thermodynamically more likely.

The diffusion rate of ions is highest at the phase transition temperature of the membrane, and this may be a reflection of the structural discontinuity of the membrane at this temperature. For example, channels could be present through which the ions could travel. Because ions are polar, direct movement through the lipid bilayer is unlikely. Instead, this function is performed by intrinsic proteins, which form pores within the lipid bilayer. The diameter of these channels can be inferred by testing the permeabilities of different ions with different hydrated radii. For a wide variety of membranes, diffusion occurs much less readily if the ion is greater than 0.5 nm in diameter (Table 2–1); thus the pore size is about 0.6 nm.

Cells contain many molecules (e.g., proteins, nucleic acids) that are charged but cannot cross the plasma membrane because of their large size. Essentially, these are trapped anions within the cell. When smaller ions diffuse into and out of the cell, the gradient of charge set up by the trapped large ions must be taken into account. A theoretical basis for this phenomenon was worked out by F. Donnan (Figure 2–3). To balance out the trapped negative charges, an equal number of cations must remain inside the cell. If, for example, K^+ stays inside the cell to balance out the macromolecules, more K^+ will be inside than outside of the cell. In addition, the diffusible anions such as Cl^-, will also be imbalanced.

TABLE 2–1 Permeant Diameter and Diffusion Rate

Permeant	Diameter (nm)	Relative Diffusion Rate
Water	0.30	5×10^7
Urea	0.36	4×10^7
Chloride ion	0.39	4×10^7
Potassium ion	0.41	1×10^2
Sodium ion	0.51	1
Lactate ion	0.52	0.9
Glycerol	0.62	0.6
Ribose	0.74	0.5
Glucose	0.86	0.3
Sucrose	1.04	<0.1

ADAPTED FROM: Giese, A. C. *Cell physiology*. Philadelphia: W. B. Saunders, 1973.

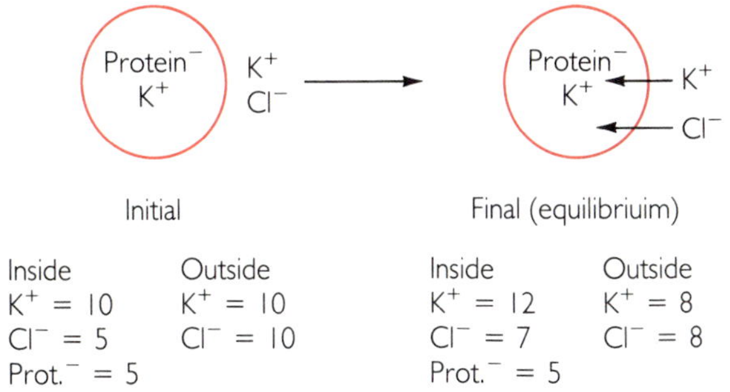
Protein⁻
K⁺
K⁺
Cl⁻
Initial
Inside
K⁺ = 10
Cl⁻ = 5
Prot.⁻ = 5
Outside
K⁺ = 10
Cl⁻ = 10
Protein⁻
K⁺
K⁺
Cl⁻
Final (equilibriuim)
Inside
K⁺ = 12
Cl⁻ = 7
Prot.⁻ = 5
Outside
K⁺ = 8
Cl⁻ = 8

FIGURE 2–3 The Donnan equilibrium. Balancing out protein anions with diffusible ions causes an imbalance in the latter.

Facilitated Transport: Ionophores

Membrane permeability to diffusible cations is quite low, even with the postulated presence of hydrophilic pores, but it can be dramatically increased by the use of ionophores. These naturally occurring and synthetic antibiotics transport ions across membranes. Three classes of ionophores are those that carry metal cations, those that carry protons, and those that form membrane channels (Figure 2–4).

Cation carriers are specific in their ion preferences. For example, valinomycin carries K^+, enniatin A carries Rb^+, and monesin carries

FIGURE 2–4 Structures of ionophores, which assist in the transport of ions across membranes.

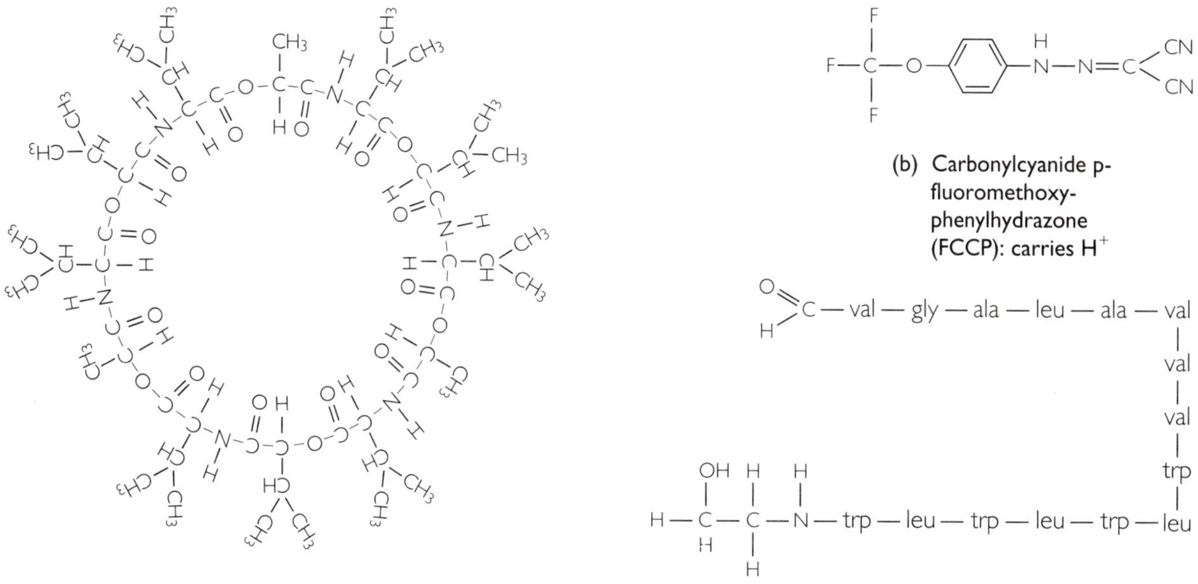
O
H
C — val — gly — ala — leu — ala — val
val
val
trp
OH H H
H — C — C — N — trp — leu — trp — leu — trp — leu
H H

(a) Valinomycin: Carries monovalent cations, especially K^+

(b) Carbonylcyanide p-fluoromethoxy-phenylhydrazone (FCCP): carries H^+

(c) Gramicidin A: forms membrane channels for monovalent cations

Na^+. Because the carriers must interact with both a polar ion and a nonpolar bilayer, they usually have a hydrophobic outer region surrounding a hydrophilic center. In valinomycin, for example, the carbonyl groups face inward after the ion binds (Figure 2–4). The hydrophobic region interacts with membranes much less efficiently when the lipids are highly ordered at temperatures below the phase transition. This shows that the carrier ionophore diffuses through the fluid lipids to transport its ion across the membrane.

Proton carriers are generally lipid-soluble weak acids. Often, the dissociation constant of the acid (pK_a) is the same as the pH of the medium, thus ensuring protonation. As would be expected, proton carrier ability is highly dependent on pH, and, indeed, a difference in pH across the membrane is needed to permit this facilitated transport. Most of these molecules are halo-, nitro-, or oxygenated phenols (Figure 2–4).

Channel formers, such as gramicidin, contain either long hydrocarbon chains or hydrophobic amino acids. In both cases, the chains are long enough for the molecule to span the lipid bilayer at least once to form a pore through which ions can pass. That the pores are charged is indicated by the fact that some channel formers facilitate cation diffusion and others facilitate anion diffusion. Because the ionophore itself is not diffusing through the membrane, fluidity is not important. Thus the action of these molecules is undiminished at temperatures below the phase transition. Channel diameters are estimated to be between 0.4 and 0.7 nm, sufficient space for the passage of many diffusible ions of biological importance.

Ionophores have a wide variety of applied uses. Their major use in biology has been to study processes involving membrane transport. The disruption of electrochemical gradients by these molecules is used as a criterion for deducing the necessity of these gradients in certain biochemical processes. For example, ionophore use was important in establishing the nature of the membrane-mediated gradient used to drive the synthesis of ATP (see Chapters 3 and 4).

A second use is in the organic chemistry laboratory, when it is necessary to dissolve a polar molecule that is highly reactive. The ionophore here acts as a protective coating. Third, an artificial lipid bilayer with an ionophore can act as a sensor-transducer where ionic current is involved. In electronics and information processing, it is often desirable to have channels that can detect ion currents and allow them to pass. Finally, ionophores have potential as drugs. When injected into dogs, the calcium ionophore, X-537A, raises cardiac Ca^{++}, leading to greater heart muscle contractility and increased blood flow. Studies are underway to test its effectiveness in treating heart disease in people.

Facilitated Transport: Gated Channels

During the 1950s, A. Hodgkin and A. Huxley did a series of experiments in which they measured the ionic currents across the membrane of a large nerve cell axon in the squid. They did this by inserting a tiny glass electrode into the cytoplasm while the voltage across the membrane was measured by a second electrode. Initially, they found that the membrane at rest was polarized, with the inside of the axon negatively charged relative to the outside. This phenomenon was due in part to the Donnan effect (Figure 2–3) and largely to the fact that the cations inside the cell did not balance the anions trapped inside. Instead, the cell's interior had a somewhat lower concentration of cations than the exterior (Figure 2–3).

However, when the nerve cell was stimulated, it became depolarized (less negative inside) because Na^+ entered the cell at a greatly increased rate. Then, K^+ left the cell to restore the original polarity. Calculations of diffusion rates needed to bring about these changes showed that the ions were entering the cells at rates far greater than their naturally slow diffusion through the nonpolar membrane. Indeed, these rates were even higher than those of transport via binding to carriers (see below). This led to the hypothesis that excitable membranes have naturally occurring channel formers similar to ionophores. In the axon, it was voltage that determined whether the channel was open, and when it was open the ions could rush across the membrane.

An important technical advance led to the identification of these channels. A major problem with interior electrode experiments to measure ionic permeability is that they must be done on large cells, and only a broad average of many individual events is actually observed. The **patch-clamp** technique overcomes these difficulties (Figure 2–5). A tiny (diameter 1 μm) glass electrode is applied to the surface of a membrane-bound organelle or cell. Slight suction results in a tight seal of a small part of the membrane around the pipet tip. If the membrane is ruptured, it quickly reseals around the tip, leaving a patch within the tip with the rest of the membrane in contact with the outside medium. In either the whole-cell or patch configuration, the electrode can be used to measure the electrical potential difference across the membrane, as well as to control the medium in which the membrane is bathed. Thus the effects of various substances on ion transport can be studied in isolation or combination, and in favorable instances the properties of single channels can be examined.

Not unexpectedly, the hydrophilic channels turn out to be lined with intrinsic membrane proteins that completely span the membrane. However, the channels are not always open. Apparently, the proteins

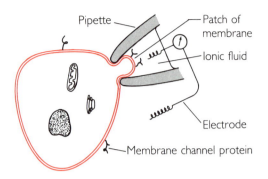

FIGURE 2–5

The patch-clamp technique. *Top:* A tiny pipette forms a seal around a patch of membrane. This can be studied in situ, or the patch can be broken off and bathed in intracellular *(bottom left)* or extracellular *(bottom right)* fluid. (From the National Science Foundation, USA.)

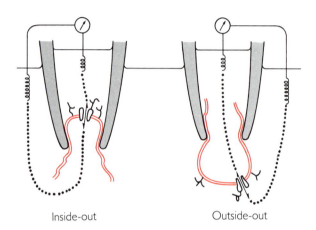

exist in two configurations, one of which permits an open channel and the other a closed one, with the two states in a dynamic equilibrium (Figure 2–6). Studies of such diverse tissues as neuronal axons, the postsynaptic receptor for acetylcholine (see below), the visual system and other sensory organs, the gap junction between animal cells (see Chapter 14), and protoplasts from legume plant leaves show that the equilibrium can be shifted toward the open configuration by certain ions or ligands. In the neurons, electrical activity unlocks the channel's "gate"; in the axon, the ligand binds directly to the membrane protein to alter its configuration. In both cases, gating allows the cell to regulate channel opening for physiological function. There are many examples of this phenomenon (Table 2–2).

A striking example of a **voltage-gated** channel is the sodium ion channel in the nerve cell axon. The nerve impulse is transmitted along this long cellular process via the depolarization of the membrane. The axonal membrane is normally polarized with the interior negative, but during cell activity it becomes transiently depolarized because of a

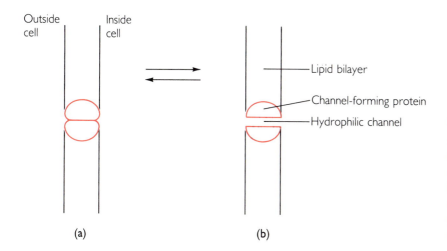

Outside cell | Inside cell

Lipid bilayer
Channel-forming protein
Hydrophilic channel

(a)   (b)

FIGURE 2–6
A gated channel. An equilibrium exists between the closed **(a)** and open **(b)** states. Binding of a ligand to the protein or ionic changes shift the equilibrium to the right, so that ions freely flow through the channel.

greatly increased permeability to Na^+, which enters the cell via a gated channel.

The **Na^+ channel** is believed to have two gates, one at the exterior surface of the membrane and the other at the interior surface. When the cell is at its typical resting potential, the outer gate is closed and the inner gate is open, and thus Na^+ does not enter through the gate. When the depolarization arrives at the axonal region, the voltage induces the outer gate to open, and Na^+ rushes into the cell through the channel. After a very brief period (milliseconds), the inner gate closes; this sets up a refractory period, during which depolarization cannot occur. Opening of the inner gate then occurs at the same time as the closing of the outer gate, and the resting state is restored.

The neurotoxins tetrodotoxin (from the puffer fish) and saxotoxin (from paralytic shellfish) bind with high affinity to the sodium channel,

TABLE 2–2 **Types of Gated Ion Channels**

Gating Effector	Ion(s)	Example
Voltage	Na^+, K^+, Ca^{++}	Neuron
Mechanical	K^+	Auditory hairs
Odor	K^+	Olfactory epithelium
Taste	K^+, Na^+	Taste receptor
Neurotransmitter	K^+, Na^+	Acetylcholine receptor
Ion	K^+	Ca^{++}-activated K^+ channel
Cyclic nucleotide	Na^+	cGMP retinal rod channel
G protein	K^+	Muscarinic acetylcholine receptor

and this fact has permitted its isolation from the neuronal membrane. Gene cloning has allowed the entire protein to be studied in some detail. The protein is quite large (over 1800 amino acids), with four homologous domains that surround a central channel through which the Na^+ flows. Each of the four domains has six membrane-spanning regions. Five of these regions are totally hydrophobic, and the sixth (termed S4) contains positively charged amino acids at every third position, with hydrophobic residues in between.

The S4 region, which occurs in most voltage-gated channels, is apparently the voltage sensor. Evidence for this role has come from molecular biological experiments, in which the S4 region's amino acid sequence was altered. When the positively charged amino acids were replaced with neutral ones, the amount of voltage needed to open the channel increased. Indeed, the magnitude of this increase was correlated with the degree of reduction of positive charge in the S4 region.

In addition to the Na^+ channel, K^+- and Ca^{++}-gated channels have important roles in the nervous system. The so-called A-type of **K^+ channel** is suppressed by the neurotransmitter acetylcholine; when this occurs, the effect is to render the inside of the cell more positively charged. Another neurotransmitter, noradrenaline, suppresses a different K^+ channel with similar effects. In both cases, the neuron is more likely to "fire" (produce an action potential due to depolarization) than it normally would be.

The isolation of K^+ channels was made possible through a fruit fly mutation called "shaker" because of the trembling characteristic of the strain. Physiological experiments indicated that this trembling was due to a defect in K^+ flow. After the mutation was mapped, molecular biology methods were used to isolate and clone the gene involved. The amino acid sequence predicted from the DNA sequence shows a channel that spans the hydrophobic part of the membrane six times.

Mutation analyses have led to the isolation of three additional K^+ channel genes and proteins from fruit flies. These channels differ from each other in terms of both amino acid sequence (overall 40% similarity) and physiological function (e.g., sensitivity to voltage). An analogous gene family has been found in mammals. An individual channel has four subunits, and these may come from more than one member of the family.

When an action potential reaches the synaptic ending of a neuron, the voltage stimulates the opening of a **Ca^{++} channel.** This ion rushes into the cell, stimulating vesicles to secrete their neurotransmitter contents into the synaptic cleft. The likelihood of the postsynaptic neuron firing is related to the amount of neurotransmitter released, and this in turn depends on the depolarization voltage (controlled in part by the K^+ channel) and the intracellular concentration of Ca^{++} (controlled by the

Ca^{++} channel). Studies by E. Kandel of learning and neural pathways in the mollusc *Aplysia* indicate that modulation of these two ion channels may be the molecular mechanism for simple learning.

In addition to their role at the synapse, Ca^{++} channels are important in the contraction of cardiac and smooth muscle cells. These channels open when the muscles are stimulated, allowing the entry of extracellular Ca^{++}. At the same time, Ca^{++} within the cells is also increased via release from intracellular stores. The combined effect of the increases in cell Ca^{++} results in the stimulation of myosin kinase, which binds actin to myosin, and muscle cell contraction follows. People with hypertension (high blood pressure) often have too strong a heartbeat as well as increased resistance in the smooth muscle in the lining of blood vessels. Several drugs, notably verapamil and nifedipine, block Ca^{++} channels from opening; this reduces the levels of intracellular Ca^{++} in the heart and blood vessels, leading to a reduction in blood pressure.

The nicotinic acetylcholine receptor is an example of a **ligand-gated** channel. It is found on the muscle cell plasma membrane at the neuromuscular junction (Figure 2–25). When the neuronal action potential reaches the junction, the neuron releases the transmitter, acetylcholine, into the synaptic cleft. The transmitter then binds to the acetylcholine receptor on the muscle cell membrane, where it acts as a ligand to open an ion channel. The inflow of cations (e.g., Na^+) then depolarizes the cell. The receptor channel is called nicotinic because nicotinic acid can activate it.

The nicotinic receptor has been purified and is composed of five polypeptide subunits of four types of chains, with one chain present twice (Figure 2–7). The subunits are transmembrane glycoproteins, which have hydrophobic bilayer-spanning domains as well as prominent extracellular and cytoplasmic regions. The five subunits form around a charged pore, and the negatively charged amino acids apparently exert

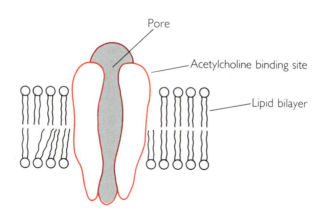

Pore

Acetylcholine binding site

Lipid bilayer

FIGURE 2–7

Cross section of an open nicotinic acetylcholine receptor channel. Only three of the five subunits that form the channel pore are shown.

the selectivity for cations. Other ligand-gated channels have similar structures, including receptors for the neurotransmitters glutamate, glycine, and gamma-amino butyric acid.

The epithelial cells that line the respiratory tract in humans have ligand-gated **chloride channels.** In response to the ligand, cyclic AMP, these channels open, allowing Cl^- to leave the cells, and this is followed by water so that the airway is lined with a moist mucus. Patients with the genetic disease cystic fibrosis have a single amino acid missing in their channel protein such that it is insensitive to cAMP and therefore does not open properly. As a result, their mucus is very thick and dry. This provides a good environment for pathogenic organisms, and there are repeated infections of the respiratory tract.

In the pancreas of such patients, blockage of the duct to the intestine leads to atrophy of the tissues that produce digestive enzymes. Death is usually due to the pulmonary factors and occurs before age 30. Current treatment for this defect in ion channels is to loosen the mucus mechanically and apply antibiotics to kill off the infecting organisms. But the determination of the exact mutation (a triumph of molecular genetics) gives hope that rational therapies directed at the channel will be devised to help treat the disease, which affects 1 in 2000 Caucasians in the United States.

In addition to voltage and ligand-gating, some ion channels act as transducers of sensory information and so are **sense-stimulus gated** (Table 2–3). For example, the ear contains auditory hair cells, which transduce sound waves into electrical signals. Displacement of cilia in the hair cells opens K^+ channels in the cell membrane, which leads to de-

---

TABLE 2–3 **Characteristics of Facilitated and Active Transport**

Both Mechanisms

1. Substrate specificity
2. Saturable kinetics
3. Competitive inhibition
4. Modification by protein-modifying reagents
5. Unregulated synthesis of carrier
6. Genetic mutability

Facilitated Transport Only	Active Transport Only
1. Travel with concentration gradient	1. Travel against concentration gradient
2. Bidirectional	2. Unidirectional
3. No requirement for metabolic energy	3. Requires metabolic energy

polarization. In a similar fashion, cilia in the olfactory epithelium use odorants to open ion channel gates. Finally, there are mechanosensitive gated ion channels in many cell types, with both positive and negative mechanical stimuli being effective.

Ion channels are widespread in nonnervous tissues. For example, a Ca^{++} channel is important in the regulation of the release of insulin from pancreatic cells by glucose. In the epithelium that lines the colon, ion channels for Na^+, K^+, and Cl^- act to ensure electrolyte balance. In oocytes, a K^+ channel opens right after fertilization, and this causes the oocyte membrane to resist further sperm binding (polyspermy). In plant leaves, movement of K^+ through a gated channel is in part responsible for the movement of the leaves to face the sun and the opening and closing of stomata (the pores in the leaf that regulate gas exchange with the atmosphere).

Facilitated Transport: Carriers

Cell membranes contain specific carrier proteins that increase the rate of diffusion of a large number of molecules (Figure 2–8). Most animal tissues have this facilitated transport for molecules they receive from blood plasma, such as sugars, amino acids, and purines. Constant breakdown of these molecules within the cells maintains the concentration gradient with the exterior.

Many of the characteristics of facilitated transport are similar to those of enzymes: substrate specificity, saturable kinetics, and so on (Table 2–3). These characteristics distinguish this transport mechanism from simple diffusion, which is concentration independent and does not directly involve membrane proteins. Transport proteins exhibit Michaelis constants (K_m) for their ligands. These concentrations at which the transport rate is half maximal (Table 2–4) are considerably higher than those of typical substrates for enzymes, showing the relative inefficiency of the systems. Binding and transport are two distinct steps.

Among the best-studied facilitated transport carriers are the **glucose transporters.** Several types are present in specific tissues. The human erythrocyte glucose transporter (Figure 1–13) is a transmembrane protein of 492 amino acids, which has 12 membrane-spanning hydrophobic regions. It is also present in many fetal tissues as well as the adult brain. Closely related to this protein in sequence (55% identity) is the glucose transporter of liver. This organ plays a key role in the regulation of blood glucose and can store glucose as glycogen. The topology of this carrier in the membrane is the same as that of the red cell carrier.

Still another related glucose transporter occurs in cells that are responsive to the hormone insulin. A major effect of this hormone is to

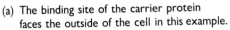

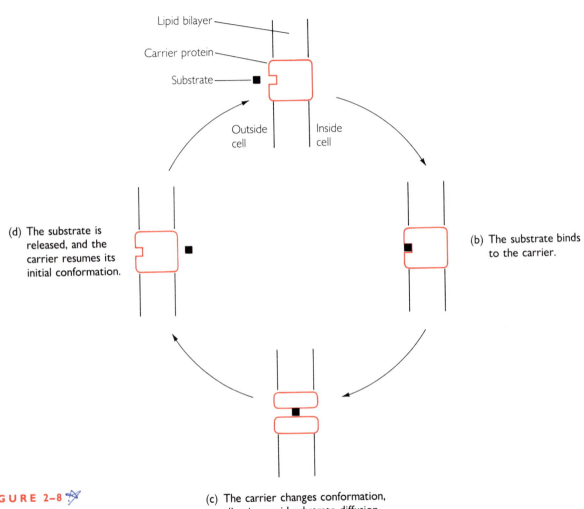

FIGURE 2-8

Facilitated diffusion.

stimulate the transport of glucose into skeletal muscle and adipocytes (fat cells). One way this happens is that the activity of glucose transporters in the plasma membrane of the cells increases. In diabetes mellitus type II, cells are much less responsive to the typical levels of insulin and instead fail to transport glucose from the blood across the plasma membranes into their cytoplasm. As a result, blood glucose levels rise to dangerously high levels. It is not surprising that the number of glucose transporters at the membrane is at least 50% lower in diabetics compared to nondiabetics. This reduction is apparently due to a reduction in the mRNA for the transporter.

TABLE 2–4 Kinetics of Facilitated Transport

Permeant	Cell Type	K_m(mM)
Glucose	Adipocyte	20
Glucose	Erythrocyte	7
Glucose	Fibroblast	2
Glucose	Hepatocyte	30
Glucose	Tumor cell	3
Lactate	Erythrocyte	66
Leucine	Erythrocyte	13
Lysine	Erythrocyte	0.07
Chloride ion	Erythrocyte	25
Phosphate ion	Erythrocyte	80
Choline	Synaptosome	0.08
Glucosamine	Synaptosome	2
Ammonia	Liverwort (plant)	0.02

DATA FROM: Jain, M. *Introduction to biological membranes,* 2nd ed. New York: Wiley, 1988, and other sources.

Alteration of the protein carriers of the cell membrane can alter the transport characteristics of that membrane. This phenomenon is illustrated by the agent of malaria, *Plasmodium falciparum.* This is currently among the most widespread human infectious diseases with over 300 million annual cases, 3 million deaths, and over 2 billion people at risk. The parasite is transmitted to humans via the bite of a female anopheline mosquito. Eventually, the parasite makes its way into erythrocytes, where it exerts its clinically devastating effects, with a single organism multiplying to produce up to 32 progeny within two days. About six hours after a red cell is invaded, its surface transport activities are changed: New and more efficient carriers for carbohydrates and ions are synthesized and inserted into the erythrocyte membrane. This provides the substrates for the intense metabolism needed to produce the new parasites. The antimalarial drug phlorizin appears to enter the cells via these new pathways and then bind to one of the carriers at its cytoplasmic surface. This effectively inhibits further parasite multiplication.

Active Transport

Transport Against a Concentration Gradient

The intracellular pH of acid-secreting parietal cells of the stomach is approximately 7; this means that the hydrogen ion concentration is 10^{-7} M. On the other hand, the pH of gastric juice external to these cells is

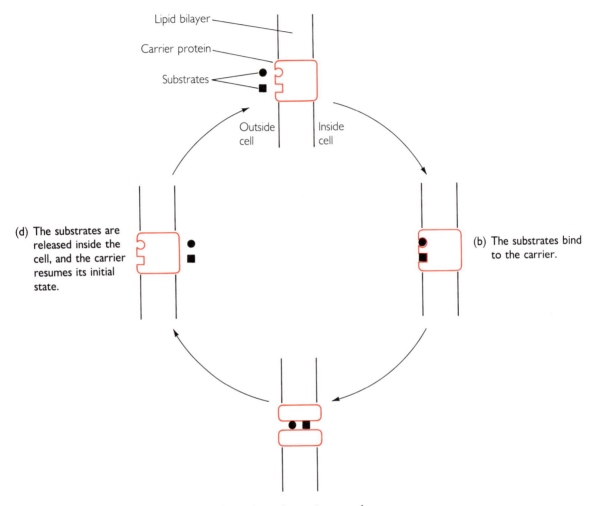

(a) The binding sites on a carrier protein for two substrates face the outside of the cell in this example. One substance (e.q., Na^+) is in a favorable diffusion gradient for entry into the cell, whereas the other substance (e.g., glucose) is not.

Lipid bilayer

Carrier protein

Substrates

Outside cell

Inside cell

(d) The substrates are released inside the cell, and the carrier resumes its initial state.

(b) The substrates bind to the carrier.

(c) Binding induces the carrier to undergo a conformational change, allowing rapid transport of the two substrates. The potential energy from the diffusion of one (e.g., Na^+) drives the movement of the other (e.g., glucose) against its gradient.

FIGURE 2–9 Co-transport (symport).

about 1, giving a hydrogen ion concentration of 0.1M. According to the equation:

$$\Delta G = RT \log_e (c_2/c_1)$$

if c_2 is 10^{-1} and c_1 is 10^{-7}, ΔG is 95 Kcal/mole, considering only the concentration gradient. Clearly, to maintain this gradient is thermodynamically unfavorable and requires massive inputs of energy. For example, the hydrolysis of 12 moles of ATP (at 8 Kcal/mole released energy) would be the minimum needed to pump out a mole of hydrogen ions through the parietal cell membrane. It is not surprising that this type of "uphill" active transport consumes 30% to 60% of a typical cell's production of energy-rich compounds.

A common phenomenon in biochemistry is the coupling of an energy-releasing reaction (the "driver") with an energy-requiring one (the "driven"). Clearly, ATP hydrolysis (as in the example in the previous paragraph) would be one way to do this coupling. Another way is to couple the potential energy released by diffusion along a concentration gradient with the energy required for transport of a second substance against its gradient. This can be achieved by a single protein (Figure 2–9) in **co-transport.**

Many co-transport systems couple an ion with an organic molecule (Table 2–5). The best studied of these are the Na^+-linked systems for sugars and amino acids in the intestinal epithelial cell. This cell has a much higher glucose concentration (>5 mM) than that of the intestinal lumen (< 1 mM). On the other hand, Na^+ is higher (120 mM) in the lumen than in the cell (10 mM). A single protein carrier in the mucosal cell membrane transports one Na^+ (the "driver") with each glucose molecule (the "driven"). These requirements are specific: No other ion or sugar can bind to this carrier, and sugar transport does not occur in the absence of ion transport. The membrane protein responsible for this co-transport has been sequenced via gene cloning and is a 662 amino acid single chain that spans the lipid bilayer 11 times.

A co-transport system involving only ions is the Na^+-K^+-Cl^- transporter of intestinal and kidney cells as well as erythrocytes. In this

T A B L E 2–5 **Some Co-Transport Systems**

Permeants	Organism, Tissue
Alanine/Na^+	Mouse, intestine
Amino acids/Na^+	Mouse, ascites tumor
Glucose/Na^+	Mammal, intestine, kidney
Glycine/Na^+	Pigeon, erythrocyte

case, the major "driver" is Na^+, with Cl^- and K^+ the "driven," especially the latter. An important role for this system is volume regulation. If the tissues are placed in a medium whose osmotic concentration exceeds that of the cells, water tends to leave the cells and they shrink. However, often after a while the cells are observed to return to their normal size. What has happened is that the triple ion transporter has been induced to take in the three ions, and water then follows. In the kidney, this phenomenon also occurs in that ions are reabsorbed into the bloodstream, and water follows.

In certain diseases (for example, congestive heart failure) it is desirable to lower the amount of fluid in the blood. This can be done by decreasing the reabsorption of water by the blood vessels in the kidney. The diuretic furosemide does exactly that by inhibiting the Na^+-K^+-Cl^- transporter. This results in more water in the urine rather than in the blood.

In addition to co-transport, another way to couple an energy-releasing reaction to transport of a substance against its concentration gradient is to use the free energy released from the hydrolysis of ATP to ADP and inorganic phosphate. This is termed **active transport.** Several examples will be considered, the first being the multidrug transporter or P-glycoprotein.

This transporter was discovered when clinicians noticed that some cancers became resistant to a cytotoxic drug and to many other drugs as well. The drugs involved—vinblastine, doxorubicin, and so on—share few common structural attributes except for being rather hydrophobic. The resistant cells, but not the susceptible cells, have increased amounts of a membrane protein called the P-glycoprotein (Pgp). If the gene for this protein is put into a drug-susceptible tumor cell, the latter becomes resistant. This is strong evidence that Pgp is the cause of resistance.

Pgp is a 170 kd membrane protein that occurs in two adjacent copies, each of which has six membrane-spanning hydrophobic domains, as well as prominent cytoplasmic and extracellular regions. The drugs bind on the cytoplasmic side after they enter the cell via diffusion through the lipid bilayer. Binding of the drug is followed by binding and hydrolysis of ATP.

The free energy released from ATP hydrolysis has one of two possible effects. The most widely accepted model is that it drives a conformational change in the transporter protein so that it forms a pore and expels the drug, against its concentration gradient (Figure 2–10). Alternatively, a conformational change in the protein may "flip" the drug from the inner leaflet of the lipid bilayer to the outer one, and then the drug diffuses away from the cell. In either case, the mode of resistance to the chemotherapeutic drugs is that the cells pump them from their cytoplasm across the plasma membrane into the extracellular medium.

This, then, is the mode of resistance: The cells pump the drugs out from their cytoplasm across the plasma membrane into the extracellular medium.

How a single glycoprotein recognizes and pumps out a number of different molecules is not known. Also a mystery is how the presence of a drug induces the synthesis of the transporter, since its expression is greatly increased in resistant cells. Answers to these questions are both of basic interest in the study of active transport and of practical interest in the treatment of cancer.

The Sodium Pump

In virtually all cells, the concentration of Na^+ is greater outside than inside the cell, and the concentration of K^+ is higher inside than outside. Typical concentrations (in mM) are:

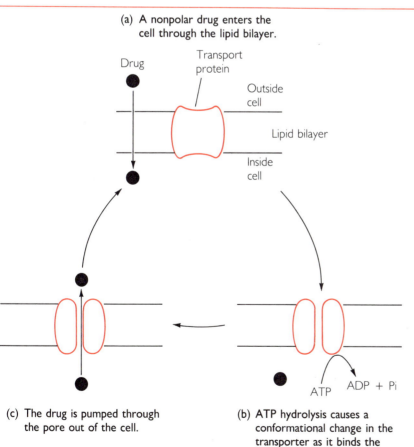

(a) A nonpolar drug enters the cell through the lipid bilayer.

Drug

Transport protein

Outside cell

Lipid bilayer

Inside cell

(c) The drug is pumped through the pore out of the cell.

(b) ATP hydrolysis causes a conformational change in the transporter as it binds the drug so that a pore is opened.

ATP   ADP + Pi

FIGURE 2–10
The multidrug transporter.

$$Na^+: 150 \text{ outside vs. } 10 \text{ inside}$$
$$K^+: \quad 5 \text{ outside vs. } 140 \text{ inside}$$

Co-transport and the slow diffusion of these ions through the membrane would tend to lower these concentration imbalances. However, they are held constant by the sodium pump.

This membrane protein couples the transport of $3Na^+$ out of the cell and $2K^+$ into the cell. Because these movements are clearly against their concentration gradients, there must be a "driver" to provide energy, and this is achieved by the hydrolysis of one molecule of ATP. The ratio of ions transported leads to an electrogenic potential gradient across the membrane. The two ions actually activate the hydrolysis of ATP, which can occur even in the absence of membranes.

Since its discovery in 1957 by F. Skou, the Na^+-K^+ ATPase has been intensively studied. It has been purified to homogeneity, its genes have been cloned and sequenced, and the protein's amino acid sequence has been derived from the gene sequence. It is composed of two subunits. The larger of the subunits is a transmembrane protein, which crosses the bilayer seven times and acts as the catalytic center, and the other one is a smaller polypeptide that crosses only once and is important in some unknown way to the function of the pump. Apparently, the active form in the membrane is composed of two copies of each of the subunits.

The isolated protein is not active, and activity is restored only when the protein is incorporated into a membrane. Boundary lipids, especially cholesterol, are essential to the functioning of the system. The importance of membrane fluidity for the activity of the pump was shown by an experiment in which polarized epithelial cells were used. Although there were Na^+-K^+ ATPase molecules in the apical region of the cell, they were not active, probably because of restricted lipid fluidity in this region (see Chapter 1). When the membrane was treated with an agent that resulted in greater membrane fluidity, the pump was activated.

The actual mechanism is a sequence of reactions (Figure 2–11):

1. At the inner membrane surface, the enzyme binds both phosphate (from ATP hydrolysis) and Na^+ to form an E_1-Na^+-P complex.

2. This complex, via the obligate cofactor Mg^{++}, releases Na^+ to the outside, leaving E_2-P.

3. This complex picks up K^+ on the outside and releases P.

4. The E_2-K^+ complex releases K^+ on the inside, returning to the E_1 configuration.

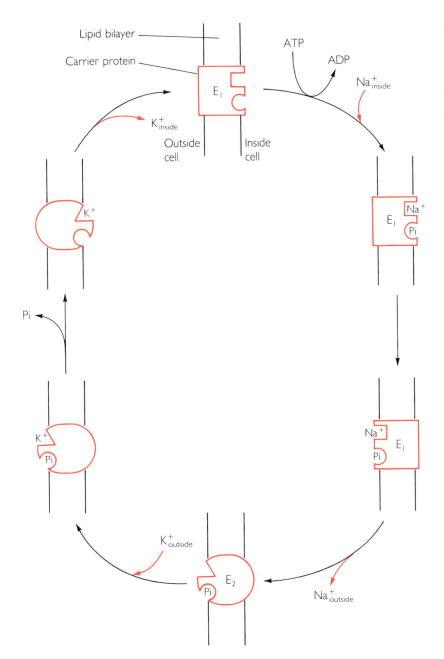

FIGURE 2–11

Active transport: The Na^+-K^+ pump. See text for details. E_1 and E_2 are different conformational states of the carrier.

The Na^+-K^+ ATPase has many important functions. Most obviously, it is largely responsible for the gradients of the two ions in excitable cells such as nerve and muscle. These provide the basis for the generation of action potentials and their propagation along the cell surface. In addition, the sodium pump helps to maintain cells in osmotic

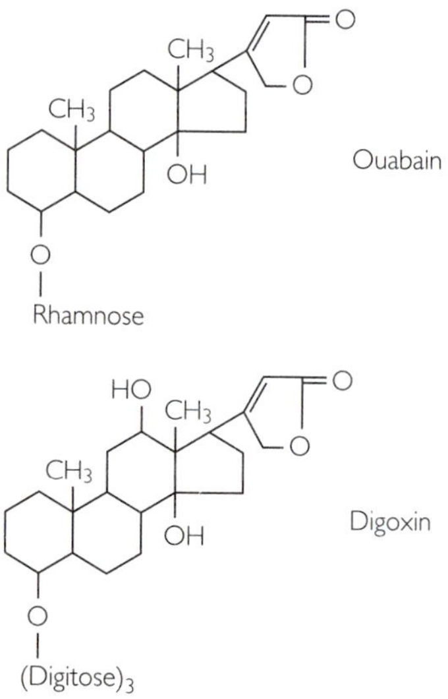

Ouabain

Rhamnose

Digoxin

(Digitose)$_3$

FIGURE 2–12

Cardiac glycosides that block the Na$^+$-K$^+$ pump.

equilibrium with their environment. Finally, the pump is essential in generating ion gradients for co-transport of metabolites (see Figure 2–9).

In a typical mammalian cell, about 30% of the ATP produced is hydrolyzed to drive this one pump. A reduction in the efficiency of the pump would increase its metabolic requirements. This forms the basis of a theory proposed by E. Racker to account for the high glycolytic rate in some tumors. Alteration of the pump via gene products endogenous to the cell (from oncogenes) or from cancer viruses could cause it to need increased amounts of cellular ATP, produced by the tumor cell.

A striking property of the sodium pump is its potent inhibition by certain plant steroids (Figure 2–12). One of these, ouabain, has been used extensively to demonstrate that the pump causes certain physiological effects. Ouabain-sensitive processes have the Na$^+$-K$^+$-ATPase as a necessary component. A related family of steroid glycosides has been isolated from the foxglove plant, *Digitalis.* Extracts of this plant have been used for at least two centuries to treat patients with dropsy, an old term for the fluid accumulation that occurs when the heart does not pump with sufficient vigor.

Today, molecules purified from these foxglove extracts are widely used. These cardiac glycosides act by binding to the outside surface of the pump, preventing it from picking up K$^+$ (step 3, above) and increasing intracellular Na$^+$. A carrier in the cardiac cell membrane facilitates the diffusion of Na$^+$ and Ca^{++}. Thus, an increase in intracellular Na$^+$ leads to a concomitant increase in intracellular Ca^{++}. The latter ion is a potent activator of the contractile proteins (see below), and its increase results in increased myocardial contractile force. Thus this sequence of events, beginning with glycoside inhibition of the Na$^+$-K$^+$ pump and ending with a more vigorous heartbeat, can alleviate the adverse effects of congestive heart failure. Millions of patients take these drugs daily.

The Calcium and Proton Pumps

The contraction and relaxation of muscle are controlled by the concentration of calcium ions. During contraction, Ca^{++} around the myofilaments is about 10 μM. For relaxation to occur, this concentration must be reduced to 0.1 μM in less than a tenth of a second. This hundredfold reduction is achieved by sequestering the Ca^{++} inside the sarcoplasmic reticulum, a membrane-surrounded structure in the muscle cell (see Chapter 6). Within the reticulum, the Ca^{++} concentration reaches 10 mM. This means that the movement of Ca^{++} from the cytoplasm into the reticulum is strongly against its concentration gradient. Once again, the "driver" for this movement is the hydrolysis of ATP. In this case, one

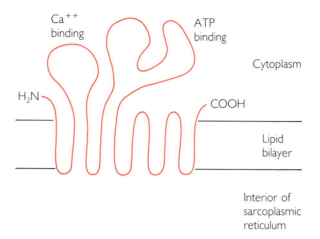

FIGURE 2–13

Structure of the Ca^{++} pump in the membrane of the sarcoplasmic reticulum. The plasma membrane Na^+K^+ pump has a similar design.

molecule of ATP hydrolysis powers the movement of $2Ca^{++}$ across the membrane into the reticulum.

The protein responsible for calcium movement is termed the **Ca^{++} pump.** It shares a number of properties with the sodium pump—specificity, subunits, requirement for Mg^{++} and requirement for membrane when reconstituted—and its mechanism is probably similar to that of the sodium pump (Figure 2–13). In the sarcoplasmic reticulum, the Ca^{++} pump accounts for about 60% of the total membrane protein. The pump and calmodulin, the receptor protein for Ca^{++}, have been localized in a wide variety of cells. This system appears to have many actions, including stimulating neurotransmitter release at synapses and activation of the mitotic apparatus.

A H^+-ATPase **(proton pump)** is responsible for energy generation in chloroplasts, mitochondria, and other membranes. Essentially, it does the pumping, using a proton gradient set up by asymmetric electron transport carriers in the membrane to make ATP. This chemiosmotic hypothesis will be described in detail in the context of these organelles (see Chapters 3 and 4). A second type of H^+-ATPase is found at the plasma membrane, where its function is to maintain the cytoplasmic H^+ concentration (pH) and, in the case of some plant cells, to acidify the cell wall during hormone-stimulated cell elongation (see Chapter 13). The third class of H^+ pumps occurs in the membranes of vacuoles, including the plant vacuole, lysosome, and coated vesicle. Its role is to acidify the contents of the cellular organelle (see Chapter 8).

A remarkable aspect of the cation pumps is their homology. Their sizes are similar (around 1000 residues), their amino acid sequences are identical to the extent of about 20%, and their three-dimensional structures are similar in the locations of membrane-spanning regions and ATP binding site. By inference from amino acid sequences, a K^+-ATPase from the bacterium *Streptococcus faecalis* appears to be nearest the ancestral form, and a "family tree" leading to the many current Na^+-K^+, H^+, and Ca^+ pumps can be drawn. Divergence of the catalytic alpha subunit of the sodium pump seems to have occurred about 500 to 300 million years ago.

Genetics of Transport

Both active and facilitated transport are mediated by membrane proteins, which are coded for by genes. It is not surprising, therefore, that some human genetic diseases involving transport have been described (Table 2–6). One place where these diseases are most manifest is the kidney, where amino acids filtered out are returned to the blood by facilitated transport carriers. If reabsorption into the blood does not occur, the amino acids are excreted in the urine. There are four separate facilitated transport proteins for amino acids, and failure of three of them can result from the genetically recessive diseases cystinuria, glycinuria, and Hartnup disease. In the latter, intestinal transport is also defective, showing possible identity between the kidney and intestinal carriers. In the three diseases, several amino acids become more concentrated in the urine, indicating a relatively broad specificity for the carriers.

Cystinuria is easily detected. In Massachusetts, as well as in other locations, it has been screened for in all newborns, and its frequency of incidence is 1 in 14,000 live births. Cystine (the dimerized form of cysteine) is the least soluble of the common amino acids. When excessive amounts of it remain in the renal tubule because of a lack of a functioning transport protein to get it back into the blood, it will precipitate

TABLE 2–6 Inherited Transport Diseases in Humans

Disease	Transport Deficiency
Cystinuria	Cystine, lysine, arginine
Glycinuria	Glycine, proline, hydroxyproline, alanine, serine, threonine, leucine, isoleucine, valine, tyrosine, tryptophan, phenylalanine
Malabsorption of vitamin B12	Vitamin B12
Pellagra (hereditary)	Tryptophan
Renal glycosuria	Glucose
Rickets (hereditary)	Phosphate
Spherocytosis (hereditary)	Na^+, K^+

and form crystals. These crystals can grow to block the urinary tract and damage the kidney. Treatment of this disease includes a low protein diet and penicillamine, a molecule that complexes with cysteine to form a more soluble dimer than cystine.

Bulk Transport

The passage of macromolecules such as proteins through the cell membrane is limited because of their large size and ionic charge. As was explained above, certain proteins can pass through the bilayer by means of a hydrophobic signal sequence (Figure 1–18). But in most cases, large substances moving in and out of cells do not traverse the bilayer directly but do so indirectly via membrane-bound vesicles.

In **endocytosis,** the cell membrane invaginates and fuses around an extracellular substance, internalizing it into the cytoplasm. In **exocytosis,** an intracellular vesicle containing a substance targeted for extracellular release fuses with the cell membrane, releasing its contents to the medium. This process is described in Chapter 7.

When extracellular particles greater than 0.2 μm in diameter are taken up by cells, the endocytotic vesicle forms tightly around the particle, excluding most of the extracellular fluid. This process is phagocytosis ("eating"). Uptake of smaller particles (e.g., proteins) also involves vesicles, and some extracellular fluid enters the cell. This process is pinocytosis ("drinking").

Phagocytosis may involve entire bacteria or even larger structures, as in the familiar protozoan *Amoeba.* In this single-celled organism, ingestion of food occurs almost entirely by this mechanism. In mammals, macrophages phagocytose inert particles, bacteria, antigen-antibody complexes, and even dead cells (Figure 2–14). In the liver, these cells line the blood sinusoids, clearing them of unwanted materials. In the lung, they line the airways and alveoli, helping to maintain a sterility often compromised by foreign material in the inhaled air. Polymorphonuclear leukocytes travel the bloodstream and are the major defense mechanism against infecting bacteria. Most of these phagocytic cells in animals are selective for the materials they digest, since otherwise they might digest host tissues. The nature of this selectivity is not clear, but it may involve pH (dying or dead tissues have a higher surface pH than living ones) or recognition markers.

Pinocytosis occurs in many cell types and in many organisms. Indeed, the formation of small vacuoles at the cell surface was first noted for *Amoeba* in 1925. Electron microscopy has revealed the process in many mammalian tissues, most notably the endothelium that lines blood vessels, kidney tubule cells, and the placenta. There is also some

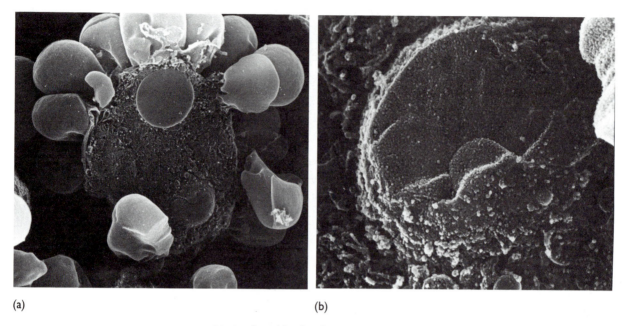

(a)   (b)

FIGURE 2–14   **(a)** A white blood cell (macrophage) endocytosing erythrocytes in the rat lymph node. **(b)** Magnified view of the endocytotic event. Note the many processes emanating from the macrophage plasma membrane. (Photographs courtesy of Dr. K. Sasaki.)

evidence that it occurs in the nectaries and stigmas of flowers of higher plants.

A well-studied pinocytotic tissue is the endothelium (Figure 2–15). Here, a single cell layer is often all that separates the blood from the surrounding tissue. The cells often appear to be filled with small vesicles, 70 nm in diameter, most of them attached by small "necks" to the plasma membrane.

A wide variety of electron-dense tracers have been introduced into blood and enter the endothelial cells via the pinocytotic vesicles. There is considerable selectivity: For instance, serum albumin, a protein important in the maintenance of blood osmotic pressure, is excluded, but albumin glycosylated with glucose does enter the cells. This observation may be clinically important because the high blood glucose levels in diabetes can lead to nonenzymatic glycosylation of proteins such as hemoglobin and albumin. Aberrant protein pinocytosis by endothelial cells could lead to the capillary damage that is often seen in diabetic patients.

A second, and more specific, type of pinocytosis involves **coated pits.** These plasma membrane pouches with a fuzzy, electron-dense lining on the cytoplasmic side were first observed by T. Roth and K. Porter in developing mosquito oocytes. These cells selectively take up yolk protein, and electron micrographs clearly showed the protein first in coated

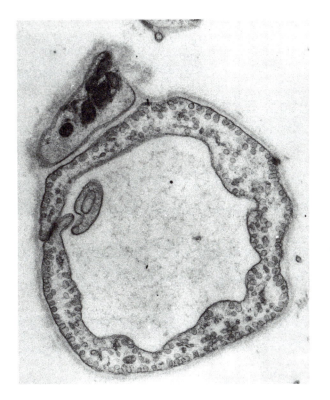

FIGURE 2–15
Electron micrograph of a cross section of a small-caliber capillary composed of a single endothelial cell conjoined to itself. Note the numerous vesicles at both cell surfaces. (Photograph courtesy of Dr. J. Casley-Smith.)

pits and later in vesicles within the cells. Later, similar observations were made on chicken oocytes, where a single cell takes up over 1 gram of yolk protein per day, and placenta, where maternal antibodies are taken in. Coated pits have now been extensively studied in fibroblasts, cells that are easily manipulated in the laboratory (Figure 2–16). The fuzzy coating on the cytoplasmic side of the membrane has been characterized and it is composed of multiple units of a single protein, clathrin. The intracellular role of coated vesicles is discussed later (see Chapter 7). Here, the function of the coated pit will be described.

Coated pits mediate the endocytosis of specific structures or molecules for which receptors exist on the cell surface. Many substances enter cells via this **receptor-mediated endocytosis** pathway. These range from proteins (e.g., the iron-binding protein transferrin) to viruses (e.g., Semliki Forest virus). Fibroblasts have been useful in the working out of the pathway. The scheme that has emerged from these studies (Figure 2–17) envisions the molecule or ligand to be taken up, binding to randomly diffusing (see Chapter 1) receptors on the cell surface. Calculations of the number of receptors, their rate of diffusion, and number of pits indicate that the average receptor will enter a pit every few seconds.

Once a ligand-bound or unbound receptor enters a pit, it becomes immobilized and then internalized. Why only these receptor proteins,

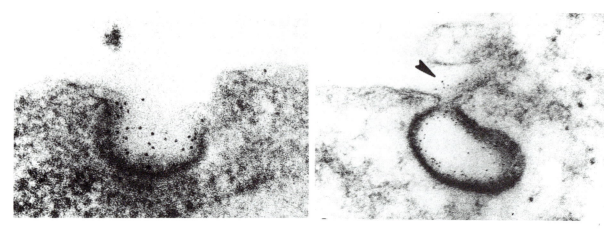

F I G U R E   2–16   Coated-pit–mediated endocytosis of low-density lipoprotein (LDL) by a fibro-blast. The left electron micrograph shows a ligand (LDL coupled to electron-dense ferritin) as dots bound in the coated pit. In the right micrograph, the membrane is presumed to have pinched off into a coated vesicle, with the LDL inside. (Photographs courtesy of Drs. M. Brown and J. Goldstein.)

and not other proteins that pass through the pit, are so treated involves the interaction of the cytoplasmic domain of the receptor with adaptor proteins (adaptins) of the pits. The relevant receptor domains have been identified in several cases (e.g., transferrin receptor in vertebrates). They share the properties of being short (4–6 amino acids), and have a strong tendency to exhibit a sharp turn in the three-dimensional structure of the protein. With the structures of the adaptins now known from DNA analyses, it will soon be possible to determine how the receptor and pit interact for internalization.

Soon after internalization, an uncoated vesicle containing internalized ligand-receptor complexes appears in the cytoplasm just below the plasma membrane. The origin of this vesicle may be a pinching off of the coated pit and its subsequent uncoating (an ATPase at the vesicle membrane has been proposed to do this) or the pinching off of a special uncoated area of the pit. At any rate, soon after the endocytotic event, the coated pits and receptors are once again on the plasma membrane.

A cell surface receptor is specific for a given ligand, and a particular cell type may have many different kinds of receptors. Because endocytosis is a general pathway, a cell can regulate which molecules are taken up by regulating the synthesis and degradation of the surface receptors. This can be done in the normal course of cellular differentiation when only certain receptors are made. For example, hepatocytes in the liver have many more insulin receptors than do nerve cells, although both cell types are normally incubated in the hormone as it flows by in the blood.

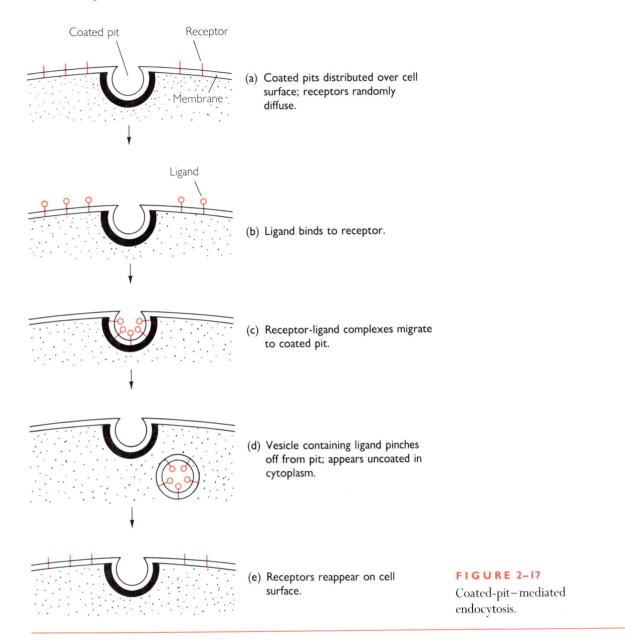

(a) Coated pits distributed over cell surface; receptors randomly diffuse.

(b) Ligand binds to receptor.

(c) Receptor-ligand complexes migrate to coated pit.

(d) Vesicle containing ligand pinches off from pit; appears uncoated in cytoplasm.

(e) Receptors reappear on cell surface.

FIGURE 2–17
Coated-pit–mediated endocytosis.

A second method of receptor regulation is to reduce the number of receptors for a ligand when it accumulates in the cell. This occurs by a combination of allowing those receptors already present to break down and not replacing them by resynthesis.

Internalization takes about a minute and constantly occurs in fibroblasts. Because about 2% of the fibroblast cell surface is normally in coated pits, this means that, if the pit forms the pinocytotic vesicle, the

TABLE 2–7   **Internalization of Peptide Hormones by Mammalian Cells**

Hormone	Cell Type	Site of Accumulation
Chorionic gonadotropin	Ovary	Nucleus, Golgi, lysosome
Insulin	Hepatocyte	Nucleus, mitochondrion
Luteinizing hormone releasing hormone	Pituitary	Golgi
Prolactin	Mammary gland	Golgi
Thrombin	Fibroblast	Nucleus
Thyroxine	Hepatocyte	Mitochondrion

entire cell membrane is internalized every hour! Even if the vesicle comes from part of the pit, the amount of membrane invagination is extensive. Because the fibroblast retains its size, new plasma membrane must be added to replace what was internalized.

The fate of the vesicle, called an endosome, varies, but it usually transports its ligand to a membrane-bound organelle (Table 2–7). Following deposition of the ligand, the endosome may fuse with a lysosome (see Chapter 8), where its contents are degraded. Ultimately, the endosome ends up at the Golgi, from where it can be reinserted to the membrane by exocytosis. This phenomenon of internalization and shuttling of the endosome essentially recycles the membrane lost during endocytosis (Figure 2–18).

The round trip of a receptor from and to the plasma membrane takes about 15 minutes. This means that many of the receptor molecules at a given moment are not on the cell surface but within the cell, taking part in the recycling pathway. Dissociation of ligand from receptor occurs in the endosome, probably via acidification of the endosome, which actively transports protons into its lumen via its membrane H^+-ATPase. Then the receptor and ligand travel via different pathways, with the receptor in vesicles that avoid the lysosome and the ligand in vesicles that fuse with the lysosome. This allows recycling of the receptor and degradation of the ligand.

A clinically important endocytotic system is that for **low-density lipoprotein (LDL).** This macromolecular complex carries some of dietary cholesterol in the bloodstream and delivers it to cells. Uptake of LDL occurs via receptor-mediated endocytosis in coated pits (Figure 2–19). The resulting endosome fuses with a lysosome, where the cholesterol is released for use by the cell in making its membranes. The presence of this new cholesterol inhibits endogenous cholesterol synthesis by the cell, thus saving cellular energy and maintaining a constant cholesterol level.

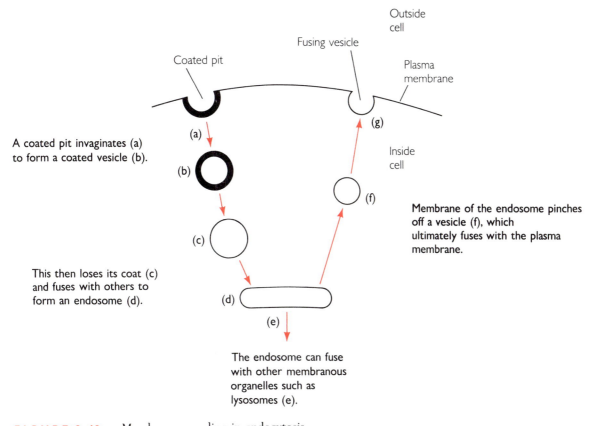

Outside cell

Fusing vesicle

Coated pit

Plasma membrane

(a)

A coated pit invaginates (a) to form a coated vesicle (b).

(b)

Inside cell

(g)

(f)

Membrane of the endosome pinches off a vesicle (f), which ultimately fuses with the plasma membrane.

(c)

This then loses its coat (c) and fuses with others to form an endosome (d).

(d)

(e)

The endosome can fuse with other membranous organelles such as lysosomes (e).

FIGURE 2–18   Membrane recycling in endocytosis.

The genetic disease familial hypercholesterolemia occurs in some form in 1 in every 500 people in most countries. Affected people have high blood levels of LDL and, as a result, develop cholesterol deposits on the inside of their arteries. This atherosclerosis often leads to myocardial infarctions (heart attacks) before the age of 50; in genetic homozygotes, this can even occur by the age of three. Studies of these patients' fibroblasts by M. Brown and J. Goldstein have shown endocytotic defects.

In most cases, the patients have a mutant, nonfunctional LDL receptor, either for its binding to LDL or for its internalization. For instance, one mutant is a deletion for the LDL-binding extracellular domain (see Figure 2–19). Another is a deletion in the bilayer-spanning region, and LDL binds but is not internalized. A third type is mutated in the cytoplasmic tail, and once again internalization is prevented. These mutations have been invaluable in mapping of the functional architecture of the LDL receptor. Rarely, there is a defect in the insertion of the receptor into the plasma membrane. In most cases, the lack of internalization of LDL results in abundant cholesterol outside the cells, and

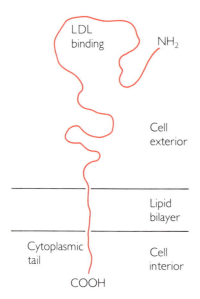

FIGURE 2–19

Structure of the low-density lipoprotein (LDL) receptor. The cytoplasmic domain is about 50 amino acids long, and the extracellular domain has about 700 amino acids.

it may be deposited on the inner surface of arteries. Because LDL cholesterol does not enter the cells, endogenous cholesterol synthesis goes on unabated and contributes to the accumulation.

Peptide hormones enter cells via coated pits and accumulate at various locations in the cell (Table 2–7). Discovery of this phenomenon has opened up a whole new area of endocrinology not considered before the 1970s. Previously, it was thought that these hormones could not cross the membrane and therefore exerted their effects solely by binding to the cell surface and then activating membrane proteins (e.g., adenyl cyclase for cyclic AMP synthesis). It is important to note that the internalization of these hormone-receptor complexes occurs only after the hormone is bound. This contrasts with the typical case of ligands such as LDL, where there is constant internalization of the receptor, whether bound or not.

There are a number of disorders in which high circulating hormone levels are coupled with a reduction in receptors for that hormone on the cell surface (e.g., insulin-resistant diabetes). By analogy to the LDL defects mentioned above, the lack of either hormone receptors or internalization in coated pits could be the cause of the metabolic errors in these patients.

The Cell Membrane and Recognition

The cell membrane separates the interior of the cell from its surrounding environment. As a boundary, it has an important role in determining with which other cells a given cell interacts. These interactions may be direct, with cells touching each other via glycoproteins and glycolipids of the cell surface. Cell-cell interactions will be described in Chapter 14. Or, the interactions between cells may be indirect, as occurs when one cell sends out chemical messengers that affect a distant cell. The present discussion focuses on these indirect interactions.

Hormones and the G Protein Cycle

Hormones are substances that are made in one location of an organism and that travel through the circulation to another location(s) where they exert specific effects. Implicit in this definition is the concept of specific receptors on the target tissue that allow the hormone to exert its effects there and not in nontarget tissues. Some hormones (e.g., steroids) are sufficiently hydrophobic that they can travel through the lipid bilayer. In such cases, the hormone will enter any cell, and receptor differentiation is within the cell. Other hormones (e.g., peptides) do not cross the bilayer. In these cases, a specific receptor on the cell surface determines whether the hormone will bind and then exert its effects. These are the hormones that will be discussed below.

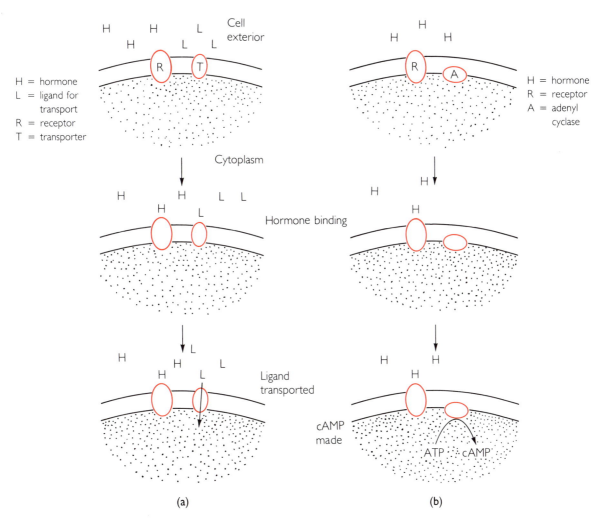

H = hormone
L = ligand for
    transport
R = receptor
T = transporter

Cell exterior

Cytoplasm

Hormone binding

Ligand transported

H = hormone
R = receptor
A = adenyl
    cyclase

cAMP made

(a)   (b)

FIGURE 2–20 Direct **(a)** and indirect **(b)** action by a hormone at the cell surface.

Following binding to the cell surface receptor, the hormone exerts its effects either directly or indirectly (Figure 2–20). In the direct scheme, hormone binding induces a membrane transport protein to perform its function. The receptor either may be the transporter itself or is linked in some way to the transporter. An example of this type of hormone is the plant hormone, indole acetic acid (auxin), which induces a H^+-ATPase at the membrane. The pumping of protons into the cell wall compartment apparently leads to the breakage of bonds in the wall. The loosening of the wall allows the cell to expand in response to the hormone (see Chapter 13).

In the indirect scheme, the hormone binding may activate adenylate cyclase, a protein located on the inner surface of the plasma membrane that catalyzes the synthesis of the "second messenger," cyclic AMP. This small molecule then exerts a number of metabolic effects. An example of these hormones is mammalian epinephrine (adrenalin). When it binds to its receptor on the surface of adipose tissue cells, binding stimulates adenylate cyclase to produce cyclic AMP on the inner surface of the membrane (Figure 2–21). The cAMP then acts as an activator of a protein kinase, an enzyme that phosphorylates serine residues of the enzyme lipase. This activates the enzyme, and lipids are hydrolyzed to release fatty acids for energy. The general pathway for this type of cAMP action is:

Hormone stimulates cAMP formation → cAMP binds to protein
kinase → protein kinase phosphorylates target proteins →
target proteins cause physiological changes.

The way that epinephrine and similar hormones actually activate an effector such as adenylase cyclase involves a second indirect pathway (Figure 2–22). This pathway, involving guanine nucleotide-binding **G proteins,** has been shown to exist also for the photoreceptor rhodopsin and muscarinic acetylcholine receptors. Three membrane proteins are involved in the transduction of information from extracellular environment to the cytoplasm: a receptor for the ligand (or light), an intermediary G protein, and the effector. The sequence occurs in five steps (see Figure 2–22):

1. The ligand binds to the receptor. This causes a conformational change so it can bind the G protein, which is complexed to GDP.

2. Binding to the receptor causes the G protein to undergo a conformational change, so it can now bind GTP instead of GDP. The ligand is released.

3. The G protein diffuses through the plane of the membrane to bind to the effector.

4. This binding causes the effector to undergo a conformational change that activates its function.

5. GTPase catalyzes the hydrolysis of the G-protein–bound GTP to GDP. This causes the protein to leave its site on the effector, and the situation returns to its initial state.

In smooth and cardiac muscle, the receptor for epinephrine is termed a beta-adrenergic receptor, and it is a transmembrane protein,

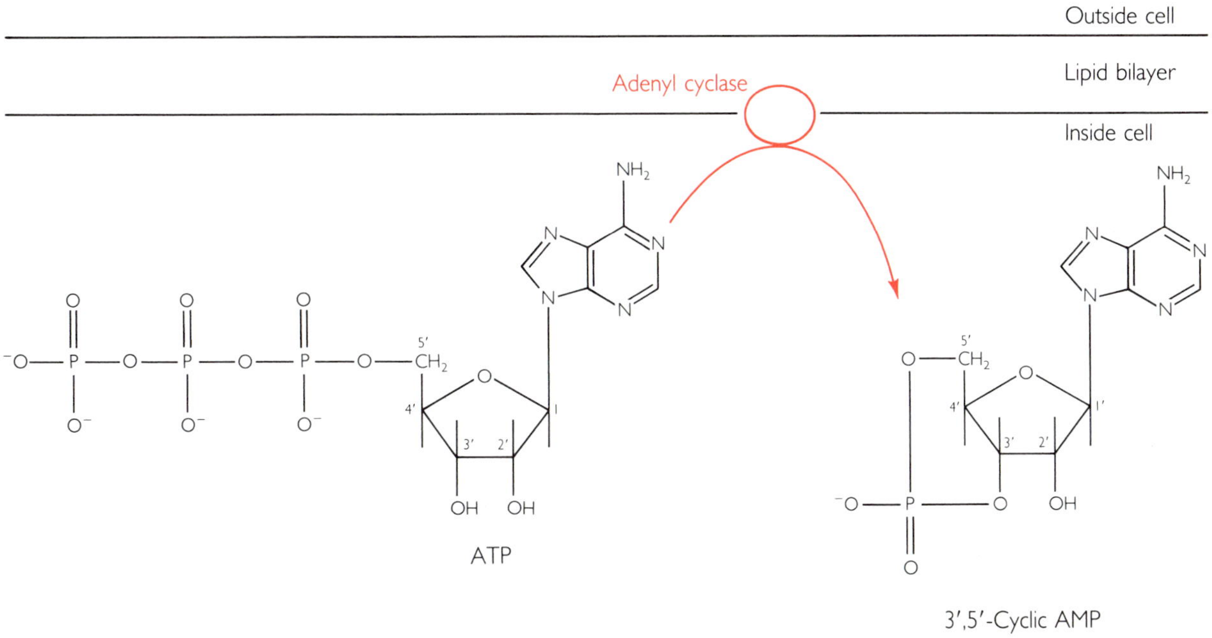

(a)

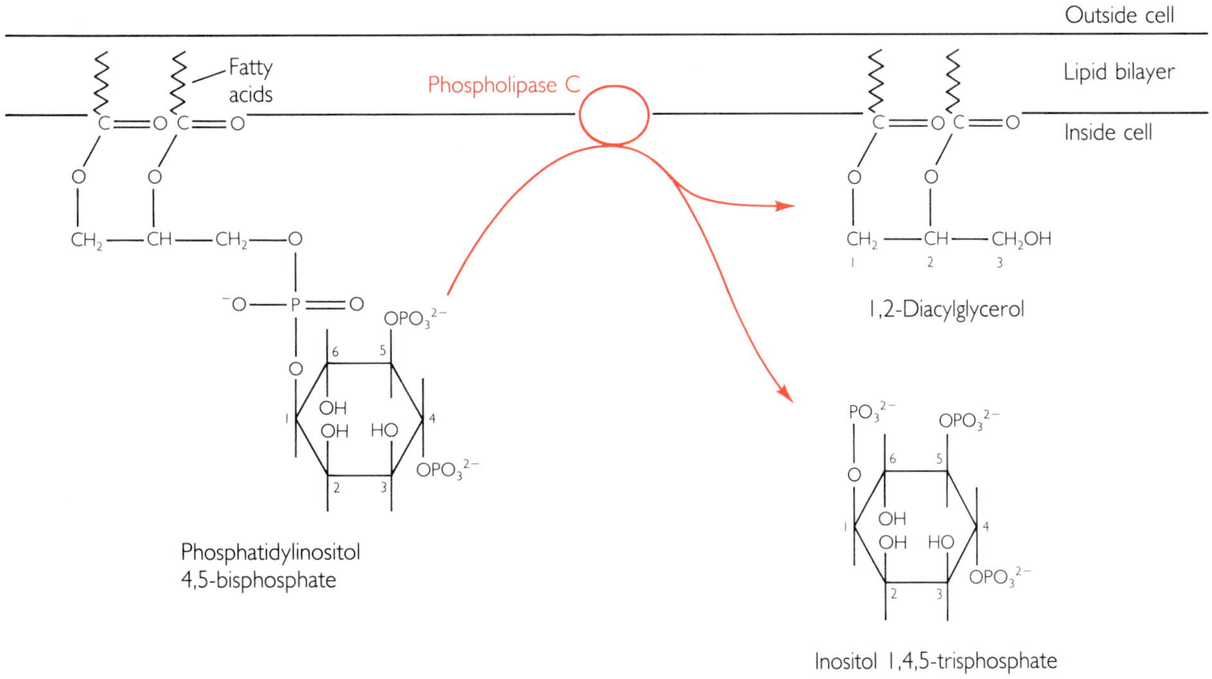

(b)

FIGURE 2–21 Formation of two kinds of intracellular "second messengers" from membrane events via adenyl cyclase **(a)** and phospholipase C **(b).**

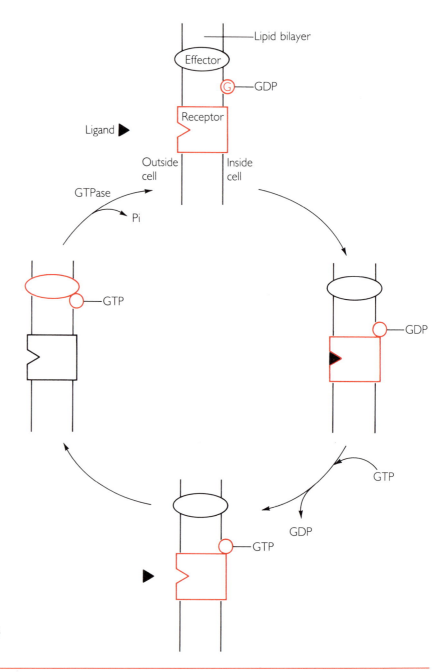

FIGURE 2–22

Signal transduction by a G protein. See text for details. "G": G protein.

with seven membrane-spanning regions and binding sites for both epinephrine and the G protein (Figure 2–23). The cAMP produced after the G protein cycle leads to the phosphorylation of Ca^{++} and K^+ channels, which stimulates the muscle to contract. A number of drugs can bind to the hormone site on the receptor. Some (e.g., isoproterenol) not only bind but also stimulate the G protein cycle; they are called beta-

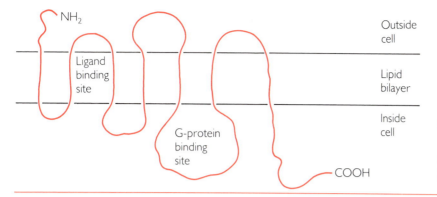

NH$_2$

Ligand binding site

G-protein binding site

COOH

Outside cell

Lipid bilayer

Inside cell

FIGURE 2–23

Structure of the beta-adrenergic receptor, a typical G-protein linked receptor.

agonists and are used to speed up the beat rate and contractility of the heart. Others, such as propranolol, bind but do not stimulate and are called beta-antagonists. They are used to reduce the heart rate and contractility.

If epinephrine binds to an alpha-adrenergic receptor, the effect is dramatically opposite. In this case, a different G protein (termed G_i for inhibitory to contrast it with the stimulatory G_s) acts to inhibit the adenylate cyclase. The recognition that there are different G proteins for different messages in the cell has led to the discovery of a number of G-related signaling pathways (Table 2–8). These can be irreversibly inhibited by pertussis or cholera toxins, which covalently modify the G proteins (see below).

The G_T protein, also called transducin, mediates the reception of light by the eye pigment, rhodopsin. When it absorbs light, rhodopsin activates G_T, which then interacts with the effector cGMP phosphodiesterase. This enzyme hydrolyzes the nucleotide cGMP by acting as a phosphodiesterase. Ordinarily, cGMP acts as a ligand to open the Na$^+$ ion channels in the photoreceptor membrane, and the cell is usually stimulated (i.e., in the dark it is depolarized and secretes an inhibitory neurotransmitter). But in the light, when the ligand's level falls, the channels are closed. This repolarization of the receptor cell (a rod) then

TABLE 2–8 Some G Proteins and Their Effects

Receptor	G Protein	Effector	Inhibitory Toxin
Beta-adrenergic	G_s	Adenylate cyclase	Cholera
Alpha-adrenergic	G_i	Adenylate cyclase	Pertussis
Opiate	G_o	Ion channels	Pertussis
Rhodopsin	G_T	cGMP phosphodiesterase	Pertussis
Many	G_P	Phospholipase C	Pertussis

results in the electrical stimulation of the membrane of the optic nerve, which leads to the brain.

The G_P protein carries information between many receptors and the enzyme effector phospholipase C. Over 30 hormones (e.g., vasopressin), growth factors (e.g., nerve growth factor), and other stimuli (e.g., sperm binding to egg) have been shown to respond via their receptors by stimulating the G_P protein to activate a form of phospholipase C. This enzyme catalyzes the partial hydrolysis of inositol phospholipids in the membrane to form inositol triphosphate and the remaining diacylglycerol (Figure 2–21).

These two products have profound effects on the cell. Inositol triphosphate binds to vesicles and the membrane of the endoplasmic reticulum (see Chapter 6) to cause the efflux of stored Ca^{++} into the cytoplasm. This cation acts to stimulate such processes as cell division and secretion. Diacylglycerol activates protein kinase C, an enzyme that puts phosphate groups on target proteins to activate them. The targets include ion channels, mRNA transcription factors, and various enzymes.

A remarkable family of proteins, called "ras," are present on the cytoplasmic side of the plasma membranes of many mammalian cells. These proteins are similar in structure to the G proteins, bind GTP, and have GTPase activity. But they are not clearly linked to receptors, and their intracellular targets are not known. What is known is that ras proteins are coded for by certain oncogenic (cancer-causing) viruses. In these cases, the ras genes are usually mutated so that the proteins bind GTP but do not hydrolyze it. This "always-active" ras protein appears to cause the infected cell to become cancerous.

Of even greater importance to people (since few human tumors are caused by viruses) is the fact that endogenous ras genes are commonly mutated in human cancers. Indeed, if one compares ras genes of bladder cancer tissue with the same genes in either normal tissue of the same person or normal bladder in a different person, only the tumor-derived ras gene is mutated. This observation can be used to identify tumors. Moreover, it clearly shows that a tumor can arise from a single cell that becomes mutated.

Growth Factors and Tyrosine Kinase Receptors

While epinephrine binding to a cell induces the synthesis of cAMP to produce physiological responses, there is another major mechanism used by some membrane-binding hormones. This involves the tyrosine kinase family of receptors, and a good example of this phenomenon is the action of insulin.

Insulin is produced by the pancreas in response to high levels of blood glucose. Its immediate effect is to increase the uptake of glucose

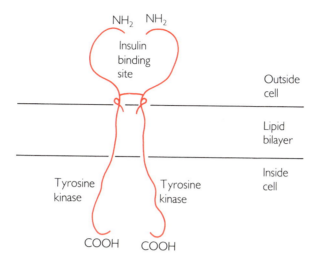

FIGURE 2–24
Structure of the insulin receptor, a typical tyrosine kinase receptor.

into cells, specifically muscle and adipocytes. The insulin receptor (Figure 2–24) has two subunits: an extracellular one, which has the binding site for the hormone, and a second subunit, which spans the membrane once and has a prominent cytoplasmic domain. The latter has tyrosine protein kinase activity. Generally, two copies of each subunit form the intact receptor.

Tyrosine protein kinase catalyzes the reaction:

$$\text{protein with tyrosine} + \text{ATP} \rightarrow$$
$$\text{protein with phosphotyrosine} + \text{ADP}.$$

As is the case with the cAMP cascade systems, phosphorylation changes the three-dimensional structure and function of target proteins. It is clear that when insulin binds to the receptor, the protein kinase activity is increased and is essential for the immediate physiological effect of increased glucose transport. This can be shown by the lack of insulin effect on cells whose receptors lack kinase activity.

The precise identification of the targets for phosphorylation remains to be achieved. One hypothesis involves a protein kinase cascade, where the tyrosine kinase of the receptor is just the first member. Another involves direct phosphorylation of an intracellular pool of glucose transporters, which are then inserted into the plasma membrane to enhance glucose transport. In addition, the endocytosis of insulin may be responsible for many of its long-term metabolic effects.

Remarkably, one of the targets of phosphorylation of the insulin receptor is the receptor itself! Indeed, most receptors can be phosphorylated, and this is presumed to result in a modulation of the receptor's activity. A well-known physiological phenomenon is down-regulation, in which prolonged exposure to a hormone causes a target cell to become less sensitive to that hormone. This could occur via receptor phosphorylation and inactivation. Or, it could occur by an actual decrease in the number of receptors on the cell surface via endocytosis.

A special class of molecules, **growth factors,** are similar to hormones in that they usually travel through the blood serum in low concentrations to exert their effects, but they differ in that they can affect the cells that produce them. Most of these factors are mitogens; that is, they stimulate cell division. A number of them, including epidermal growth factor (EGF) and platelet derived growth factor (PDGF), interact with cell membranes by the tyrosine kinase receptors.

Like ras, tyrosine protein kinase can be coded for by oncogenic viruses. Indeed, it was the first enzyme activity to be identified with a known oncogene. Comparison of the viral tyrosine kinase genes with cellular DNA revealed that many were homologous, although not identical to, cell membrane growth factor receptors. For example, the oncogene v-erbB of avian erythroblastosis virus is similar to the receptor for epidermal growth factor in chicken plasma membranes, except that the ligand-binding domain is missing. From this and other similar examples, it has been hypothesized that the cancer-causing ability of these oncogenes is intimately linked to the growth-promoting sequelae of membrane-bound tyrosine protein kinase activity.

Neurotransmitters

Neurotransmitters are small molecules that mediate the transmission of nerve impulses across synapses, the 50 nm gaps that mediate the unidirectional flow of information between either presynaptic and postsynaptic neurons or neurons and muscles. In the presynaptic element, a typical synapse contains many vesicles (Figure 2–25), which contain neurotransmitter. On stimulation, the latter is released by exocytosis and diffuses across the synapse. At the postsynaptic element, the neurotransmitter binds to a specific receptor, resulting in changes in ion flow and depolarization of the cell.

The events linking binding to ion flow are beginning to be characterized. In skeletal muscle, acetylcholine binding occurs on the outside of the membrane to a receptor that is an oligomer of five polypeptide chains. The receptor is actually a ligand-gated hydrophilic channel (see above and Figures 2–6 and 2–7). Binding of the neurotransmitter induces the gate to open, and positively charged ions (Na^+) flow along their electrochemical gradient into the cell, causing depolarization. Neurotoxins such as tetrodotoxin bind tightly to the acetylcholine binding site on the receptor, keeping it permanently open with disastrous results

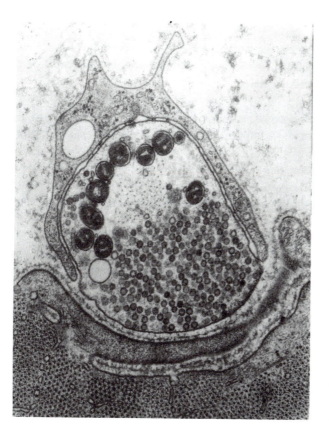

FIGURE 2–25
Synapse from a motor nerve ending. The neuron ending, with its many vesicles filled with neurotransmitter, is in the upper half of the photograph. There is a synapse (gap) between the neuron and the muscle cell in the bottom half. ($\times$70,000. Photograph courtesy of Dr. John Heuser.)

for nervous and muscular coordination. In the disease myasthenia gravis, many patients develop antibodies to their acetylcholine receptors. This autoimmune reaction leads to blockage of the synaptic event in muscles, and the patients become fatigued very easily.

The skeletal muscle nicotinic receptor (so called because it is stimulated by the drug nicotine) provides for rapid changes in ionic permeability and depolarization. However, in the heart, where changes occur over a longer period (seconds instead of milliseconds), a slower system is used. The muscarinic acetylcholine receptors (stimulated by the drug muscarine) of heart muscle are linked to a K^+ channel via a G protein cycle (Figure 2–22). The ligand (G-protein)-gated K^+ channel opens and K^+ rushes out of the muscle cells; this makes them hyperpolarized (even more negative inside), and they are inhibited from contracting.

Dopamine, an important transmitter in the brain, acts at the synapse indirectly by binding to its receptor and inducing adenylase cyclase at the inner membrane surface to make cyclic AMP. This second messenger apparently activates an enzyme (protein kinase) that phosphorylates a membrane protein. This event increases the negativity on the inside surface of the membrane, making membrane depolarization less likely. Thus, dopamine acts as an inhibitory neurotransmitter.

Toxins, Antigens, and Lectins

Infection of mammals by the bacterium *Vibrio cholerae* leads to the release of cholera **toxin.** This protein molecule has two subunits, one of which binds specifically to a cell membrane sphingolipid (GM_1 ganglioside). This binding results in a G-protein cycle that is defective, because the second subunit of the toxin catalyzes the covalent addition of the ligand, poly-ADP-ribose, to the G_s protein (Table 2–8). This modified G protein then enters the cycle (Figure 2–22), binding to the GM_1 receptor and adding GTP as usual. In this state, it binds to and activates adenyl cyclase to produce cyclic AMP. But the modified G protein cannot hydrolyze GTP to GDP and thus inactivate itself. Instead, it permanently binds GTP and permanently and excessively activates adenyl cyclase. This leads to a great excess of intracellular cAMP. Changes in membrane permeability rapidly occur, and in the intestine, water and electrolytes are released, and severe diarrhea and dehydration result. Pertussis toxin acts on different G proteins (Table 2–8) with similar results.

The bacterium *Corynebacterium diphtheriae,* when harboring a certain viral genome, produces diphtheria toxin. This protein has two regions. One interacts with specific receptors on the cell surface and contains a hydrophobic signal sequence that allows the toxin to cross the lipid bilayer. The other region, once in the cell, blocks a specific elongation factor on the ribosome, thus arresting protein synthesis.

Antigens are molecules foreign to an animal that can bind specifically to the surfaces of only certain cells, the lymphocytes. These white blood cells produce and may secrete to the bloodstream antibodies, proteins that bind specifically to the antigen. The antigen-antibody complex may then be endocytosed by phagocytes, thus eliminating the foreign molecule. Because there are many different antigens, there are many different classes of antibody, and usually a given lymphocyte will synthesize only a single type.

Which cells make which antibodies is signaled by the presence of the antibody on the cell surface. Thus an antigen selectively binds to those lymphocytes that will make its cognate antibody. Initially, binding is to antibodies that are diffusely scattered over the cell surface. These complexes then patch and cap (Figure 1–16). This is followed by a host of membrane changes, including altered permeabilities to ions and molecules. The lymphocyte then divides and differentiates to form a group of cells secreting the specific antibody.

Lectins are glycoproteins that bind to specific oligosaccharides on

T A B L E 2–9 **Some Plant Lectins and Their Specificities**

Lectin Name	Source Plant	Sugar Specificity
Concanavalin A	*Canavalia ensiformis*	Glucose, mannose
Castor bean lectin	*Ricinus communis*	Galactose, Acetylgalactosamine
Mung bean lectin	*Phaseolus aureus*	Galactose, xylose
Lentil lectin	*Lens culinaris*	Glucose, mannose
Soybean lectin	*Glycine max*	Galactose, Acetylgalactosamine
Wheat germ lectin	*Triticum vulgare*	Acetylglucosamine, Acetylmuramic acid
Blood group O lectin	*Ulex europeus*	Fucose
Blood group A lectin	*Vicia cracca*	Acetylgalactosamine
Blood group B lectin	*Bandeirea simplicifolia*	Galactose

the surface of the cell membrane. Over 60 lectins have been isolated, mainly from plants and invertebrates but also from cells as diverse as bacteria, slime molds, and rat white blood cells. Some of the plant lectins are noted in Table 2–9. Many of these are storage proteins that the plants synthesize as they make seeds and then break down when the seeds germinate, supplying the embryonic plant with amino acids. Others are secreted outside of the plant cell to its cell wall, where they have roles in defense against invading pathogens (see Chapter 13).

In terms of their membrane-binding properties, lectins are usually multivalent; that is, they have several of the specific sugar-binding sites per molecule, and this property allows them to induce receptor clustering on single cells and to cross-link between adjacent cells. The latter phenomenon has been observed to be more common in cancer cells than in normal cells. Although this is a useful observation in classifying cells, its significance in terms of cancer is not clear.

Differences in lectin-binding properties have been used to categorize the changing cell surfaces of virus-infected cells, host-parasite interactions, and embryonic cells. Lectins have even been used to purify different cell types from mixtures. The binding of a lectin to its cell surface receptor can lead to no affect, permeability, and then metabolic changes, or even cell death, depending on the cell type, the lectin, and its concentration. In the organisms that make them, lectins may be involved in defense against pathogens and symbiotic associations with other organisms. These roles will be discussed in Chapter 14.

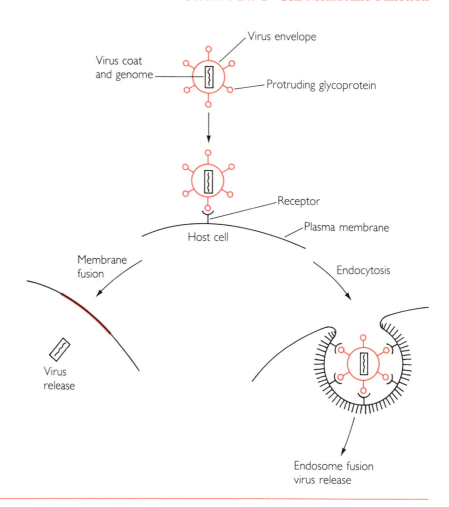

FIGURE 2–26
Mechanisms of enveloped virus
entry into cells.

Viruses

Viruses are often specific for the cells that they infect, and this specificity usually resides in cell surface receptors. Many viruses important to
people are enveloped (have a membrane coat surrounding a nucleopro-
tein core). These include the viruses causing rubella, rabies, mumps, influenza, and hepatitis. Protruding from the viral envelope is a glycoprotein that can bind to a different glycoprotein on a host cell. Binding is
followed by internalization of the virus, either by membrane fusion or by
receptor-mediated endocytosis (Figure 2–26).

Certain white blood cells, termed T helper lymphocytes, have a
CD4 receptor glycoprotein on their cell membranes. An important role
for this receptor is to bind to invading cells that contain an antigen on
their surface. Actual recognition is achieved not only by the presence of
the antigen but also by the presence of a histocompatibility complex

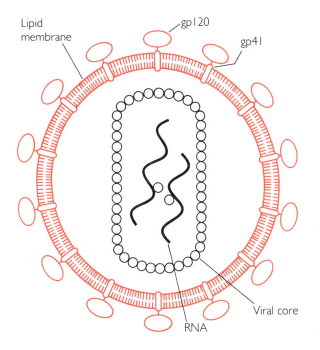

F I G U R E 2–27

Structure of an enveloped virus, the human immunodeficiency virus (HIV). The glycoprotein involved in recognition of the host cell receptor is gp120.

molecule. People with the disease acquired immunodeficiency syndrome (AIDS) have defective immune systems and hence impaired ability to fight off diseases such as infections and cancers. The main defect is in the T helper cells that have CD4 on their surfaces.

These observations led to the idea that the agent that causes AIDS affects cells with CD4 on their surface because it is the receptor for the infective agent. When the human immunodeficiency virus (HIV) was isolated (Figure 2–27), it was shown to have an envelope with two glycoprotein subunits: gp120 and gp41 (so named because their approximate molecular weights are 120,000 and 41,000 daltons, respectively). Because gp120 protrudes from the membranes, it not unexpectedly is the binding site to CD4. Indeed, the three-dimensional structure of gp120 is quite similar in its interactions with CD4 with the latter's natural substrate, a cell surface with an antigen and histocompatibility complex.

The role of gp41 is to help internalize the virus. After HIV binds to the target cell, gp41 apparently penetrates the host plasma membrane (note that it is an intrinsic transmembrane protein—see Figure 2–27). This event is essential for the fusion of the two membranes, viral and host cell. As a result, the envelope-free virus enters the cell and then can exert its lethal effects. One way to block the infection of this virus would be to upset the virus-receptor interaction. Clinical trials are underway using soluble CD4 receptors to tie up HIV in the bloodstream, as well as vaccines that bind to gp120.

Summary: The Erythrocyte Membrane

No cell membrane has been better characterized than that of the mammalian erythrocyte (Figure 1–1). This is not surprising, since the tissue is readily available in pure form and the plasma membrane is the only membrane of the cell. Many of the properties of membranes were first demonstrated on these cells and then generalized to other tissues.

Chemically, the red cell membrane is 40% lipid and 60% protein (Table 1–2). The major **lipids** are phospholipids, with much smaller amounts of the other lipid species (Table 1–3). About half of the fatty acid chains are saturated, with the remainder unsaturated. The phospholipids retain an asymmetry in the bilayer, with choline phospholipids (phosphatidylcholine and sphingomyelin) in the external half and amino phospholipids (phosphatidylserine and phosphatidylethanolamine) in the interior half. There is some evidence that membrane cholesterol may also be distributed asymmetrically, with more of it at the periphery of the cell.

Erythrocytes are biconcave disks, with a "doughnut" image under the microscope. The high cholesterol content in the regions peripheral to the "doughnut hole" contributes to the rigidity of this region. Structural alterations in red cell lipids can occur in several diseases (Table 2–10). In addition, an increase in the ratio of cholesterol to phospholipids, which reduces membrane fluidity, alters cell shape with adverse hematological consequences.

Erythrocyte membrane **proteins** were first characterized on acrylamide gel electrophoresis in the presence of sodium dodecyl sulfate (SDS). This technique solubilizes both hydrophilic and hydrophobic proteins and separates them on the basis of molecular weight. The result was a remarkably small number of major polypeptides (Figure 2–28). These have subsequently been individually characterized, mapped on the membrane, and assigned functions (Table 2–11).

T A B L E 2–10   **Alterations of Erythrocyte Lipids in Human Disorders**

Condition	Cholesterol: Phospholipid Ratio	Hematological Effect
Abeta lipoproteinemia	Elevated	Spiny cell shape
Lecithin-cholesterol acyl transferase deficiency	Elevated	Mild hemolysis
Severe hepatocellular disease	Elevated	Anemia

ADAPTED FROM: Ballas, S., and Krasnow, S. Structure and function of erythrocyte membrane and its transport functions. *Ann Clin Lab Sci* 10 (1980):209–219.

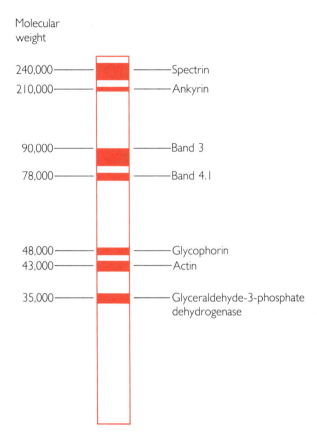

Molecular
weight

240,000 —————— Spectrin
210,000 —————— Ankyrin

90,000 —————— Band 3
78,000 —————— Band 4.1

48,000 —————— Glycophorin
43,000 —————— Actin

35,000 —————— Glyceraldehyde-3-phosphate
dehydrogenase

FIGURE 2–28

Electrophoretic separation of
erythrocyte membrane proteins
on an SDS-poly-acrylamide gel.

The major **intrinsic proteins,** glycophorin and band 3, span the
bilayer. Glycophorin is a "typical" intrinsic protein, with its amino terminal half glycosylated and sticking out on the exterior surface, as might
be expected from its mode of synthesis on the endoplasmic reticulum.
The part of the protein embedded in the bilayer is composed of hydrophobic amino acids (Figure 1–8). Although large in amount, the precise
function of glycophorin is not known; it contains the MN blood group
antigens on its extensive carbohydrate chains. The band 3 protein, on the
other hand, has a well-defined function—anion (Cl^- and HCO_3^-)
transport—but a more complex structure. This molecule spans the
membrane six times and has its amino terminus sticking out of the cytoplasmic side of the bilayer. To account for this orientation, multiple
internal hydrophobic insertion regions are present, and there is no signal
peptide.

The major **extrinsic proteins** lie on the cytoplasmic face of the
bilayer and form a submembranous skeleton. Electron microscopy reveals a mass of fibers in a hexagonal array. These fibers are composed of
spectrin, a large (molecular weight 450,000) rodlike protein with two
subunits with repetitive helical folds. Tetramers of spectrin are nonco-

TABLE 2–11   Major Erythrocyte Membrane Proteins

Protein (copies per cell)	Function
Intrinsic Proteins	
Glycophorin (600,000)	MN blood group
Band 3 (800,000)	Anion transport, ABO blood group
Na^+-K^+-ATPase (500,000)	Sodium pump
Extrinsic Proteins/Cytoplasmic Face	
Spectrin (200,000)	Filamentous network
Actin (400,000)	Attaches spectrin filaments
Ankyrin (400,000)	Binds spectrin to bilayer
Band 6 (148,000)	Glyceraldehyde-3-phosphate dehydrogenase
Adducin (200,000)	Binds actin filaments
Band 4.1 (200,000)	Binds actin filaments
Band 4.9 (43,000)	Binds actin filaments
Extrinsic Proteins/Extracellular Face	
Acetylcholinesterase	
Blood group antigens	

valently linked to short bundles of actin, the major component of cytoplasmic microfilaments (see Chapter 9) at junctional complexes. These regions also contain three proteins, adducin and bands 4.1 and 4.9, which act to promote the spectrin-actin association. Tropomyosin also associates with the actin filaments. The regulation of bundling of these proteins into a skeletal structure appears to be via a cAMP-dependent protein kinase. This enzyme can phosphorylate protein 4.9, and when this occurs the protein no longer binds to actin to promote actin-spectrin association.

The skeleton is connected to the inner surface of the membrane by ankyrin, which links spectrin to the anion transporter. Because spectrin has also been detected in neural, muscle, gastrointestinal, and respiratory tissues, as well as in lymphocytes and platelets, the membrane skeleton is apparently not a unique component of erythrocytes.

This submembranous structure gives the erythrocyte its integrity, elasticity, and disc shape. Two lines of evidence indicate this important role: First, chemical dissociation of the extrinsic proteins with solutions of low ionic strength leads to a disintegration of the skeleton and formation of a vesicular shape. Second, a number of abnormalities of red cell shape can be traced to abnormalities of the extrinsic protein spectrin (Table 2–12).

TABLE 2–12 Clinical Spectrin Abnormalities

Clinical Syndrome	Spectrin Abnormality
Hereditary elliptocytosis	Altered tryptic peptides; heat sensitive
Hereditary spherocytosis	Diminished stability
Muscular dystrophy	Abnormal phosphorylation
Sickle cell anemia	Altered filament organization

The best studied of these abnormalities is hereditary spherocytosis. This is the most common cause of hemolytic anemia in people of Northern European descent and is inherited most often as an autosomal dominant. Whereas in normal people, erythrocytes have about 240,000 spectrin dimers per cell, people with this disease have as low as 75,000 dimers. The result is a defective red cell membrane skeleton, and the cells appear spherical in shape (Figure 2–29). This results in increased deformability and lysis, and the cells have a lifespan much reduced from the normal 120 days. Severe anemia occurs that can sometimes be lethal.

Patients with hereditary elliptocytosis have a predominance of elliptical-shaped erythrocytes. This condition maintains itself even in red

FIGURE 2–29 Scanning electron micrographs of erythrocytes from normal **(a)** and hereditary spherocytosis **(b)** patients. (Courtesy of Dr. P. Agre.)

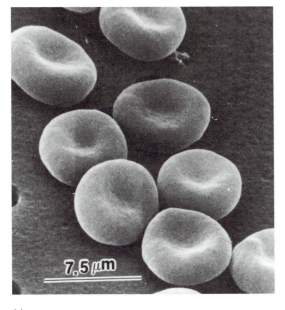
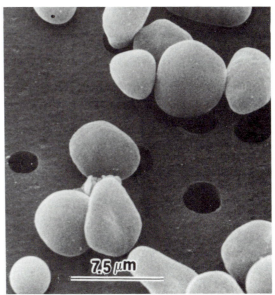

(a)   (b)

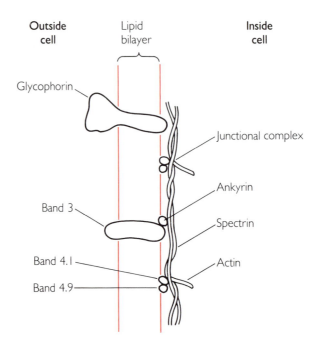

Outside cell · Lipid bilayer · Inside cell

Glycophorin

Junctional complex

Ankyrin

Band 3

Spectrin

Band 4.1

Actin

Band 4.9

FIGURE 2–30

Model for erythrocyte membrane protein organization.

cell ghosts (empty cells), indicating that the abnormal membrane shape is not due to the cytoplasm but is a true membrane property. The red cells are unusually heat-sensitive and fragile, and patients usually suffer from anemia. Spectrin isolated from their erythrocytes is also heat-sensitive; the amino acids sequence of the protein has been altered. Other patients with this disorder have abnormalities in the ability of spectrin to self-associate or in the relative amounts of the two subunits.

Some other people with anemias have defects in ankyrin and protein 4.1, with predictable damage to the red cell skeleton. An interesting abnormality that does not lead to clinical problems is ovalocytosis, a dominantly inherited characteristic common in Southeast Asia in which the red blood cells have an oval shape. Here, the defect is once again in a membrane protein, in this case the major transmembrane protein, band 3 (the anion transporter). Normally, this protein binds loosely to the submembrane skeleton via ankyrin. But in these people, the cytoplasmic region of band 3 protein is changed so that it binds very tightly. As a result, its lateral mobility in the membrane is restricted. This adds another factor, cytoplasmic skeletal interactions, to the list of determinants of membrane protein mobility.

A model for the arrangement of erythrocyte cell membrane proteins is shown in Figure 2–30. The major intrinsic proteins are band 3 and glycophorin. The main component of the skeleton is spectrin, present in tetrameric form, with the four monomers laid parallel to each other. At their ends, they associate with polymerized actin (the muscle protein) via bands 4.1 and 4.9 proteins in a junctional complex. The complex appears to cross-link the spectrin monomers; ankyrin, binding to a different part of the spectrin molecule, links with band 3, the trans-membrane intrinsic protein. Control of cell shape by this system is believed to reside in a protease that can specifically hydrolyze the various components of the membrane skeleton.

In addition to the major proteins, a number of enzymes are present in the erythrocyte membrane in catalytically important but quantitatively minor amounts. These include 6 ATPases, including the Na^+-K^+ pump and Ca^{++} pump, which span the membrane; 14 enzymes of carbohydrate metabolism; 4 enzymes of nucleotide metabolism; and 10 other enzymes and transport proteins. Thus the plasma membrane represents a metabolic compartment of some importance.

Further Reading

Agnew, W., Claudio, T., and Sigworth, F. Molecular biology of ionic channels. *Curr Top Membr Transp* 33 (1989).

Agre, P. Hereditary spherocytosis. *JAMA* 262 (1989):2887–2890.

Amino, N. Receptors in disease: An overview. *Clin Biochem* 23 (1990):31–36.

Antman, E., and Smith, T. Current concepts in the use of digitalis. *Adv Int Med* 34 (1989):425–454.

Apell, H.-J. Electrogenic activities of the Na, K pump. *J Membr Biol* 110 (1989): 103–114.

Aronson, P. The renal proximal tubule: A model for the diversity of anion exchangers. *Ann Rev Physiol* 51 (1990):419–441.

Baldwin, S. Uniporters and anion antiporters. *Curr Op Cell Biol* 2 (1990):714–721.

Bennett, V. Spectrin-based membrane skeleton: A multipotential adaptor between plasma membrane and cytoplasm. *Physiol Rev* 70 (1990):1029–1959.

Berard, J., Bourhis, J., and Riou, G. Clinical significance of multiple drug resistance in human cancers. *Anticancer Res* 10 (1990):1297–1302.

Berridge, M. J. Inositol triphosphate, calcium, lithium and cell signaling. *JAMA* 262 (1989):1834–1841.

Birnbaumer, L., Abramowitz, J., and Borwn, A. Receptor-effector coupling by G proteins. *Biochim Biophys Acta* 1031 (1990):163–224.

Borst, P. Genetic mechanisms of drug resistance. *Rev Oncol* 4 (1991):87–99.

Bourne, H., Sanders, D., and McCormick, F. The GTPase superfamily: Conserved structure and molecular mechanism. *Nature* 349 (1991):117–126.

Bretscher, M. Endocytosis: Relation to capping and cell locomotion. *Science* 224 (1984):681–685.

Brown, A. A cellular logic for G protein coupled ion channel pathways. *FASEB J* 5 (1991):2175–2179.

Brown, M., and Goldstein, J. A receptor-mediated pathway for cholesterol homeostasis. *Science* 232 (1986):34–47.

Cadena, D., and Gill, G. The nuclear pore: at the crossroads. *FASEB J* 6 (1992):2288–2295.

Cantley, L., Auger, K., Carpenter, C., Duckworth, B., Graziani, A., Kapeller, R., and Soltoff, S. Oncogenes and signal transduction. *Cell* 64 (1991):281–302.

Carafoli, E. The calcium pumping ATPase of the plasma membrane. *Ann Rev Physiol* 53 (1991):531–547.

Carafoli, E. The calcium pump of the plasma membrane. *J Biol Chem* 267 (1992):2115–2118.

Catt, K., Hunyady, L., and Balla, T. Second messengers derived from inositol lipids. *J Bioenerg Biomembr* 23 (1991):7–12.

Cheesman, C. Molecular mechanisms involved in the regulation of amino acid transport. *Prog Biophys Molec Biol* 55 (1991):71–84.

Chrispeels, M., and Raikhel, N. Lectins, lectin genes and their role in plant defense. *Plant Cell* 3 (1991):1–9.

Cornelius, F. Functional reconstitution of the sodium pump, kinetics of exchange reactions performed by reconstituted Na/K-ATPase. *Biochim Biophys Acta* 1071 (1991):19–66.

Dawson, D. Ion channels and colonic salt transport. *Ann Rev Physiol* 53 (1991):321–339.

Delaunay, J., Alliosio, N., Morie, L., and Pothier, B. The red cell skeleton and its genetic disorders. *Mol Asp Med* 11 (1990):116–241.

Field, M., Rao, M., and Chang, E. Intestinal electrolyte transport and diarrheal disease. *New Engl J Med* 321 (1989):879–883.

Franciolini, F., and Petris, A. Chloride channels of biological membranes. *Biochim Biophys Acta* 1031 (1990):247–259.

Frolich, O. The tunnelling mode of biological carrier mediated transport. *J Membr Biol* 101 (1988):189–198.

Geering, K. The functional role of the beta-subunit in the maturation and intracellular transport of Na,K-ATPase. *FEBS Lett* 285 (1991):189–193.

Gilman, A. G proteins and regulation of adenyl cyclase. *JAMA* 262 (1989):1819–1825.

Ginsburg, H., and Stein, W. New permeability pathways induced by the malarial parasite in the membrane of host erythrocytes. *Biosci Rep* 7 (1987):455–467.

Gordon, D. Ion channels in nerve and muscle cells. *Curr Op Cell Biol* 2 (1990):695–707.

Gould, G., and Bell, G. Facilitative glucose transporters: An expanding family. *TIBS* 15 (1990):18–23.

Grover, A. K. Calcium pump isoforms: diversity, selectivity and plasticity. *Cell Calcium.* 13 (1992):9–17.

Hammersen, F. *Endothelial cell vesicles.* Berlin: Karger, 1985.

Harris, R. A., and Allan, A. Alcohol intoxication: Ion channels and genetics. *FASEB J* 3 (1990):1689–1695.

Herbert, D., and Carruthers, A. Uniporters and anion antiporters. *Curr Op Cell Biol* 3 (1991):702–709.

Higgins, C., and Gottesman, M. Is the multidrug transporter a flippase? *Trends Biochem Sci* 17 (1992):17–21.

Hoekstra, D., and Kok, W. Entry mechanisms of enveloped viruses: Implications for fusion of intracellular membranes. *Biosci Rep* 9 (1989):273–305.

Hoffman, J., and Giebisch, G. *Current topics in membranes and transport.* Multivolume series. New York: Academic Press, 1983–.

Horisberger, J.-D., Lemas, V., Kraehenbuhl, J.-P., and Rossier, B. Structure-function relationship of Na,K-ATPase. *Ann Rev Physiol* 53 (1991):565–584.

Hunter, T. Protein modification: Phosphorylation on tyrosine residues. *Curr Op Cell Biol* 1 (1989):1168–1181.

Jan, L., and Jan, Y. Voltage-sensitive ion channels. *Cell* 56 (1989):13–25.

Jencks, W. How does a calcium pump pump calcium? *J Biol Chem* 264 (1989):18855–18858.

Jorgensen, P., and Andersen, J. Structural basis for conformational transitions in the Na, K-pump and Ca-pump proteins. *J Membr Biol* 103 (1988):95–120.

Juranka, P., Zastawny, R., and Ling, V. P-glycoprotein: Multidrug resistance and a superfamily of membrane-associated transport proteins. *FASEB J* 3 (1989):2583–2592.

Kane, S. E., Pastan, I., and Gottesman, M. Genetic basis of multidrug resistance of tumor cells. *J Bioenerg Biomembr* 22 (1990):593–618.

Kieber-Emmons, T., Jameson, B. A., and Morrow, W. The gp120–CD4 interface: Structural, immunological and pathological considerations. *Biochim Biophys Acta* 989 (1989):281–300.

Kobilka, B. Adrenergic receptors as models for G protein coupled receptors. *Ann Rev Neurosci* 15 (1992):87–114.

Krueger, B. Toward an understanding of structure and function of ion channels. *FASEB J* 3 (1989):1906–1914.

Lazarides, E., and Woods, C. Biogenesis of the red cell membrane skeleton and the control of erythroid morphogenesis. *Ann Rev Cell Biol* 5 (1989):427–452.

Lester, H. A. Strategies for studying permeation at voltage gated ion channels. *Ann Rev Physiol* 53 (1991):477–496.

Levitzki, A. From epinephrine to cyclic AMP. *Science* 241 (1988):800–805.

Libert, F., Vassart, G., and Parmentier, M. Current developments in G protein coupled receptors. *Curr Op Cell Biol* 3 (1991):218–223.

Lingrel, J., Orlowski, J., Shull, M., and Price, M. Molecular genetics of Na/K ATPase. *Prog Nucl Acid Res* 38 (1990):37–63.

Lolley, R., and Lee, R. Cyclic GMP and photoreceptor function. *FASEB J* 4 (1990):3001–3008.

Lux, S., and Becker, P. Disorders of the red cell skeleton. In *The metabolic basis of inherited disease,* 6th ed. (pp. 2367–2408). Scriver, C. M. et al. New York: McGraw-Hill, 1989.

McPherson, M. A., and Dormer, R. Molecular and cellular biology of cystic fibrosis. *Molec Asp Med* 12 (1991):1–81.

Mellman, I., Fuchs, R., Schmid, S., and Helenius, A. Control of intracellular membrane traffic: Implications for calcium and phosphate. *Prog Clin Biol Res* 252 (1988):101–108.

Milligan, G., Wakelam, M., and Kay, J. (Eds.). G proteins and signal transduction. *Biochem Soc Symp* 56 (1990).

Mooseker, M., and Morrow, J. (eds.) Ordering the membrane-cytoskeleton trilayer. *Curr Top Dev Biol* 38 (1992).

Morris, C. Mechanosensitive ion channels. *J Membr Biol* 113 (1990):93–107.

Morse, M., Satter, R. L., Crain, R., and Cote, G. Signal transduction and phosphatidylinositol turnover in plants. *Physiol Plant* 76 (1989):118–121.

Narahashi, T. *Ion channels.* New York: Plenum, 1989.

Nelson, N. Structure, function and evolution of proton ATPases. *Plant Physiol* 86 (1988):1–3.

Ohnishi, S., and Ohnishi, T. *Membrane-linked disorders.* Philadelphia: Taylor and Francis, 1989.

Palmgren, M. Regulation of plant plasma membrane proton-ATPase activity. *Physiol Plant* 83 (1992):314–323.

Parker, E., and Ross, E. G protein-coupled receptors: Structure and function of signal-transducing proteins. *Curr Top Membr Trans* 36 (1990):131–160.

Pastan, I., and Willingham, M. (Eds.). *Endocytosis.* New York: Plenum, 1986.

Pasternak, C. A. Membrane transport and disease. *Mol Cell Biochem* 91 (1989):3–11.

Pauza, C. The endocytic pathway for human immunodeficiency virus infection. *Adv Exp Med Biol* 300 (1991):111–138.

Quinton, P. Cystic fibrosis: A disease of electrolyte transport. *FASEB J* 4 (1990):2709–2717.

Rana, R., and Hokin, L. Role of phosphoinositides in transmembrane signalling. *Physiol Rev* 70 (1990):115–150.

Reid, E., Cook, G., and Luzio, J. *Cell membranes and disease.* New York: Plenum, 1987.

Roper, S. The cell biology of vertebrate taste receptors. *Ann Rev Neurosci* 12 (1989): 329–353.

Senwen, K., and Pouyssegur, J. G protein signal transduction pathways and the regulation of cell proliferation. *Adv Cancer Res* 58 (1992):75–93.

Serrano, R. *Plasma membrane ATPase of plants and fungi.* Boca Raton, FL: CRC Press, 1985.

Sharon, N. Lectins as cell recognition molecules. *Science* 246 (1989):227–234.

Silverman, M. Molecular biology of the Na^+-D-glucose transporter. *Hosp Prac* (1989):180–204.

Skou, J. The energy-coupled exchange of Na for K across the cell membrane. The Na/K pump. *FEBS Lett* 268 (1990):314–324.

Smith, T. Digitalis: Mechanisms of action and clinical use. *New Engl J Med* 318 (1988):358–365.

Smith, V. L., Kaetzel, M., and Dedman, J. R. Stimulus-response coupling: The search for intracellular calcium mediator proteins. *Cell Regul* 1 (1990):165–172.

Spach, G., Duclothier, H., Molle, G., and Valleton, J. Structure and supramolecular architecture of membrane channel-forming peptides. *Biochimie* 71 (1989):11–21.

Standen, N. Potassium channels, metabolism and muscle. *Exp Physiol* 77 (1992): 1–26.

Stein, W. D. The cotransport systems. *Curr Op Cell Biol* 1 (1989):739–745.

Strehler, E. Recent advances in the molecular characterization of plasma membrane Ca pumps. *J Membr Biol* 120 (1991):1–15.

Taylor, C. The role of G proteins in transmembrane signalling. *Biochem J* 272 (1990): 1–13.

Von Heijne, G. The signal peptide. *J Membr Biol* 115 (1990):195–201.

Wagner, R., and Casley-Smith, J. Endothelial vesicles. *Microvascular Res* 21 (1981): 267–298.

Wang, K., Villalobo, A., and Roufogalis, B. The plasma membrane calcium pump. *Trends Cell Biol* 2 (1992):46–51.

White, M. Structure and function of tyrosine kinase receptors. *J Bioenerg Biomembr* 23 (1991):63–80.

Yamamoto, H., and Kanaide, H. Release of intracellularly stored Ca ions by inositol triphosphate: An overview. *Gen Pharmacol* 21 (1990):387–393.

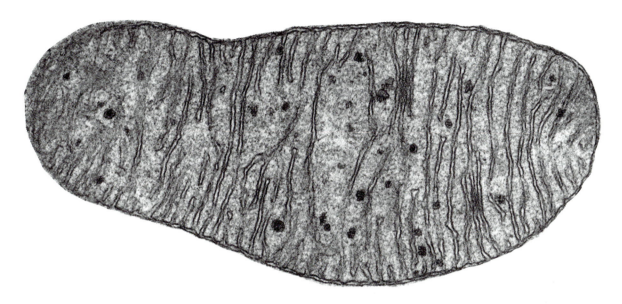

Mitochondria

An Energy-Transducing Compartment

Membranes are not used only to separate cells from their outside environment. Within the cell, separate compartments are surrounded by membranes to form defined structures. These **membrane-bound organelles** *include mitochondria, plastids, the endoplasmic reticulum, the Golgi complex, lysosomes, peroxisomes and other vacuoles, and the nucleus. They will be the subjects of this and succeeding chapters.*

The membrane that surrounds an organelle has the same roles as the plasma membrane. First, it separates the organelle from its surroundings (the cytoplasm), creating a cellular compartment in which certain biochemical reactions can occur away from others. This implies a second role of the organelle's membrane in exerting some control over what crosses it in both directions. Not surprisingly, the basic fluid mosaic model holds for organelle membranes, and their transport properties are often quite similar to those of the plasma membrane. For instance, as will be shown in this chapter, mitochondrial membranes perform co-transport and active transport.

Mitochondria and plastids (see Chapter 4) have internal membranes that have a third, important, property, that of **energy transduction.** *As will be made clear in the next two chapters, this process occurs largely as a result of the other two membrane properties: compartmentation and transport. In the case of the mitochondrion, the transduction is from the energy of oxidation-reduction to the phosphorylation of ADP to synthesize ATP. This provides usable chemical energy for processes elsewhere in the cell, such as protein synthesis in the cytoplasm and active transport at the plasma membrane.*

Structure

Light Microscopy

The development of the cell theory and improvements in the optics of the light microscope set late 19-century biologists on a search for structures inside the cell. These early cell biologists called their science **cytology,** a term now used primarily by pathologists when examining the overall structures of cells in diseased tissues.

By the late 1800s, cytologists had identified tiny granules as constituents of most cells of higher organisms. From their threadlike appearance during spermatogenesis, C. Benda termed the granules mitochondria (from the Greek "mitos" [thread] and "chondros" [granule]).

In 1900, L. Michaelis (who later became famous for his studies of enzyme kinetics) found that mitochondria in living cells could be specifically stained by the dye Janus Green B (Figure 3–1). Because this dye must be oxidized to become colored, Michaelis proposed that mitochondria are cellular oxidizing agents. This observable dye reaction became the criterion for identification of the organelle. Under the microscope, stained mitochondria were observed to move, probably in response to cytoplasmic streaming (Chapter 9), and to change shape, coalesce, and divide.

Mitochondria were first isolated in 1888, when A. Kolliker teased them out of insect flight muscle tissue. He showed that they were osmotically active and inferred that they are surrounded by a membrane. But it was not until the late 1940s that methods were developed for the large-scale isolation of mitochondria from tissues by differential centrifugation (see Appendix I). This allowed the identification of fatty acid breakdown, the citric acid cycle, electron transport, and ATP synthesis as biochemical functions compartmentalized in the organelle. To come full circle, in 1953 it was shown that the dye Janus Green B was structurally similar to the coenzyme FAD and that it was actually oxidized by the mitochondrial electron carrier cytochrome oxidase.

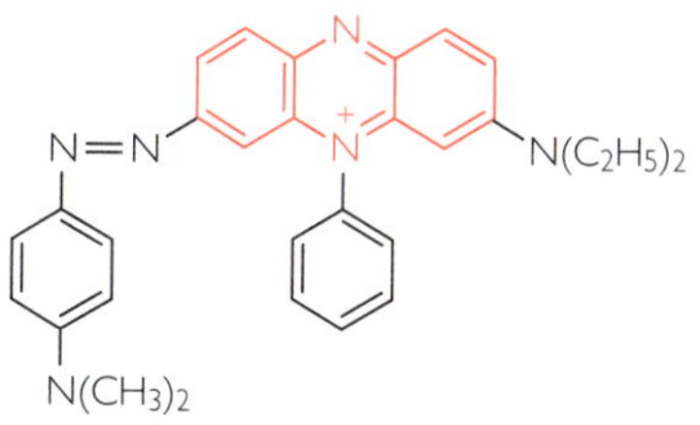

Electron Microscopy

When the electron microscope was used to examine the fine structure of cells in the early 1950s, it was soon revealed that the mitochondrion is a complex organelle (Figure 3–2). In 1956, G. Palade gave this still valid structural definition: "Two spaces or chambers are outlined by the mitochondrial membranes, an outer chamber contained between the two membranes, and an inner chamber bounded by the inner membrane. The inner chamber is penetrated, and, in most cases, incompletely partitioned by laminated structures which are anchored with their bases in the inner membrane and terminated in a free margin after projecting more or less deeply inside the mitochondrion."

The "laminated structures" are termed **cristae** (folds); the interior of the organelle is the **matrix.** The outer and inner membranes are 10 nm thick, and the space between the two membranes is approximately 8 to 10 nm.

Cristae vary in shape, size, and number. They are often arranged in parallel, perpendicular to the longitudinal axis of the mitochondrion, and appear as plates (Figure 3–2). In contrast, in some tissues, such as the adrenal cortex, they are tubular. This arrangement is also common in protozoans, insects, and algae, leading to the hypothesis that it represents a phylogenetically primitive structure. Cells active in energy metabolism (e.g., muscle) often have many cristae with a low matrix volume, and this situation is reversed in less active tissues (e.g., cells that

(a) Janus Green B

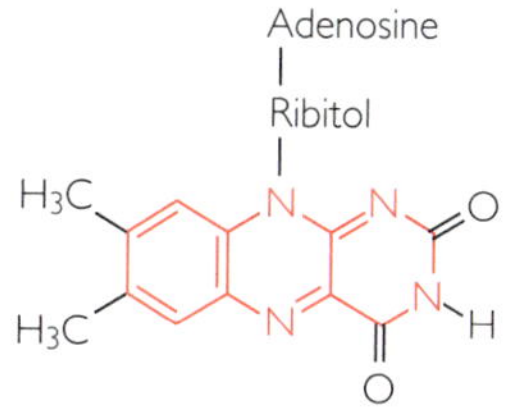

(b) FAD

FIGURE 3–1

Structure of Janus Green B **(a),** the redox dye used to stain mitochondria in living cells. Note the similarity to the structure of the flavin group of the coenzyme FAD **(b).**

FIGURE 3–2

Scanning electron micrograph of a mitochondrion from a rat epididymal cell. The outer membrane and inner membrane folds (cristae) are evident. (×35,000. Photograph courtesy of Dr. K. Tanaka.)

form eggs and sperm). However, the relationship between cristae and oxidative metabolism does not always hold. For example, if rice is germinated anaerobically, mitochondria develop abundant cristae but have no electron transport activity.

When visualized with the electron microscope, mitochondria characteristically appear as spheres or rods, 1 to 4 μm long and 0.3 to 1 μm wide. Their *number* per cell varies considerably, usually depending on the metabolic activity of the cell, so that active cells have more mitochondria than less active ones (Table 3–1). But counts of mitochondria can be misleading. In yeast and the rat nephron, reconstructions of over 150 serial sections of entire cells have revealed the existence of a single, giant, branched mitochondrion. When rapidly growing yeast slow down to stationary growth, numerous small mitochondria are observed, raising the possibility that the large structure fragments to form the small ones. The reverse has been observed to occur in spermatogenesis in mammals: Here, the midpiece of the mature sperm contains a fused mitochondrion that comes from the many separate ones in the spermatocyte.

In some tissues, mitochondria have a nonrandom **distribution** in the cell. In the sperm midpiece, the mitochondrion lies next to the tail, where its energy is used. Similar arrangements occur in muscle (mitochondria near the fibrils), epithelial cells (near the ciliated border), and kidney tubules (near the plasma membrane). In cells involved in ion and water transport, the plasma membrane has many folds to increase its

TABLE 3–1 Numbers of Mitochondria in Cells

Cell	Number of Mitochondria	% Cell Volume Occupied by Mitochondria
Frog oocyte	300,000	5
Liver hepatocyte	1,300	20
Plant meristem	300	10
Spermatocyte	100	—
Yeast	1(?)	10

surface area, and these folds form narrow cytoplasmic channels in which the mitochondria lie to supply energy for active transport. In the parietal cells that line the stomach, mitochondria lie adjacent to the smooth endoplasmic reticulum system used to secrete acid to the lumen of the stomach. Thus, mitochondria may lie near the location where they are needed, presumably to lower the time needed for ATP diffusion to its site of hydrolysis. But in most cells, mitochondria appear to be randomly distributed in the cytoplasm.

Mitochondria are dynamic organelles, which move about the cell and **change their shape.** Time-lapse cinematography has been used to follow their movements, and, although some patterns have been seen, their physiological significance is unknown. Since they were originally shown to exhibit osmotic swelling by Kolliker, mitochondrial volume changes have been extensively characterized. Swelling is promoted in an isotonic medium by such diverse agents as P_i, Ca^{++}, short-chain fatty acids, and thyroxine. It is reversed by Mg^{++} and ATP. Because these agents also promote muscle contraction, it has been proposed that mitochondrial swelling and contraction involve muscle-like proteins.

These changes may have physiological significance. If mitochondrial size is monitored (by light scattering and electron microscopy) during respiration, changes occur depending on the metabolic state of the cell. When a molecule that can be used for cellular respiration (e.g., pyruvate, which reduces NAD to NADH when oxidized) is absent, the inner membrane is largely detached from the outer one and is collapsed at the middle of the organelle, much like a plasmolyzed plant cell (Figure 2–1) except that there are a few points of attachment to the outer membrane. If substrate and phosphate are added, the inner membrane expands to a more typical configuration, lying next to the outer membrane. This change may be important in allowing more efficient translocation of metabolites between the cytoplasm, across the two membranes, and the matrix.

Chemical Composition

Isolated mitochondria are composed almost entirely of protein (70% by weight) and lipid (30%). Subfractionation involves first solubilizing the outer membrane in the detergent digitonin. This leaves the inner membrane surrounding the matrix (mitoplast). Further detergent treatment solubilizes this membrane.

The outer and inner membranes differ strikingly in **composition.** On an overall basis, the outer membrane is richer in lipid (40% vs. 20%) and poorer in protein (60% vs. 80%). These differences extend to a qualitative analysis (Table 3–2). The outer membrane is relatively rich in phosphatidyl inositol, while the inner one has a high cardiolipin content that serves as a marker during its isolation. Outer membrane lipids tend to be less fluid than those of the inner membrane, with more saturated fatty acids in the former and more unsaturated fatty acids in the latter. An additional factor that contributes to the rigidity of the outer membrane is its high content of galacturonic and other sugar acids. Finally, qualitative analysis of membrane proteins on gel electrophoresis shows only a few major species in the outer envelope but many in the inner membrane.

The two membranes are quite different in their osmotic behavior and **permeability.** The former is borne out by the swelling and contraction cycle mentioned above, where the outer membrane appears rigid while the inner membrane expands or shrinks. The inner membrane is similar to the plasma membrane in that it is impermeable except to very small nonpolar molecules. The only way for small polar molecules to get through the inner membrane is via specific carriers, which are intrinsic membrane proteins.

On the other hand, the outer membrane is permeable to practically all molecules up to a molecular weight of about 5000. Vesicles made of isolated outer membranes are permeable to sucrose, nucleotides, and small peptides but not to cytochrome c (molecular weight 13,000). Favorable electron micrographs and X-ray diffraction studies show pores in the membrane, which could act as a molecular sieve. The role of this leakiness to small but not large molecules is not clear. It probably acts to block cytoplasmic enzymes from reacting with inner membrane components.

If the major outer membrane protein is extracted and added to artificial lipid vesicles, they attain the sieving properties of intact outer membranes. This pore protein, or **porin,** disrupts the rather rigid bilayer to form channels about 2 nm in diameter through which molecules with molecular weights of less than 6000 can pass. The pore appears to be voltage-gated, opening only when there is a potential across the outer membrane, and is also sensitive to anions. In some situations, porin acts to connect the inner and outer membranes.

TABLE 3–2 **Lipid Composition of Cauliflower Mitochondrial Membranes**

	Outer Membrane	Inner Membrane
Lipids (% wt)		
Phosphatidylcholine	42	41
Phosphatidylethanolamine	24	37
Phosphatidylglycerol	10	3
Phosphatidylinositol	21	5
Cardiolipin	3	14
Fatty Acids (%)		
$C_{16:0}$	50	10
$C_{18:0}$	4	1
$C_{18:1}$	20	7
$C_{18:2}$	8	13
$C_{18:3}$	18	69

FROM: Hanson, J., and Day, D. Plant mitochondria. In Stumpf, P., and Conn, E. (Eds.). *The biochemistry of plants,* Vol. 1. New York: Academic Press, 1980.

The connection of the outer and inner membranes appears to be necessary for the import of proteins into the organelle that are made in the cytoplasm (see Chapter 5). ADP stimulates membrane contacts, and a number of kinases that produce ADP (e.g., creatine kinase, hexokinase, glycerol kinase) bind to porin. A function of this binding of normally cytoplasmic enzymes may be to indirectly promote contact between the two membranes.

In electron micrographs, the mitochondrial matrix usually contains regions with low electron density, containing fibrils that are 5 nm in diameter. These disappear when the tissue section is treated with deoxyribonuclease, indicating the presence of mitochondrial DNA. Ribosomes, usually smaller in appearance than those in the cytoplasm, are also present.

The two membranes essentially divide the organelle into four compartments: outer membrane, intermembrane space, inner membrane, and matrix. In the liver hepatocyte, the proportion of total mitochondrial protein in the compartments is about 8%, 4%, 21%, and 67%, respectively. The four compartments delineate different functions and enzyme contents (Table 3–3).

The marker enzyme that uniquely identifies the outer membrane in mammals is **monamine oxidase,** which catalyzes the oxidative removal of amino groups from the so-called biogenic amines: epinephrine, norepinephrine, dopamine, and serotonin. The resulting products are ultimately converted into molecules that are excreted in the urine. Because these amines are important in hormonal and neural communication as

---

TABLE 3–3 Compartmentation of Enzymes in
Mitochondria

Outer Membrane

Choline phosphotransferase

Cytochrome b_5 reductase

Fatty acyl CoA synthetase

Fatty acid elongation, C_{14} and C_{16}

Glycerophosphate acyl transferase

Monamine oxidase

Intermembrane Space

Adenylate kinase

Creatine kinase (heart, skeletal muscle, brain)

Nucleoside phosphokinases

Sulfite oxidase

Inner Membrane

ATPase (reversible)

Anion and cation translocases

Carnitine–fatty acyl transferase

Respiratory chain system

Succinate dehydrogenase

Matrix

Citric cycle enzymes (except succinate dehydrogenase)

Fatty acid β-oxidation system

Fatty acyl CoA synthetase

Nucleic acids and protein synthesis systems

Urea cycle (part)

---

either hormones (e.g., epinephrine) or synaptic transmitters (e.g., dopamine), the activity of the oxidase on the outer membrane can regulate the amounts of these substances a cell can secrete for effects on target cells.

A number of psychoactive drugs act by inhibiting this enzyme. For instance, when a person drinks an alcoholic beverage, the ethanol is initially converted to acetaldehyde. The latter is an inhibitor of monamine oxidase for serotonin, and apparently the resulting accumulation of the latter is responsible for some of the intoxicating effects. Other monamine oxidase inhibitors are used to treat depression (e.g., isocarboxazid and phenelzine).

The enzymes marking the intermembrane space are nucleoside phosphokinases. These phosphorylating enzymes allow the nucleotides formed to be translocated into the mitochondria from the cytoplasm, where they are made.

The mitoplast (inner membrane and matrix) contains the enzymes of energy metabolism. While glycolytic breakdown of glucose occurs in the cytoplasm, most of the citric acid cycle occurs in the mitochondrial matrix. The transduction of energy by oxidative phosphorylation occurs in the inner membrane. The mitoplast is also involved in fatty acid metabolism. Transport of fatty acids for breakdown occurs via the carrier, carnitine, in the inner membrane, and once in the matrix, further breakdown occurs by beta-oxidation enzymes.

Function: Translocation Across the Inner Membrane

Ions

Mitochondria have an important role in the regulation of the ionic composition of the cell, and this role is performed primarily by the inner membrane. These ionic movements have three general requirements:

1. Because the inner membrane is generally impermeable to ions, specific translocase proteins are needed.

2. The ions often move against their concentration gradient, implying the need for metabolic energy (active transport).

3. The movement of an ion into or out of the mitochondrion upsets the electrical potential across the membrane and this is involved in the synthesis of ATP. An ion of equal and like charge may be expelled for each ion taken up into the matrix, or one of opposite charge is taken up at the same time.

In animal mitochondria, the ion taken up in the greatest concentration is Ca^{++}, especially if the organelles are active in electron transport. In fact, if a mixture of Ca^{++} and ADP is added to mitochondria in the presence of phosphate ion, they preferentially accumulate Ca^{++} and phosphate over ADP and phosphate. The influx of Ca^{++} is accompanied by an efflux of protons and, usually, an influx of an anion such as phosphate. Because these movements are against their concentration gradients, they must be driven by an energy-releasing process. Usually, this is via a Ca^{++}-activated membrane-bound ATPase (a calcium pump), but in some instances the energy needed comes directly from electron transport.

When mitochondria accumulate Ca^{++} and phosphate, dense granules of insoluble calcium phosphate appear in the matrix. These may be in storage for future calcification reactions, such as egg shell or bone formation. The decrease in cytoplasmic Ca^{++} mediated by mitochondria has physiological significance because this cation has a multitude of effects in the cytoplasm, including promotion of glycogen breakdown, pyruvate oxidation, and muscle contraction. Thus, maintenance of a proper cytoplasmic concentration of Ca^{++} is important to the cell. Release of Ca^{++} by the mitochondria can occur because of hormonal (e.g., ACTH) stimulation or an inhibition of electron transport.

Extensive Ca^{++} transport does not occur across the inner membrane of plant mitochondria. Instead, the predominant ion is phosphate, which enters along with counterions such as K^+ and Mg^{++}. Again, movements are usually against concentration gradients and are driven by either ATP hydrolysis or electron transport.

The uptake of monovalent cations into mitochondria is considerably slower than that of Ca^{++} but can be greatly increased by ionophores (Figure 2–4). These hydrophobic antibiotics carry ions such as K^+ across the lipid bilayer. Ionophoretic transport upsets ionic gradients across the mitochondrial membrane and uses energy otherwise destined for ATP synthesis (see below).

Metabolites

The matrix of the mitochondrion is a cellular metabolic compartment. In it, pyruvate is broken down to CO_2, fatty acyl-CoA's are oxidized to acetyl-CoA, ATP is made, and substrates are present for the synthesis and breakdown of amino acids. But the inner mitochondrial membrane is impermeable to pyruvate, fatty acids, ATP, and glutamate, thus making transport through the membrane by diffusion difficult. This difficulty is overcome by the presence of specific protein **translocases** in the inner membrane.

Translocation was first found during the 1960s during studies of osmotically driven mitochondrial swelling. If an ammonium salt of a permeant anion (e.g., acetate) is tested, the NH_4^+ will cross the inner bilayer as nonpolar NH_3 and the anion as its conjugate acid (acetic acid). Once inside the matrix, the salt (ammonium acetate) forms once again. But in the process, a proton has also been translocated. This causes an imbalance of protons across the membrane and leads to an osmotic swelling of the mitochondrion, which is measured by light scattering. Other methods to detect anion entry include isotopic labeling of the permeant with rapid separation of organelle and medium and the use of specific inhibitors of anion transport.

Some of the translocation systems discovered by these techniques are shown in Table 3–4. Each carrier is vital to the metabolic roles of the mitochondrion (see Appendix II for a review of the biochemistry involved):

TABLE 3–4 Translocases on the Inner Mitochondrial Membrane

Name	Molecules Translocated (in ↔ out)
Pyruvate	Pyruvate^{-1} ↔ OH$^-$
Fatty acid	Fatty acyl carnitine ↔ carnitine
Phosphate	Phosphate$^-$ ↔ OH$^-$
Adenine nucleotide	ADP ↔ ATP
Oxoglutarate	Malate$^-$ ↔ α-ketoglutarate$^-$
Glycerophosphate	Glycerophosphate$^-$ ↔ dihydroxyacetone-P$^-$
Aspartate	Glutamate$^-$ ↔ aspartate$^-$
Glutamate	Glutamate$^-$ ↔ OH$^-$
Dicarboxylates	Succinate$^-$, malate$^-$ ↔ phosphate$^-$
Tricarboxylates	Citrate$^-$, isocitrate$^-$ ↔ malate$^-$

The **pyruvate** carrier imports pyruvate, the product of aerobic glycolysis.

The **fatty acid** carrier imports fatty acids coupled to the carrier carnitine, since the inner membrane is relatively impermeable to fatty acyl-CoA, which is the initial product in the oxidation of fatty acids. Once across the inner membrane, carnitine is released and transported out, and the fatty acid is oxidized by enzymes in the matrix.

The **phosphate** carrier imports phosphate needed to synthesize ATP.

The **adenine nucleotide** carrier imports ADP so that it can be converted to ATP at the inner membrane. The ATP is then translocated out to be used throughout the cell.

The **oxoglutarate** carrier imports malate, a citric acid cycle intermediate, and exports α-ketoglutarate, which is important in the anabolism of amino acids by cytoplasmic enzymes.

The **glycerophosphate** carrier is involved in a shuttle system for the import of reducing equivalents into the mitochondrion.

The **aspartate** carrier and the oxoglutarate carrier form a shuttle system for the import of reducing equivalents into the mitochondrion.

The **glutamate** carrier imports this product of amine (from protein) metabolism to the mitochondrial matrix, where it can be deaminated and the ammonium formed into urea.

The **dicarboxylate** and **tricarboxylate** carriers import citric acid cycle intermediates that have been formed in the cytoplasm from certain amino acids.

Like other membrane transport systems (see Chapter 2), the mitochondrial translocases are proteins. They exhibit substrate specificity and saturable kinetics and have specific inhibitors. Perhaps the best studied system is the adenine nucleotide carrier, which catalyzes a 1:1 exchange between ATP and ADP and is specific for those nucleotides only. During active respiration, the translocation is directional, with ADP entry into the mitochondrion and ATP exit being favored. Thus, the ratio of ATP/ADP is much higher in the cytoplasm than in the mitochondrial matrix. Because ATP has three charged groups and ADP two, this asymmetry is electrogenic; that is, it sets up a potential gradient across the membrane. This gradient is fueled by some of the protons pumped across the inner membrane during electron transport (see below).

The nucleotide carrier has been isolated and is a protein dimer that undergoes a conformational change when the nucleotide binds to it. There are separate binding sites that face the cytoplasm and matrix. Binding of ADP is competitively inhibited by atractyloside, a toxin isolated from the thistle *Atractyles gummifera;* ATP binding is inhibited by the bacterial toxin bongkreic acid.

The glycerophosphate carrier is the key component in a **shuttle** for the transport of reducing equivalents from the cytoplasm into the mitochondrion. Glycolysis occurs in the cytoplasm and results in the formation of some NADH via the reduction of NAD while a glucose metabolite is oxidized. In an overall sense:

$$\text{glucose} + \text{NAD} \rightarrow \text{pyruvate} + \text{NADH}.$$

In order for glycolysis to continue, the NADH formed must be reoxidized to NAD, and, at the same time, this reoxidation can provide energy for transduction to ATP formation (see below). Typically, in the presence of adequate O_2, these reactions occur in the mitochondrion, and the NADH enters the organelle to be oxidized, leaving as NAD.

But in liver cells, the inner mitochondrial membrane is not permeable to NADH. Instead, the NADH is oxidized in the cytoplasm, and the reducing equivalents, now present in a "proxy," are translocated into the mitochondrion. When the "proxy" is oxidized there, the reducing equivalents are transferred to NAD (or FAD) and then enter the electron transport chain. The oxidized "proxy" is then translocated out into the cytoplasm, where the process can begin again.

The shuttling process is simply shown by the glycerol phosphate shuttle (Figure 3–3). Here, the oxidized "proxy" is dihydroxyacetone phosphate, and when reduced, it is α-glycerophosphate. The mitochondrial acceptor is FAD. Another shuttle, presented more schematically, is the malate-aspartate shuttle (Figure 3–4), which is somewhat more indirect but also requires the use of inner membrane translocases.

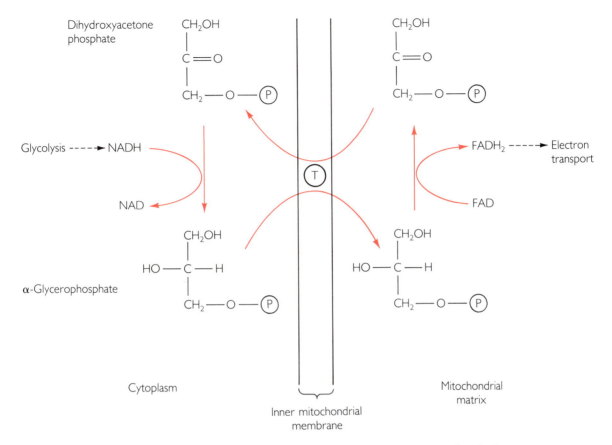

FIGURE 3–3 The α-glycerophosphate shuttle for the oxidation of extramitochondrial NADH in liver cells. The reducing equivalents of NADH are transferred to dihydroxyacetone phosphate, forming α-glycerophosphate. The latter is translocated (via T) into the mitochondrial matrix, where a second enzyme converts it back to dihydroxyacetone phosphate. In the process, FAD is reduced to $FADH_2$, which enters electron transport. The dihydroxyacetone phosphate is now translocated back to the cytoplasm, so the process may begin again.

Function: Energy Transduction

Biochemical Background

The mitochondrion transduces chemical energy for use by the cell in synthetic reactions and active transport. This energy is derived from the three major cellular foodstuffs—carbohydrates, lipids, and proteins. These reduced, energy-rich compounds, or their breakdown intermediates, are translocated into the mitochondrial matrix, where, through the

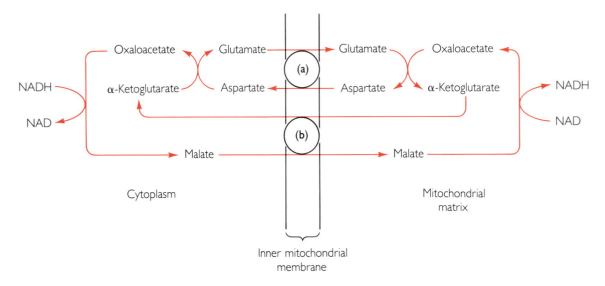

FIGURE 3–4   The malate-aspartate shuttle for transferring cytoplasmic-reducing equivalents into the mitochondrion in liver. Note the involvement of two translocases: **(a)** α-ketoglutarate-malate transporter and **(b)** glutamate-aspartate transporter.

citric acid cycle, they are oxidized to CO_2 (Figure 3–5). This oxidation is coupled with the reduction of coenzymes such as NAD, which pass their reducing energy to oxygen in the electron transport chain, located in the inner membrane. The oxidation of the components of this chain is coupled with the phosphorylation of ADP to form ATP, the molecule that "stores" energy for use by the cell. Thus, the "flow" of energy in the mitochondrion is:

$$\text{foodstuffs} \rightarrow \text{citric acid cycle} \rightarrow \text{coenzymes} \rightarrow$$
$$\text{electron transport chain} \rightarrow \text{ATP}.$$

Carbohydrate breakdown begins in the cytoplasm with glycolysis:

$$\text{Glucose} + 2\text{ADP} + 2\text{P}_i + 2\text{NAD} \rightarrow$$
$$2 \text{ pyruvate} + 2\text{ATP} + 2\text{NADH}.$$

The pyruvate is translocated through the inner membrane to the mitochondrial matrix, where it is first metabolized by the pyruvate dehydrogenase complex:

$$\text{Pyruvate} + \text{CoA-SH} + \text{NAD} \rightarrow \text{acetyl-CoA} + \text{NADH} + CO_2.$$

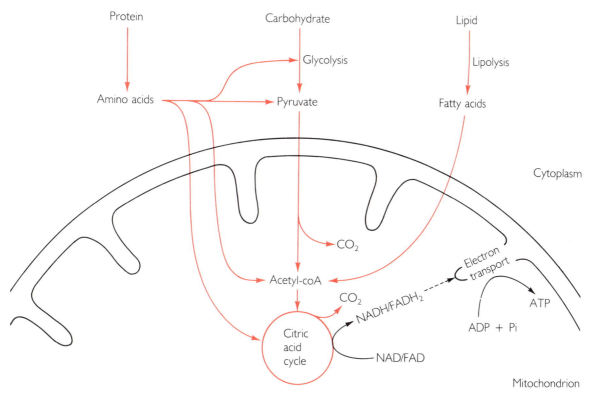

FIGURE 3-5 Overview of the breakdown of the three major fuels—carbohydrates, lipids, and proteins. Note that in each case the metabolites flow into the mitochondrion.

The acetyl-CoA is then fully oxidized by the citric acid cycle:

$$\text{Acetyl-CoA} + 3\text{NAD} + \text{FAD} + \text{GDP} + \text{Pi} \rightarrow$$
$$2\text{CO}_2 + \text{CoA-SH} + 3\text{NADH} + \text{FADH}_2 + \text{GTP}.$$

It is important to note that reactions in which the glucose is being transformed from a reduced to an oxidized state are coupled with coenzyme reactions in which they are being reduced (e.g., NAD to NADH). The cytoplasmic NADH formed during glycolysis can be reoxidized and its reducing equivalents translocated into the matrix via the glycerophosphate and malate-aspartate shuttles (Figures 3–3 and 3–4). Thus, carbohydrate catabolism leads to CO_2 and reduced coenzymes in the mitochondrial matrix.

Lipid catabolism generates fatty acids and glycerol. The latter can be converted into the glycolytic intermediate dihydroxyacetone phosphate and then can be converted to pyruvate and enter the citric acid cycle. Fatty acids are first acylated with CoA in the cytoplasm and are then translocated across the inner membrane via the carrier carnitine. In the matrix, they are oxidized to acetyl-CoA with concomitant coenzyme reduction: For example,

$$\text{palmitic acid } (C_{16:0}) + 8\text{CoA-SH} + 7\text{NAD} + 7\text{FAD} \rightarrow$$
$$8\text{acetyl-CoA} + 7\text{NADH} + 7\text{FADH}_2.$$

The acetyl-CoA then enters the Krebs cycle, as above. Once again, the principal products are CO_2 and reduced coenzymes in the mitochondrial matrix.

The carbon atoms of **amino acids** enter the citric acid cycle either directly, via one of its intermediates, or indirectly. Some (e.g., aspartate, glutamate) must be translocated through the inner membrane, whereas others enter via a catabolic product. The ketogenic amino acids leucine, lysine, phenylalanine, and tyrosine are converted into acetoacetyl-CoA, which in turn forms acetyl-CoA. Isoleucine, leucine, and tryptophan form acetyl-CoA directly. Alanine, cysteine, glycine, and serine are converted to pryuvate. The remaining amino acids are transformed into one or another of the intermediates in the citric acid cycle. Therefore, catabolism of amino acids' carbon skeletons also ends in the formation of CO_2 and reduced coenzymes through the citric acid cycle.

Electron Transport

Carbohydrates, lipids, and proteins are catabolized from energy-rich reduced states to CO_2, a relatively energy-poor oxidized compound. Their reducing energy is stored in the coenzymes NADH and $FADH_2$ in the mitochondrial matrix. These represent a source of potential chemical energy.

The tendency of a reduced compound to lose its reducing electrons, as in the reaction:

$$\text{NADH} \rightarrow \text{NAD}^+ + \text{H}^+ + 2e^-$$

is given by the standard redox potential E_0'. This is the force generated by the flow of electrons from the compound to a suitable electron acceptor under standard (1.0M, pH 7.0, 25°C) conditions.

In biochemical convention, these potentials are termed reduction potentials, so that compounds can be compared in terms of their abilities to accept electrons:

$$\text{NAD}^+ + \text{H}^+ + 2e^- \rightarrow \text{NADH} \qquad E_0' = -0.32 \text{ volts}$$
$$\tfrac{1}{2}\text{O}_2 + 2\text{H}^+ + 2e^- \rightarrow \text{H}_2\text{O} \qquad E_0' = +0.82 \text{ volts}$$

Additional potentials relevant to mitochondrial function are listed in Table 3–5.

These potentials are usually measured in electrochemical cells and compared to hydrogen, but the situation in living cells is different. Instead of an acceptor electrode, the mitochondrion has electron accepting molecules. Electrons tend to flow from more electronegative (tend to give up electrons and become oxidized) to electropositive (tend to accept electrons and become reduced) molecules. The free energy change for this coupled reaction is:

$$\Delta G_0' = -nF\Delta E_0'$$

where $\Delta G_0'$ is the standard free energy change, n is the number of electrons transferred, F is Faraday's constant (23,062 calories/volt/mole), and $\Delta E_0'$ is the standard reduction potential change.

These relationships can be used to calculate the free energy released when electrons are transferred between any two compounds. The $\Delta E_0'$ is calculated by writing the reduction as shown in Table 3–5 and the oxidation in reverse. Thus, the sign of the potential of the oxidative reaction is changed. For instance, for the coupled transfer of electrons from NADH to O_2,

$$\Delta E_0' \text{ is } (0.82-(-0.32)) = 1.14 \text{ volts.}$$

Thus,

$$\Delta G_0' = (-2)(23062)(1.14) = -52,600 \text{ cal/mole.}$$

This is clearly a large free energy change that is thermodynamically highly favorable and could be used to fuel active transport and synthesis.

TABLE 3–5 Reduction Potentials

Reaction	Red. Pot. (E_0', volts)
$1/2 O_2 + 2H^+ + 2e^- \rightarrow H_2O$	+0.82
Cytochrome a_3-Fe^{+3} + $e^- \rightarrow$ cytochrome a_3-Fe^{+2}	0.55
Cytochrome a-Fe^{+3} + $e^- \rightarrow$ cytochrome a-Fe^{+2}	0.29
Cytochrome c-Fe^{+3} + $e^- \rightarrow$ cytochrome c-Fe^{+2}	0.25
Ubiquinone + $2H^+$ + $2e^- \rightarrow$ ubiquinone H_2	0.10
Cytochrome b-Fe^{+3} + $e^- \rightarrow$ cytochrome b-Fe^{+2}	0.08
FMN + $2H^+$ + $2e^- \rightarrow FMNH_2$	−0.12
FAD + $2H^+$ + $2e^- \rightarrow FADH_2$	−0.18
NAD + $2H^+$ + $2e^- \rightarrow NADH + H^+$	−0.32

Unfortunately, these reactions require much less free energy input. For example, the conformational change in the Na^+-K^+ pump requires less than 5 Kcal/mole. Therefore, direct use of the free energy from direct transfer of electrons from NADH to O_2 would be wasteful to the cell. Instead, the energy is "parceled out" into smaller chemical energy portions. This is achieved by transporting the electrons from NADH (and FADH) down the electron transport chain.

The carriers of the electron transport chain are part of the inner mitochondrial membrane. They represent a series of **redox carriers** intermediate in reduction potential between NADH and O_2. Table 3–5 shows these carriers' reduction potentials, and it is clear that there is a gradual increase in positivity from NADH to O_2. The free energy released in each step is illustrated in Figure 3–6.

Each oxidation-reduction reaction is catalyzed by a specific protein (e.g., a cytochrome) that has a prosthetic group that can accept electrons from the electron donor. A carrier with a less positive E_0' gives electrons to one with a more positive E_0'. Some of the potential changes ($\Delta E_0'$) are of sufficient magnitude that considerable free energy ($\Delta G_0'$) is released during the electron transfer step. These changes are shown as heavy lines in Figure 3–6. Passage of electrons from E-FMN to coenzyme Q yields a $\Delta G_0'$ of -12.2 Kcal/mole; from cytochromes b to c_1 $\Delta G_0'$ is -9.9 Kcal/mole; from cytochrome a to O_2, $\Delta G_0'$ is -23.8 Kcal/mole. The "parceling out" of reducing energy from NADH is clearly seen from these steps.

The order of the electron carriers in the inner membrane is specific

FIGURE 3–6

Energy relationships in the flow of electrons in the mitochondrial respiratory chain. The E_0's are redox potentials, and G_0's represent free energies based on these potentials. (E-FMN = flavin nucleotide linked to NADH dehydrogenase; Q = coenzyme Q [ubiquinone]; b, c_1, c, a = cytochromes.)

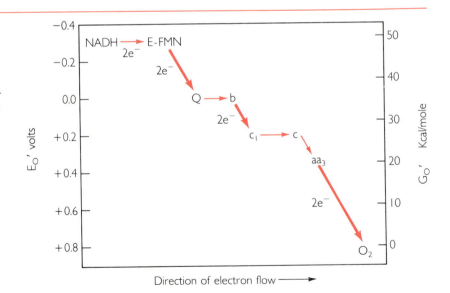

and invariant. This makes thermodynamic sense, since the carriers, beginning with NADH dehydrogenase, should be arranged in order of increasingly positive E_0'. Experimental confirmation of this order depends on a sensitive assay of the oxidized or reduced state of each carrier. Fortunately, each carrier has a specific spectrophotometric absorption spectrum that differs when the molecule is oxidized or reduced. For example, oxidized cytochrome c has low absorption at 550 nm, whereas the reduced carrier absorbs strongly at this wavelength. This has led to the development of optical techniques to determine with high precision the redox state of each carrier in intact mitochondria.

These techniques can be used along with specific **inhibitors** of the electron transport chain (Figure 3–7). The fish poison rotenone blocks transfer of electrons from E-FMN to coenzyme Q, while the antibiotic antimycin A blocks the cytochrome b to c_1 transfer, and cyanide blocks the final transfer to O_2. Thus, if mitochondria are incubated with pyruvate, NADH is formed through pyruvate dehydrogenase and enters the electron transport pathway. If rotenone is present, only E-FMN will show a reduced spectrum; if antimycin A is present, E-FMN, coenzyme Q and cytochrome b will be reduced. Therefore, coenzyme Q and cytochrome b must follow E-FMN in the chain. Experiments such as this not only confirm the thermodynamically predicted order of carriers but also delineate the different entry points for different substrates into the system.

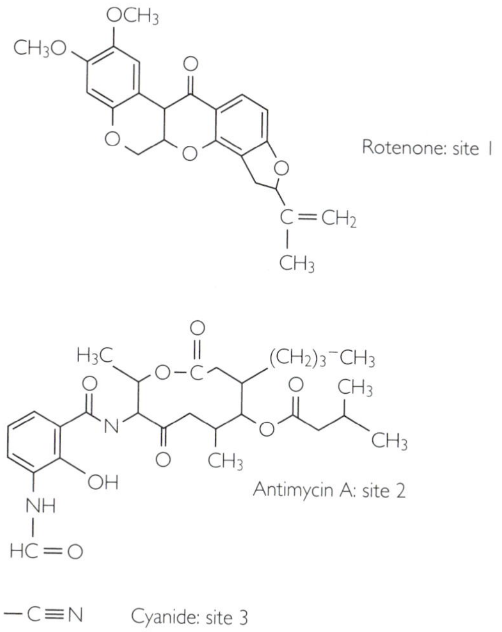

FIGURE 3–7

Inhibitors of electron transport and their sites of action.

FIGURE 3-8

FIGURE 3-8

An experiment on the effects of various electron transport inhibitors on oxygen consumption by mitochondria from liver. The Y-axis shows the oxygen content of the medium surrounding the mitochondria. A downward slope indicates oxygen uptake (and use) by the organelles, and a horizontal line indicates that no uptake of oxygen is occurring. See text for explanation.

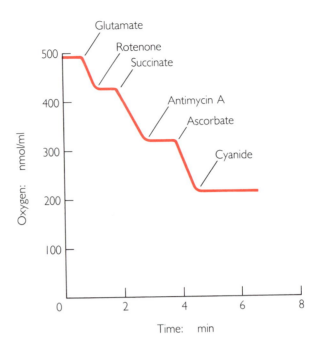

Another approach is the use of the oxygen electrode or manometry to measure **oxygen uptake** by isolated mitochondria. The consumption of O_2 is directly proportional to electron transport, since O_2 is the final electron acceptor in the scheme. A typical experiment is shown in Figure 3-8. Mitochondria using endogenous substrates have a low rate of O_2 uptake. If substrate is added (with ADP, see below), oxygen consumption rapidly increases. In the experiment, added glutamate leads to reduced NAD (i.e., NADH), which then enters the electron transport scheme to ultimately reduce O_2. Rotenone stops electron transport somewhere between glutamate and O_2. Addition of succinate leads to a resumption of transport, indicating that succinate enters the chain between the rotenone block point and O_2. By similar reasoning, ascorbate enters between the succinate entry point and O_2.

Direct evidence for the order of electron carriers comes from their **isolation** from the inner membrane. Treatment of mitochondria with mild detergents yields the carriers in three lipid-protein complexes. Lipid is essential for activity (e.g., cytochrome aa_3 has an absolute requirement for cardiolipin), an indication that the proteins are embedded in the lipid bilayer. These complexes will carry out their cognate reactions if supplied with electrons of suitable E_0'. They are indeed complex (Table 3-6).

The electron transport chain with its substrates, inhibitors, and complexes is summarized in Figure 3-9. The chain is similar in both

TABLE 3–6 Complexes of the Electron Transport Chain*

Complex	Mol. Wt.	Subunits	Components	Function
I NADH-Q-Reductase	850,000	16	E-FMN, 20 nonheme irons	Transfers e^- from NADH to Q
III Q-Cyt. c-Reductase	280,000	8	Two cyt. b, one cyt. C_1, 3 heme irons	Transfers e^- from Q to cyt. c
IV Cyt. c oxidase	200,000	9	Cyt. a, cyt. a_3, 2 heme irons, 2 Cu centers	Transfers e^- from cyt. c to O_2

*Complex II, succinate dehydrogenase, is present in the inner membrane and reduces complex III but is not part of the electron transport chain proper.

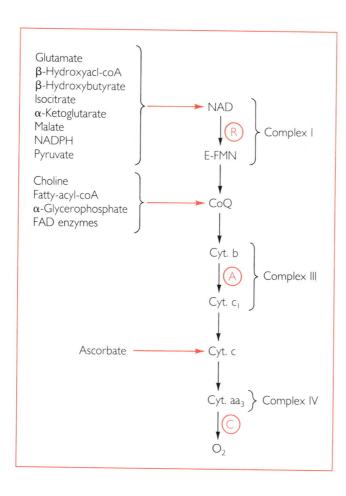

FIGURE 3–9
The electron transport chain in the inner mitochondrial membrane. Sites of input of reducing electrons from various substrates are shown. Inhibitors rotenone (R), antimycin A (A), and cyanide (C) act at the sites indicated. Complexes I, III, and IV have been isolated from the membrane.

plant and animal cells with one exception. Plant mitochondria have an NADH dehydrogenase that can oxidize both external and matrix NADH. This obviates the need for shuttle systems in plants to oxidize extramitochondrial NADH.

Oxidative Phosphorylation

The successful "parceling out" of the reducing energy in NADH by electron transport does not complete the process of mitochondrial energy transduction. First, not all energy-using reactions in the cell are reductions and hence use reducing power. For example, a change in protein conformation is what leads to the functioning of the Na^+-K^+ pump, which uses at least a third of the cell's energy. Such a change requires free energy input but not from oxidation of some reduced coenzyme. Second, the free energy-yielding electron transport reactions occur in the inner membrane of the mitochondrion, but chemical energy is needed all over the cell. Again, there must be some way to get the high-energy compound that drives the Na^+-K^+ pump to the plasma membrane. Clearly, a transportable, versatile chemical store for the free energy is needed, and that store is ATP.

The storage of free energy in ATP is via the phosphorylation of ADP. This reaction is energy-requiring:

$$ADP + Pi \rightarrow ATP + H_2O \qquad \Delta G_0' = +8 \text{ Kcal/mole.}$$

Conversely, its hydrolysis is free energy yielding under suitable conditions:

$$ATP + H_2O \rightarrow ADP + Pi \qquad \Delta G_0' = -8 \text{ Kcal/mole.}$$

During electron transport, three reactions have a G_0' sufficiently above 8 Kcal/mole (Figure 3–6). Therefore, there are three potential sites of ADP phosphorylation in the oxidative transfer of electrons from NADH to O_2.

Verification of these three sites of ATP production comes from determination of the **P:O ratio** of respiring mitochondria (Figure 3–10). Electron transport is closely coupled with ATP synthesis, and if there is an inadequate supply of ADP in the mitochondrial matrix, oxygen consumption (due to electron transport reducing O_2 to H_2O) is low. But if ADP is added to the medium, it is translocated into the matrix by the adenine nucleotide carrier. There, the ADP is phosphorylated to ATP, concomitant with a measured amount of O_2 consumption. The P:O ratio with glutamate is about 3, indicating that there are three sites of phosphorylation between NADH and O_2. For succinate, the ratio is 2, and for ascorbate, the ratio is 1. This confirms the pre-

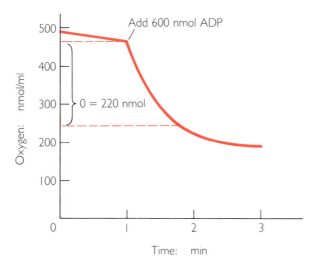

FIGURE 3–10
Determination of the P/O ratio. Liver mitochondria are incubated in the presence of glutamate. The rate of O_2 uptake from the medium, measured by oxygen electrode, is initially low. If ADP is added, respiration speeds up until the ADP is phosphorylated to ATP. The latter can be measured as esterification of P_i. If all of the ADP is esterified, the P/O ratio is $600/220 = 2.7$.

dictions based on thermodynamics of the locations of the phosphorylation sites.

Once the existence of the electron transport chain and its coupling to ATP synthesis were established about 1960, research turned to the mechanism by which the two are coupled. In 1961, P. Mitchell proposed such a mechanism, the **chemiosmotic hypothesis.** Since he was awarded the Nobel Prize for this proposal in 1978, it has gained the force of a "biological central dogma." Sociology aside, there is considerable evidence for Mitchell's idea, which involves proton electrochemical coupling.

The hypothesis is described in Figure 3–11. Stated another way, it envisages a vectorial transport of protons out of the mitochondrial matrix through a membrane otherwise impermeable to protons. This transport occurs as the electrons of NADH are passed to O_2. The pumping out of protons sets up an electrochemical gradient with the pH outside the matrix lower (more H^+) than inside:

$$\Delta\mu_{H+} = \Delta\psi - Z\Delta pH$$

where $\Delta\mu_{H+}$ is the electrochemical gradient, $\Delta\psi$ is the membrane potential with the outside more positive than the inside, and ΔpH is the pH gradient. The term Z is used to convert ΔpH into millivolts, the units of the other two terms; at 30° C, $Z = 60$.

The relationship of the free energy change $\Delta G_0{}'$ to the electrochemical gradient approximates:

$$\Delta G_0{}'/F = n\Delta u_{H+}$$

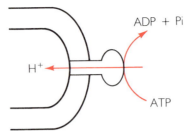

(a) One proton pump: ATP hydrolysis drives protons across the pump. As H^+ builds up on the other side, it sets up a potential gradient to drive ATP synthesis.

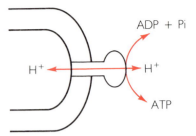

(b) Equilibrium: between energy derived from ATP hydrolysis and energy needed to pump protons against an increasing gradient.

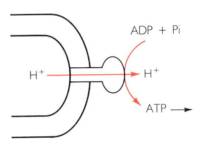

(c) ATP removed as it is made: the proton gradient drives ATP synthesis. This depletes the gradient, unless it is constantly replenished.

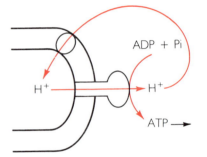

(d) Second proton pump: This replenishes the proton gradient so that ATP synthesis can continue.

FIGURE 3–11

The chemiosmotic theory of proton electrochemical coupling. (Adapted from: Nicholls, D. *Bioenergetics: An introduction to chemiosmotic theory.* New York: Academic Press, 1983.)

where F is Faraday's constant and n is the number of protons pumped per phosphorylation site. This means that a gradient of H^+ across a membrane results in available free energy. This energy can be used for the synthesis of ATP from ADP by reversing the flow of protons across the membrane. Recalling that the original flow was out of the matrix, the reverse flow should be inward. This inward movement of H^+ is achieved via the ATP synthase, a proton pump located at the inner mitochondrial membrane.

In order to keep ATP synthesis going, two things must occur:

1. The proton gradient must be continuously replenished by the electron transport carriers.

2. ATP must be continuously removed by translocation so that its synthesis is favored.

The dependence of phosphorylation by mitochondria on these two events is well established.

The chemiosmotic theory is experimentally testable. First, it predicts that respiring mitochondria will set up a **pH gradient** with more protons on the outside and fewer in the matrix. On a crude basis, this can be done by supplying mitoplasts with substrate and then measuring the external pH, which indeed goes down.

More sophisticated techniques are used to measure both ΔpH and $\Delta\psi$. These involve permeant fluorescent amines, which show reduced fluorescence in a higher pH, and fluorescent membrane probes, which are sensitive to potential changes across the membrane. Experiments using these compounds give μ_{H+} values of about 0.2 volts for a number of mitochondrial systems. Thus:

$$\Delta G_0' = nF\Delta H^+ = n\,(23062)(0.2) = n\,(4600)\ \text{cal.}$$

So, for the synthesis of one ATP molecule ($\Delta G_0' = 8$ Kcal/mole), a minimum of two protons must cross the membrane. Careful measurements indicate that the actual number is three to four, of which the potential energy of one is used in ATP translocation out of the matrix (see above).

Two other kinds of experiments support the idea that an electrochemical gradient is important for ATP synthesis. First, imposition of such a gradient on mitochondria in the absence of electron transport leads to immediate ATP synthesis. If NADH is added to the organelles, there is a consistent 5- to 10-millisecond lag before ATP is made; presumably, this is sufficient time to set up the pH gradient. Second, ionophores that upset the potential gradient uncouple electron transport from phosphorylation. Most notable among these are the K^+-carrier valinomycin and various H^+-carriers such as FCCP and dinitrophenol (Figure 2–4).

A second prediction of the chemiosmotic theory is that the electron transport carriers should be arranged **asymmetrically** in the inner membrane, to transport electrons in one direction (inward) and protons in the other (outward). Moreover, the three complexes involved in ATP synthesis should span the inner membrane to perform this task.

The topology of the inner membrane has been studied in two ways. First, impermeant reagents (see Chapter 1) or specific antibodies to carriers have been used to **resolve** on which surface they are located (Figure 3–12). These antibodies do not penetrate the inner mitochondrial membrane. Therefore, in mitochondria stripped of outer membranes, they will bind to their antigenic target (e.g., one of the electron carriers) if it is exposed on the outside of the inner membrane.

If mitochondria are treated with ultrasound (sonicated), the membranes pinch off to form vesicles. In the case of the inner membrane,

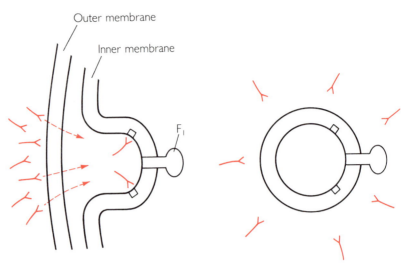

Outer membrane

Inner membrane

F_1

(a) Intact mitochondria:
anticytochrome C antibodies
bind to their target.

(b) Submitochondrial particles:
anticytochrome C antibodies
cannot penetrate the inner
membrane and so do not bind to
their target. The F_1 "knob" is a
protein extending into the
matrix in intact mitochondria.

FIGURE 3–12

Using antibodies to investigate
the topology of the inner mitochondrial membrane.

these vesicles form by membrane fusion at the junction of the crista with
the main part of the inner membrane. These vesicles, termed submitochondrial particles, have as their outer surface the side that faced the
matrix; thus they are "inside-out." Proteins on this surface, inaccessible
to the impermeant antibody in the intact organelle, are now accessible.
Thus the binding of an antibody or impermeant reagent to a membrane
protein in submitochondrial particles indicates that the carrier faces the
matrix.

The second topographical method for the inner membrane is **reconstitution** (Figure 3–13). Essentially, this involves taking the inner
membrane apart and putting it back together again. Initial success at this
technique came with the analysis of the 8-nm–diameter knobs or "lol-
lipops" seen projecting from the inner membrane into the matrix in
electron micrographs. These were initially proposed to contain both
electron transport and ATP synthesizing activities, but elegant experiments by E. Racker disproved this.

When the knobs (termed F_1 particles) were removed from the inner membrane by treating the membrane with an agent that disrupts
protein-lipid interactions, the membrane retained electron transport activity but did not synthesize ATP. The isolated F_1 particles had ATP-

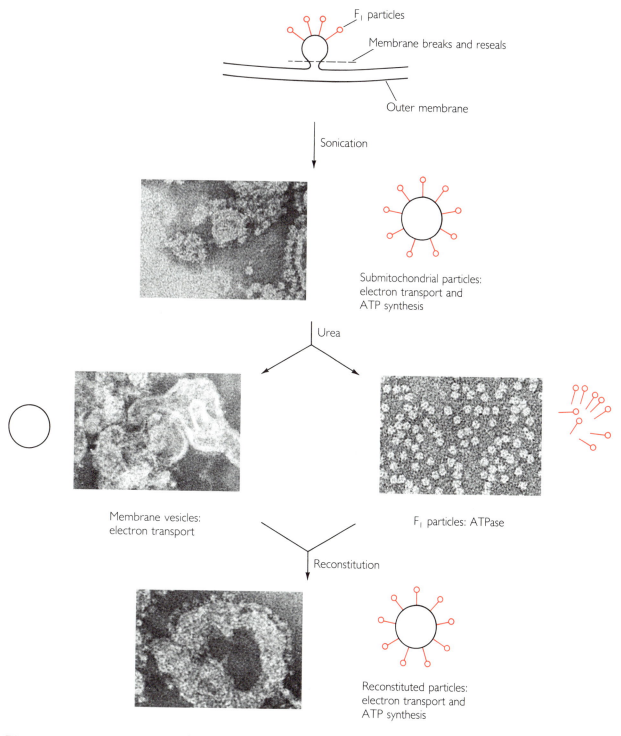

F₁ particles

Membrane breaks and reseals

Outer membrane

Sonication

Submitochondrial particles:
electron transport and
ATP synthesis

Urea

Membrane vesicles:
electron transport

F₁ particles: ATPase

Reconstitution

Reconstituted particles:
electron transport and
ATP synthesis

FIGURE 3–13   Disruption and reconstitution of the inner mitochondrial membrane. (×175,000. Photographs courtesy of Dr. E. Racker.)

hydrolyzing activity. Reconstitution of the membrane and knobs restored oxidative phosphorylation. This experiment not only identified F_1 as the ATP synthesizing enzyme and located it as, appropriately, facing the matrix, but it also established the necessity of a membrane for ATP synthesis. Subsequently, the three major electron transporting complexes were isolated from, and reconstituted into, the inner membrane. In order for oxidative phosphorylation to occur after reconstitution, asymmetry of the carriers was essential.

The topography of the electron carriers and their function as a proton pump are shown in Figure 3–14. This is a simplified diagram, since there are many other proteins involved whose locations are uncertain, but it does give the picture of asymmetry predicted by the chemiosmotic hypothesis. One major problem that remains to be solved in this regard is exactly how the carriers pump protons.

ATP Synthesis

The enzyme responsible for ATP synthesis in mitochondria is a large complex, visualized under the electron microscope as a knob or "lollipop" projecting from the inner membrane into the matrix (Figures 3–11 and 3–12). It is composed of two main parts: the membrane

FIGURE 3–14

Topography of proton translocation and the electron transport chain in liver mitochondria.

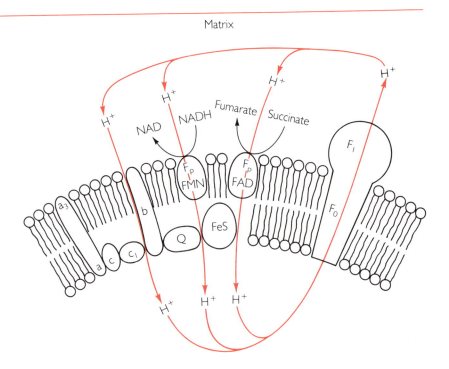

bound F_0, and the projecting F_1; thus the complex is referred to as F_1F_0. The F_1 projection is further resolvable into three copies each of two subunits, α and β, and single copies of three additional subunits. Analyses of the DNA sequences of the subunits from various species indicate that the two major ones have quite similar (40%) amino acid sequences, with more diversity in the minor subunits.

The F_0 portion in the membrane typically has three subunits: one, a hydrophobic protein, binds to F_1; another, a proteolipid, is the proton translocator. Evidence for the latter comes from mutants of the fungi *Neurospora* and yeast that are resistant to the antibiotic oligomycin. This compound, from *Streptomyces,* uncouples oxidative phosphorylation by blocking the passage of protons through F_0. Mutant strains that are unaffected by oligomycin have an altered F_0 proteolipid. If proteolipid from normal mitochondria is inserted into artificial lipid bilayer vesicles, they will generate $\Delta\mu_{H+}$ in proportion to the F_0 present.

The F_1 component synthesizes ATP via the chemiosmotic proton flux coming through F_0 in the membrane. As shown above (Figure 3–13), F_1 is easily detached from the membrane. Although soluble, it cannot synthesize ATP, only hydrolyze it. Because it catalyzes a reversible reaction, when $\Delta\mu_{H+}$ is low or ATP is not removed (Figure 3–11), the hydrolysis of ATP should be favored. This hydrolysis reaction is inhibited by an additional subunit, an ATPase inhibitor.

There is also some evidence that ATP synthesis can be regulated by an inhibitory anion channel, which, when open, severely reduces the electrochemical gradient. The three sites on F_1 that actually bind the adenine nucleotide and do the catalysis are apparently on the three beta subunits. Actual formation of ATP occurs when ADP displaces an hydroxyl group on inorganic phosphate.

The function of F_1 is to use $\Delta\mu_{H+}$ set up by the electron transport chain to do chemiosmotic work, that is, to synthesize ATP. The mechanism by which this occurs is under study, and two proposals have been made.

1. The first is a direct mechanism in which two protons pass through F_0 to the active site of F_1 where they act on one of the oxygen atoms of phosphate to form water. The highly reactive species that remains reacts with ADP to form ATP.

2. The second idea involves an indirect mechanism, whereby ΔH^+ induces a conformational change in F_0, and this change is transmitted to F_1 via the interactions of the many subunits. At this time, nucleotides are bound to the active sites in F_1. Once the conformational change reaches these sites, ATP is released.

Uncoupling and Thermogenesis

The proton gradient set up by electron transport can be used for another purpose in addition to ATP synthesis. Newborn babies without much body hair, hibernating animals, and cold-stressed animals use it to produce heat, a process called metabolic thermogenesis.

This process occurs in a specialized tissue, brown fat, and in its constituent cells, the adipocytes. Most of the time, animals use other means to keep warm, such as a hairy or furry coat, blood circulatory mechanisms, and heat generated by physical activity such as shivering. But at certain times, these mechanisms are either undeveloped (as in a newborn baby) or inadequate (as in an animal kept in very cold conditions). It is here that the adipocyte comes into prominence.

The key observation on these cells is that their mitochondria are unusually permeable to protons, much more so than those of other tissues. Their two typical proton pumps—the electron transport chain and the F_1F_0-ATP synthetase—are normal in activity. However, these cells have a third mitochondrial inner membrane proton pump. This protein, called **thermogenin,** spans the inner membrane six times to form a channel through which protons can flow in the same direction as the ATP synthetase.

This situation means that there are now two pumps that harness the energy inherent in the electrochemical gradient set up by the carriers of electron transport, but only one of them (F_1F_0) harnesses the free energy released as the protons pass into the matrix to make ATP. The other, thermogenin, couples the energy not into a chemical reaction but into heat. Because this has the net effect of reducing the coupling of electron transport and ATP synthesis, it is not surprising that thermogenin has also been called "uncoupling protein" (Figure 3–15).

The thermogenin system develops when it is needed. For example, in humans, adipose tissue of a fetus at 28 weeks' gestation (full term is about 40 weeks) contains twice the thermogenin content as the same tissue in an adult. In children up to their teenage years, the thermogenin content is about seven times the adult value. This indicates that metabolic thermogenesis is important in the newborn and child, who have not fully developed the other thermogenic mechanisms.

Once an animal reaches adulthood, brown fat adipocytes gradually reduce in number. But they can reappear if the animal is under consistent cold stress. Studies of animals under such conditions have revealed control mechanisms for metabolic thermogenesis. Not only is there more thermogenin in fat cells under cold stress, but the channel proteins already present are unmasked. Masking of the channel apparently involves purine nucleotide (ADP, ATP, GDP, GTP) binding, and unmasking requires free fatty acids. Hormones that bind to the β-adrenergic

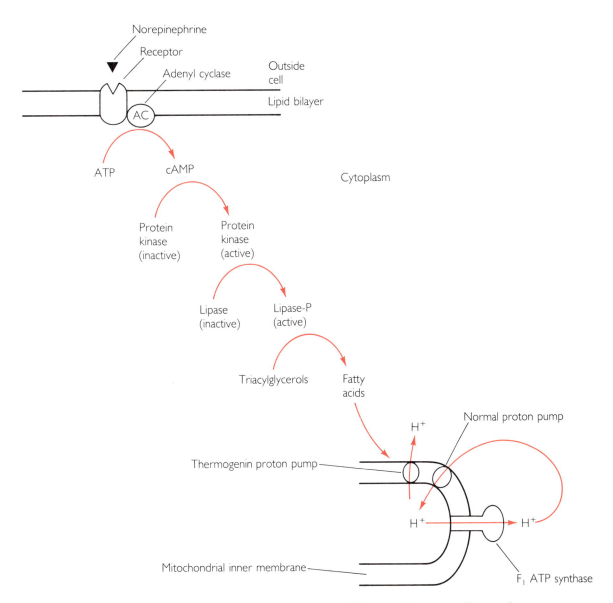

FIGURE 3–15   The cascade of events in the generation of heat in adipocytes of brown fat. The binding of a hormone, norepinephrine, to its receptor on the adipocyte leads to the formation of cyclic AMP inside the cell. This second messenger indirectly leads to the release of fatty acids, which activate the thermogenic pump in the inner mitochondrial membrane. This pump upsets the H^+ gradient set up by the electron carriers. Instead of ATP synthesis, thermogenin uses the free energy of the electrochemical gradient to produce heat. Compare to the chemiosmotic theory, Figure 3–11.

receptor (see Chapter 2) stimulate thermogenin via adenyl cyclase (Figure 3–15).

Mitochondria and Disease

Alterations of mitochondrial structure and function occur in a number of human diseases. In the **mitochondrial myopathies,** staining characteristics of the organelle are used as a sign in clinical diagnosis. Muscle biopsies of these patients can be taken and when examined by light microscopy show "ragged-red fibers," which represent aggregates of abnormal mitochondria. Under the electron microscope, these abnormalities are apparent as crystals in the matrix, concentric cristae, and so on (Figure 3–16). The relationship of these structural aberrations to mitochondrial function is not known.

Patients with mitochondrial myopathies show a wide variety of muscular disorders, ranging from inadequate contraction to fatigue. Because these symptoms are thought to be caused by the abnormal mitochondria, their functioning has been examined in some cases. One obvious problem in examining these mitochondria is tissue availability: A muscle biopsy from a patient is usually less than a gram. The development of techniques for studying small amounts of mitochondria from tissue culture cells has overcome this difficulty.

Different abnormalities in mitochondrial function have been found in different patients with these myopathies. The first patient described in this manner (by R. Luft in 1959) had loosely coupled mitochondria. Respiration was insensitive to oligomycin, and F_1 ATPase activity was quite high. In addition, there was little respiratory control: Oxygen consumption was high, regardless of the presence of ADP (Figure 3–10). Thus, the patient maintained a high metabolic rate to promote a high rate of electron transport so that her inefficient ATP synthesis could proceed. This could explain her muscle weakness.

Since the original case report, a number of myopathies and encephalopathies (central nervous system disorders) have been shown to have abnormal mitochondrial function (Table 3–7). These include defects in all four complexes of the electron transport chain, as well as the citric acid cycle, use of certain substrates, and uptake of metabolites. In many instances, the clinical picture is grim, with severely adverse brain and muscle function evident from birth, and death at an early age. In other cases, the only symptom is an inability to exercise vigorously.

A well-studied case of mitochondrial disease is Leber's hereditary neuropathy, which causes blindness in young adults, especially men. In this case, the protein that is altered as a result of the mutation is subunit 4 of the NADH-coenzyme Q oxidoreductase, which is part of respiratory complex I (Table 3–6). This has been shown both by the specific electron transport defect and by DNA analysis.

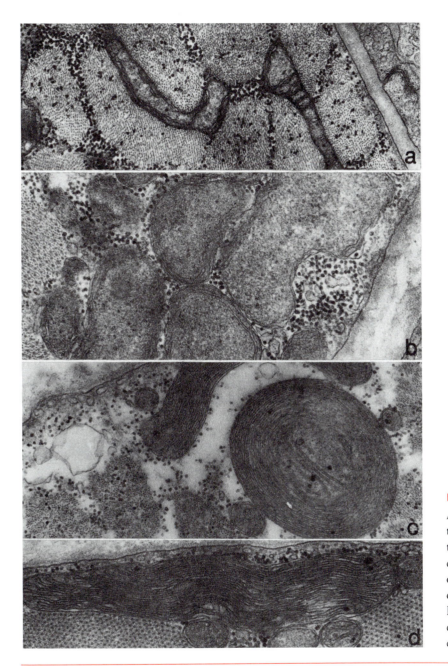

F I G U R E 3–16
Abnormal mitochondria in patients with mitochondrial myopathies. **(a)** Increased matrix and crystals. **(b)** Increased matrix, only one crista. **(c)** Large mitochondrion, concentric cristae. **(d)** Large mitochondrion, parallel cristae. (×40,000. Photographs courtesy of Dr. H. Schmulbruch.)

The relationships between the mitochondrial defects and the clinical picture are not clear in any case but have led to some interesting hypotheses. For example, in the genetic disease porphyria, heme breakdown products such as protoporphyrin accumulate, and these have been shown to uncouple oxidative phosphorylation in isolated mitochondria.

TABLE 3-7 **Some Mitochondrial Diseases**

Disease	Mitochondrial Abnormality
Acyl-CoA dehydrogenase deficiency	Reduced fatty acid oxidation
Carnitine deficiency	No transport of fatty acyl-CoA
Chronic alcohol intake	Reduced inner membrane fluidity
Fumarase deficiency	Reduced Krebs cycle
Kearns-Sayre syndrome	Cytochrome oxidase deficiency
Leber's optic neuropathy	NADH dehydrogenase deficiency
Menke's disease	Cytochrome aa_3 deficiency
Mitochondrial myopathy	Pyruvate dehydrogenase deficiency
Mitochondrial myopathy	Cytochrome b deficiency
Mitochondrial myopathy	Altered F_1-ATPase
NADH dehydrogenase deficiency	No oxidation of NADH
Porphyria	Uncoupling of oxidative phosphorylation
Zellweger's syndrome	No oxidation of succinate

When uncoupling occurs, the energy released during electron transport is not used to make ATP but rather, as in the case of brown fat mitochondria, comes off as heat. People with porphyria suffer from periodic high fevers, and these could be due to the uncoupling in mitochondria.

Environmental factors can also lead to mitochondrial pathologies. In the United States, most maize (corn) grown carries a gene for male sterility (T-cytoplasm), which prevents self-pollination. This is essential if a genetic outcross is desired, as in the production of hybrid corn. Unfortunately, this gene confers on the plants a high sensitivity to the toxin produced by a race of the fungus *Bipolaris,* the causative agent of southern corn leaf blight. In addition, the maize is very sensitive to the insecticide methomyl. Both the toxin and insecticide bind to a small protein on the inner mitochondrial membrane, coded for by the T-cytoplasmic system. Binding induces ion leakage across the inner membrane, with a resulting loss of the chemiosmotic gradient and, ultimately, death of the cell.

Further Reading

Adams, V., Griffin, L., Towbin, J., Gelb, B., Worley, K., and McCabe, E. Porin interaction with hexokinase and glycerol kinase: Metabolic microcompartmentation at the outer mitochondrial membrane. *Biochem Med Metab Biol* 45 (1991): 271–291.

Bereiter-Hahn, J. Behavior of mitochondria in the living cell. *Int Rev Cytol* 122 (1990):1–37.

Bergeron, M., Guerette, D., Forget, J., and Thiery, G. Three dimensional characteristics of the mitochondria of the rat nephron. *Kidney International* 17 (1980):175–185.

Bianchi, G., Carafoli, E., and Scarpa, A. Membrane pathology. *Ann NY Acad Sci* 448 (1986).

Boyer, P. The unusual enzymology of ATP synthase. *Biochemistry* 26 (1987):8503–8506.

Brand, N., and Murphy, M. Control of electron flux through the respiratory chain in mitochondria and cells. *Biol Rev* 62 (1987):141–193.

Brdiczka, D. Interaction of mitochondrial porin and cytoplasmic proteins. *Experientia* 46 (1990):161–167.

Brdiczka, D., Bucheler, K., Kottke, M., Adams, V., and Nalam, V. Characterization and metabolic function of mitochondrial contact sites. *Biochim Biophys Acta* 1018 (1990):234–238.

Carafoli, E., and Roman, I. Mitochondria and disease. *Molec Aspects Med* 3 (1980): 295–429.

Chan, S. I., and Li, M. Cytochrome c oxidase: Understanding nature's design of a proton pump. *Biochemistry* 29 (1990):1–12.

Coper, H. Depression and monoamine oxidase. *Prog Neuropsychopharm* 3 (1989):441–463.

Cooper, C., Nicholls, P., and Freedman, J. Cytochrome oxidase: structure, function and membrane topology of polypeptide subunits. *Biochem Cell Biol* 69 (1991):586–607.

Cramer, W. A., and Knaff, D. B. Energy Transduction in Biological Membranes. New York: Springer-Verlag, 1991.

DeVivo, D., and DiMauro, S. Mitochondrial defects of brain and muscle. *Biol Neonate* 58 (1990):54–69.

Dihanich, M. The biogenesis and function of mitochondrial porins. *Experientia* 42 (1990):146–153.

Dimauro, S., Bonilla, E., Zeviani, M., Servidei, S., Devivo, D., and Schon, E. Mitochondrial myopathies. *J Inher Metab Dis* 10 Suppl. 1 (1987):113–128.

Englebrecht, S., and Junge, W. Subunit delta of H+-ATPases: At the interface between proton flow and ATP synthesis. *Biochim Biophys Acta* 1015 (1990):379–390.

Futai, M., Noumi, T., and Maeda, M. ATP synthase: Results by combined biochemical and molecular biological approaches. *Ann Rev Biochem* 58 (1989):111–136.

Geny, C., Cormier, V., Meyrignac, C., Cesaro, P., Degos, J., Gherardi, R., and Rotig, A. Muscle mitochondrial DNA in encephalomyopathy and ragged red fibers. *J Neurol* 238 (1991):171–176.

Godinot, C., and DiPietro, A. Structure and function of the ATPase-ATP synthase complex of mitochondria as compared to chloroplasts and bacteria. *Biochimie* 68 (1986):367–374.

Good, N. Active transport, ion movements and pH changes. *Photosyn Res* 19 (1988): 225–250.

Halestrap, A. The regulation of matrix volume of mammalian mitochondria in vivo and in vitro and the control of mitochondrial metabolism. *Biochim Biophys Acta* 973 (1989):355–382.

Hanson, M. Plant mitochondrial mutations and male sterility. *Ann Rev Genet* 25 (1991):461–486.

Harding, A. E., and Holt, L. Mitochondrial myopathies. *Brit Med Bull* 45 (1989): 760–771.

Hatefi, Y. The mitochondrial electron transport and oxidative phosphorylation systems. *Ann Rev Biochem* 54 (1985):1015–1070.

Himms-Hagen, J. Brown adipose tissue thermogenesis: Interdisciplinary studies. *FASEB J* 4 (1990):2890–2898.

Kramer, R., and Palmieri, F. Molecular aspects of isolated and reconstituted carrier proteins from animal mitochondria. *Biochim Biophys Acta* 974 (1989):1–23.

Lemasters, J., Hackenbrock, C., Thurman, R., and Wetserhoff, H. *Integration of mitochondrial function.* New York: Plenum, 1989.

Lenaz, G. Role of mobility of redox components in the inner mitochondrial membrane. *J Membr Biol* 104 (1989):193–209.

Levings, C. S. The Texas cytoplasm of maize: Cytoplasmic male sterility and disease susceptibility. *Science* 250 (1990):942–948.

McCormack, J. G., and Denton, R. M. The role of mitochondrial Ca^{++} transport and matrix Ca^{++} in signal transduction in mammalian tissues. *Biochim Biophys Acta* 1018 (1990):287–291.

Mela-Riker, L., and Bukoski, R. Regulation of mitochondrial activity in cardiac cells. *Ann Rev Physiol* 47 (1985):645–663.

Mitchell, P. Keilin's respiratory chain concept and its chemiosmotic consequences. *Science* 206 (1979):1148–1159.

Moore, A., and Beechey, R. *Plant mitochondria.* New York: Plenum, 1987.

Morgan-Hughes, J., Schapira, A., Cooper, J., Holt, I., Harding, A., and Clark, J. The molecular pathology of respiratory chain dysfunction in human mitochondrial myopathies. *Biochim Biophys Acta* 1018 (1990):217–222.

Morris, M. Mitochondrial mutations in neuro-ophthalmological diseases: A review. *J Clin Neuro Ophthalmol* 10 (1990):159–166.

Nicholls, D. G. *Bioenergetics: An introduction to the chemiosmotic theory.* New York: Academic Press, 1982.

Nicholls, D. G., and Locke, R. Thermogenic mechanisms in brown fat. *Physiol Rev* 64 (1984):1–64.

Papa, S., Chance, B., and Ernster, L. *Cytochrome systems.* New York: Plenum, 1987.

Pietrobon, D., Virgilio, F., and Pozzan, T. Structural and functional aspects of calcium homeostasis in eukaryotic cells. *Eur J Biochem* 193 (1990):599–622.

Racker, E. *Reconstitution of transporters, receptors, and pathological states.* New York: Academic Press, 1985.

Rial, E., and Nicholls, D. The uncoupling protein from brown adipose tissue. *Revis Biol Cell* 11 (1987):75–104.

Ricquier, D., Casteilla, L., and Bouillaud, F. Molecular studies of the uncoupling protein. *FASEB J* 5 (1991):2237–2242.

Rosamond, J. Structure and function of mitochondria. *Int Rev Cytol Suppl* 17 (1987):121–145.

Tedeschi, H., Kinnally, K., and Mannella, C. Properties of channels in the mitochondrial membrane. *J Bioenerg Biomembr* 21 (1989):451–459.

Walz, B., and Baumann, O. Calcium-sequestering cell organelles. *Prog Biochem Cytochem* 20 (1989):1–47.

Williams, R. J. P. Proton circuits in biological energy interconversions. *Ann Rev Biophys Chem* 17 (1988):71–97.

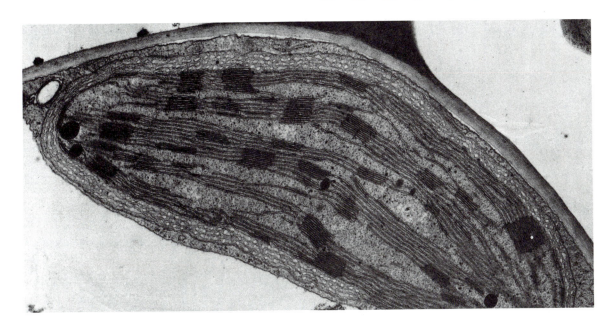

Plastids

Energy Transduction and Biosynthesis

Mitochondria are instrumental in the transduction of chemical energy of reduced molecules, such as carbohydrates, into ATP, which is used for biosynthesis and active transport. The ultimate source of this chemical energy is the sun. Green plant photosynthesis converts this solar energy into ATP and NADPH, which are then used in the biosynthesis of carbohydrates from CO_2 and water. This key process, then, involves two components: the trapping of light energy and its use in anabolism.

Both of these processes occur in a membrane-bound organelle, the chloroplast. When one considers what happens in the mitochondrion, it is not surprising that the energy-trapping events in the chloroplast occur at membranes; this constitutes the so-called "light reaction." Also not unexpectedly, the biosynthetic pathways occur in the soluble part of the organelle; these are called the "dark reaction." Once the carbohydrate is synthesized, it is often transferred to other regions of the cell (and then to the rest of the plant). This implies the need for regulation of transport into and out of the chloroplast compartment. These three functions, the light reaction, the dark reaction, and the translocation of metabolites, constitute the primary events in the ecological food chain on which all animal life depends.

Chloroplasts were first seen under the microscope by the plant anatomist N. Grew in the 17th century. Although it was evident that they were unique to plants, their role was not made clear until late in the 19th century, when they were experimentally linked to light-driven gas exchange and the production of starch.

But further study of these organelles awaited a method for their isolation. This was achieved in 1938 by S. Granick, who ground up spinach leaves in an isosmotic glucose solution and then used differential centrifugation to isolate the chloroplasts. The isolated organelles have been used to study the structure, composition, and function of chloroplasts and other plastids.

Structure

Light Microscopy

All green plant tissues have chloroplasts. In lower plants, they show great variation in size, shape, and number. For instance, a cell may have only one large chloroplast, measuring over 100 μm in diameter, or over 100 smaller ones. They may be shaped like a convex lens, a spiral, or a star. Indeed, in the algae the morphology of the chloroplasts is so diverse and species-specific that it is used as a criterion for taxonomic classification.

In higher plants, chloroplasts are usually lens shaped, 4 to 6 μm in diameter, and 5 to 10 μm in length. In a typical leaf cell, there are about 40 of them, and this means that there are about half a million per square millimeter of leaf surface area. The chloroplasts are in constant motion because of cytoplasmic streaming and do not exhibit a preferred location in the cell. However, their distribution in tissues is clearly nonrandom. They are most abundant in green photosynthesizing tissues such as leaves and less abundant or absent in the root. Often, other types of plastids predominate where chloroplasts are absent (see below).

Electron Microscopy

The first electron micrographs of isolated chloroplasts, published in 1947 by S. Granick and K. Porter, showed that the grana appeared as stacks of internal disks. When J. Finean made thin sections of chloroplasts in 1953, an ultrastructural picture emerged of an organelle with three compartments: an outer envelope, an internal lamellar membrane system, and the stroma (Figure 4–1).

The **outer envelope** is actually a double membrane. Each membrane is 6 to 8 nm thick and the space between them is 10 to 20 nm. The outer membrane is rather rigid structurally. When chloroplasts are put into a medium of high osmotic concentration, the inner membrane visibly shrinks away from the outer one in a manner reminiscent of mitochondrial contraction. In this case, the space between the two membranes may increase tenfold; restoration of osmotic equilibrium restores the inner membrane to its closely apposed position. The significance of this cycle in chloroplasts is not clear.

Some electron micrographs show vesicles associated with the inner of the two membranes, leading to the hypothesis that the inner membrane contributes to the lamellar system. In a number of instances, micrographs show the two membranes of the plastid envelope to be fused at places. This could provide a mechanism for lipid transfer, or, more speculatively, it could be a way to import proteins from the cytoplasm into the plastid (see Chapter 5).

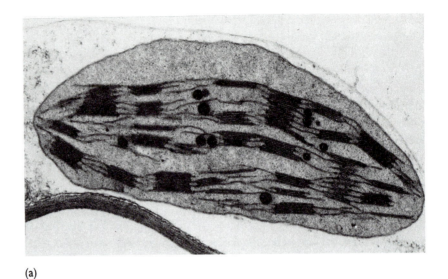

(a)

(b)

F I G U R E 4–1 Chloroplast structure. **(a)** Electron micrograph of a cross section of a spinach chloroplast. (×50,000. Courtesy of Dr. L. A. Staehelin.) **(b)** Diagram of structures within a chloroplast.

The **lamellar system** appears quite complicated. In electron micrographs of thin sections (Figure 4–2), it appears to be composed of parallel, flattened, membrane sacs, with electron-dense stacks in the grana regions and extensive but less dense stroma lamellae. In 1962, W. Menke coined the term *thylakoid* ("saclike") to describe these membranous structures. A typical chloroplast has about 50 of them, each about

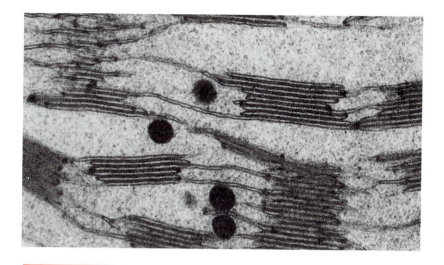

FIGURE 4–2

Thin section of spinach thylakoid membranes showing stacked grana and unstacked stroma membrane regions. The dark dots are plastoglobuli. (×95,000. Photograph courtesy of Dr. L. A. Staehelin.)

0.1 μm in diameter with 2 to 100 stacked thylakoids. Careful examination of this internal membrane system has shown that it is formed from the folding and interconnecting of a single membrane sheet that separates the stroma from the intracisternal space. Thus, the grana are not "sacs" or discs but rather appear so because of artifacts of specimen preparation for electron microscopy.

The extensive folding of the lamellar membrane to form the grana provides a great increase in surface area. A. Wellburn has calculated that a typical leaf has 600 times its surface area of photosynthetic membranes for each cell layer containing chloroplasts. While electron micrographs may give the impression of a static structure (Figure 4–2), the thylakoid membrane is quite dynamic, with constant stacking and unstacking in a given region. Like the overall morphology of the chloroplast, the substructure of the internal membrane system varies among different plants. In general, algal plastids do not contain grana, but instead lamellae appear as parallel sheets that either never become appressed or are closely associated throughout the length of the chloroplast.

The **stroma,** or matrix, of the plastid usually contains several kinds of discrete bodies. Starch grains, up to 2 μm in length, either lie freely in the stroma, as in higher plants, or are associated with the internal membranes as pyrenoids in lower plants. Plastoglobuli, lipid deposits about 0.1 μm in diameter, are a common feature of chloroplasts, and as leaves turn color, the plastoglobuli may accumulate colored pigments. Ribosomes, at 17 nm diameter somewhat smaller than their cytoplasmic counterparts, are a universal feature of the stroma. Finally, 2.5-nm–diameter fibrils, digestible with deoxyribonuclease, represent the chloroplast DNA. The remainder of the stroma is rich in proteins but microscopically structureless.

Chemical Composition

As in the mitochondrion, the outer envelope and internal lamellae of the chloroplast differ considerably in their chemical composition and function. Envelope membranes contain little **protein;** indeed, their 30% protein content is among the lowest of all biological membranes. The outer membrane of the pair that composes the envelope contains even less protein than the inner one. Qualitative analysis of envelope proteins by gel electrophoresis indicates several major proteins and a number of minor ones in terms of quantity. In contrast to the thylakoid membrane, the proteins of which are rather small, several envelope proteins are well over 100,000 in molecular weight. By inference, most of these occur in the inner membrane, where functions of metabolite transport and enzymes reside.

Envelope **lipids** (Table 4–1) are remarkable in terms of their high glycolipid content, especially digalactosyldiglyceride. The predominant fatty acid is $C_{18:3}$, a high degree of unsaturation that ensures considerable fluidity.

As noted above, the outer membrane is osmotically rather inert, while the inner membrane is differentially permeable. A number of studies show that, like the outer mitochondrial membrane, the outer envelope membrane allows the diffusion of a wide variety of substances such as phosphate, nucleotides, carboxylic acids, and sucrose. Higher molecular weight (greater than 2000) dextrans do not cross this bilayer, indicating a size limit to its porosity. A porin-like protein apparently makes channels that allow the passage of these substances across the outer membrane.

In contrast, the inner membrane is not porous. Instead, the small molecules that diffuse across the outer membrane require translocases to get across the inner one (see below).

The internal lamellar system has a relatively low lipid content. Glycolipid, specifically monogalactosyldiglyceride, is the predominant constituent, with a highly fluid $C_{16:3}$ and $C_{18:3}$ fatty acid composition (Table

---

TABLE 4–1 **Lipid Composition of Spinach Chloroplast Membranes**

	Outer Envelope	Lamellae
Total % lipid	70	30
Phosphatidylcholine (as % of total lipid)	27	10
Phosphatidylglycerol	13	16
Sulfolipid	6	7
Glycolipid	38	59

4–1). Given the abundance of cells containing plastids, these glycolipids are probably the most abundant lipids in the natural world. But the fluidity of this membrane is not uniform. Near the chlorophyll-protein complexes where light is absorbed, there is much less fluidity than between the complexes. This may be important in the transfer of electrons between the two photosystems, one of which oxidizes water (photosystem II [PS-II]) and the other of which reduces the coenzyme, NADP, to NADPH (photosystem I [PS-I]).

However, by far the most distinctive lipid components are the pigments. **Chlorophylls** are magnesium-containing molecules with hydrophilic porphyrin rings binding the Mg^{++} and a long phytol hydrocarbon chain ensuring hydrophobicity for insertion into the membrane (Figure 4–3). Thus the molecule is arranged much like a phospholipid, and it is believed to be positioned in the membrane in a similar manner, with the phytol chain perpendicular to the plane of the lamella. The porphyrin rings are associated with membrane proteins in a manner similar to the association of heme and globin to form hemoglobin.

Carotenoids (Figure 4–4) are accessory pigments to the chlorophylls. Their long hydrocarbon chains allow them to be widely distributed in the thylakoid membrane, and they are generally not bound to proteins. In addition to their role as accessory pigments in photosynthesis, carotenoids have an important protective role in that they absorb excess light that could physically damage the chlorophyll molecules.

The thylakoid membrane **proteins** and pigments can be made soluble by the use of a detergent. When they are separated by means of electrophoresis in the presence of sodium dodecyl sulfate (SDS), many components are seen. Some of them are associated with chlorophyll into

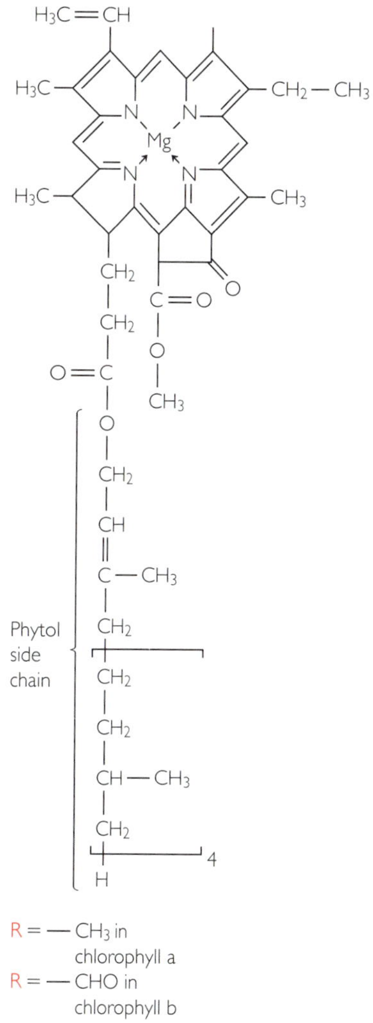

$R = —CH_3$ in chlorophyll a
$R = —CHO$ in chlorophyll b

FIGURE 4–3
Structures of chlorophylls a and b. The long phytol chain is hydrophobic and can embed in a membrane lipid bilayer. The ringed structures, on the other hand, are polar and will be at the periphery of the bilayer. Note the similarity in arrangement to a membrane phospholipid.

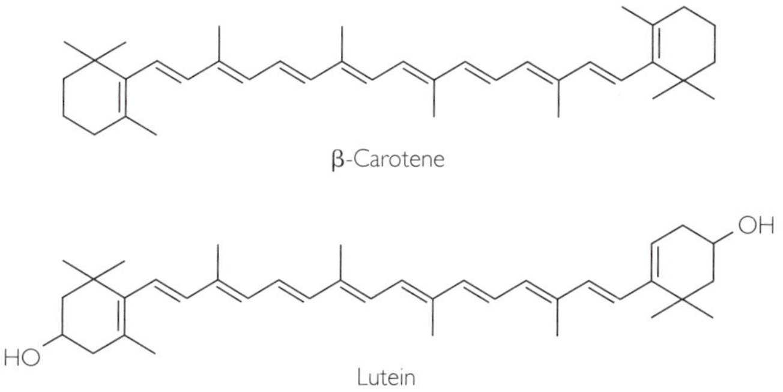

β-Carotene

Lutein

FIGURE 4–4
The two major carotenoids of chloroplast thylakoids. Lutein is also called a xanthophyll. The long, nonpolar chains permit the molecules to embed in a membrane lipid bilayer, and the double bonds ensure fluidity.

chlorophyll-protein complexes. Others make up the electron transport system that links the two photosystems. Still others make up the ATP synthesis machinery. The properties of the six molecular complexes that have been isolated and characterized from thylakoid membranes will be described below.

The distribution of enzyme activities associated with the chloroplast shows compartmentation within the organelle (Table 4–2). In its two membranes, the outer envelope contains a number of enzymes, several of which involve lipid, especially galactolipid, synthesis. The reduction of assimilated nitrogen by nitrate reductase appears to be an envelope function in some plants. The function of the Mg^{++}-ATPase (which can also use Mn^{++}) is not known. Translocases for a number of metabolically important molecules occur at the inner membrane (see below).

The thylakoid is the location of the "light" reactions of photosynthesis, where radiant energy is converted into chemical energy in the form of ATP and NADPH. The two photosystems, an electron transport system, and the oxidative phosphorylation mechanism are located in the lamellae. This chemical energy is used to fix CO_2 into sugars in the Calvin cycle, located in the stroma. In addition, this soluble phase plays an important role in the anabolism of fatty acids and amino acids. It also

T A B L E 4–2   **Compartmentation of Enzymes in the Chloroplast**

Outer Envelope

Fatty acid desaturase

Galactolipid synthesis enzymes

Long-chain acyl-CoA synthetase

Mg^{++}-ATPase

Nitrate reductase

Translocases

Thylakoid Membrane

ATP synthetase

Electron transport carriers

Glutamate dehydrogenase

Photosystems I and II

Stroma

Amino acid synthesis (aspartate, glutamine, glutamate, methionine, tryptophan)

Calvin cycle enzymes

Fatty acid synthetase

Nucleic acid and protein synthesis enzymes

Sulfate reduction enzymes

contains a molecular genetic apparatus for the synthesis of its own genome and some of its proteins.

Function: The Light Reaction

Biochemical Background

The major function of chloroplasts is in the conversion of light energy into chemical energy. In the thylakoid membrane, the primary process of energy transduction occurs:

$$\text{Light} + \text{NADP} + \text{ADP} + P_i + H_2O \rightarrow O_2 + \text{ATP} + \text{NADPH}_2.$$

The input of energy in this system is visible **light,** which has the characteristics of both a particle and a wave. The energy of the particle, the photon, can be expressed in terms of its ability to do electrical work, as in a solar cell. Physicists define the energy of a single photon as a quantum. But not all photons have the same energy content, with light at a shorter wavelength (e.g., blue light) having photons with more energy than light at a longer wavelength (e.g., red light). This is formally expressed by the equation:

$$E = hc/\lambda$$

where E is the energy of a quantum of photons in electron volts, h is Planck's constant, and c is the speed of light. When λ is expressed in nm, the product of hc is 1235. Thus

$$E = 1235/\lambda.$$

The molecules in the chloroplast that absorb sunlight are the chlorophylls (Figure 4–3), carotenoids (Figure 4–4), and other pigments. The actual photochemical reaction occurs in the 1% of the pigments termed the **reaction centers,** which are composed only of chlorophyll a. The remaining molecules of chlorophyll a, and all of the other pigments, are accessory: They absorb light and channel the energy to the reaction center. This process of light harvesting is an essential one when one considers that a single chlorophyll molecule can absorb one photon of light per second yet completes its photochemical reaction in 10^{-15} sec. Were it not for accessory pigments channeling light to the active chlorophyll, it would spend most of its time in "idle," waiting for the next photon.

When a pigment absorbs light at a certain wavelength, it often releases the energy absorbed in the form of emitted light. In a "perfect world," this energy released would be the same as that absorbed, so that if a pigment absorbed light at 500 nm it would emit energy as light at 500 nm. But the second law of thermodynamics states that in any transformation involving energy or matter, the usable energy after the reaction will be less than that at the start (*entropy,* a term for unusable energy, increases). This means that the emitted light will be lower in energy (and longer in wavelength) than the absorbed light. Chlorophyll a absorbs visible light maximally at 430 nm and 675 nm, with the latter wavelength being photochemically most effective. The **accessory pigments** absorb at wavelengths shorter than 675 nm and emit longer wavelengths, making these photons' energy available to chlorophyll a.

The usefulness of accessory pigments is dramatically illustrated by red algae, marine plants that live in water at depths well over 15 meters. At this depth, clear seawater absorbs over 99% of the light at wavelengths over 600 nm, so that virtually no light at 675 nm reaches the chlorophyll directly. Yet the plants are active photosynthetically, and their reaction centers use chlorophyll. They can do this because of a series of accessory pigments, which absorb light at shorter wavelengths and channel its energy to the reaction center (Table 4–3). Light energy is absorbed at a shorter wavelength and emitted at a longer wavelength, where it is absorbed by the next carrier. The antenna pigments of the red algae are usually aggregated to form large (25 nm) aggregates called phycobilosomes, and they are embedded in the lamellae adjacent to the chlorophyll reaction centers.

The **chemistry of chlorophyll** and the other pigments that allows the absorption of light energy is a series of alternating single and double carbon-carbon bonds. In chlorophyll a, this occurs in the porphyrin part of the molecule. Chemically, the part of this conjugated double-bond system that interacts with light is the π orbital system. When it absorbs light at 675 nm, electrons use the energy to become "excited" and move into higher electronic orbitals away from the nucleus. This excited state is unstable and lasts a very short time, on the order of 10^{-15} second.

In the absence of a continuous energy input, the excited electron will return to its more stable and less energetic ground state. The excitation energy is released as light at a longer wavelength because some energy is lost in the excitation–de-excitation process (see above). In chlorophyll a, the fluorescence emission of light absorbed at 675 nm is at 690 nm.

However, chlorophyll a has an unusual property—instead of its excited electron returning to the ground state, the electron can be used to reduce an acceptor molecule in the photosystem complex. This acceptor ultimately transfers the electron to NADP to produce NADPH, one of the two main products of the light reaction. This loss of electrons leaves the chlorophyll molecules in a highly oxidized, unstable state, and the lost electrons are replaced via the oxidation of water to molecular oxygen.

T A B L E 4–3 **Spectroscopic Properties of Accessory Pigments of Red Algae**

Pigment	Visible Absorption Maximum	Fluorescence Emission Maximum
Phycoerythrin	498, 538, 567	578
Phycocyanin	555, 615	640
Allophycocyanin	650	660
Allophycocyanin B	618, 671	680

In essence, the overall light reaction to form reduced NADP is a coupling of the oxidation of water with chlorophyll as an energy-transducing intermediary:

$$H_2O \rightarrow \tfrac{1}{2}O_2 + 2H^+ + 2e^- \qquad E_0{}' = -0.82 \text{ volts}$$

$$NADP + 2H^+ + 2e^- \rightarrow NADPH + H^+ \qquad E_0{}' = -0.32 \text{ volts}$$

The net change in $E_0{}'$ for the coupled reaction is -1.14 volts. Because $\Delta G_0{}' = -nF\Delta E_0{}'$ (see Chapter 3), this means that, on a molar basis, $\Delta G_0{}'$ for the light driven reduction in photosynthesis is $+52$ Kcal. Clearly, this reaction requires a considerable free energy input, and this is where sunlight plays its role.

In the reaction center, chlorophyll a absorbs light at about 675 nm. The energy of a quantum of light at this wavelength is 1.84 electron volts; a mole (6.02×10^{23}) of such quanta has 42.4 Kcal of energy. This is not sufficient to supply the free energy needed (52 Kcal) for the reduction of NADP coupled to water oxidation. Therefore, the minimum amount of light needed is 2 moles of quanta/mole NADP reduced. The actual ratio in isolated chloroplasts is closer to 4.

The **experimental evidence** that chloroplasts use the redox system outlined in the two equations comes from a number of sources. Perhaps the simplest yet most elegant one was done by T. Engelmann on the green alga *Spirogyra* in 1894. He put these cells under a microscope and was able to illuminate them with either the whole field or a narrow area of the cell lighted. As an oxygen sensor, he used a strain of aerobic bacteria, which swim toward that gas. In wide-field light, they clustered all over the spiral chloroplast. But when only a small area of the cell received light, they clustered there if the chloroplast was present in that region (Figure 4–5). He concluded that the chloroplast produced O_2.

Confirmation of this idea came when R. Hill demonstrated oxygen evolution directly by isolated chloroplasts in 1937. Moreover, he found that, in order for O_2 to be produced, a suitable electron acceptor (NADP or another molecule with a similar $E_0{}'$) had to be present to be reduced.

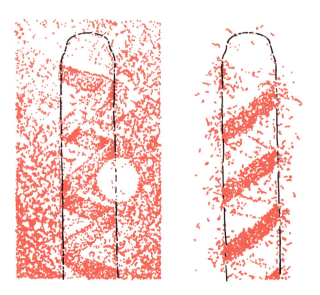

FIGURE 4–5

Engelmann's experiment showing
that the chloroplast produces O_2
when illuminated. Aerobic bacteria swam toward high O_2 concentrations over the chloroplast
only when it was illuminated. If a
nonchloroplast region of the cell
was illuminated, the bacteria did
not swim to it because oxygen
gas was not being made.

The dyes that Hill used are still employed to measure the light reaction.
Because it was the first photosystem discovered, the NADP-reducing
complex is termed photosystem I.

In studying which wavelengths of light were most efficient at promoting photosynthesis, R. Emerson found that two wavelengths (680
nm and 700 nm) were optimal. Because both wavelengths were better then either alone, he suggested that there were two coupled
photosystems.

Finally, the discovery of cytochromes in the thylakoid, along with
ATP synthesis as a second important chemical energy product, led to the
proposal of photophosphorylation. R. Hill and F. Bendall plotted all of
the redox molecules involved in terms of their E_0's and proposed the
Z-scheme for electron flow in the light reaction (Figure 4–6).

The **Z-scheme** describes the flow of electrons from water,
through the two photosystems, to NADP. Like the electron transport
scheme in mitochondria, it is a thermodynamic formulation based on the
known E_0' values for the molecular species involved. The photosystem
numbering seems out of order in this scheme—and it is. The numbering comes from the fact that photosystem I was the first to be isolated.

The Z-scheme essentially has four steps:

1. The primary reaction at **photosystem II** is the oxidation due to light
absorption of the P680 reaction center. This reduces the pigment phaeophytin, which in turn reduces plastoquinone, the carrier to the cytochrome b/f complex. The oxidized P680 extracts an electron from the
electron donor, termed carrier Z, which in turn extracts an electron
from donor M. The latter, a manganese-containing component, has the

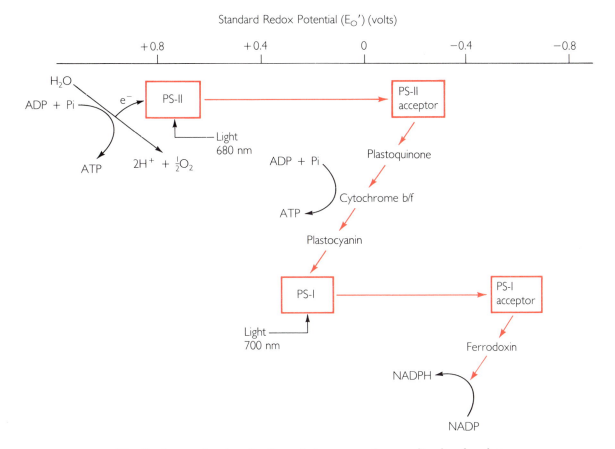

F I G U R E 4–6 The Z-scheme, showing the flow of electrons and noncyclic phosphorylation in the photosynthetic light reaction. The two photosystems are PS-I and PS-II. Reactions that go from right to left (toward increasingly positive E_0') release free energy. On the other hand, those that go from left to right (toward increasingly negative E_0') require an energy input, in this case provided by sunlight.

property of extracting an electron from water, thereby producing oxygen. Thus the flow of electrons in this photosystem is:

$$H_2O \rightarrow M \rightarrow Z \rightarrow P680 \rightarrow \text{phaeophytin}$$
$$\rightarrow \text{plastoquinone.}$$

2. The **cytochrome b/f complex** mediates the transfer of electrons between the two photosystems via an iron sulfur complex:

$$\text{Plastoquinone} \rightarrow \text{cyt } b_6 \rightarrow \text{FeS protein}$$
$$\rightarrow \text{cyt } f \rightarrow \text{plastocyanin.}$$

3. A second photochemical event is involved in the transfer of electrons through **photosystem I.** In this case, excitation of the P700 reaction center and the subsequent loss of an electron cause the P700 to reoxidize plastocyanin. While the ultimate product of this photosystem is reduced ferrodoxin, there are several intermediate steps along the way. These include a specific molecule of chlorophyll a, phylloquinone (vitamin K), and X (an iron sulfur center). The sequence of transfer is as follows:

$$\text{Plastocyanin} \rightarrow \text{P700} \rightarrow \text{Chl A}_0 \rightarrow \text{phylloquinone} \rightarrow$$
$$\rightarrow \text{X(Fe-S)} \rightarrow \text{ferrodoxin.}$$

4. The final step in this scheme is the **reduction of NADP** to form NADPH. This is done via a flavoprotein:

$$\text{Ferrodoxin} \rightarrow \text{flavoprotein} \rightarrow \text{NADP.}$$

The production of a reduced coenzyme, NADPH, is essential for the reductive synthesis of carbohydrates from CO_2 in the dark reaction. However, ATP is also needed, and this is produced through chemiosmotic coupling (see Chapter 3). Two reaction sequences produce protons that can form the basis of an electrochemical gradient: The most obvious is electron transport between the two photosystems via the cytochrome b/f complex. The second is the transport from water to the P680 reaction center of photosystem II.

The production of ATP by the Z-scheme is termed noncyclic photophosphorylation, since the electrons essentially go in one direction, from water to NADP. In **cyclic photophosphorylation,** light again causes the oxidation of chlorophyll, but instead of the electrons ultimately being lost to the chlorophyll, they return. The difference in energy between electrons in the excited state and the ground state (see above) can be used to provide energy for ATP synthesis.

The pathway for the electrons in the cyclic pathway involves some of the same molecules as the noncyclic pathway (Figure 4–7). The chlorophyll center is P700 of photosystem I, but in this case its ultimate electron acceptor, ferrodoxin, reduces not NADP but plastoquinone, which in turn reduces the cytochrome b/f complex. The latter then transfers the electrons to plastocyanin, which passes them back to P700. In the process of transfer through the b/f complex, protons are released, and this provides a chemiosmotic gradient for ATP synthesis.

ATP and NADPH must be synthesized by the light reaction in the correct proportion (1.5 ATP per NADPH) for the dark reactions. When this ratio is too low (the noncyclic scheme produces 1.3 ATP per NADPH and a minimum of 1.5 ATP per NADPH is needed), the cyclic pathway is used to supply the extra ATP. In some cells (e.g., the nitrogen-fixing heterocysts of blue-green algae and the bundle-sheath cells of

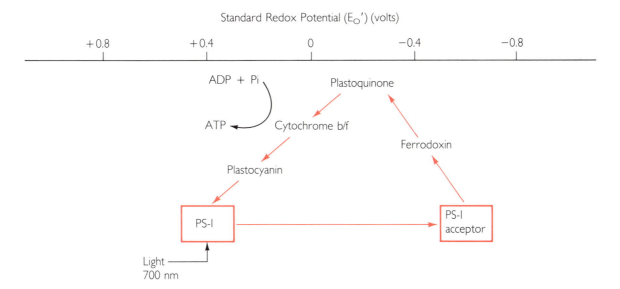

FIGURE 4–7   The flow of electrons in cyclic phosphorylation in the photosynthetic light reaction. Only ATP is produced. Compare to Figure 4–6, where NADPH is also produced.

C4 plants—see below), the noncyclic pathway does not occur, and chemical energy from the light reaction is produced only by the cyclic pathway.

Membrane Structure for the Light Reaction: Biochemistry

Taken together, the concepts of the reaction center and Z-scheme indicate a complex array of molecules performing the photosynthetic light reaction. An indication of this complexity is given in Table 4–4. Literally dozens of different components occur within the internal lamella system of the chloroplast. The challenge facing the cell biologist is to determine the functional arrangement of these molecules in the membrane. Both biochemical and structural techniques have been used for this task.

The pioneering work in this area was done by J. Deisenhofer and H. Michel, who crystallized the photosynthetic reaction center of the purple bacterium *Rhodopseudomonas,* which has only one photosystem. They found a complex consisting of two light-absorbing chlorophylls, phaeophytins (chlorophylls that do not contain magnesium) and quinones, all bound together with three transmembrane proteins. It was possible to trace the pathway of the photochemical reaction through this complex. They were awarded a Nobel prize for this research.

TABLE 4–4 Approximate Composition of the Molecular Complex Performing the Photosynthetic Light Reaction in Spinach

Component	Number of Molecules
Chlorophyll a	160
Chlorophyll b	70
Carotenoids	48
Plastoquinones	28
d-Tocopherol	10
Vitamin K	4
Phospholipids	116
Sulfolipids	48
Glycolipids	490
Ferrodoxin	5
Cytochrome b 563	1
Cytochrome b	1
Cytochrome f	1
Plastocyanin	1
Other proteins	30
Iron	12 atoms
Copper	6 atoms
Manganese	2 atoms

FROM: Metzler, D. *Biochemistry* (p. 773). New York: Academic Press, 1977.

Six intact, functional complexes have been isolated and characterized from thylakoid membranes. While nomenclature schemes have varied, Table 4–5 presents a summary of these molecular aggregates, each of which has a vital role in the light reaction.

Core complex I reduces NADP via the carrier protein ferrodoxin and contains the reaction center of photosystem I, P700. A polypeptide, 70,000 molecular weight, binds in four copies per P700 and also to 50 molecules of chlorophyll a and carotene. Mutants of the green alga *Scenedesmus,* which lack this protein, cannot carry out photosystem I. The other polypeptides are Fe-S proteins, which act as electron acceptors. This complex is associated with a light-harvesting complex **(LHC-I)** containing both chlorophylls, bound to several polypeptides. This LHC acts as an antenna, absorbing light and channeling it to the photosystem.

TABLE 4–5   **Pigment and Protein Complexes from Thylakoids**

Name	Composition
Core complex I (photosystem I)	Chlorophyll P700, 40 chlorophyll a, 5 carotene, 3 Fe-S centers, 8 polypeptides
LHC-I (light-harvesting)	Chlorophylls a and b (ratio 5:1), 3 polypeptides
Core complex II (photosystem II)	Chlorophyll P680, 5 chlorophyll a, 1 carotene, 1 phaeophytin, cytochrome b559, 20 polypeptides
LHC-II (light-harvesting)	Chlorophylls a and b (ratio 1:1), 2 polypeptides
Cytochrome b_6-f complex	Cytochromes b_6 and f, 2 other polypeptides
ATP synthetase	CF_1: 9 polypeptides; CF_0: 3 polypeptides

Core complex II actually has two variants that are almost identical but differ slightly in the amino acid sequences of their major apoprotein. In each case, a single protein, molecular weight about 45,000, binds to the P680 reaction center, 40 chlorophyll a molecules, and phaeophytin. A second set of proteins, 32,000 and 34,000 molecular weight, bind to plastoquinone and constitute the site of its reduction. This entire aggregate is very similar to the reaction center in purple bacteria (Figure 4–8) and carries out photosystem II.

Core complex II is associated with a light-harvesting complex **(LHC-II),** which also comes in two forms. The major one of these, LHC-IIb, is remarkable in that it is the most abundant protein in the thylakoid, and it has a relatively high chlorophyll b content (a:b ratio of 8:7). Indeed, barley mutants lacking chlorophyll b also lack LHC-IIb. It is estimated that LHC-II binds half of all chlorophyll in nature. This complex also acts as an energy antenna but also may be a regulator of membrane stacking (see below).

Cytochrome b_6-f complex has a number of polypeptides that are cytochromes and Fe-S proteins. It acts as the transporter of electrons between photosystems II and I, in much the same way that the electron transport system in mitochondria passes electrons from electron donors such as NADH to oxygen (see Chapter 3). In addition, this complex has a vital role in cyclic photophosphorylation (Figure 4–7), where it accepts electrons indirectly from ferrodoxin and passes them to photosystem I.

Again similar to mitochondria, the **ATP synthetase** has a membrane-embedded F_0 portion, which acts as a proton channel, and a protruding ATP synthetase, which uses the proton gradient to make ATP. This complex is described in detail below.

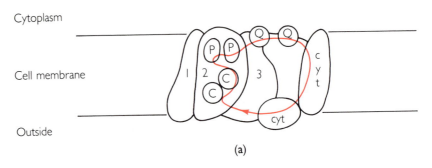

Cytoplasm

Cell membrane

Outside

(a)

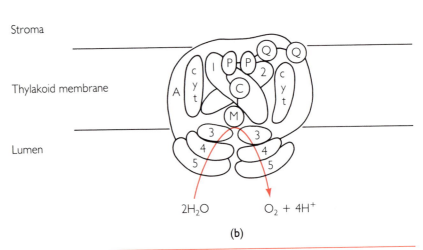

Stroma

Thylakoid membrane

Lumen

$2H_2O$ $O_2 + 4H^+$

(b)

FIGURE 4–8

Diagram of the pigments and proteins of the photosynthetic reaction center of the purple bacterium *Rhodopseudomonas* **(a)** and core complex II in the higher plant thylakoid membrane **(b).** The flow of electrons is noted. (C = active chlorophyll; A = accessory chlorophyll; P= phaeophytin; cyt = cytochrome; Q = quinone; 1–5 = proteins.)

FIGURE 4–9

Three herbicides that act by inhibiting the photosynthetic light reactions. Atrazine and Diuron block the transfer of electrons out of photosystem II, while paraquat accepts electrons from photosystem I.

Atrazine

Paraquat

Diuron

A number of important **herbicides** act by binding to these complexes (Figure 4–9). For example, atrazine (Atraton) and Diuron bind to core complex II and inhibit the transfer of electrons to plastoquinone. This essentially stops the reduction of NADP, since there is no longer a connection between the two photosystems. An important aspect of these herbicides is their selectivity. In most cases, the crop of interest (e.g., corn) has the ability to rapidly break down the herbicide so that it is resistant, whereas the target weeds cannot do this. Sometimes, weeds develop resistance to the herbicide, and when this occurs it is often due to a mutation in a gene coding for the complex II protein that normally binds the chemical.

Some herbicides have a broader target base. For example, paraquat (Figure 4−9) inhibits photosynthesis not by blocking electron transport but by acting as an electron acceptor from photosystem I. It can do this because its E_0' is about −0.5 volts, well above that needed for "capture" of the excited electrons from this photosystem (Figure 4−6). The herbicide then transfers the electrons to molecular oxygen, forming superoxide radicals, which then kill the cell:

$$PS\text{-}I \rightarrow paraquat \rightarrow O_2 \rightarrow O_2^-(superoxide).$$

Mapping the complexes in the thylakoid membrane has employed methods similar to those used to localize components in the mitochondrial membrane (see Chapter 3). The isolation of the six complexes has permitted analysis by antibodies and other impermeant reagents. For example, antibodies to cytochrome f and plastocyanin bind to their targets only if they are sonicated along with the lamellae; this indicates that they must face the inside of the thylakoid, since they are inaccessible in intact plastids or isolated lamellae. On the other hand, an antibody to ferrodoxin binds readily to isolated, nonsonicated lamellae, indicating that ferrodoxin is on the side of the lamella facing the stroma.

Radioactive impermeant reagents that bind nonspecifically to proteins have also been used. If one of these reagents is applied to an intact lamellar system it will bind to stroma-facing proteins, which can then be isolated and their radioactivity demonstrated. In this manner, cytochrome b has been shown to be stroma-facing, and core complex I and LHC-I have been shown to span the membrane. Finally, a coupling factor (F_1) for ATP synthesis has been shown to extend out into the stroma by antibody binding as well as microscopy. Figure 4−10 shows the arrangement of molecules in the membrane as determined biochemically.

Brief incubation of chloroplasts in detergents leads to a separation of the internal membranes. The membranes facing the stroma, which are **unstacked,** become detached from those composing the grana **(stacked).** Because of their different densities, these regions are then easily separated by centrifugation. Analyses of these two fractions show that the stroma membranes are considerably richer in photosystem I activity, and the grana membranes lack this complex and instead have photosystem II activity. The cytochrome b/f complex is present in both regions, and the CF_1-ATP synthetase is located on stroma membranes (Figure 4−11).

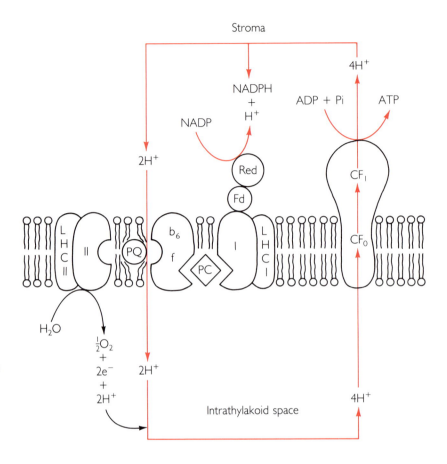

FIGURE 4–10
Diagrammatic representation of the arrangement of the photosynthetic complexes in the thylakoid membrane. The two proton pumps are shown. (I = core complex I; II = core complex II; LHC = light-harvesting complex; CF = coupling factor (ATP synthetase); b,f = cytochromes b_6 and f; PC = plastocyanin; PQ = plastoquinone; Fd = ferrodoxin; Red = NADP reductase.)

This spatial separation of the coupling factor from the electron transfer reactions is not unexpected, since a chemiosmotic mechanism is presumed to operate for ATP synthesis (see below). However, the physical separation of the two photosystems in the plane of the membrane has an important consequence: There must be carriers that shuttle through the fluid bilayer between the photosystems and the cytochrome b/f complex.

The high content of unsaturated lipids ensures considerable fluidity of the lamellae at temperatures within the typical leaf. Lipids and proteins in less fluid membranes have diffusion rates for lateral movement of about 1 μm/sec (Table 1–7). In the more fluid lamella, this could easily be an order of higher magnitude, 10 μm/sec. Assuming this rate of diffusion, and a granum diameter of 0.1 μm, it would be possible for the carrier linking the two photosystems to traverse the 50 nm of mem-

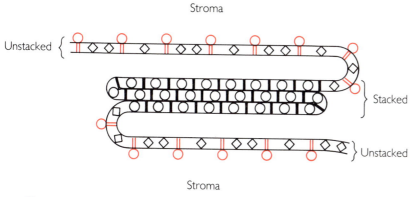

Stroma

Unstacked {

} Stacked

} Unstacked

Stroma

CF-ATPase
core complex I
core complex II
cytochrome-b/f complex

FIGURE 4–11

The separation of photosynthetic complexes in stacked versus unstacked thylakoid membranes. (After J. Anderson.)

brane between them in 5 msec. This is well within the time measured for the rate-limiting (slowest) step of the light reaction, 20 msec.

The messengers proposed to link the two photosystems and the cytochrome complex are plastoquinone, the most abundant electron transport pigment, and plastocyanin. Indeed, the photosystem II complex has a binding site for plastoquinone, the photosystem I complex binds plastocyanin, and the cytochrome b/f complex appropriately binds both plastocyanin and plastoquinone.

The **stacking and unstacking mechanism** of thylakoid membranes occurs via their surface charges. In stacked regions, the adjacent membranes have little charge, and this means that there is little electrostatic repulsion between the two surfaces so that van der Waals forces can hold the membranes together. In contrast, in unstacked regions the surface charges are much larger, and the two membranes repel each other.

Thylakoid stacking and unstacking constantly occurs within the chloroplast, so there must be a way to regulate the surface charges to achieve this. The regulator is believed to be LHC-II. A fraction of this complex is mobile in the plane of the membrane and is unassociated with core complex II. This fraction can be reversibly phosphorylated via threonine residues on the apoprotein. The increase in surface charge resulting from LHC-II phosphorylation is sufficient to cause membrane repulsion and unstacking. The identification of LHC-II as the putative membrane stacking agent has been made from developmental studies (plastids develop stacking only when LHC-II is produced) and mutants (in the barley mutant that lacks the major LHC-II apoprotein, stacking requires unphysiologically very high ion concentrations).

LHC-II–mediated stacking and unstacking is important for **light reaction regulation.** Photosystem I (P700) and photosystem II (P680) have different absorption efficiencies for incident light. Yet for the Z-scheme (Figure 4–6) to operate most efficiently, both photosystems should be equally excited. When photosystem II has excess energy compared to photosystem I, LHC-II becomes phosphorylated. This causes a localized unstacking of the thylakoid, and the LHC-II can diffuse to core complex I, where it increases energy absorption at the same time as its absence from complex II leads to a lowering of the efficiency of this photosystem. This tends to equalize the reaction efficiencies in the two photosystems. On the other hand, excess energy at photosystem I causes dephosphorylation, the LHC-II migrates back to photosystem II, and the process is reversed.

The migration of LHC-II is quite rapid (20 sec) and provides the plant with short-term adaptation to intense light. Longer-term adaptation is seen in plants adapted to sunny or shady environments. Excess light can reduce the efficiency of photosynthesis (photoinhibition), and plants that live in sunny environments prevent this by reducing their maximal photosynthetic rate by having a much reduced LHC-II activity relative to core complex II.

On the other hand, shade-adapted plants need to trap all of the light they can, and they have an extensive LHC-II. Of course, when the sun shines brightly on such plants, they are especially sensitive to photoinhibition. For example, pine trees in the northern hemisphere are typically adapted to the shade of the forest and have a high LHC-II:core complex II ratio. In the spring, bright sunlight reflected from snow on the ground causes photoinhibition and the pine needles turn yellow.

Membrane Structure for the Light Reaction: Electron Microscopy

Information on membrane structure that complements the biochemical approaches has come from the use of the **freeze-fracture** technique. In this method, isolated thylakoids are frozen rapidly in liquid freon, and the frozen membrane is fractured in a vacuum with a knife. The fractured surface is coated by metal, deposited at a 45° angle. When the plant membrane is digested away, what remains is a replica of the surface, with shadows cast by protruding particles because of the angle of metal coating.

In favorable instances, the fracture plane passes through the interior of the membrane, splitting the bilayer. Four interior faces of the lamellar network can be seen. The PF, or protoplasmic face, is the bilayer half facing the stroma. The EF, or exoplasmic face, is the bilayer half facing the intrathylakoid space. These faces are discerned in both stacked (s) and unstacked (u) regions. A complex—some say, bewildering—array of particles is seen on these fracture faces under the electron microscope (Figure 4–12). An interpretation of these particles is shown in Figure 4–13.

In the **stacked** regions are smaller, 8-nm–diameter particles (average size), on the P membrane face (PFs) and mostly larger, 16-nm but some 8-nm particles also, on the EFs face. Several lines of evidence indicate that the large particles represent PS-II complexes with LHC-II, and the smaller ones represent the two separate components. First, the isolated components have these molecular dimensions when viewed under the electron microscope after incorporation into lipid vesicles. Sec-

FIGURE 4–12   Freeze-fracture image of pea chloroplast membranes. The four fracture faces are observed. (×135,000. Photograph courtesy of Dr. L. A. Staehelin.)

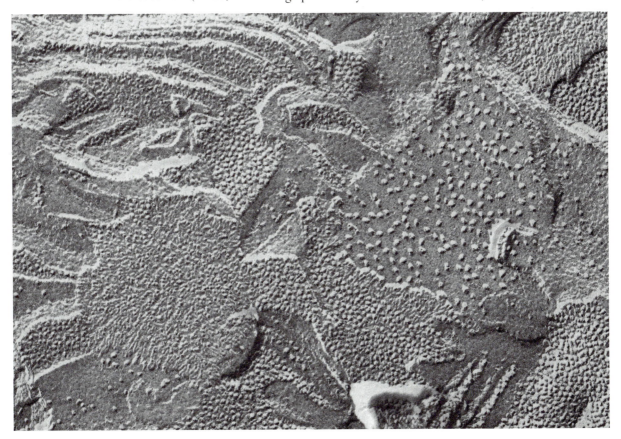

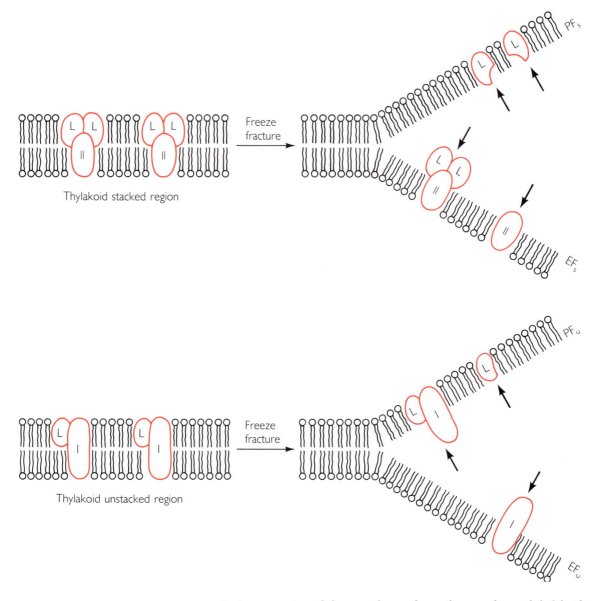

FIGURE 4–13 An interpretation of the particles on freeze-fracture faces of thylakoids (Figure 4–12). Arrows indicate particles. (I = photosystem I core; II = photosystem II core; L = light-harvesting complex.)

ond, in a developmental study, it was noted that when pea plants are grown in the dark, all the EFs particles are 8 nm. Illumination leads to an increase in the 16-nm particles. Biochemically, illumination leads to greening because of chlorophyll synthesis, and LHC-II accumulates at this time. Therefore, the LHC represents the extra 8-nm particle. Third, there is a mutant of barley that lacks chlorophyll b and part of LHC; its 12-nm EFs particles are smaller than those of the wild type barley.

In **unstacked** lamellae, the Efu (stroma facing, unstacked) contains 8-nm particles, while the PFu (facing the interior of the thylakoid, unstacked) has larger 11-nm particles along with some smaller ones. In this case, an interpretation is that the smaller 8-nm particles represent the PS-I core, and the larger ones represent the core with added LHC-I components. The evidence for this is similar to that for the stacked regions. First, the isolated PS-I core can be incorporated into lipid vesicles and shows the expected 8-nm dimension. Second, during the development of greening in pea plants, the previous 8-nm PFu particles increase in size to 11 nm. This occurs simultaneously with the production of some LHC components. Third, a mutant in maize (corn) having no PS-I activity lacks specific PS-I proteins. Its PFu particles are smaller, as are its EFu particles, indicating that both of these contain PS-I.

The combination of biochemical, structural, and genetic approaches has given a considerable amount of information on the structure of the thylakoid membrane (Figure 4–10). The flow of electrons can now be traced topographically from photosystem II to photosystem I to NADP.

ATP Synthesis

In addition to NADPH, the light reaction channels some of its energy into the synthesis of ATP. Because light is involved, this process is termed **photophosphorylation.** In the thylakoids, a chemiosmotic mechanism with proton electrochemical coupling occurs that is similar in many respects to oxidative phosphorylation in the inner mitochondrial membrane (see Chapter 3).

The synthesis of ATP can be coupled to electron transport between the photosystem II acceptor and photosystem I. The electrons move to carriers of increasingly more positive E_0' (Figure 4–6). This uphill movement in reduction potential is a downhill movement in terms of energy, since:

$$\Delta G_0' = -nF\Delta E_0'.$$

In addition to the ATP that can be made via electron transport between the two photosystems, some ATP can be made via the electron transport in photosystem II between water ($E_0' = +0.8$ volts) and the reaction center P680 ($E_0' = +1.1$ volts). Both electron transport and water photolysis release protons, and these are used to set up the chemiosmotic gradient.

The evidence for chemiosmotically driven ATP synthesis in the chloroplast is similar to that for its occurrence in mitochondria:

1. The architecture of the membrane protein is **asymmetric** (Figure 4–10). The arrangement of the photosystem complexes is such that protons are pumped into the intrathylakoid space during electron transport. The pumping is believed to occur between plastoquinone and cytochrome b. An F_1-ATP synthetase, very similar in structure and appearance to mitochondrial F_1, is used to pump the protons back out into the stroma. Microscopically, this chloroplast F_1 (CF_1) appears as a "lollipop" extending out from the lamella into the stroma.

2. As he has done with mitochondria (Figure 3–12), E. Racker has been able to dissociate CF_1 from the thylakoid. When **reconstituted** into the membrane or even into artificial membranes, it will synthesize ATP if an H^+ gradient is supplied. CF_1 is located almost exclusively on unstacked lamellae. This is consistent with chemiosmotic coupling, as the ATP synthetase need not be adjacent to the site of proton pumping into the intrathylakoid space.

3. Measurements of the pH outside of the thylakoids show an increase on illumination. This **pH increase** is due to the pumping of protons out of the stroma during electron transport. The pH gradient set up in this way is 3–4 pH units. The lamella membrane is permeable to ions, in contrast to mitochondrial inner membrane, which is not. When H^+ is pumped in, either Cl^- accompanies it or Mg^{++} is pumped out to maintain electroneutrality. This means that, for chloroplasts, the equation:

$$\Delta\mu_{H+} = \Delta\psi - 60\Delta pH$$

becomes

$$\Delta\mu_{H+} = -60\Delta pH.$$

With a pH gradient of 3.5, μ_{H+} becomes -210 mv. This figure for the proton potential is similar to that obtained for mitochondria. For the phosphorylation of one ADP molecule, $2H^+$ must cross the membrane, and indeed this is what occurs.

Thylakoids will even make ATP in the absence of electron transport if the proper pH gradient is supplied. This was demonstrated by A. Jagendorf and E. Uribe, who placed thylakoids in a succinic acid solution at pH 4, which resulted in an intrathylakoid pH of 4. The thylakoids were then placed in a dilute pH 8 solution ($\Delta pH = 4$). Electron transport was inhibited by the fact that the experiment was done in the dark and specific blockers were present. If ADP and P_i were supplied, ATP was synthesized in a rapid burst. After some time, synthesis slowed down because the proton gradient was not replenished and ATP was not removed.

4. **Ionophores** such as FCCP (Figure 2–4) transport protons across the thylakoid membrane, upsetting the H^+ gradient, and inhibit pho-

tophosphorylation. As is the case with mitochondria, this is strong evidence for the involvement of a proton gradient in ATP synthesis.

The structure and possible mechanism of CF_1 are very similar to those of mitochondrial F_1. Indeed, amino acid sequences derived from DNA sequences indicate considerable homologies between mitochondrial and chloroplast ATP synthetases. The plastid ATP-synthesizing component extends into the stroma and has five different subunits, two of which are present in three copies for a total of nine polypeptides. Both of the large α and β subunits contain nucleotide binding sites, but only the ones on beta are apparently involved in catalysis. The single γ polypeptide is essential for binding the F_1 portion to the membrane, and the other two subunits may be regulatory. The membrane-embedded CF_0 has three subunits and forms a proton channel (Figure 4–14). Conformational changes due to chemiosmotic proton transport have been implicated in the mechanism of ATP formation.

Like mitochondrial F_1, chloroplast CF_1 will act as an ATP synthetase as long as there is a continuous replenishment of the proton gradient (Figure 3–12). This occurs at all times in physiologically normal mitochondria because of the electron transport chain. But in the chloroplast, electron transport and proton pumping happen only in the presence of light. Therefore, at night, the CF_1 would tend to act as an ATPase and hydrolyze any ATP present in the plastid. For this reason, CF_1 is subject to regulatory controls. It apparently is active only when a sulfhydryl group on the gamma subunit is reduced. At night, the group is oxidized and the enzyme is inhibited in both directions, thus preventing its ATPase from functioning.

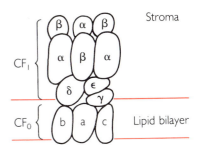

F I G U R E   4 – 1 4

The CF-ATPase/ATP synthetase in the thylakoid membrane. The protruding CF_1 portion catalyzes ATP production/hydrolysis and has three copies of each of the alpha and beta subunits and one copy each of the other three subunits. The CF_0 portion has three subunits and forms the H^+ channel in the membrane.

Function: The Dark Reaction

Biochemical Background

In the light reaction, radiant energy is converted into chemical energy in NADPH and ATP. In the dark reaction (so named because light is not necessary for it to occur), these two compounds are used to reduce atmospheric CO_2 to "fixed" carbohydrate:

$$CO_2 + NADPH + ATP \rightarrow (CH_2O)_n + NADP + ADP + P_i.$$

The enzymes that catalyze this series of reactions occur in the chloroplast stroma. This is functionally optimal, since both NADPH and ATP are formed on the stroma side of the lamella membrane (Figure 4–10).

The reactions of CO_2 fixation were worked out in the laboratory of M. Calvin. He and his colleagues incubated the unicellular green

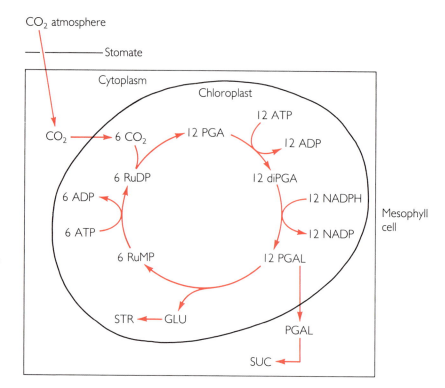

FIGURE 4–15

Metabolic relationships in C_3 photosynthetic carbon fixation (Calvin cycle). (PGA = phosphoglyceric acid; STR = starch; diPGA = diphosphoglycerate; SUC = sucrose; PGAL = phosphoglyceraldehyde; GLU = glucose-P; RuMP = ribulose monophosphate; RuDP = ribulose biphosphate.)

alga *Chlorella* in $^{14}CO_2$, and, following illumination, he identified ^{14}C-products by paper chromatography. The Calvin cycle (Figure 4–15) has three steps: carboxylation, reduction, and regeneration:

1. Carboxylation involves the reaction of CO_2 with the five-carbon ribulose-1,5-biphosphate to form two molecules of the three-carbon phosphoglyceric acid (PGA). This reaction is catalyzed by the enzyme ribulose-1,5-biphosphate carboxylase/oxygenase or RubisCO. This enzyme has a large catalytic subunit and a smaller regulatory subunit. Regulation is achieved by the stimulatory actions of Mg^{++} (transported into the stroma as H^+ is pumped out during the light reaction), alkaline pH (a consequence of the proton pump), and NADPH. Thus the light reaction is not only essential for but also promotes the dark reaction. RubisCO accounts for up to half of the total protein in the chloroplast stroma. It is probably the most abundant protein in the natural world.

2. Reduction of CO_2 requires energy inputs from the light reaction in the forms of ATP and NADPH. The product is triose phosphate (phosphoglyceraldehyde). This reduced molecule can then be converted into starch within the chloroplast or translocated out of the organelle (see below) to be a precursor for sucrose. In a typical leaf, the ratio of starch: sucrose production is about 1:2.

3. Regeneration of ribulose-1,5-biphosphate, the acceptor for CO_2, must occur for the cycle to begin anew. Of the triose phosphate formed, only one-sixth is used for sucrose or starch synthesis. The remainder is used in a series of 10 individual reactions to regenerate the CO_2 acceptors.

The Michaelis constant (K_m) measures the concentration of a substrate at which an enzyme performs its catalytic function half-maximally. If a substrate concentration falls below the K_m of its enzyme, the reaction proceeds very slowly. The K_m of CO_2 for RubisCO is 12 μM. In a typical leaf, chloroplasts are most abundant in the mesophyll cells, which lie just below the surface of the leaf. CO_2 enters the leaf via pores, or stomata, and diffuses into the mesophyll cells and chloroplasts, where carboxylation occurs. Given diffusion of CO_2 and its constant removal, estimates of its effective concentration in the stroma are about 0.0001% or 200 μM. This is well above the K_m, so that carboxylation occurs efficiently.

In the leaves of tropical grasses and their relatives such as sugarcane and maize, green chloroplasts are abundant, not near the surface but in cells surrounding the conductive tissues in the middle of the leaf (Figure 4–16). Because of the longer diffusion distance, the effective CO_2 concentration in the chloroplast stroma of these cells would be lower than that in the other leaves. Another reason for this is that, because of hot daytime temperatures where they grow, these plants tend

---

FIGURE 4–16   Comparative anatomy of the leaves of typical C_3 and C_4 plants. In C_3 plants, CO_2 fixation and the Calvin cycle occur in the mesophyll cells, which contain abundant chloroplasts. In C_4 plants, CO_2 fixation occurs in the mesophyll, but the Calvin cycle occurs in the bundle sheath cells.

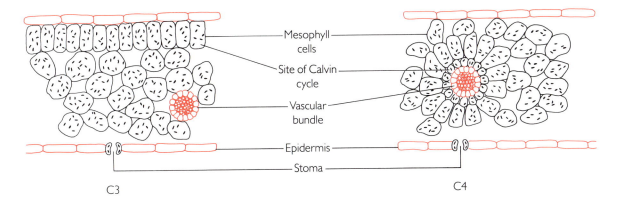

to close their stomatas to conserve water. As a result of these factors, CO_2 in the stroma of bundle sheath cells' chloroplasts would be as low as 1 μM, far below the K_m for RubisCO.

To overcome this difficulty, these plants fix CO_2 in plastids in the mesophyll cells not by RubisCO, which is absent, but via a three-carbon acceptor into a four-carbon molecule. This different fixation reaction is highly efficient at low CO_2 concentrations. The four-carbon molecule, or a metabolite, is transported to the bundle sheath chloroplasts, where it "drops off" its CO_2, producing a three-carbon molecule that regenerates the three-carbon acceptor. Figure 4–17 shows these relationships for one type of C4 plant.

There are actually three types of **C4 pathways,** distinguished by the method of converting the C4 molecule into a C3 molecule. The three reactions occur in different cell compartments:

$$\text{Malate (C4)} + \text{NADP} \rightarrow \text{pyruvate (C3)} + CO_2 + \text{NADPH}$$
$$\text{(chloroplast—see Figure 4–16)}$$

$$\text{Malate (C4)} + \text{NAD} \rightarrow \text{pyruvate (C3)} + CO_2 + \text{NADH}$$
$$\text{(mitochondrion)}$$

$$\text{Oxaloacetate (C4)} + \text{ATP} \rightarrow \text{phosphoenol pyruvate (C3)}$$
$$\text{(cytoplasm)} \qquad + CO_2 + \text{ADP}$$

The C4 pathway results in a CO_2 concentration of 10 μM in the bundle sheath stroma. This is a tenfold improvement over the possible situation with the C3 pathway alone, and the Calvin cycle can begin efficiently.

There is an important feature associated with the C4 pathway. At high O_2 concentrations, RubisCO does not fix CO_2 but acts on some of the ribulose-biphosphate as an oxygenase, forming PGA and phosphoglycolic acid. Ultimately, CO_2 is released and ATP is used up by these reactions, termed **photorespiration:**

$$\text{RuBP} + 15\ O_2 + 5\ \text{NADH} + 29\ \text{ATP} + 15\ \text{NADPH} \rightarrow$$
$$5\ CO_2 + 25\ H_2O + 5\ \text{NAD} + 15\ \text{NADP} + 29\ \text{ADP} + 31\ P_i.$$

In C4 plants, mesophyll and bundle sheath cells can only carry out cyclic photophosphorylation. This generates ATP, but an important side effect is that it does not generate O_2. This means that in the mesophyll cell, where CO_2 arrives from the air, the ratio of O_2 to CO_2 is quite low. This is not important for the C4 fixation reaction. But in the bundle sheath cell, where the fixation is by RubisCO, the low ratio means that

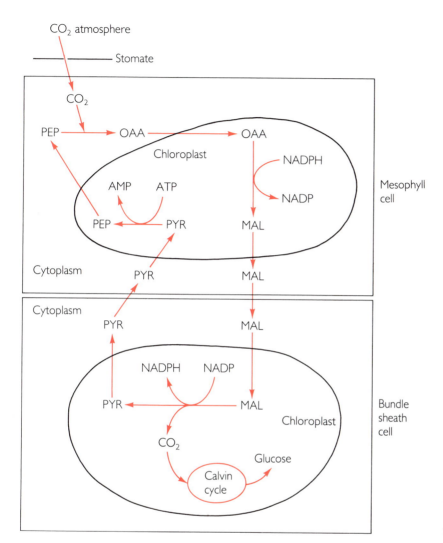

FIGURE 4–17
Metabolic relationships in C_4 photosynthetic carbon fixation. (PEP = phosphoenol pyruvate; OAA = oxaloacetate; MAL = malate.)

photorespiration will not occur to any appreciable extent. This is a reason why C4 plants are highly productive in terms of their production of fixed carbon. Calculations by M. D. Hatch put the quantum yield (moles of CO_2 fixed per mole of light quanta absorbed) of C4 plants at 0.065, while the number for C3 plants is 0.053.

The most important food crops are C3 plants, whose productivity could increase if photorespiration losses were reduced. Agricultural scientists are pursuing methods to do this chemically and genetically. However, some recent evidence indicates that photorespiration acts to prevent the formation of harmful free radicals when intense light interacts with plant cells. If this is so, elimination of this activity of RubisCO could be harmful to the plant.

In some C4 plants, CO_2 fixation occurs primarily at night, and the C4 product is stored until the next day, when it is decarboxylated. This storage can be easily detected by the sour taste (due to malic acid) of such plants in the early morning. Storage actually occurs inside the vacuole (see Chapter 8), which protects the cytoplasm from a reduction in pH because of the high concentrations of malate.

These plants are mostly succulent cacti of the family *Crassulaceae* (thus the name **Crassulacean acid metabolism** [CAM]) and live in hot, dry regions. Because of this, they keep their leaf stomatas closed during the day to conserve water. Of course, this means that CO_2 cannot enter the leaf to get fixed at this time but instead does so at night when the stomates can open. The next day, the malate is transported out of the vacuole and the C4 pathway continues, leading to fixation of CO_2 by RubisCO. A look at the CAM pathway (Figure 4−18) shows that it is a typical C4 pathway except for the fact that the "night shift−day shift" separation has replaced the cellular separations of most C4 plants.

The initial enzymes of CO_2 fixation in C3 and C4 plants differ in their preferences for the two naturally occurring isotopes of carbon, ^{12}C and ^{13}C. RubisCO has a decided preference for $^{12}CO_2$, while PEP carboxylase has a much reduced preference. This allows the biochemist to evaluate the relative contributions of the two pathways by simply analyzing the isotopic content of fixed carbon. It also can permit one to distinguish the source of sugar: Beet sugar is extracted from a C3 plant, and cane sugar is from a C4 plant.

C4 photosynthesis occurs in a wide variety of genera from at least 16 plant families, all of which also include members with C3 photosynthesis. Thus, there is no simple theory to account for the evolution of C4 photosynthesis, except that it probably appeared on many separate occasions. With their similar cellular anatomy and biochemistry, the C4 plants represent an interesting case of convergent evolution.

Translocation of Metabolites

The photosynthetic dark reactions require considerable transport of metabolites in and out of the chloroplast. The envelope is the barrier between stroma and cytoplasm, and of the two membranes in the envelope, the outer one is quite permeable to most small molecules, and the inner one is relatively impermeable to hydrophilic solutes. These molecules cross the inner membrane by means of specific translocases. Some of these are summarized in Table 4−6.

The **phosphate** translocator exchanges incoming P_i for outgoing triose phosphate and other fixed carbon phosphates. It is most efficient

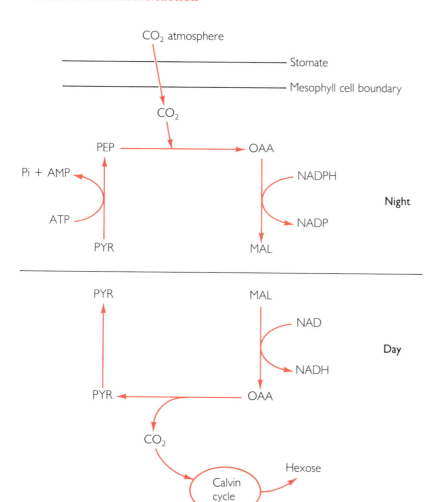

FIGURE 4–18

Crassulacean acid metabolism. At night, plants fix CO_2 into C_4 acids. These are stored in the vacuole until the next day, when the C_4 acids are decarboxylated and the Calvin cycle occurs. Note the similarity in overall scheme to the C_4 cycle, Figure 4–17.

with the former, the most important product of the Calvin cycle. The transport protein exhibits saturable kinetics (K_m 0.1 mM), competitive inhibition (pyrophosphate) and is inhibited by reagents that bind to certain amino acids. The phosphate translocator, an integral protein in the inner membrane, plays an important physiological role, in that it imports Pi into the stroma for ADP phosphorylation and exports fixed carbon to the cytoplasm.

Dicarboxylate transport is also an exchange across the envelope. Because the inner membrane is impermeable to NADP and NAD, the malate-aspartate shuttle (Figure 3–4) may operate in chloroplasts to transfer reducing equivalents from the stroma to the cytoplasm. In

TABLE 4–6   **Some Translocases in the Inner Chloroplast Membrane**

Name	Molecules Transported
Phosphate	Phosphate-3-phosphoglycerate, dihydroxyacetone-P; less efficiently: glycerol-1-P, erythrose-4-P, ribose-5-P, phosphoenol-pyruvate
Dicarboxylate	Malate, oxaloacetate, α-ketoglutarate, aspartate, glutamate
Adenine nucleotide	ATP-ADP
Sugars	Glucose, mannose, fructose, ribose, xylose
Amino acids	Glycine, serine
Glycolate	Glycolate–glycerate
Pyruvate	Pyruvate (C4 plants only)

this case, the shuttle would operate in reverse to that seen in the mitochondrion.

Another way to obtain reduced coenzymes in the cytoplasm for anabolism is via the 3-phosphoglycerate shuttle. This is catalyzed by the **phosphate** translocator, which exchanges incoming cytoplasmic 3-phosphoglycerate with outgoing stroma dihydroxyacetone phosphate (Figure 4–19). This shuttle also carries out stroma ATP, the other product of the light reaction. Such an indirect method for ATP export is necessary because the chloroplast adenine nucleotide carrier is a rather weak one. It appears to have a role in ATP entry to the stroma, supplying the molecule to the dark reaction at night when the light reaction does not occur.

The **glucose** carrier exports this product of the breakdown of starch from the stroma. Its kinetic and molecular properties are remarkably similar to the carrier in mammalian cells. **Glycolate** is a product of photorespiration in C3 plants and is removed by a specific carrier, while C4 plants contain a **pyruvate** carrier for its export from the bundle sheath cell chloroplast and import into the organelle in the mesophyll cell (Figure 4–17).

Nongreen Plastids

Many plant parts, such as roots, buds, and flowers, contain nongreen plastids. Most of these are derived from chloroplasts. Table 4–7 summarizes some of the properties of these organelles, and Figure 4–20 shows some of their structures.

Amyloplasts occur in storage tissues such as potato tubers and seeds, meristems, and the very tips of roots. They lack an internal membrane system and contain abundant starch deposits in the form of a sin-

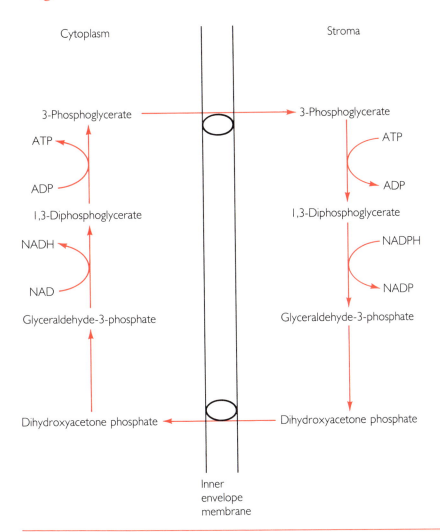

Cytoplasm   Stroma

3-Phosphoglycerate ⟶ ⟶ 3-Phosphoglycerate

ATP ⟵
ADP

1,3-Diphosphoglycerate   1,3-Diphosphoglycerate

NADH ⟵
NAD

Glyceraldehyde-3-phosphate   Glyceraldehyde-3-phosphate

Dihydroxyacetone phosphate ⟵ ⟵ Dihydroxyacetone phosphate

ATP
ADP

NADPH
NADP

Inner
envelope
membrane

FIGURE 4–19

The 3-phosphoglycerate shuttle
for transporting ATP and reducing equivalents from chloroplast
stroma to cytoplasm.

gle grain or multiple grains. The starch grains appear to be formed in
concentric layers. In storage tissues, their role is to store fixed carbon,
and as such amyloplasts are important in human nutrition. In root caps
and a layer of cells in shoots, amyloplasts lie within membrane-delimited
bodies called statoliths. These plastids have a small layer of surrounding
cytoplasm, so that their membranes do not touch the vacuolar membrane. Because starch is denser than the surrounding medium, statoliths
tend to sediment to the bottom of the cell under the influence of gravity.
This apparently is essential in the perception of gravity by the plant (e.g.,
roots tend to grow down, and shoots grow up). Indeed, if the starch
grains are removed from these cells, the plant cannot sense gravity. The
biochemical connection between statolith sedimentation and growth is
not clear.

TABLE 4–7 **Nongreen Plastids**

Name	Characteristics
Amyloplast	Starch grains
Chromoplast	Nongreen pigments
Elaioplast	Lipid deposits
Eoplast	Fragmented thylakoids
Etioplast	Prolamellar body
Proteinoplast	Protein crystals

Chromoplasts are responsible for the orange, red, and yellow colors of senescing leaves, flowers, and fruits. They have a variable morphology, ranging from round to tubular to ameboid, some with internal membranes but none with thylakoids. Studies of autumn leaves show that their chromoplasts derive from chloroplasts in which the thylakoids are broken down and lipid deposits appear. The dominant biochemical characteristic of these organelles is the presence of pigments, usually carotenoids and xanthophylls, and these are synthesized in large amounts via enzymes in the plastid envelope as the chromoplasts develop. The function of chromoplasts lies in the attraction or repulsion to colors of many animals, such as pollinators. This does not, however, explain their presence in underground organs, such as radishes and carrots.

The conversion of green chloroplasts into colored chromoplasts is only part of the general process of senescence, a degenerative process that leads to the death of the organelle (and of the cell). Ultrastructurally, the first sign of plastid senescence is unstacking of the thylakoids, which severely reduces photosynthetic efficiency. This is followed by a breakdown of the lamellae with a concomitant appearance of plastoglobuli to store the lipids released. Finally, the envelope membranes rupture. However, these ultrastructural changes represent the effects of biochemical events. Studies of senescing leaves indicate that the first biochemical change is a loss of activity by RubisCO. Because the light reactions are still taking place and NADPH cannot be reoxidized, the electrons produced by the oxidation of chlorophyll may generate harmful free radicals, which then damage thylakoid lipids.

Elaioplasts and **proteinoplasts** are chloroplasts modified for the storage of lipid and protein, respectively. Both lack thylakoids, but internal tubular membranes are often present. Lipid is deposited in rounded plastoglobuli, and protein is deposited as crystals. Often, the protein deposits are crystalline RubisCO. **Eoplasts** contain small, fragmented thylakoids. They are often formed in meristems and seeds, and

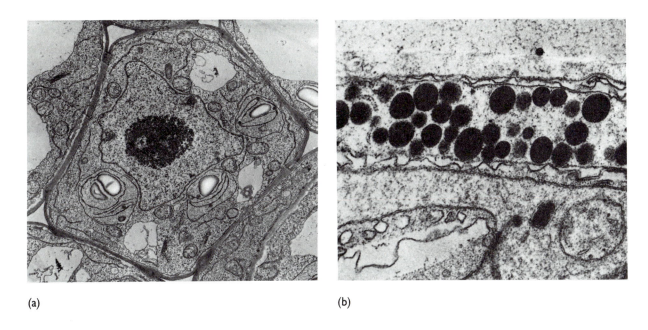

(a)   (b)

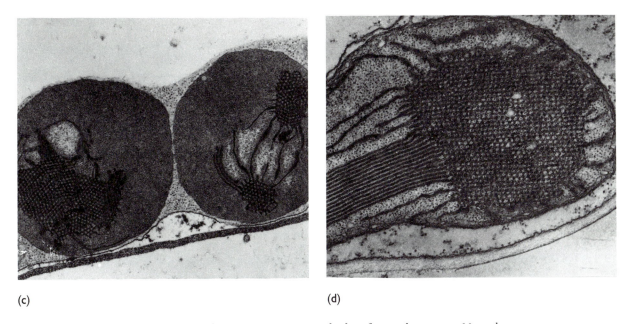

(c)   (d)

FIGURE 4–20   Nongreen plastids (×40,000). **(a)** Amyloplast from a bean root. Note the two starch grains. **(b)** Chromoplast from a fruit of the red tomato. The dense granules are carotenoid pigments. **(c)** Etioplasts from a bean seedling kept in the dark. **(d)** Proplastids from a bean root tip. Courtesy of Dr. W. Harris.

their function is clear, although it has been suggested that they represent the progenitor of all other plastid types.

Etioplasts are formed in plants grown in the dark. They are usually smaller than chloroplasts, with a spherical shape. Their most conspicuous feature is a three-dimensional lattice of tubules, the prolamellar body. Membranes resembling thylakoids can be observed seemingly emerging from the crystal. This has led to the idea that the prolamellar body is a repository for stored membranes. Brief (10-min) illumination results in a disintegration of the prolamellar body. Later, thylakoids and grana appear, and within a day in light, the etioplast has been converted into a mature chloroplast. During this period, the LHC is synthesized and added to the developing membrane system. There is considerable reorganization of the membrane components of the prolamellar body as the mature lamellae develop, and this can be observed at both the microscopic and biochemical levels. Molecular studies indicate that among the earliest events on illumination is the synthesis of mRNAs corresponding to the various proteins of the LHC and photosystems. The primary receptor that mediates the response to light is the pigment phytochrome.

Further Reading

Allen, J. How does protein phosphorylation regulate photosynthesis? *Trends Biochem Sci* 17 (1992):12–16.

Anderson, J. Consequences of spatial separation of photosystem I and 2 in thylakoid membranes of higher plant chloroplasts. *FEBS Let* 124 (1981):1–8.

Anderson, J., and Anderson, B. The dynamic photosynthetic membrane and regulation of solar energy conversion. *Trends Biochem Sci* 13 (1988):351–355.

Bennett, J. Regulation of photosynthesis by protein phosphorylation. In Cohen, P. (Ed.), *Molecular aspects of cell regulation,* Vol e (pp. 227–246). New York: Academic Press, 1984.

Biswal, U., and Biswal, B. Ultrastructural modifications and biochemical changes during senescence of chloroplasts. *Int Rev Cytol* 113 (1988):172–320.

Bonner, J., and Varner, J. (Eds). *Plant biochemistry.* New York: Academic Press, 1976.

Chitnis, P., and Thornber, J. P. The major light-harvesting complex of photosystem II. *Photosyn Res* 16 (1988):41–63.

Creed, D., and Caldwell, R. Photochemical electron transfer reactions. *Photochem Photobiol* 41 (1985):715–740.

Deisehhofer, J., and Chau, V. Structures of bacterial photosynthetic reaction centers. *Ann Rev Cell Biol* 7 (1991):1–24.

Dodge, A. D. (Ed.). *Herbicides and plant metabolism.* Cambridge, UK: Cambridge University Press, 1989.

Douce, R., and Joyard, J. Biochemistry and function of the plastid envelope. *Ann Rev Cell Biol* 6 (1990):173–216.

Evans, M., and Bredenkamp, G. The structure and function of the photosystem I reaction center. *Physiol Plant* 79 (1990):415–420.

Friesner, R. A., and Won, Y. Photochemical charge separation in photosynthetic reaction centers. *Photochem Photobiol* 50 (1989):831–839.

Gantt, E. Phycobilosomes. *Ann Rev Plant Physiol* 32 (1981):327–347.

Gounaris, K., Barber, J., and Harwood, J. The thylakoid membranes of higher plant chloroplasts. *Biochem J* 237 (1986):313–326.

Green, B., Pichersky, E., and Kloppstech, K. Chlorophyll a/b binding proteins: An extended family. *Trends Biochem Res* 16 (1991):181–186.

Gregory, R. P. F. *Biochemistry of photosynthesis,* 3rd ed. New York: Wiley, 1989.

Hansson, O., and Wydrzynski, T. Current perceptions of photosystem II. *Photosyn Res* 23 (1990):131–162.

Hatch, M. D. C4 Photosynthesis: A unique blend of modified biochemistry, anatomy and ultrastructure. *Biochim Biophys Acta* 895 (1989):81–106.

Hatch, M. D., and Boardman, N. K. *Photosynthesis: Biochemistry of plants,* Vol. 10. New York: Academic Press, 1988.

Hober, J. K. *Chloroplasts.* New York: Plenum, 1984.

Holzwarth, A. Structure-function relationships and energy transfer in phycobiliprotein antennae. *Physiol Plant* 83 (1992):518–528.

Huber, R. Structural basis of light energy and electron transfer in biology. *Eur J Biochem* 187 (1990):283–305.

Margulies, M. Photosystem I core. *Plant Sci* 64 (1989):1–13.

Mauzerall, D., and Greenbaum, N. The absolute size of a photosynthetic unit. *Biochim Biophys Acta* 974 (1989):119–140.

Nugent, J. Photosynthetic electron transport in plants and bacteria. *Trends Biochem Sci* 9 (1984):354–357.

O'Keefe, D. Structure and assembly of the chloroplast cytochrome b-f complex. *Photosyn Res* 17 (1988):189–216.

Reilly, P., and Nelson, N. Photosystem I complex. *Photosyn Res* 19 (1988):73–84.

Rutherford, A. W. Photosystem II, the water splitting enzyme. *Trends Biochem Sci* 14 (1989):227–232.

Scheller, H., and Moller, B. Photosystem I polypeptides. *Physiol Plant* 78 (1990): 484–494.

Schnepf, E. Types of plastids: Their development and interconversions. In Reinert, J. (Ed.), *Results and problems in cell differentiation,* Vol. 10 (pp. 1–28). Berlin: Springer-Verlag, 1980.

Siefermann-Harms, D. Carotenoids in photosynthesis. *Biochim Biophys Acta* 811 (1989):325–355.

Staehelin, L. A., and Arntzen, C. A. (Eds.). Photosynthesis III: Photosynthetic membranes and light-harvesting systems. In *Encyclopedia of Plant Physiology,* Vol. 19. Berlin: Springer-Verlag, 1986.

Thomson, W., and Whatley, J. Development of non-green plastids. *Ann Rev Plant Physiol* 31 (1980):375–394.

Thornber, J. P., Peter, G. F., and Nechustal, R. Biochemical composition and structure of photosynthetic pigment-proteins from higher plants. *Physiol Plant* 71 (1987):236–240.

Thornber, J. P., Staehelin, L. A., and Hallick, R. (Eds.). *Biosynthesis of the photosynthetic apparatus.* New York: A. R. Liss, 1984.

Wellburn, A. R. Plastids. *Int Rev Cytol Suppl* 17 (1987):149–206.

Wiley, D., and Gray, J. Synthesis and assembly of the cytochrome b-f complex in higher plants. *Photosyn Res* 17 (1989):125–144.

CHAPTER 5

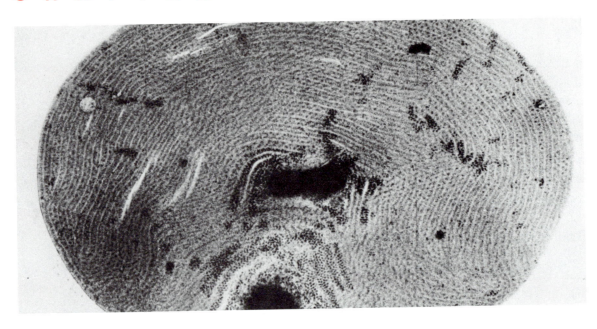

Biogenesis of Plastids and Mitochondria

Semiautonomous Organelles

O V E R V I E W

Mitochondria and plastids are cellular compartments where specialized processes occur, and each organelle has its own set of enzymes and membrane carriers to carry them out. The cell has been called a factory, with the mitochondrion as its "powerhouse," the plastid as part of the plant cell's "manufacturing center," and the nucleus as the "management."

But styles of management change and so have ideas about the roles of these organelles. In manufacturing, the idea that an all-important central management team should make the decisions for all parts of a company is being changed to include more participation by people in the divisions in the decision-making process. Likewise in cell biology, the idea that the nucleus has total control over what goes on in mitochondria and plastids has been modified as evidence has accumulated that these organelles have their own DNA and internally synthesize some of their proteins. As this chapter will show, organelle independence (as that of the factory worker) is only partial, and the two organelles can be called semiautonomous rather than fully independent.

Organelle Division

Microscopy

Many algae have large **plastids,** and under a microscope these can be seen to divide by fission into two more or less equal parts. Such observations led A. Schimper to propose that plastids are autonomous organelles, or "cells within cells." Patient microscopy, or, more conveniently, a microscope fitted with a motion picture camera, can follow a plastid from its initial formation by division to its own division, the whole process being a plastid cycle (Figure 5–1). For both algae and higher plants, the division cycle times range from 10 to 22 hours, a period rather similar to the division cycle time of the whole cell (see Chapter 12).

The patterns of plastid division have been followed in two ways. First, the division plane of the plastid can be experimentally modified. For instance, if a large algal cell is stretched by applying a physical force, the plastid will align itself to the stress field and divide along this direction. Then, if the force field changes, so does the plastid division orientation. This shows that the geometry of an organelle's division responds to conditions in the surrounding cytoplasm. Second, the rate of

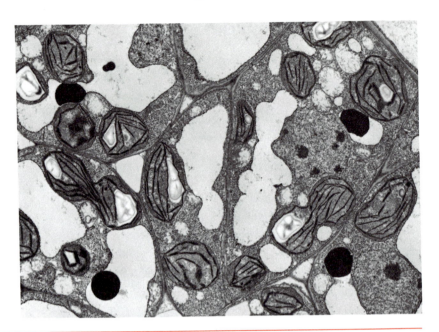

FIGURE 5–1

Dividing chloroplasts in cells of a young leaf of a sesame plant. (×8300. Photograph courtesy of K. Platt-Aloia and W. Thompson.)

plastid division varies, especially during development. When a leaf is formed, there is considerable cell division, but the plastids divide faster, leading to a great increase in the number of plastids per cell (Table 5–1). How this rate of division is regulated is not known.

The actual division of a plastid is preceded by an elongation of the organelle. During this period, the chloroplast DNA divides and the two molecules formed are distributed to opposite ends of the organelle. A constriction of the outer envelope then occurs, followed by membrane fusion by pinching off. In green tissues, low light intensity and reduced temperature slow down chloroplast division, but light is not an absolute requirement, since nongreen plastids can divide without it.

In some tissues, such as germinating seeds, chloroplasts come not from division but from the light-stimulated differentiation of preexisting, immature plastids. These proplastids contain little, if any, internal membranes (Figure 4–20). When the seedling emerges from the ground, the proplastids form etioplasts, which then form thylakoids and become chloroplasts.

Early electron micrographs taken during the 1950s showed **mitochondria** associated with the nuclear membrane and endoplasmic reticulum, and this led to the hypothesis that these two organelles form mitochondria. But more careful microscopy showed these associations to be artifactual or coincidental. Instead, mitochondria are formed from division of other mitochondria. This is hard to see clearly under the light

TABLE 5-1 Chloroplast Numbers in Developing Leaves

Species	Leaf Age (Days after Emergence)	Cell Number per Leaf ($\times 10^3$)	Chloroplasts per Cell
Bean	1	2,000	8
	6	50,000	50
Spinach	1	7,580	10–50
	14	15,850	200
Sunflower	1	90	19
	29	5,021	50
Tobacco	1	170	10–20
	8	50,630	200
Wheat	1	991	46
	8	4,295	150

FROM: Possingham, J. Plastid replication and development in the life cycle of higher plants. *Ann Rev Plant Physiol* 31 (1980):113–129.

microscope, since these organelles are often very small (about $1\,\mu\text{m}$) and move about the cell quite rapidly. Nevertheless, time-lapse films of living cells clearly show mitochondria dividing in a manner similar to plastids. This has been most clearly shown in the ciliated protozoan *Tetrahymena*.

In synchronized *Tetrahymena,* there is a mitochondrial division cycle with divisions similar to the overall cell division cycle (see Chapter 12). The mitochondria have an overall division cycle time of 14 hours, which is the same as that of the nucleus, and exhibit cycle subdivisions of G1, S, G2, and M.

Also like the plastids, mitochondria can arise from preexisting, undifferentiated organelles. These promitochondria occur, for example, when yeast cells are grown in the absence of oxygen; when it is restored, mitochondrial differentiation ensues with the formation of cristae. Electron micrographs taken during cell division of mature tissues often show dumbbell-shaped mitochondria, providing circumstantial evidence for their biogenesis by division. But more compelling evidence has come from the use of autoradiography.

Autoradiographic Experiments

D. Luck did a series of elegant experiments on the biogenesis of **mitochondria** in the mold *Neurospora.* He used this organism because it is easily and reproducibly grown in the laboratory. An important factor was that its genetics had been extensively studied, and there was a single gene mutant that required choline in the medium in order to survive and grow. If radioactive choline was supplied in the medium, the organism's

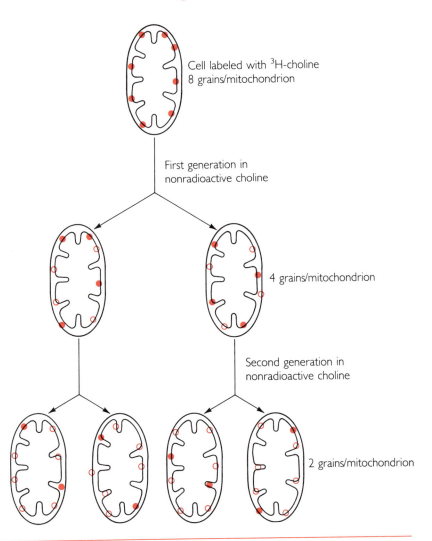

Cell labeled with ³H-choline
8 grains/mitochondrion

First generation in
nonradioactive choline

4 grains/mitochondrion

Second generation in
nonradioactive choline

2 grains/mitochondrion

FIGURE 5–2

Experiment showing growth and division of mitochondria. *Neurospora* were grown in radioactive choline, which was incorporated into the mitochondrial membrane lipids. After one or two growth cycles in nonradioactive choline, mitochondria were isolated and their relative radioactivity determined by autoradiography. *Closed dots:* silver grains due to radioactive choline. *Open dots:* nonradioactive choline (for illustration).

newly synthesized lipids would be highly radioactive. A sample of cells was taken immediately after the labeling period and their mitochondria were isolated. As expected, mitochondrial membranes were radioactive. Electron microscope autoradiography showed that all mitochondria were labeled, with an average silver grain count (proportional to radioactivity) of 8.

The labeled *Neurospora* was subsequently incubated in nonradioactive choline for a sufficient time for the cell mass and the mitochondria to double (one generation time). Sampling of mitochondria then showed that although the total amount of lipid label was the same, it was now evenly distributed over twice the number of mitochondria: Each one had four grains. Still another generation in nonradioactive choline led to a further halving of the label per organelle to two.

The interpretation of this experiment (Figure 5–2) is that during the first nonradioactive generation, the mitochondria divided evenly into two smaller organelles with an equal amount of labeled lipid in their membranes. These then grew, building their membranes out of unlabeled choline, and when they reached mature size, their membranes had a mixture of labeled and unlabeled phospholipid. Because all mitochondria were dividing, all had the same amount of radioactivity. A similar situation occurred in the next generation. This experiment leads to the conclusion that when a mitochondrion divides, its lipids are evenly distributed to its "offspring."

A somewhat analogous autoradiographic experiment has been performed on **chloroplasts.** In this case, another rather stable component, DNA, has been used. The strategy is similar to Luck's experiment on mitochondria: Chloroplast DNA was heavily labeled with the use of the radioactive precursor thymidine. The cells were then placed in a medium containing nonradioactive thymidine for a generation or two, and the chloroplasts were assayed for DNA radioactivity by autoradiography.

In synchronously dividing cultures of algae, labeled DNA was distributed equally among the progeny chloroplasts. In disks from spinach leaves placed in culture, the number of chloroplasts increased sixfold, whereas the number of autoradiographic grains per plastid decreased sixfold. Thus, DNA from the chloroplast divides during plastid division and is evenly distributed to the two new plastids.

Genetic Evidence for Organelle Autonomy

Maternal Inheritance

If chloroplasts and mitochondria can divide and have their own genetic material (DNA), what degree of genetic autonomy do these organelles possess? One way to approach this question is to examine strains with abnormal organelles and to determine whether the abnormality is inherited through the nuclear or organelle genomes. Fortunately, these modes of inheritance give strikingly different results in genetic crosses.

When sperm are formed in animals, mitochondria often form an aggregate and provide the energy for the beating of the sperm tail. When the sperm interacts with the egg, however, only the nucleus enters the egg cytoplasm for fertilization. Thus, the new organism inherits nuclear material from both parents, but cytoplasm (including the mitochondria) inherits material from the mother only (Figure 5–3).

A similar situation occurs in higher plants. When the generative cell in a pollen grain divides to form the precursor of the sperm cell and the auxiliary tube nucleus, only the region around the latter contains other organelles. This means that mitochondria and plastids are essentially excluded from the sperm. On the other hand, the egg cell contains

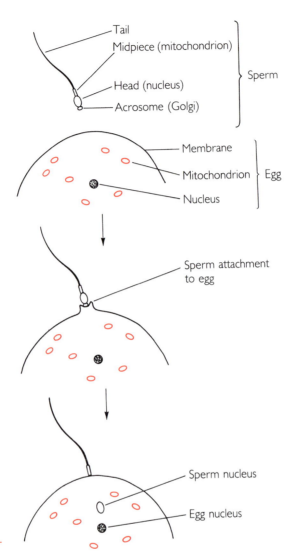

FIGURE 5–3

The cellular basis of maternal inheritance in higher animals. The sperm does not contribute mitochondria to the fertilized egg, while the egg cell does. This means that characteristics coded for by the mitochondrial genome will be inherited through the female only.

an abundance of all organelles. Therefore, the fertilized egg has only organelles from the female parent (Figure 5–4).

These cellular aspects of sexual reproduction ensure that genetic characteristics carried on organelle DNA will be inherited through the female parent only. This provides a rather simple method for testing for organelle genes (Table 5–2). If **reciprocal crosses** are done — making the female parent homozygous mutant and the male homozygous normal and vice versa — a gene inherited through the typical nuclear system will show simple Mendelian inheritance: A dominant gene will show up in the offspring no matter which parent carries it. On the other hand, an

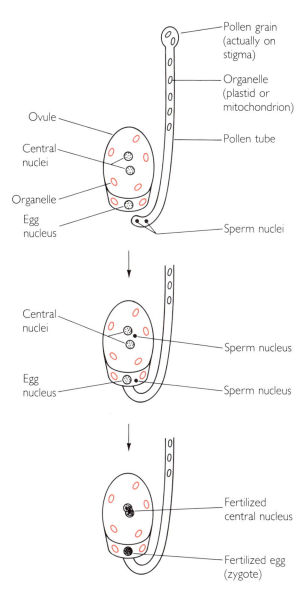

Ovule

Central nuclei

Organelle

Egg nucleus

Pollen grain (actually on stigma)

Organelle (plastid or mitochondrion)

Pollen tube

Sperm nuclei

Central nuclei

Egg nucleus

Sperm nucleus

Sperm nucleus

Fertilized central nucleus

Fertilized egg (zygote)

FIGURE 5–4

The cellular basis of maternal inheritance in higher plants. The fertilized egg contains plastids and mitochondria only from the female parent. Therefore characteristics coded for by the genomes of these two organelles will be inherited through the female only.

organelle gene will be transmitted only if it is carried by the female parent. The study of organelle mutations not only gives some clues as to the specific functions controlled by the organelle but is also important in establishing that it is indeed somewhat autonomous.

Plastid Mutations

In the Japanese four o'clock plant, *Mirabilis japonica,* leaves of different branches have different patterns of green coloring. Some branches have normal green leaves; others have white leaves; still others have leaves that have intermixed blotches of green and white tissues and are termed

TABLE 5-2 **Maternal Inheritance**

| Male Parent | Female Parent | Offspring Phenotype if: | |
		Mendelian	Maternal
AA	aa	Normal	Mutant
aa	AA	Normal	Normal

A is a dominant allele for normal phenotype.

a is a recessive allele for mutant phenotype.

(Note that in maternal inheritance, there is usually not a true diploid genotype.)

variegated. Reciprocal crosses of flowers from white and green branches result in offspring that always look like the female parent (Table 5–3), a clear demonstration of maternal inheritance.

Examination of the different leaves under the light microscope shows that white regions lack recognizable chloroplasts, whereas they are abundant in green areas. This leads to a simple explanation of the results of the crosses with variegated females: By chance during cell division, or by some other means, one cell could get all of the white ones, and another could get all of the green ones. By cell division, these cells could then give rise to the white and green areas.

In the unicellular green alga *Chlamydomonas,* cells can be either haploid (single copy of all chromosomes) or diploid (two copies). The latter come from a fusion of two haploid cells, one of mating type (+) and the other of mating type (−). Because the two cell types (gametes) are the

TABLE 5-3 **Reciprocal Crosses of Japanese Four O'Clock Plants**

Phenotype of Branch Bearing Male Gamete	Phenotype of Branch Bearing Female Gamete	Phenotype of Offspring
Green	Green	Green
White	Green	Green
Variegated	Green	Green
Green	White	White
White	White	White
Variegated	White	White
Green	Variegated	Green, white, or variegated
White	Variegated	Green, white, or variegated
Variegated	Variegated	Green, white, or variegated

same size and equal amounts of cytoplasm are contributed by each to the zygote, one might expect that the maternal pattern of inheritance could not occur. But it does: Plastid mutations for antibiotic resistance and lack of green color have been described and clearly show uniparental inheritance. This can happen because in the zygote, organelle DNA of the ($-$) parent is degraded, while the ($+$) DNA survives. Why this occurs is a mystery, but at any rate the situation is analogous to that in higher plants where one parent contributes organelles to the offspring.

Mitochondrial Mutations

The fungus *Neurospora* is usually haploid. But under certain conditions, two haploid cells can fuse to form a diploid zygote, which then divides by meiosis to form haploid spores. Only one of the two fusing strains (the "female") contributes cytoplasm to the zygote, and therefore, maternal inheritance can be followed in this mold. Several slow-growing, maternally inherited mutants termed "poky" have been isolated, and these strains are deficient in cytochromes a and b (Table 5–4). Mitochondria from a similar strain, "abnormal," have been injected into wild type *Neurospora*. After several generations, cultures with the "abnormal" phenotype appear, indicating that the mutant mitochondria have divided and "taken over" the overall mitochondrial population of the cells. This shows that the "abnormal" gene is carried on the mitochondria.

Yeast, like *Neurospora,* is a fungus with haploid and diploid phases of the life cycle. When yeast cells fuse to form a zygote, however, both strains contribute cytoplasm, and so simple reciprocal crosses are not very informative. Nevertheless, mitochondrial mutants have been detected in this organism. When yeast cells are placed in an anaerobic environment, they grow very slowly and form a small colony. But some strains, termed petites (from the French word meaning small), have small colonies even when oxygen is present. They are unable to perform

T A B L E 5–4 **Organelle Mutations**

Organism	Mutations	Organelle	Defect
Japanese four o'clock	White leaf	Chloroplast	
Chlamydomonas	Streptomycin resistance	Chloroplast	Ribosome
Corn	Striped leaf	Chloroplast	Ribosome
Neurospora	Poky	Mitochondrion	Cytochromes a,b
Yeast	Petite	Mitochondrion	Mitochondrial DNA
Maize	Male sterility	Mitochondrion	Membrane

mitochondrial energy transduction and receive energy only from inefficient fermentation.

Some petite strains show inheritance patterns that are Mendelian and so are due to mutant nuclear genes coding for mitochondrial functions. Others are aberrant when crossed with a wild type, as the petite phenotype disappears and all progeny are normal. This is interpreted as a situation in which the wild type cytoplasm completely corrects for the defect in the mutant. This could happen (and does) if the petite strain lacks mitochondrial DNA entirely.

Cytoplasmic male sterility in over 140 different plant species, including maize, is inherited maternally and is due to a mitochondrial gene. Molecular cloning indicates that the gene product is a small mitochondrial protein, and exactly how it causes pollen grains to die is not known. Unfortunately, plants with this protein are sensitive to attack by the fungus *Bipolaris maydis*. The mitochondrial protein binds to the fungal toxin, and the complex then binds to the inner mitochondrial membrane, making it leaky to ions (see Chapter 3) and ultimately killing the plant. In addition, the same protein makes the plant susceptible to the herbicide Lannate (methomyl), and so when this chemical is used to kill pests such as the corn borer, it will also damage the corn plant.

Despite the risks of disease and herbicide susceptibility, the search for similar genes for male sterility in self-pollinating plants, such as cereal grains, is an active area of plant-breeding research. A major effort of plant breeders is to take two inbred strains of corn, for instance, and cross them to make a hybrid. Such hybrids have grain-producing abilities much greater than the sum of the two strains that went into making them, a phenomenon called hybrid vigor. But corn has both sexes on the same plant, and just putting the two different strains in a field near each other can lead to self-fertilization, as well as hybridization. To ensure that only hybrids are formed, the breeder used to have to detassel (remove the male reproductive organs of) one of the two strains. To do this manually is very tedious, and the gene for male sterility solves the problem.

Organelle Protein Synthesis

Strategies for Study

Mitochondria and plastids contain many proteins (Tables 3–3 and 4–2). The existence of mutations that lead to organelle abnormalities, some of which come from the nuclear genome and some from the organelle genome, shows that some organelle proteins are coded for by the DNA of the nucleus and some by the DNA of the organelles. Determining the intracellular site of synthesis of a protein is tantamount to determining the site of its DNA code, since there is little evidence that large nucleic acids can cross the chloroplast or mitochondrial membranes. Three experimental methods have been used to determine where organelle proteins are made: genetics, inhibitors, and in vitro protein synthesis.

Behind the **genetic** approach is the assumption that organelle genes are maternally inherited. If a mitochondrial or chloroplast protein has mutant forms, organisms with one of these can be reciprocally crossed with the wild type. Following the reasoning of Table 5–2, if the protein's inheritance pattern is through both parents, it is a product of the nuclear genome. If it is through the female parent only, it is coded for by the organelle. In a strict sense, this method does not give information on which organelle is responsible: A chloroplast protein that is maternally inherited could be made in the mitochondrion and then transported to the plastid. But no example of this phenomenon has been found.

The major protein of the chloroplast stroma is RubisCO, which is composed of a large and small subunit. S. Wildman described how mutants of the two subunits are inherited in tobacco plants. These mutants were detectable as differences in chromatographic behavior of pieces of the protein generated by limited digestion by the protease trypsin. The extra tryptic peptide of the large subunit of the tobacco species *Nicotiana gossei* appeared in crosses with *Nicotiana tabacum* only when the former was the female parent. This indicates that the large subunit is a product of the organelle genome. On the other hand, an extra small subunit peptide present in *N. tabacum* but not in *N. glauca* appeared in both reciprocal crosses of these two species, indicating nuclear inheritance (Figure 5–5).

A somewhat different approach has been taken in studies of mitochondrial diseases (see Chapter 3). Analyses of human pedigrees in which some people suffer from familial mitochondrial encephalomyopathy have shown maternal inheritance. This rare disease results in a form of epilepsy, deafness, and dementia. Biochemical studies of isolated mitochondria of these patients have shown that the defect is in one of the subunits of complex I (NADH dehydrogenase). This indicates that this subunit is coded for by the mitochondrion. A similar result has come from studies of a rare vision defect, Leber's neuropathy, which affects a different subunit of complex I.

Inhibitor studies depend on the fact that certain antibiotics blocking protein synthesis at the ribosome have different effects on ribosomes that are in the cytoplasm than on the ribosomes that are present in mitochondria and plastids. This occurs because the ribosomal proteins of the two systems differ, and these proteins are the targets for the anti-

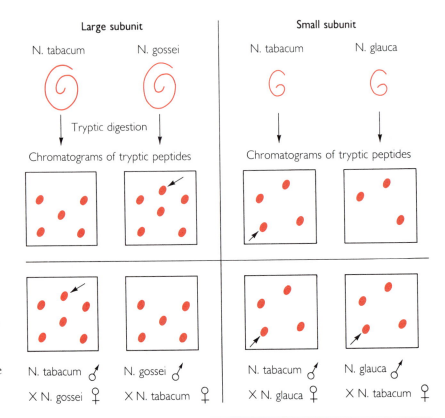

FIGURE 5-5

Diagram of the analysis of mutants of tobacco RubisCO. The protein was cleaved by trypsin, and the fragments then were separated by paper electrophoresis. "Extra" peptides are indicated by an arrow. The results show maternal inheritance of the large subunit and Mendelian inheritance of the small subunit.

biotics. For example, chloramphenicol inhibits protein assembly on chloroplast and mitochondrial ribosomes, while cycloheximide inhibits protein synthesis on cytoplasmic ribosomes.

To perform these analyses, separate groups of cells are incubated in each inhibitor in the presence of a radioactive amino acid. Organelles are isolated, and the radioactive proteins made are then analyzed. If an organelle protein is made in the presence of cycloheximide but not in the presence of chloramphenicol, it must be made on organelle ribosomes. On the other hand, synthesis in the presence of chloramphenicol but not in the presence of cycloheximide indicates assembly on cytoplasmic ribosomes followed by transport into the organelle. Again, synthesis by one organelle and transfer to another cannot be ruled out.

In yeast mitochondria, cytochrome oxidase is composed of seven subunits, resolvable by gel electrophoresis. A. Tzagaloff incubated yeast cells in radioactive leucine in the presence of chloramphenicol. Mitochondria were isolated, and electrophoresis showed that four of the

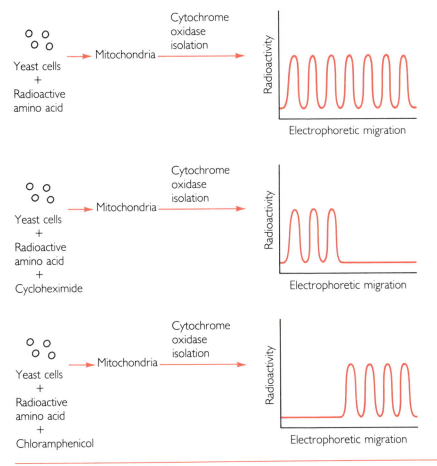

FIGURE 5-6
How inhibitors of protein synthesis can be used to elucidate the site of synthesis of an organelle protein. Cycloheximide blocks protein synthesis on 80S (cytoplasmic) ribosomes, while chloramphenicol blocks it on 70S (organelle) ribosomes. In these experiments, mitochondrial cytochrome oxidase was studied.

seven subunits were labeled. However, if yeast cells were similarly incubated in the presence of cycloheximide, these four were unlabeled and the other three subunits were labeled. Therefore, three of the seven subunits of cytochrome oxidase are made by the mitochondrion, and the other four are made in the cytoplasm (Figure 5–6).

The third method used to analyze the site of synthesis of organelle proteins is **in vitro protein synthesis** by isolated organelles. In this case, chloroplasts or mitochondria in the test tube are supplied with the necessary cofactors and substrates (usually ions, GTP, ATP, and amino acids) for protein synthesis. If one or more of the amino acids is radioactive, the resulting proteins are assayed by their radioactivity.

In chloroplasts, the ATP synthetase on the thylakoid is composed of eight different subunits: five make up the CF_1 synthetase, and three make up the membrane-embedded CF_0 proton channel. G. Schatz isolated spinach chloroplasts and incubated them in radioactive methionine as well as the other amino acids and cofactors. The resulting radioactive proteins were assayed for reaction against specific antibodies for each of the subunits. Four of the CF_1 and one CF_0 subunits were labeled, indicating that they are made by the chloroplast. Verification of this came from the fact that these subunits were not made in the presence of chloramphenicol.

Organelle-Coded Proteins

A combination of the three techniques has shown that most organelle proteins are synthesized in the cytoplasm and only a small number inside the organelles. These experimental data have been confirmed and extended by the sequencing of organelle DNA and derivations of proteins coded from these sequences.

It is estimated that the mitochondrion needs at least 400 different proteins for its metabolic capabilities. Of these, only a few components of the electron transport chain and part of the F_1-ATP synthetase are made locally (Table 5–5). The enzymes of the citric acid cycle, fatty acid metabolism, translocases, and the bulk of the energy-transducing apparatus are imported from the cytoplasm.

In the chloroplast, about 700 proteins are needed for its functions, and of these about 10% are made locally (Table 5–6). These include some components of the light reaction complexes, part of the ATP synthetase, and the large subunit of RubisCO. But most of the light reaction components and almost all of the enzymes of carbon dioxide fixation are coded for by the nucleus.

An interesting pattern is seen in core complex II for photosystem II (Figure 4–8). Here, the polypeptide subunits of the complex and cytochrome b559 that are embedded in the thylakoid membrane are coded for by the chloroplast genome. But the three polypeptides that are extrinsic and lie on the luminal side of the membrane are coded for by the

---

TABLE 5–5   **Products of the Mammalian Mitochondrial Genome**

Electron Transport Complexes
I: NADH reductase: 7 subunits of 27
III: cytochrome c reductase: 1 subunit of 10
IV: cytochrome oxidase: 3 subunits of 13

Oxidative Phosphorylation
F_0-proton channel: 2 subunits of 3

Molecular Genetic Apparatus
Ribosomal RNA
All transfer RNAs

T A B L E 5–6 Products of the Chloroplast Genome

Light Reaction

Core complex I: 5 subunits of 11

Core complex II: 6 subunits of 9

Cytochrome b_6-f complex: 2 subunits of 5

CF_1 ATPase: 3 subunits of 5

CF_0 ATPase: 2 subunits of 3

Dark Reaction

RubisCO: large subunit

Molecular Genetic Apparatus

Ribosomal RNA

All transfer RNAs

One-third of ribosomal proteins

Elongation factors T and G for protein synthesis

---

nucleus. Such a spatial pattern of proteins coded for by the genomes is the exception rather than the rule, however. In most cases, cooperatively made oligomeric proteins show no particular pattern as to the genome coding for their subunits.

Much of the organelle genome is taken up by the production of RNA molecules and some molecules needed for protein synthesis (see below and Tables 5–5 and 5–6). In addition, the nuclear-cytoplasmic system contributes over 80 proteins to the apparatus of organelle protein synthesis. The reason for this considerable effort by the cell to set up a special compartment just to make a few proteins is not known and may be a reflection of the possibly endosymbiotic origin of these organelles.

Transport from Cytoplasm to Organelle

Most mitochondrial and plastid proteins are synthesized on cytoplasmic ribosomes and then cross the organelle membranes on their way to their destinations. In some cases, the protein is destined for the matrix or stroma; in others, it is inserted into a membrane such as the crista or thylakoid; in still others, it is a subunit of a larger oligomeric protein and joins with organelle-synthesized subunits to form a final product. Studies on both organelles reveal common themes in synthesis and transport.

Proteins destined for organelles are typically made on free ribosomes in the cytoplasm, and transport occurs following completion of the polypeptide chain. This is in marked contrast to the syntheses of membrane and secreted proteins, which generally occur on membrane-bound ribosomes and in which transport is coupled to synthesis; that is, transmembrane movement is cotranslational. That the uncoupling of synthesis and transport occurs in the organelle proteins is shown by the lag time of several minutes between the synthesis of a protein and its arrival at the organelle.

A second characteristic is that most organelle proteins are made on cytoplasmic ribosomes as precursors of higher molecular weight than the mature forms. The amino-terminal "signal" sequence is cleaved off at the organelle in a manner similar to membrane protein processing (Figure 1–18). Although most organelle proteins are made with this terminal sequence (Table 5–7), some, such as the adenine nucleotide translocator of mitochondria, are not and apparently have an internal signal. Proteins that go to the outer mitochondrial membrane (e.g., porin) have a targeting sequence, usually of hydrophobic amino acids for membrane insertion, but no cleavable signal.

Organelle **signal sequences** are usually 20 to 70 amino acids long and do not necessarily contain an abundance of hydrophobic amino acids. The most common situation is a prevalence of positively charged and neutral hydroxyl-containing amino acids (e.g., serine and threonine) and few negatively charged residues. The signal sequences tend to form helices with the charged residues on one surface and hydrophobic residues on the other surface. This makes it unlikely that they act as a "leader," forming a channel through the hydrophobic bilayer. Indeed, many of the precursors travel through the cytoplasm to their destination as more hydrophilic species. This is especially true of membrane-bound proteins

TABLE 5–7   Molecular Weights of Precursors and Mature Forms of Some Cytoplasmically Made Organelle Proteins

Organelle	Protein	Molecular Weight of:	
		Cytoplasmic Precursor	Organelle Product
Chloroplast	CF_1 subunit	26,000	18,000
Chloroplast	LHC protein	32,000	28,000
Chloroplast	RubisCO small subunit	16,000	12,000
Mitochondrion	CF_1, α subunit	64,000	58,000
Mitochondrion	Cytochrome c_1 apoprotein	37,000	31,000
Mitochondrion	Ornithine transcarbamylase	40,000	36,000

such as the cytochromes or LHC subunits. The helices may interact with the surface of the organelle membrane via electrostatic interaction with the negatively charged phospholipids.

The importance and role of the signal in protein targeting to the organelle have been demonstrated experimentally, and the entire process can be studied in the test tube. If mRNA for an imported protein is translated in vitro, the precursor protein can be synthesized. It can then be added to the appropriate organelle, and uptake, cleavage, and intraorganellar destination can be studied. Molecular biological techniques make it possible to synthesize altered precursor proteins, or chimeric precursors, with the signal of one protein attached to the main body of another.

A major conclusion from targeting studies is that the signal sequence is both necessary and sufficient for transport of a protein into the organelle. If the cytoplasmic precursor of the small subunit of RubisCO is deleted or sufficiently mutated, the protein will not be taken up by plastids. Studies of the precursor of mitochondrial ornithine transcarbamylase (OTCase) show that it has a 32 amino acid leader. Deletion of amino acids 8–22, or mutation of a single arginine to glycine, completely abolishes import into the mitochondria. These and many other studies show the necessity of the leader peptide.

Chimeric protein studies show that it has sufficient information for proper targeting. For example, the RubisCO small subunit leader peptide has been fused to a bacterial protein, neomycin phosphotransferase; the chimera is properly directed to the plastid and ends up properly cleaved and in the stroma. Likewise, the OTCase leader has been fused to a cytoplasmic protein, dihydrofolate reductase; in this case, the latter ends up in the mitochondrial matrix.

The question of organellar specificity has been examined by an in vivo test using the bacterial protein, chloramphenicol acetyltransferase. (This is easily detected in cells because it enables them to metabolize the antibiotic chloramphenicol.) The protein was fused to either chloroplast or mitochondrial leaders and then introduced into plant cells by a genetic transformation method. When the bacterial gene was translated, its protein was directed to the plastid if it had the plastid leader, or it was directed to the mitochondrion if it had the mitochondrial leader. This shows that the leader peptide has information for specific organelle targeting.

On the other hand, the same bacterial gene has been fused to the gene for the leader of mitochondrial cytochrome oxidase, and in this case the protein goes to both mitochondria and plastids of tobacco cells. This means that the leader in this case is not organelle-specific and raises the possibility that the same protein can be made in the cytoplasm and imported by both organelles.

Once the signal is translated, a series of events occurs, which, for yeast mitochondria, are summarized in Figure 5–7. An important initial event is the association of the newly made protein with another protein whose role is to prevent the organelle-bound protein from assuming its normal three-dimensional shape. This keeps the targeted protein unfolded, with its signal available for binding to the organelle, and prevents the protein from associating with itself.

The unfolding protein is a "**molecular chaperone**," and gene sequencing shows that it is a member of the family of stress proteins that occur in both prokaryotic and eukaryotic cells. These proteins always act to prevent target proteins from folding up inappropriately. For instance, other stress proteins are made when a cell is exposed to elevated temperature (heat shock), an environmental stressor that would tend to denature important enzymes. The chaperones here act to save the life of the cell. In bacteria, a chaperone protein is required for the assembly of the head of infecting bacteriophage (viruses), where it keeps the subunits of the head apart until they are ready to self-assemble.

The chaperone-organelle protein complex now arrives at the surface of the organelle. The **recognition** of the signal with the organelle occurs at the organelle outer membrane surface and can be specific. For instance, the precursor of the small subunit of RubisCO from algae does not enter pea plastids, possibly because its signal peptide differs considerably from that of the pea subunit. On the other hand, the precursor of a *Neurospora* mitochondrial protein can be imported into yeast mitochondria, implying a rather nonspecific transport mechanism.

The actual receptor for targeted proteins on the organelle surface can be isolated if the protein to be imported is attached to one end of a bifunctional organic reagent. When this protein is presented to the appropriate organelle, the reagent will attach covalently to the surface receptor, immobilizing the complex. The receptor on yeast mitochondria has been studied in this fashion and identified. Another approach is to use an antibody to the signal peptide, in the hope that this antibody will recognize the receptor as well (and it does).

In yeast, these methods have identified two receptor proteins, both embedded in the outer mitochondrial membrane with the bulk of the proteins exposed to the cytoplasm. If either protein is disrupted, protein import into mitochondria is lowered by 50–80%. This indicates that these receptors are functionally redundant and that a single one is not responsible for all import.

On plastids, the receptor protein appears to be the site of action of a fungal toxin, tentoxin. Besides allowing an easier purification of the receptor for imported proteins, toxin sensitivity has possibilities for the design of herbicides.

Once binding occurs, the precursor is **translocated** across the membrane in an energy-requiring step. This requirement can be proved

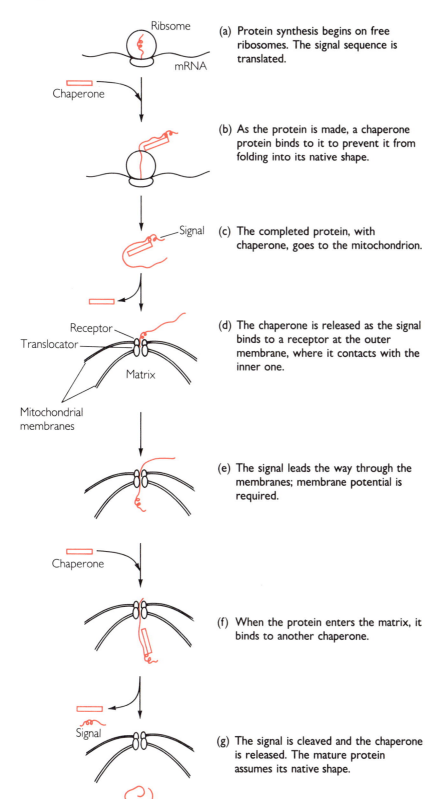

(a) Protein synthesis begins on free ribosomes. The signal sequence is translated.

(b) As the protein is made, a chaperone protein binds to it to prevent it from folding into its native shape.

(c) The completed protein, with chaperone, goes to the mitochondrion.

(d) The chaperone is released as the signal binds to a receptor at the outer membrane, where it contacts with the inner one.

(e) The signal leads the way through the membranes; membrane potential is required.

(f) When the protein enters the matrix, it binds to another chaperone.

(g) The signal is cleaved and the chaperone is released. The mature protein assumes its native shape.

FIGURE 5–7

The role of molecular chaperones in the import of yeast mitochondrial proteins.

if cells are treated with uncouplers of oxidative phosphorylation or are mutated such that cellular ATP is depleted. In either case, unprocessed precursors of organelles will accumulate in the cytoplasm. One reason for the ATP requirement for entry into the organelle is that the protein loses its chaperone on arrival at the receptor and must be kept in an elongated shape so that it can get through the organelle membrane(s).

In mitochondria, but not in plastids, maintenance of membrane potential also appears to be essential for protein import, and this, coupled with hydrolysis of ATP, provides the energy for translocation through the membrane. The necessary potential does not involve H^+, since ionophores that disrupt the proton gradient (e.g., FCCP, see Figure 2−4) do not inhibit protein import. On the other hand, K^+ ionophores such as valinomycin do block import, so the gradient involves that cation. Given the negative charge on the membrane surface and positive charge on the helical region of the leader peptide, it is possible that an electrophoretic translocation through the outer membrane (envelope) could occur. Embedding in the membrane may occur via hydrophobic regions in the signal.

Electron microscopy shows that at intervals the two plastid membranes appear in close apposition, and the same thing happens with the two mitochondrial membranes. Apparently, this is where proteins are translocated, and they probably span both membranes at once during the process. There are separate translocating channels across both membranes. If antibodies to a protein are applied as it is in transit across the plastid envelope or the two mitochondrial membranes, they can be visualized under the electron microscope by coupling them with, for example, a heavy metal such as gold. When this is done, staining is observed at the dual membrane contact points, which appear at numerous places around the surface of the organelle.

Once inside the stroma or matrix, the transit signal is cleaved from the protein by a soluble **protease.** Again, this contrasts with the cleavage of plasma membrane and ER proteins, where cleavage happens at the inner surface of the membrane that is crossed. The proteases from different organisms and organelles, although different in amino acid sequence, have similar properties (e.g., they are not serine proteases and have neutral pH optima). In addition, the protein may be inhibited from folding inappropriately in the matrix or stroma by associating with a second chaperonin.

At this point, the proteins can be envisioned as being in a traffic circle. Their ultimate destinations are determined by the presence (or absence) of additional signals in their amino acid sequences. Remarkably elegant and similar plans operate in plastids (Figure 5−8) and mitochondria (Figure 5−9):

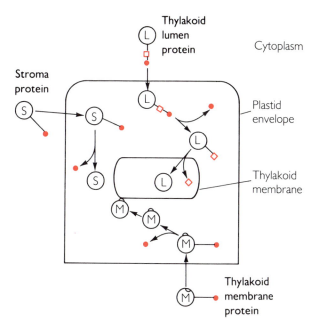

FIGURE 5–8

Three routes for directing a cytoplasmically made plastid protein to its proper location within the organelle. All of the proteins have a plastid-targeting signal, which is cleaved off once they are in the stroma. Thylakoid lumen-bound proteins have a second cleavable signal. Thylakoid membrane proteins have a noncleavable signal.

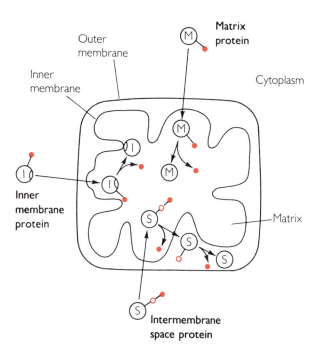

FIGURE 5–9

Three routes for directing a cytoplasmically made mitochondrial protein to its proper location within the organelle. All of the proteins have a mitochondrion-targeting signal, which is cleaved off once they are in the matrix. Intermembrane space-bound proteins have an additional cleavable signal. Inner membrane proteins have a noncleavable signal. The general scheme is similar to the one in plastids (Figure 5–8).

1. Proteins destined to stay in the stroma or matrix have no additional signals.

2. Proteins destined to cross another membrane to get to the thylakoid lumen or intermembrane space have an additional signal, generally just behind the amino-terminal organelle-targeting signal. Cleavage of the general signal exposes this second one, and it is used for membrane traverse into the second compartment. The second signal is cleaved by a second protease, this time on the inner surface of the second membrane.

3. Proteins destined to be inserted into the hydrophobic milieu of the thylakoid or inner mitochondrial membrane also have a second signal in addition to the organelle-targeting one. In this case, the signal is a non-cleavable one and is a hydrophobic part of the protein that embeds into the membrane.

There are exceptions to these pathways. A notable one is mitochondrial cytochrome c, which lacks the general organelle amino-terminal signal and instead appears to have its own binding sites on the mitochondrial surface. It requires neither ATP nor a membrane potential for translocation but instead uses the attachment of the protein to heme, which refolds the protein, to drive its transit through the membrane. Indeed, the enzyme that catalyzes the heme addition appears to be the surface binding site for the protein.

Another interesting exception, this time in the plastids, occurs in a number of classes of algae. S. Gibbs has shown that in species of brown algae and diatoms, the chloroplasts are completely enclosed by a sheet of endoplasmic reticulum with ribosomes on its outer surface. In some cases, this reticulum is continuous with the outer membrane of the nuclear envelope. In these plants, plastid proteins are probably synthesized on the "chloroplast ER" and enter the lumen, from which they are transported to the plastid by vesicles.

Interactions Between Cytoplasmic and Organelle Proteins

Many proteins and structures in plastids and mitochondria are cooperatively made, with some synthesized in the cytoplasm and the rest within the organelle (Tables 5–5 and 5–6). The two are usually coordinated such that both are made and come together in the correct proportions. The mitochondrion does not import four subunits of cytochrome oxidase only to have an inadequate mitochondrially made pool of the other three.

Coordination appears to occur at several levels. In yeast, a number of nuclear genes appear to control mitochondrial gene transcription. By mutational analysis, separate nuclear genes have been identified that act as positive transcriptional regulators of the mitochondrial genes for cytochrome oxidase subunits and cytochrome b. These regulators appear to be formed at the same time as ones for the nuclear subunits of the same protein. Thus, the cell has a way of ensuring equal amounts of the nuclear and mitochondrially coded subunits. In the plastids of the green alga *Chlamydomonas,* a nuclear-coded protein acts to stabilize the mRNA for a polypeptide subunit of core complex II. Presumably this translational control is yet another way for the nucleus to exert some influence on organelle protein synthesis. Even mitochondrial division is under nuclear control. A nuclear-coded RNA, MRP-RNA, actually enters the mitochondrial matrix and acts as part of the replicating machinery of mitochondrial DNA.

A well-studied cooperatively made protein is RubisCO of chloroplasts (Figure 5–10). The small subunit is a product of the nuclear genome. It is made on free cytoplasmic ribosomes as a precursor with a signal peptide that allows it to bind to the plastid and enter the stroma, where the signal is cleaved off. A chaperonin protein, also a product of the nuclear-cytoplasmic system, enters the plastid and forms an aggre-

FIGURE 5–10 Cooperation of nucleus and chloroplast in the synthesis of RubisCO. Note the role of the binding protein, which keeps large subunits from aggregating with each other before they can bind to the imported small subunits.

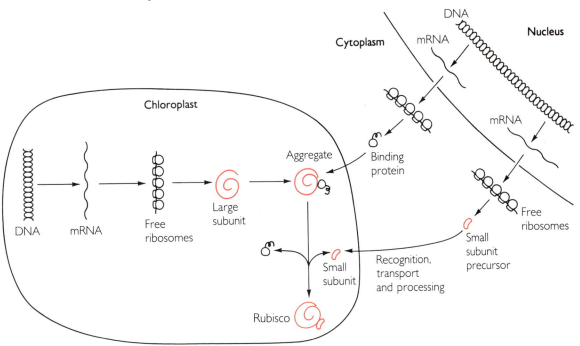

gate with the large subunit, which is made in the plastid. This aggregation appears to have an important role by rendering the large subunit soluble and by preventing improper associations by the large subunit with itself and other proteins.

The necessity of the chaperone for RubisCO has been shown in the course of genetic engineering applications. Because RubisCO is a key enzyme in photosynthetic carbon fixation (see Chapter 4), it has been the subject of intense research with the aim to make it more efficient. When the genes for the small and large subunit are expressed in bacteria, functional enzyme does not form; instead, the two subunits associate among themselves rather than with each other. This means that the gene for the plant chaperone must also be present and expressed for the enzyme to be properly formed.

Following binding of the molecular chaperone, ATP hydrolysis causes the near-simultaneous release of the large subunit and its association with the small subunit. The actual finished product (functional RubisCO in the stroma) has eight large and eight small subunits.

Organelle Molecular Genetics

Plastids

The existence of plastid DNA (ptDNA) was first indicated in 1962 by the report of deoxyribonuclease-sensitive, 2.5-nm−diameter fibrils in the stroma. It was first isolated from algae, where its different % G + C content from nuclear DNA, which leads to a different density, provided a convenient marker. Most ptDNAs, however, have the same % G + C content as the bulk DNA of the cell.

All ptDNAs exist as covalently closed circular molecules, with a molecular weight of about 10^8 (150,000 base pairs). There are 10−60 such circles in each plastid, making the plastid genetically polyploid. Two unique features that distinguish ptDNA from nuclear DNA are the absence of the unusual base, 5-methylcytosine, which occurs in nuclear DNA, and the presence of some ribonucleotides, which do not occur in nuclear DNA.

DNA sequences for plastids from two evolutionarily divergent plants (the liverwort, *Marchantia* and the higher plant, tobacco) show considerable similarity in the genes present and their organization. Indeed, the major differences between the two are a larger size in tobacco (155,844 base pairs as compared to 121,024 in the liverwort, due mostly to a long-repeated sequence in the former) and an inversion of the order of several genes.

A map derived from the DNA sequence (Figure 5−11) shows some coding sequences on one DNA strand and others on the complementary

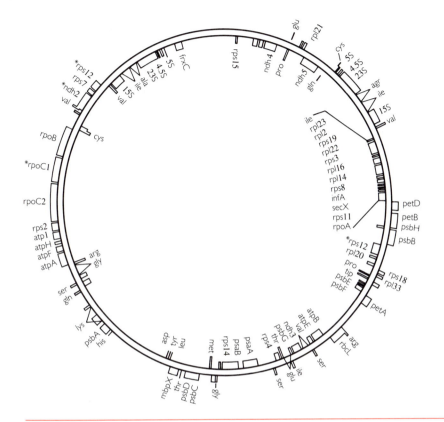

FIGURE 5–11

Map of chloroplast DNA from the liverwort plant *Marchantia*. Amino acid abbreviations refer to tRNA genes. (rps = ribosomal protein, small subunit; rpl = ribosomal protein, large subunit; 4.5S, 5S, 15S, 23S = rRNAs; inf = initiation factor for translation; secX = large ribosomal subunit protein; rpo = RNA polymerase; mbpX = membrane permease; rbc = RubisCO subunit; psa = core complex I; psb = core complex II; pet = cytochrome b_6/f complex; ndh = NADPH oxidase; frx = Fe-S protein.)

strand. In addition, the two sequences reveal six tRNA and nine protein-coding genes with intervening sequences (introns—regions within the coding part of a gene that are noncoding). In one case, a gene for one of the ribosomal proteins has one exon on one DNA strand and the other two exons with accompanying intron on the other strand. To make this protein, the two loci are transcribed separately and then spliced together. This situation is highly unusual in eukaryotes.

Chloroplasts have their own protein synthesis machinery, and often the components differ from their analogues in the nucleus and cytoplasm. For instance, chloroplast RNA polymerase, some of the subunits of which are plastid encoded, is usually sensitive to inhibition by the drug rifampin but not by amanatin; this situation is the converse in the nucleus. Chloroplast ribosomes are 70S in sedimentation and composed of 30S and 50S subunits, whereas cytoplasmic ribosomes are typically 80S, with 40S and 60S subunits. Different antibiotic sensitivities of these ribosomes have been mentioned above: Chloramphenicol blocks plastid ribosomes, while those in the cytoplasm are inhibited by cycloheximide. Finally, the organelle has its own unique transfer RNAs and associated enzymes and cofactors for protein assembly. Proteins synthesized in the organelle begin with N-formyl-methionine, a situation that does not occur in the cytoplasm.

Sequence analyses of ct-16S rRNA and the plastid-coded mRNA for the large subunit of RubisCO reveal complementary base stretches. In the rRNA is CCUCC, and in the leader (prior to the coding region) of the mRNA is GGAGG. This complementarity provides for specific recognition of the plastid mRNA and plastid ribosome. Because cytoplasmic mRNAs do not have this particular recognition sequence, this makes it unlikely that a cytoplasmic mRNA could be translated in the plastid.

Mitochondria

Fibrils identified as mitochondrial DNA (mtDNA) were first discerned in electron micrographs in 1963. Later, DNA was isolated from *Neurospora* by virtue of its different buoyant density from nuclear DNA. The relationship of this DNA to mitochondrial biogenesis is dramatically demonstrated by some petite mutants of yeast in which mitochondrial DNA is completely lacking. The result is severely defective mitochondria, since they do not have any of the mitochondrially coded proteins.

MtDNAs are usually closed circular molecules, except in certain protists, where they are linear. Their sizes vary greatly in different species: Mammalian mtDNA is usually 15,000 to 18,000 base pairs (kb), fungi range from 18 to 78 kb, and higher plants are from 200 to 2500 kb. Even within a plant family, there can be a considerable size range: Watermelon mtDNA is 330 kB, while muskmelon mtDA is 2500 kb. Much of this "extra" DNA is taken up by presumably nonfunctional rearrangements of DNA sequences, since both large and small mtDNAs produce mostly the same gene products (Table 5–5).

The best studied mtDNAs are those from yeast and humans, and both have been sequenced. Yeast mtDNA is much larger (70,000 base pairs compared to 16,000 base pairs in humans). It has a most unusual base composition, with 82% of the residues adenine and thymine and only 18% guanine and cytosine. Even regions of this DNA have unusual compositions: Half of yeast mtDNA has a base composition of 95% A + T and is termed "spacer" DNA. In addition, there are many introns among the coding regions of the genes.

Human mtDNA, in contrast, has more of a typical nucleotide content. There is little spacer, and introns are absent. Indeed, the economy of this mtDNA is such that genes often do not have complete translation stop codons, and nucleotides to complete the stop signal are added to the mRNA after transcription. The gene for mitochondrially synthesized cytochrome b illustrates the difference in size between humans and yeast

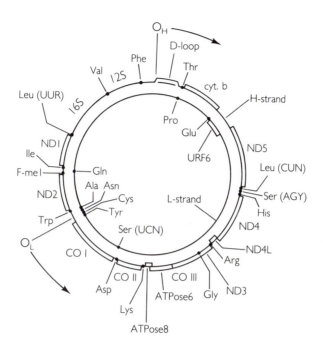

FIGURE 5–12

Map of human mitochondrial DNA. Amino acid abbreviations refer to tRNA genes. (O = origin of DNA replication; ND = NADH dehydrogenase subunit; CO = cytochrome oxidase subunit; ATPase = F_1-F_0 subunit; 12S, 16S = rRNAs; URF = unidentified reading frame.)

mtDNAs. Its coding region is the same in both (1150 base pairs), yet the yeast gene is seven times longer than the human gene because of the presence of introns. In addition, in human mtDNA many genes are separated on the genome, not by noncoding spacers but by tRNA genes (Figure 5–12). Thus, although these two mtDNAs code for a similar set of molecules (Table 5–5), their organizations are quite different.

Like chloroplasts, mitochondria contain their own protein synthesis apparatus. It has similar properties to that of the plastid and contrasts with the nuclear-cytoplasmic scheme in the same ways (see above). The patterns of mitochondrial gene transcription are quite variable. Where intervening sequences are present, they are transcribed as part of a large mRNA precursor and then excised. In human mitochondria, a number of genes are transcribed onto a single mRNA molecule, often with tRNA sequences separating protein-coding sequences. Excision of the tRNAs thus releases individual mRNAs. Also in humans, both strands of mtDNA are transcribed (Figure 5–12), thereby considerably increasing the coding capacity of this molecule.

A remarkable aspect of mitochondrial molecular biology emerged when mtDNA-coded proteins were sequenced and compared with gene sequences: Animal and fungal mitochondria (but not plant ones) have a somewhat different genetic code than other organelles or DNAs. Table 5–8 lists several of the unique features of the mitochondrial code. Most notable is that the "universal" termination codon, UGA, is read as tryptophan in mitochondria. The organelle tRNA for tryptophan has a

TABLE 5-8 Mitochondrial Exceptions to the Genetic Code

Codon	"Normal" Assignment	Mitochondrial Code			
		Mammals	Yeast	Neurospora	Plants
UGA	Stop	Trp	Trp	Trp	Stop
CUN	Leu	Leu	Thr	Leu	Leu
AUA	Ile	Met	Met	Ile	Ile
AGA, AGG	Arg	Stop	Arg	Arg	Arg
CGG	Arg	Arg	Arg	Arg	Trp

unique base sequence in its anticodon region that can read both the tryptophan codon, UGG, and the UGA codon. This makes it unlikely that a cytoplasmic mRNA, where UGA means "stop," could be accurately translated in the mitochondrion.

The D-loop of mtDNA (see Figure 5–12) contains the origin of DNA replication, as well as of transcription. Depending on the organism, it is 400 to 800 base pairs in length, and there is considerable sequence diversity between organisms. Indeed, comparisons of the D-loops of mtDNAs of a wide variety of organisms have shown that the rate of DNA sequence changes over evolutionary time is quite high (about 2% per million years). This has led to studies, especially by A. Wilson, of the evolutionary relationships between organisms as revealed by similarities (close relatives) or differences (distant relatives) in D-loop sequence.

Comparisons of several human groups show that there are even enough differences between individuals' D-loop sequences to relate humans in an evolutionary sense. For example, individual Africans have more differences when compared to each other than do Asians. This means that the Africans as a group have had more time to accumulate these differences, and so they are a more ancient group than Asians. If the rate of change of the D-loop sequence is constant and due only to mutation (recombination is low because mitochondria are inherited only from the female), the number of differences between Africans can be used to infer when their common ancestor lived. On these bases, it has been inferred that the "mitochondrial Eve" lived in Africa about 200,000 years ago.

Kinetoplasts: Modified Mitochondria

The suborder *Trypanosomatina* is composed of flagellated protozoans such as trypanosomes, which cause sleeping sickness in animals, and leishmanias, which cause skin lesions. As shown in Figure 5–13, the mitochondrion of these single-celled organisms is a single large reticulated structure. The central area of this organelle is differentiated into a specialized region called the kinetoplast. Its distinguishing feature under electron microscopy is the presence of abundant 3 nm DNA fibrils.

Kinetoplasts are semiautonomous organelles. Because they are relatively large (about 5 μm in diameter), they can be observed to divide by binary fission. Like typical mitochondria, they are enclosed within a double membrane, the inner one often differentiated into cristae.

However, the ultrastructure of the kinetoplast changes profoundly during the rather complex life cycle of the trypanosomid. When present in vertebrate blood, these parasites derive energy by a unique oxidation of glycerol phosphate; this is accompanied by an absence of their Krebs cycle and cytochrome system. The kinetoplast at this stage is modified with its DNA greatly reduced, and the mitochondrion shows few cristae. The mtDNA is active in transcription at this stage, but the translation of mitochondrial proteins is inhibited. On the parasite's returning to its alternate invertebrate host (e.g., a tse-tse fly), mitochondrial activity and kinetoplast morphology are restored.

The explanation for these cyclic changes in the kinetoplast apparently lies in the properties of kinetoplast DNA (kDNA). This accounts for up to one-third of total cellular DNA when the kinetoplast is fully differentiated. It is composed of 5000–10,000 minicircles (each 500–2000 base pairs long) and 20–50 maxicircles (each 20,000 to 40,000 base pairs), connected together in a network.

The maxicircles are typical mtDNA, with rRNA genes and various protein-coding regions. In contrast, the minicircles are apparently not transcribed and can have considerable sequence diversity that is often strain specific. Molecular hybridization probes have been successfully used to identify leishmanias on the basis of minicircle sequences. The increase and decrease in mitochondrial function in trypanosomids can be explained by the selective amplification and reduction of kDNA and its transcripts. But exactly how this occurs is not known.

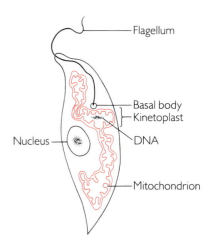

FIGURE 5–13

The protozoan *Trypanosoma* showing its large reticulated mitochondrion with kinetoplast.

Origin of Plastids and Mitochondria

Serial Endosymbiosis Theory

Plastids and mitochondria are semiautonomous organelles. They divide, contain their own genomes, and synthesize some of their own proteins. But in the broad sweep of cellular history, this was not always so. The fossil imprints of the most ancient cells, found in rocks 3.4 billion years old, are clearly prokaryotic, with little discernible internal structures. Organelle-like structures only appear in more recent imprints, about 1.5 billion years old. The remarkable transition from a solely prokaryotic world to one also populated with eukaryotic cells occurred over a long period of time. An interesting idea for how this transition happened is the serial endosymbiosis theory (Figure 5–14).

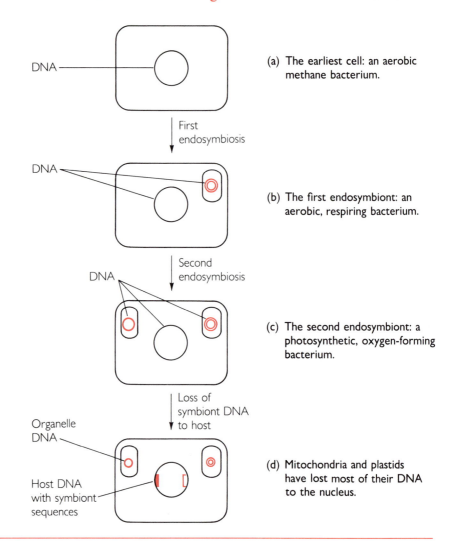

DNA ——

(a) The earliest cell: an aerobic methane bacterium.

First endosymbiosis

DNA ——

(b) The first endosymbiont: an aerobic, respiring bacterium.

Second endosymbiosis

DNA

(c) The second endosymbiont: a photosynthetic, oxygen-forming bacterium.

Loss of symbiont DNA to host

Organelle DNA ——

Host DNA with symbiont sequences

(d) Mitochondria and plastids have lost most of their DNA to the nucleus.

FIGURE 5–14

The serial endosymbiosis theory for the evolutionary origin of mitochondria and plastids.

This theory was first proposed by K. S. Mereschkowsky in 1905 and has been expanded on more recently by L. Margulis. It proposes that nuclear and organelle genomes originally were in different prokaryotic cells. The atmosphere at the time of cellular evolution is supposed to have lacked oxygen, and cells were similar to the current *Archaebacteria,* which live in extreme environments (for instance, thermal vents in the ocean). In time, photosynthetic bacteria and blue-green algae evolved, and oxygen appeared in the air as a result of their activities. The fact that oxygen gas is generally poisonous to anaerobes gave a selective advantage to aerobic microbes, which also could extract more energy via oxidative phosphorylation than could their anaerobic counterparts. According to the theory, an anaerobic cell, the precursor to the eukaryotes, ingested an aerobic microbe, which became the mitochondrion. Then, it ingested

TABLE 5–9 Some Common Properties of Prokaryotes and Organelles

Property	Eukaryotic Cell	Prokaryotes, Mitochondria, Plastids
Ribosomes	80S type	70S type
Sensitivity of protein synthesis to chloramphenicol	No	Yes
Sensitivity of protein synthesis to cycloheximide	Yes	No
DNA genome	Linear	Circular
Sensitivity of RNA polymerase to rifampin	No	Yes

a photosynthetic one, which became the chloroplast. Subsequent evolution led to a transfer of many of the symbiont's genes to the nucleus and the loss of independence of the symbiont. As a result, the organelles became semiautonomous, dependent on the nucleus for many of the proteins that they once made themselves.

Circumstantial evidence commonly used to support the endosymbiont theory is the similarity between the two organelles and the bacteria. Table 5–9 describes the most striking of these similarities. The presence of circular DNA, 70S ribosomes, and an RNA polymerase sensitive to rifampin, as well as unique antibiotic sensitivities of protein synthesis, is common to bacteria, plastids, and mitochondria but absent from the nuclear-cytoplasmic system. Unfortunately, the fact that some of these characteristics are coded for by nuclear genes (Tables 5–5 and 5–6) clouds the issue. The endosymbiont proponents envision a transfer of genes from symbiont to nucleus, but it could be argued that organelle genes are the slowly evolving remnants of primitive nuclear genes.

More rigorous evidence for or against either theory should come from an examination of the evolutionary relationships of organelle and nuclear genomes and proteins. The endosymbiont theory would be supported if it is shown that the nuclear and organelle genomes derived from separate ancestors.

Plastids and Mitochondria

In many respects, the molecular biology of plastids is very similar to that of bacteria. The genome has a relatively constant size, and there are few intervening sequences and few repetitive genes. The synthesis of plastid mRNAs does not generally involve the posttranscriptional addition of polyadenylate, a characteristic of the nuclear system. As noted above, plastid rRNA and mRNA exhibit recognition sequence complementarity, and this sequence is almost identical to the one in bacteria. The various factors involved in the initiation, elongation, and termination of protein synthesis are so similar between plastids and prokaryotes that they can be interchanged in test tube protein synthesis. Such is not the case for the plastid and cytoplasmic systems.

Gene and protein sequences point to an independent prokaryotic origin for plastids. For example, in nuclei the genes for 18S and 28S rRNA are clustered and separate from the 5S rRNA gene, but in prokaryotes and most plastids, the three genes are clustered together. Comparisons of the nucleotide sequences of plastid and bacterial and nuclear rRNAs and tRNAs show many more similarities between plastids and bacteria than between plastids and nuclei. Taken together, this type of data consistently indicates a bacterial relationship for plastids.

Perhaps the most dramatic evidence in support of the endosymbiotic origin of plastids comes from the current existence of intracellular photosynthetic symbionts. For instance, a sea slug, *Tridachia crispata,* lives on coral reefs by sucking out the cytoplasm from algae, taking out the chloroplasts, and inserting them in its cells. For a while, these plastids are fully functional and help the slug to survive, but ultimately the lack of nuclear genes for plastid proteins causes the plastids to die. Then the slug takes in another dose of plant cytoplasm, and the process repeats itself.

On a more permanent level, the single-celled flagellate *Cyanophora paradoxa* (Figure 5–15) harbors a photosynthetic organelle, the cyanelle, inside of a vacuole. This structure strongly resembles a photosynthetic

F I G U R E   5–15

Electron micrograph of *Cyanophora paradoxa* cell, with cyanelle inclusions. (×21,000. Photograph courtesy of Dr. Leo Vernon.)

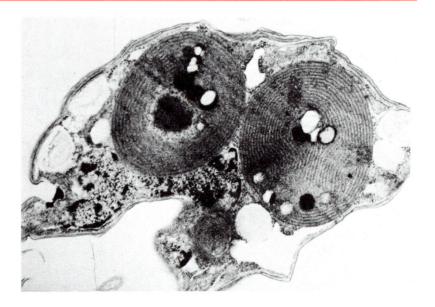

cyanobacterium but has lost the ability to live independently of its host. Morphologically, it resembles a bacterium in that it has a bacterial-like cell wall, no grana, and phycobilosomes and lacks plastocyanin. However, its DNA content is distinctly plastidlike. Because the cyanobacteria are closest phylogenetically to modern plastids of red algae, the cyanelle represents a probable step on the way to plastid formation.

Ascidians, or sea squirts, are marine chordates that form colonies. In their cloacal system they often have unicellular green algae living symbiotically. What is remarkable about these symbionts is that they have characteristics of both prokaryotic cyanobacteria and eukaryotic green algal plastids. On the bacterial side, they lack internal, membrane-bound organelles; their photosynthetic lamellae are in concentric circles; their cell wall is bacterial; and they divide by fission. They would be considered as blue-green algae, except for their distinctive green algal features. These include the presence of chlorophylls a and b (prokaryotes lack chlorophyll b) and grana and the absence of phycobilosomes. The unique characteristics of the alga *Prochloron* (Figure 5–16) have led to its being placed in a separate division of algae. It possibly represents a survivor of the progenitor of the symbiotic green algal chloroplast, although DNA sequence data indicate that it is most closely related to cyanobacteria.

As with plastids, sequence comparisons have been made of mitochondrial DNA and nuclear and bacterial DNAs. In the case of rRNAs, once again the mitochondria have more homologies to prokaryotes and nuclei. The same is true for protein-coding regions. For example, the gene sequence for the enzyme superoxide dismutase in mitochondria is

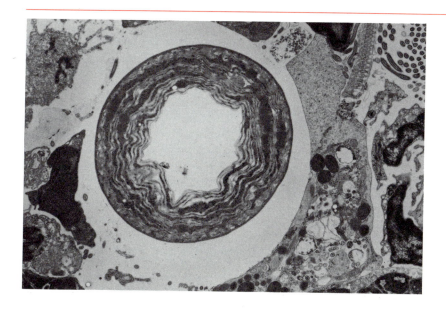

FIGURE 5–16

A cell of *Prochloron* surrounded by sea squirt host tissue. (×8700. Photograph courtesy of Dr. K. Lee.)

quite similar to the sequence of the same enzyme in bacteria. When different bacteria are examined, the genus *Rhodospirillum* seems to be closest to mitochondria and the organelle; this indicates that this bacterium and the organelle had a common ancestor.

Sequence analyses of the DNAs of the nucleus, chloroplast, and mitochondria reveal some surprising homologies. In the sea urchin, nuclear DNA contains, but does not express, sequences for part of mitochondrial 16S rRNA and a mitochondrial-coded cytochrome oxidase subunit. In maize, both the mitochondria and chloroplasts contain homologous DNA sequences for 16S rRNA, several plastid tRNAs, and even part of RubisCO. However, most of these sequences are expressed only in the chloroplast. The mechanism for movement of genes among the three cellular compartments is not known, but its existence indicates that the transfer of most of a symbiont's genes to the nucleus is not an unlikely possibility.

To summarize, the serial endosymbiosis theory can account for the evolutionary origin of plastids and mitochondria because:

1. Actual endosymbioses occur now in nature.

2. DNA can move from the organelle to the nucleus.

3. DNA sequences for the organelles are more related to bacteria than to the nuclear genome.

Further Reading

Attardi, G., and Schatz, G. Biogenesis of mitochondria. *Ann Rev Cell Biol* 4 (1988):289–333.

Baker, K., and Schatz, G. Mitochondrial proteins essential for viability mediate protein import into yeast mitochondria. *Nature* 349 (1991):205–208.

Cann, R., Stoneking, M., and Wilson, M. Mitochondrial DNA and human evolution. *Nature* 325 (1987):31–36.

Cantatore, P., and Saccone, C. Organization, structure and evolution of mammalian mitochondrial genes. *Int Rev Cytol* 108 (1987):149–201.

Clayton, C. The molecular biology of the kinetoplastidae. *Genet Engin* 7 (1988):1–52.

Clayton, D. Replication and transcription of vertebrate mitochondrial DNA. *Ann Rev Cell Biol* 7 (1991):453–478.

Craig, E., Kang, P. J., and Boorstein, W. A review of the role of 70 kDa heat shock proteins in protein translocation across membranes. *Ant van Leeuwen* 58 (1990):137–146.

Dratch, M. Mitochondrial DNA defects: Diseases from outside the nuclear family. *J NIH Res* 1 (Dec 1989):73–78.

Ellis, R. J. Molecular chaperones: The plant connection. *Science* 250 (1990):954–958.

Flugge, U.-I. Import of proteins into chloroplasts. *J Cell Sci* 96 (1990):351–354.

Forsburg, S., and Guarente, L. Communication between mitochondria and the nucleus in regulation of cytochrome genes in yeast. *Ann Rev Cell Biol* 5 (1989):153–160.

Gatenby, A., and Ellis, R. J. Chaperonin function: The assembly of RubisCO. *Ann Rev Cell Biol* 6 (1990):125–149.

Geny, C., Cormier, V., Meyrignac, C., Cesaro, P., Degos, J., Gherardi, R., and Rotig, A. Muscle mitochondrial DNA in encephalomyopathy and ragged red fibers. *J Neurol* 238 (1991):171–176.

Gibbs, S. The chloroplast endoplasmic reticulum—structure, function and evolutionary significance. *Internat Rev Cytol* 72 (1981):49–100.

Glick, B., and Schatz, G. Import of proteins into mitochondria. *Ann Rev Genet* 25 (1991):21–44.

Grace, S. C. Phylogenetic distribution of superoxide dismutase supports an endosymbiotic origin for chloroplasts and mitochondria. *Life Sci* 47 (1990):1875–1886.

Gray, M. Organelle origins and ribosomal RNA. *Biochem Cell Biol* 66 (1989):325–348.

Gruissem, W. Chloroplast gene expression: How plants turn their plastids on. *Cell* 56 (1989):161–170.

Hartl, F.-U., Pfanner, N., Nicholson, D., and Neupert, W. Mitochondrial protein import. *Biochim Biophys Acta* 988 (1989):1–45.

Hemmingsen, S., Woolford, C., van der Vies, S., Tilly, K., Dennis, D., Georgopolous, C., Hendrix, R., and Ellis, R. J. Homologous plant and bacterial proteins chaperone oligomeric protein assembly. *Nature* 333 (1988):330–335.

Jacobs, H. T., and Lonsdale, D. The selfish organelle. *Trends Genet* 3 (1987):337–340.

Kabnick, K., and Peattie, D. Giardia: A missing link between prokaryotes and eukaryotes. *Am Sci* 79 (1991):34–44.

Kawano, S. The life cycle of mitochondria in the true slime mold, *Physarum polycephalum. Bot Mag Tokyo* 104 (1991):97–113.

Kawashima, N., and Wildman, S. Studies on fraction I protein. *Biochim Biophysica Acta* 262 (1972):42–49.

Keegstra, K. Transport and routing of proteins into chloroplasts. *Cell* 56 (1989):247–253.

Kuriowa, T. The replication, differentiation and inheritance of plastids with emphasis on the concept of organelle nuclei. *Int Rev Cytol* 128 (1991):1–62.

Landry, S., and Gierasch, L. Recognition of nascent polypeptides for targeting and folding. *Trends Biochem Sci* 16 (1991):159–163.

Lestienne, P. Mitochondrial and nuclear DNA complementation in the respiratory chain function and defects. *Biochimie* 71 (1989):1115–1123.

Levings, C. S. Molecular biology of plant mitochondria. *Cell* 56 (1989):171–179.

Lewin, R., and Cheng, L. (Eds.). *Prochloron: A microbial enigma.* New York: Routledge and Chapman, 1989.

Lonsdale, D., Brears, T., Hodge, T., Melville, S., and Rottman, W. H. The plant mitochondrial genome: Homologous recombination as a mechanism for generating heterogeneity. *Phil Trans R Soc Lond* B319 (1988):149–163.

Luck, D. Genesis of mitochondria in *Neurospora crassa. Proc Nat Acad Sci* 49 (1963):233–240.

Margulis, L. *Symbiosis in cell evolution.* San Francisco: W. H. Freeman, 1981.

Neupert, W., Hartl, F., Craig, E., and Pfanner, N. How do polypeptides cross mitochondrial membranes? *Cell* 63 (1990):447–450.

Palmer, J. Contrasting modes and tempos of genome evolution in land plant organelles. *Trends Genet* 6 (1990):115–120.

Parker, W., Oley, C., and Parks, J. A defect in mitochondrial electron transport activity in Leber's hereditary optic neuropathy. *New Engl J Med* 320 (1989):1331–1333.

Pfanner, N., and Neupert, W. The mitochondrial protein import apparatus. *Ann Rev Cell Biol* 59 (1990):331–353.

Possingham, J. Control of plastid division. *Int Rev Cytol* 84 (1983):1–56.

Rothman, J. E. Polypeptide chain binding proteins: Catalysts of protein folding and related processes in cells. *Cell* 59 (1989):591–601.

Roy, H. RubisCO assembly: A model system for studying the mechanism of chaperonin action. *Plant Cell* 1 (1989):1035–1042.

Schatz, G. Signals guiding proteins to their correct locations in mitochondria. *Eur J Biochem* 165 (1987):1–6.

Simpson, L. The mitochondrial genome of kinetoplastid protozoa. *Ann Rev Microbiol* 41 (1987):363–382.

Smeekens, S., Weisbeek, P., and Robinson, C. Protein translocation into and within chloroplasts. *Trends Biochem Sci* 15 (1990):73–77.

Suguira, M. The chloroplast chromosomes in land plants. *Ann Rev Cell Biol* 3 (1989):51–70.

Umesono, K., and Ozeki, H. Chloroplast gene organization in higher plants. *Trends Genet* 3 (1987):281–287.

Wallace, D. C. Mitochondrial DNA mutations and neuromuscular disease. *Trends Genet* 5 (1989):1–6.

Wienhaus, V., and Neupert, W. Protein translocation across mitochondrial membranes. *Bioessays* 14 (1992):17–24.

Wiesbeek, P., Hageman, J., De Boer, D., Pilion, R., and Smeekens, S. Import of proteins into the chloroplast lumen. *J Cell Sci Suppl* 11 (1989):199–223.

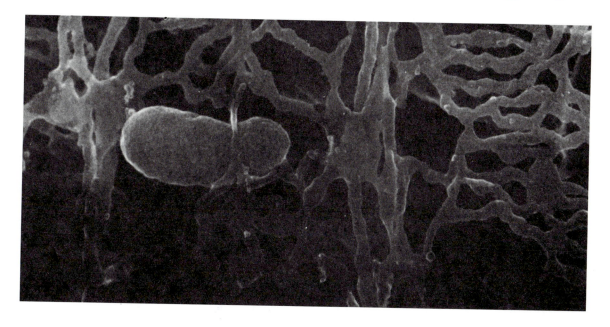

Endoplasmic Reticulum

The Largest Membrane Surface

Membranes provide a cell with a "workbench" for biochemical functions such as energy transduction, and a surface across which the cell can import needed substances and export wastes. As cells got larger over evolutionary time, the need to expand their membrane surface area became more and more urgent. The reason for this is related to the ratio of surface area to volume.

If a cell is considered as a cube, its total surface area is six times the area of one surface, or six times the square of the unit distance along one side of the cube. That is, if the cube is 1 μm on a side, the total surface area is 6 μm^2; for 2 μm on a side, the surface area is 24 μm^2. However, the internal volume of the hypothetical cell is the cube of the unit length. Thus, the volumes of these two cells are 1 μm^3 and 8 μm^3. Therefore, the surface area:volume ratios of these cells are 6.0 and 3.0.

The important point of these calculations is that as the surface area of a cell increases, its surface area:volume ratio decreases. This means that as eukaryotic cells became larger, they had to develop more extensive membrane surfaces to carry out their essential functions. One way to do this was serial endosymbiosis of prokaryotes to form internal membranous organelles such as plastids and mitochondria (Chapter 5). Another way was to form internal membranous structures out of lipids and proteins already in the cell. The endoplasmic reticulum (ER) is the most extensive of these structures.

Like the mitochondria and plastids, the ER provides both a membrane surface for biochemistry and an internal compartment to segregate certain functions. Its membrane is used for the synthesis of membrane and secretory proteins, as well as many lipids. Within the ER lumen, proteins are modified (for instance by glycosylation) and sent on their way to their final destination inside or outside the cell. In addition, the lumen acts as a storage compartment for Ca^{++}, especially in muscle cells.

Structure

Microscopy

Light microscopists observed structural organization in the seemingly "empty" cytoplasm during the first half of the 20th century. Polarizing microscopy revealed parallel, thin cytoplasmic structures that under dark field showed up as filaments, stainable in certain cell types with cationic dyes. By the 1940s, J. Brachet had demonstrated that this staining was due to RNA. These structures were commonly seen in nerve cells, where they were called Nissl bodies, and in secretory cells, where they were termed ergastoplasm.

When K. Porter examined this cytoplasmic system under the electron microscope during the late 1940s, he observed an elaborate network of membrane-delimited channels (Figure 6–1). In cultured cells, the network was less abundant in the periphery of the cell ("ectoplasm") than in the interior ("endoplasm"), and so he called the network the endoplasmic reticulum.

FIGURE 6–1 Endoplasmic reticulum from a radish root hair. The dilated region contains protein. (×38,000. Photograph courtesy of Dr. E. H. Newcomb.)

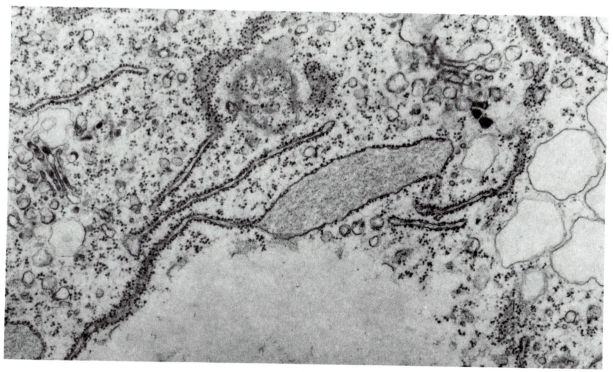

The ER consists of a system of interconnected tubules with the membranes surrounding an interior cisternal space. ER membranes, at 6 nm thickness, are generally thinner than other cellular membranes. Typically, the cisternal space is 30 nm in diameter but can increase considerably in certain situations, such as induction by drugs. As was mentioned above, the huge surface area of the ER makes it the largest membrane in the cell. For instance, in the liver hepatocyte, ER occupies 15% of the cell volume, and its surface area of 63,000 μm^2 is 37 times that of the plasma membrane.

In most cells, the ER appears interconnected with other membrane-bound organelles, most commonly with the outer membrane of the nuclear envelope (Chapter 10). In fact, this observation is so frequent that the ER has been proposed to come from the "budding off" of the envelope. ER connections with the Golgi complex, the mitochondria, and the cytoskeleton have also been reported. Time-lapse motion pictures show the ER to be quite dynamic, with tubules being extended and shrunk, as well as realigned. Although the significance of these changes is not known, they may be mediated by protein "motors" such as kinesin (see Chapter 9).

The outstanding morphological differentiation of this organelle is between rough (RER) and smooth (SER) components. The former are studded on their outer surface with numerous, 18- to 22-nm–diameter ribosomes, whereas the latter lack these structures.

RER (Figure 6–2) is commonly a network of parallel, interconnected sheets, surrounding the nucleus and radiating out toward the periphery of the cell. Its cisternae are regularly 30 nm across. As the occurrence of its ribosomes indicates, RER is most abundant in cells active in protein synthesis, especially membrane and secretory proteins. During hormone-induced cell differentiation, there is a proliferation of RER. Examples of this phenomenon include thyroid hormone-induced metamorphosis in amphibian tadpoles, testosterone-induced maturation of the seminal vesicle in mammals, and the induction of hydrolytic enzyme synthesis by gibberellic acid in the endosperm of cereal grains. The massive protein synthesis accompanying liver regeneration and the differentiation of antibody-forming cells also are preceded by RER formation.

In addition to lacking ribosomes, **SER** often appears more vesicular, with swollen cisternae (Figure 6–3). It occurs along with RER in some cell types, such as liver hepatocytes, but is more commonly proliferated in specialized cells lacking RER. These include endocrine glands, where it is correlated with steroid hormone formation, and flower petal cells, where it may be involved in the synthesis of pigments.

When an animal is treated with a xenobiotic agent (a substance foreign to the body, such as a drug), liver SER proliferates in association

FIGURE 6–2

Scanning electron micrograph of rough endoplasmic reticulum from an intestinal epithelial cell. Note the presence of ribosomes on the surface and the parallel array. (×99,000. Photograph courtesy of Dr. K. Tanaka.)

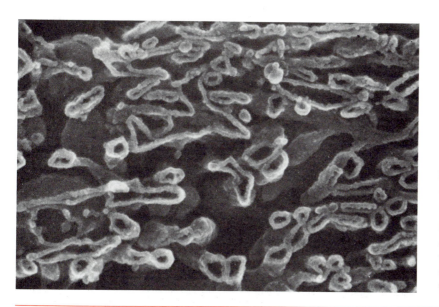

FIGURE 6–3

Scanning electron micrograph of smooth endoplasmic reticulum from a kidney cell. The cisternae are in a random array (compare to Figure 6–2). (×96,000. Photograph courtesy of Dr. K. Tanaka.)

with the metabolism of the chemical. This phenomenon of induction will be discussed in detail later in the chapter. In striated muscle cells, an extensive SER surrounds the muscle fibers, where it is complexed with tubules running internally from the plasma membrane. This sarcoplasmic reticulum is essential to the regulation of muscle contraction (see below).

During cell division, the ER appears to fragment and then reassemble. This **assembly-disassembly** cycle can be studied in cultured cells that are treated with the compactin. This drug inhibits an enzyme (HMG-CoA reductase) that is required for endogenous cholesterol synthesis, and if the cells have no outside source of cholesterol, they compensate for the inhibitor by oversynthesizing the enzyme. Because this enzyme occurs in the ER, there is a massive proliferation of the ER into a so-called crystalloid arrangement, which then collapses when cholesterol is added to the medium around the cells. This makes this system a convenient one for studies of ER dynamics. These studies have clearly shown that the crystalloid smooth ER comes from rough ER, which in turn derives from the outer membrane of the nuclear envelope.

Chemical Composition

Even the most gentle procedures for cell disruption destroy the fragile interconnections of the ER, and the membranes break and reseal, forming vesicles called **microsomes** (Figure 6–4). These can be separated from the other cell components by equilibrium density gradient centrifugation. The ER fraction can be identified by the presence of an integral membrane protein, NADH-cytochrome c reductase. Detergents can be used to remove ribosomes from the ER, an indication that these two organelles are held together by hydrophobic forces.

With regard to the microsomes that remain after ribosome removal, a major difference between RER and SER is the presence of ri-

FIGURE 6–4

Isolated microsomes from bean cotyledons. The ER membranes have been broken during isolation and sealed up to form vesicles. (×40,000. Photograph courtesy of Dr. M. Chrispeels.)

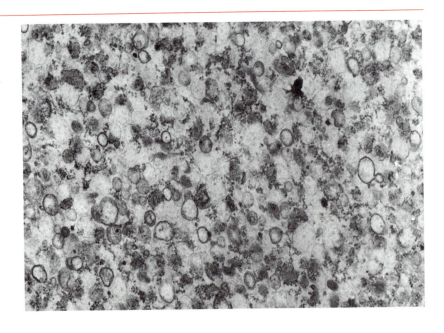

TABLE 6–1 Comparison of Rough and Smooth Microsomes

Composition	Rough	Smooth
RNA	High	Low
Ribosomes	Present	Absent
Ribophorins	Present	Absent
SRP receptor	Present	Absent
Dolichol phosphate	High	Low
Galactose, sialic acid	Low	High
Cholesterol	Lower	Higher
Physical Properties		
Net negative charge	High	Low
Monovalent cation affinity	High	Low
Density	High	Low
Tendency to aggregate	Less	More

bophorins in the membrane of the RER. These two proteins, 63,000 and 65,000 molecular weight, appear to be able to bind ribosomes to the microsomal membrane. Electron micrographs of algal cells after freeze-fracturing reveal 11-nm particles on the outer surface of the ER membrane that are associated with the imprints of ribosomes. These may be the ribophorins. Other proteins localized in the RER membrane include the receptor for the signal recognition particle (the docking protein). The remaining differences between the two ER types are summarized in Table 6–1.

The microsomal membrane is approximately 40% **lipid** (mostly phospholipid) and 60% protein by weight. Assuming molecular weights for these components of 800 and 50,000, respectively, this indicates that there are about 35 lipid molecules per protein molecule. The lipid composition (Table 6–2) shows little or no cholesterol and a fair amount of unsaturated fatty acids, indicating considerable membrane fluidity. Calorimetric studies of isolated microsomes confirm this.

Analysis of microsomal **proteins** is difficult because the negative charge of the organelle tends to absorb cytoplasmic molecules during isolation. In addition, microsomal vesicles often contain newly synthesized proteins in transit to other cell compartments. These two facts mean that an experimenter is never entirely certain if a microsomal protein is truly an ER component.

Nevertheless, electrophoresis consistently shows about 30 different proteins in the membrane and an equal number in the lumen. Some of the ones in the membrane are enzymes (Table 6–3). Their activities

TABLE 6–2 Phospholipid Composition of Castor Bean Endosperm Microsomes

	Percent Composition
Phospholipids	
Phosphatidylcholine	40
Phosphatidylethanolamine	33
Phosphatidylinositol	17
Phosphatidylglycerol	3
Fatty Acids in Phospholipids	
$C_{16:0}$	40
$C_{18:0}$	8
$C_{18:1}$	12
$C_{18:2}$	36
$C_{18:3}$	4

ADAPTED FROM: Donaldson, R., and Beevers, H. Lipid composition of organelles from germinating castor bean endosperm. *Plant Physiol* 59 (1977):259–263.

depend on the presence of lipid and in some cases show absolute specificity for boundary lipid. For example, phosphatidylcholine is essential for the activity of the cytochrome P450 system, which metabolizes xenobiotics.

Several proteins are permanent residents of the cisterna (Table 6–4). These so-called "**reticuloplasmins**" include several enzymes that modify proteins after they are made (e.g., protein disulfide isomerase, which catalyzes disulfide bond formation), proteins involved in the stabilization of protein structure within the lumen (e.g., the binding protein, BiP), and Ca^{++} binding proteins. The roles of these proteins will be described below. Their presence belies the idea that the lumen is occupied only by proteins on their way to be secreted.

An interesting reticuloplasmin in higher plant cells is one that specifically binds the hormone indoleacetic acid (auxin). In many plants, this hormone is responsible for cell elongation and, therefore, much of the growth of the plant (see Chapter 13). Although it has not been rigorously proven that the ER protein is the auxin receptor, there is suggestive evidence for it. If so, the ER in these plant cells must play an important role in growth.

The density of microsomes is determined by their RNA content, with rough microsomes being denser than smooth ones. This permits a separation of RER from SER, and analyses have revealed that although the two fractions have much in common (except for ribosomes), there

TABLE 6–3 Some Enzyme Activities of Microsomes

Carbohydrates

Ascorbic acid synthesis

Glucose-6-phosphatase

UDP-glucose dephosphorylase

UDP-glucuronic acid metabolism

Lipids

Acylation of fatty acids

Cholesterol synthesis and sterol conversions

Fatty acid synthesis and desaturation

Glycolipid synthesis

Phosphatide synthesis

Prostaglandin synthetase

Nucleotides

NADH and NADPH cytochrome c reductases

NADase

Nucleotidase-5′

Proteins

Hydroxylation of proline and lysine

Iodination of amino acids

Protein synthesis and processing

Protein disulfide isomerase

Xenobiotics

Drug, dye, and carcinogen metabolism (Table 6–8)

TABLE 6–4 Some Proteins that Reside in the ER Lumen

Protein	Function
PDI	Makes disulfide bonds
BiP	Binds unfolded proteins
GRP94	Glucose-regulated protein
Prolylhydroxylase	Hydroxyproline synthesis
Reticulin	Binds calcium ions
Glucuronyltransferase	Adds glucuronic acid

TABLE 6–5 **Heterogeneous Distribution of Microsomal Molecules**

Associated with Lighter Microsomes (mean density 1.15 g/cc)

NADH cytochrome c reductase

NADPH cytochrome c reductase

Cytochrome b_5

Cytochrome P450

Associated with Heavier Microsomes (mean density 1.22 g/cc)

RNA

Associated with All Microsomes (density range 1.13–1.26 g/cc)

Glucose-6-phosphatase

Esterase

Nucleoside diphosphatase

Glucuronyltransferase

DATA FROM: Amar-Costesc, A., and Beaufay, H. A structural basis of enzymic heterogeneity within liver endoplasmic reticulum. *J Theoret Biol* 89 (1981):217–230.

are some enzymatic differences in terms of quantity (Table 6–5). But this distribution is not strict, since cytochemical staining shows the presence of all ER enzymes throughout the ER system in cells such as liver where both RER and SER are present.

In addition to lateral protein asymmetry, the ER membrane has considerable transverse asymmetry, with some proteins facing the cytoplasm and others facing the lumen. The techniques used to reveal this phenomenon are similar to those used on the plasma membrane, mitochondrion, and plastid. Microsomal vesicles are impermeable to antibodies, proteases, and impermeant reagents; thus the binding of any of these to a microsomal protein in intact vesicles indicates that the protein is embedded in the cytoplasmic side of the membrane. On the other hand, low detergent concentrations make the vesicles permeable to these reagents, so that attachment in this case but not in intact vesicles indicates that a protein is part of the luminal side of the membrane.

These studies have revealed considerable protein asymmetry (Table 6–6). Phospholipase incubations indicate lipid asymmetry as well, with most of the phosphatidylcholine and sphingomyelin in the cytoplasmic half of the bilayer and phosphatidylserine and phosphatidylethanolamine in the inner (luminal) leaflet.

The enzymes of xenobiotic metabolism face the cytoplasm. This makes sense, since the foreign molecules to be metabolized are located there. Cytochrome b_5, NADH-cytochrome b_5-reductase, and NADPH-

TABLE 6–6 Topology of Enzymes in the Liver Endoplasmic Reticulum Membrane

Cytoplasmic Surface	Luminal Surface
ATPase	Acetanilid-hydrolyzing esterase
Cytochrome b_5	β-glucuronidase
Cytochrome P450	Cytochrome P450
GDP-mannosyltransferase	Glucose-6-phosphatase
NADH-cytochrome b_5 and NADH-cytochrome c reductase	Nucleoside diphosphatase
Nucleoside pyrophosphatase	
Nucleotidase-5′	

cytochrome c reductase are anchored in the bilayer by small hydrophobic regions, leaving the bulk of each molecule exposed. There is also evidence that the ribophorins face the cytoplasmic surface. The bulkiness of the xenobiotic metabolizing system could exclude the ribosome-binding proteins, providing a mechanism for the lateral membrane asymmetry between SER and RER.

Function: Protein Synthesis and Segregation

Secretory and Storage Proteins

The role of the RER in the synthesis and processing of integral membrane proteins has been outlined in Chapter 1. These proteins are made on RER with a hydrophobic signal, by which they can embed in the lipid bilayer. In addition, the ER is the site of synthesis of a group of proteins destined for storage within another membrane-bound organelle in the cell (e.g., protein bodies in seeds and lysosomes in animal cells) or for export outside of the cell (secretion). These will be the primary focus of the discussion in this chapter.

G. Palade described the importance of the ER in secretory protein biogenesis through studies of tissues of the exocrine pancreas, which synthesizes digestive enzymes and secretes them into the intestinal lumen via the pancreatic duct. If this tissue is incubated in a radioactive amino acid, the proteins it makes will be radioactively tagged. If the incubation time is very short, the tagged proteins can be caught in the act of being assembled, and when this happens, they are associated with ribosomes on the RER surface. RER synthesized proteins also include those destined for other membrane-bound cell compartments such as the lysosome, membrane proteins, and storage proteins in seeds. This is in contrast to cytoplasmic proteins such as the glycolytic enzymes, which are usually synthesized on unattached ribosomes.

The ribosome attaches firmly to the surface of the ER membrane by its large subunit (60S in cytoplasmic eukaryotic ribosomes). The ribophorins, possibly aggregated into an 11-nm particle that spans the ER membrane, are at or near the site where the actual attachment occurs. Evidence for this supposition comes from experiments:

1. If microsomes are treated with ionic detergents, the ribophorins and ribosomes detach from the membrane together, and the remaining ER membrane cannot bind ribosomes unless the ribophorins are added back.

2. As noted above, SER membranes generally lack ribophorins, but if they are added to smooth microsomes, they bind ribosomes.

3. Antiribophorin antibodies block the binding of the protein-synthesizing complex to the RER.

The initial binding of the ribosome to the RER is mediated by the nascent polypeptide chain. At their initially translated amino termini, almost all proteins made on the ER have a stretch of largely hydrophobic amino acids. This acts as a **signal** with which the growing chain, and its accompanying ribosome complex, can be attached to the membrane. (See Figure 1–18 for an account of the signal hypothesis for membrane protein synthesis.) Because it takes about 20 amino acids to make a single membrane-spanning α-helix, it is not surprising that ER signal sequences are all about this length (Table 6–7).

That the signal sequence is essential for protein targeting to the ER has been demonstrated by genetic manipulations similar to those used to study the signals for mitochondrial and plastid proteins (see Chapter 5). For instance, the DNA coding for the ER signal from a secreted protein can be added to the gene for β-globin; if this gene is put into a cell, the normally cytoplasmic globin is put into the ER and secreted. On the other hand, deletion of the signal from a polypeptide causes it to be localized in the cytoplasm.

Given the importance of this process, it might be expected that the signal has been evolutionarily conserved, and a common amino acid sequence might exist. Such is not the case. Table 6–8 gives the sequences of four signal peptides. Generally, the amino-terminal region is positively charged, the central region is hydrophobic, and the carboxy-terminus (nearest the start of the protein proper) is polar. Molecular models indicate that the hydrophobic signal exists as a regular α-helix, and this has been confirmed by circular dichroism spectroscopy of the isolated peptides in nonpolar environments.

TABLE 6–7 Some Secretory Proteins with Signal Peptides

Protein	Signal Size (amino acids)
Serum Proteins	
Mouse immunoglobulin, κ light chain	20
Mouse immunoglobulin, λ light chain	19
Mouse immunoglobulin, μ heavy chain	18
Bovine serum albumin	18
Bovine prothrombin	22
Hormones	
Human chorionic gonadotropin	24
Human placental lactogen	25
Human growth hormone	26
Human proinsulin	23
Bovine parathyroid hormone	25
Milk Proteins	
Sheep β-lactoglobulin	18
Sheep α-lactalbumin	19
Sheep casein	15
Egg Proteins and Others	
Chicken lysozyme	18
Chicken ovomucoid	23
Human interferon	23
Honeybee promelittin	21

TABLE 6–8 Amino Acid Sequences of Signal Peptides of Secreted Proteins

Sequence→	1	5	10	15	20	→protein
Pre-conalbumin	met-lys-leu-ile-leu-cys-thr-val-leu-ser-leu-gly-ile-ala-ala-val-cys-phe-ala-ala-pro-					pro-
Pre-λ-immunoglobulin	met-ala-trp-ile-ser-leu-ile-leu-ser-leu-leu-ala-leu-ser-ser-gly-ala-ile-ser-gly-ala-					val-
Pre-lysozyme	met-arg-ser-leu-leu-ile-leu-val-leu-cys-phe-leu-pro-leu-ala-ala-leu-gly-lys-val-phe-					gly-
Pre-proalbumin	met-lys-trp-val-thr-phe-leu-leu-leu-leu-phe-ile-ser-gly-ser-ala-phe-ser-arg-gly-val-					phe-

FROM: Von Heijne, G., and Blomberg, C. Trans-membrane translocation of proteins. *European J Biochem* 97 (1979):175–181.

But the synthesis of these proteins does not start on the ER. Instead, the signal sequence is translated on a free ribosome, and then it binds the synthesizing complex to the ER. The intermediary between translation and binding is a ribonucleoprotein complex termed the **signal recognition particle** (SRP). This is composed of six nonidentical polypeptide chains and a molecule of RNA with about 300 nucleotides. In the absence of ER, the complex as a whole functions to arrest secretory protein synthesis after the nascent signal sequence is translated. Generally, about 70 amino acids are translated before the SRP stops it. The SRP does this by having a subunit that recognizes the signal as it emerges from a pocket in the ribosome.

The signal recognition particle with its bound protein-synthesizing complex binds to ER via a docking protein on the ER membrane surface. The docking protein has an overall basic charge on its surface, perhaps allowing its interaction with the negatively charged RNA of the SRP. The docking protein also binds guanosine triphosphate (GTP), which is hydrolyzed to guanosine diphosphate (GDP) during binding. It has been proposed that the energy released by this hydrolysis "fixes" the SRP to the ER and that there is an opportunity prior to this event for the system to "proofread" and debind incorrectly matched peptides. At any rate, binding releases the SRP, and this allows translation and translocation into the ER to proceed (Figure 6–5).

Proteins destined for insertion into the ER membrane must be prevented from following the secretory pathway. To this end, they have a sequence of amino acids just past the N-terminal signal termed a **"stop-transfer"** signal. This essentially acts as an anchor to keep the protein in the membrane lipid environment, much like similar signals in thylakoid proteins in plastids and inner membrane proteins in mitochondria

---

FIGURE 6–5

The SRP cycle. (SRP = signal recognition particle; SP = signal peptide; SPase = signal peptidase; SSR = signal sequence receptor; DP = docking protein.)

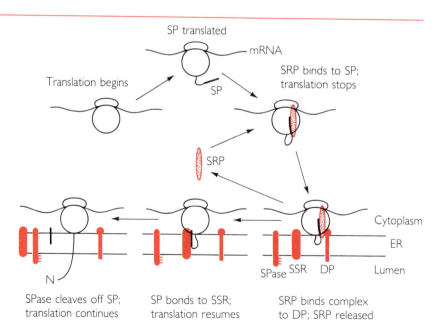

SP translated

mRNA

Translation begins

SRP binds to SP; translation stops

SP

SRP

Cytoplasm

ER

Lumen

SPase   SSR   DP

N

SPase cleaves off SP; translation continues

SP bonds to SSR; translation resumes

SRP binds complex to DP; SRP released

(Chapter 5). Obviously, the stop-transfer amino acids should be hydrophobic, and they are.

Some proteins, such as cytochrome b_5 and yeast mating pheromone, do not go to the ER as nascent polypeptides via the SRP. Instead they are fully translated first. But they must still be inserted into the membrane and so must have a mechanism for targeting them to the ER and preventing them from prematurely folding into their native shape. This is apparently achieved by posttranslational binding to a molecular "chaperone" protein, similar to those used in the assembly of RubisCO and the translocation of proteins into mitochondria (see Chapter 5). DNA sequencing has shown that the molecule is a member of the family of the so-called "stress proteins," made in response to heat shock of all organisms. Yeast mutants that lack these proteins also lack the ability to target and translocate synthesized proteins to the ER as well as the mitochondria. This indicates that translocation is similar in the two situations.

Co-Translational Processing

Approximately three minutes after its synthesis is first detected on the surface of the RER, a secretory protein from the exocrine pancreas is located inside the RER cisterna. If a protease is applied to isolated microsomes, the newly synthesized proteins are not digested because of the protection afforded by the microsomal membrane. Even nascent chains are protected from the protease by the ribosome.

The antibiotic puromycin inhibits protein synthesis in cells and organelles by acting as a false tRNA and binding covalently to the growing polypeptide chain. Cells supplied with radioactive puromycin will synthesize only nascent radioactive chains of varying length. C. Redman found that all of these radioactive chains in a secretory cell are inside the RER lumen, and this led to the concept of a **vectorial discharge** into the lumen of a secretory protein as it is made.

In order for the protein to be **translocated** through a membrane, it must be in an extended conformation. Clearly, if it is in a three-dimensional globular form, translocation will be much more difficult. But because the protein is being made on free ribosomes in the cytoplasm, it might be expected to adopt its native three-dimensional conformation as it is assembled. The SRP prevents this by acting as a molecular chaperone and keeping the polypeptide as an extended chain. Thus the SRP has two roles: to direct the complex to the ribosome and to keep it in the correct shape for translocation.

The unidirectional passage of a secretory protein into the RER could involve a hydrophobic protein "tunnel" in the membrane. The initial amino acids (signal sequence) are hydrophobic and would interact well with this membrane pore. Microsomes treated with very low concentrations of the protease elastase no longer translocate proteins across their membranes. This treatment releases a basic protein (translocase?), which, when added back to the microsomes, permits translocation to occur.

Some indications of the properties of the tunnel have come from experiments in which rough microsome vesicles are attached to a lipid bilayer separating two aqueous compartments. This has permitted electrophysiological measurements, and these have clearly revealed a channel that is fully occupied by the transitting polypeptide chain (ions are excluded when it is present). The channel closes if ribosomes are detached from the microsome, indicating that the ribosome itself may be the "ligand" that opens the channel and that the tunnel in the ribosome through which the nascent chain exits may be lined up with the ER tunnel.

Soon after the signal sequence has traversed the ER membrane, and while the rest of the secretory or storage protein is still being translated, a peptidase associated with the membrane cleaves off the signal (Figure 6–6). The amino acid sequences at the **cleavage** site appear to be somewhat specific, with small neutral amino acids (e.g., ala, gly) at one side of the cut. The loss of these 20-odd amino acids leads to a situation in which the mature finished protein has fewer amino acids than would be expected from the complete translation of its mRNA. Direct evidence for this model comes from experiments on immunoglobulins by G. Blobel:

1. Pure mRNA for immunoglobulin can be supplied with cofactors for test-tube protein synthesis. The resulting protein is 19 amino acids longer than mature immunoglobulin.

2. If RER from an immunoglobulin-synthesizing cell is used for the test-tube experiment, the mature (shorter) protein is made.

3. If the protein from (1) is added to ER during synthesis, only the mature species is synthesized.

The best explanation for these data is that the ER supplies the enzyme that cuts off the extra signal from the protein.

Cleavage of the signal peptide appears to be an essential event for the functioning of the protein translated after it. The signal, being hydrophobic, alters the three-dimensional structure of a protein existing in an aqueous environment. Indeed, if unprocessed preproteins are injected into an oocyte, their half-lives are much shorter than those of processed proteins of the same molecular species. Signals can even be internal in a polypeptide chain, and these too are cleaved for maximal activity. In

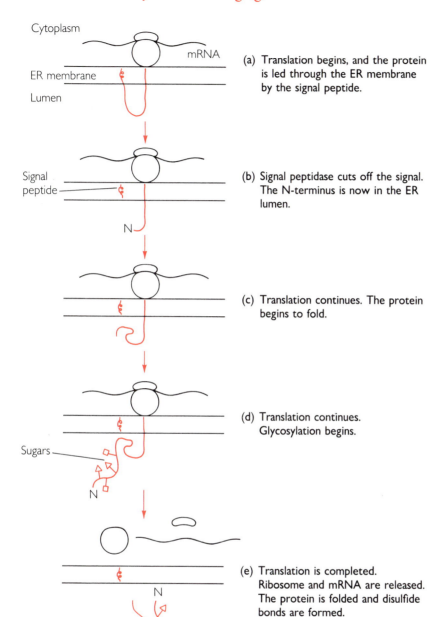

Cytoplasm

mRNA

ER membrane

Lumen

(a) Translation begins, and the protein is led through the ER membrane by the signal peptide.

Signal peptide

N

(b) Signal peptidase cuts off the signal. The N-terminus is now in the ER lumen.

(c) Translation continues. The protein begins to fold.

(d) Translation continues. Glycosylation begins.

Sugars

N

(e) Translation is completed. Ribosome and mRNA are released. The protein is folded and disulfide bonds are formed.

N

FIGURE 6–6

The completion of a protein that is destined for insertion into the ER lumen.

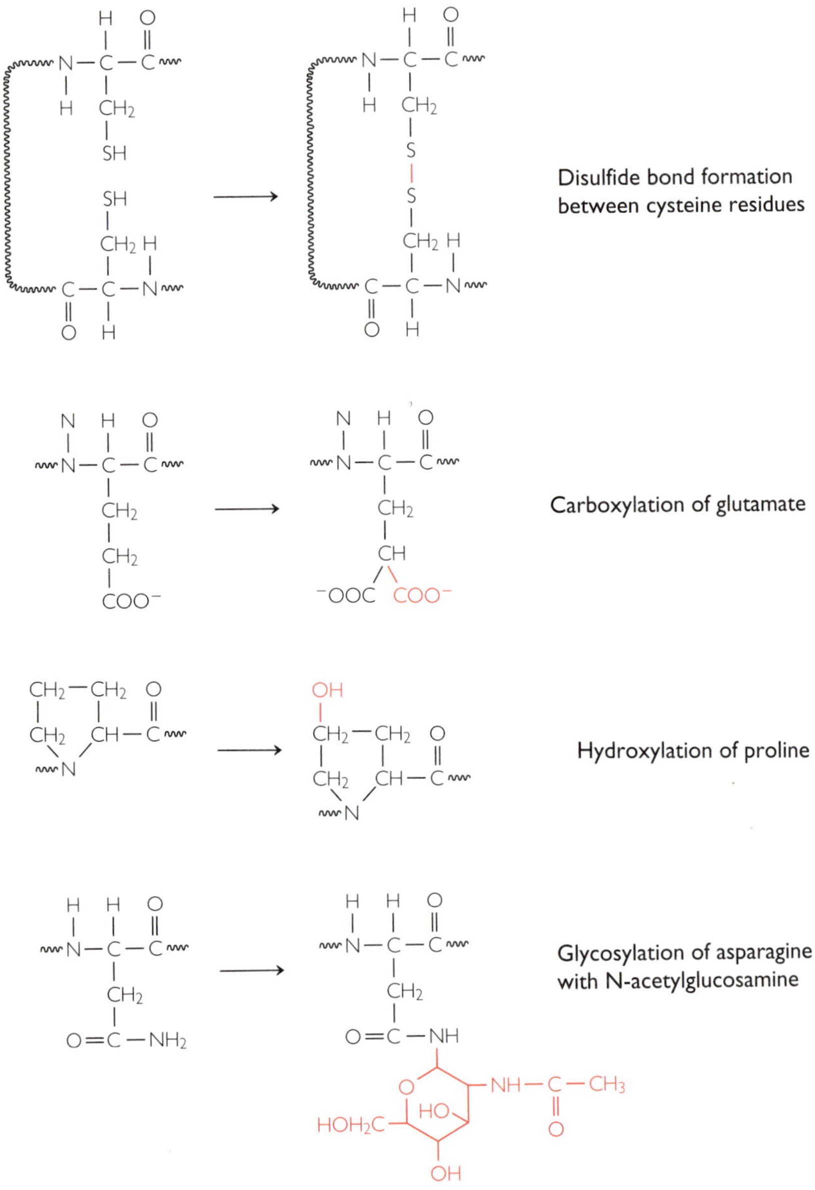

FIGURE 6–7

Four covalent modifications of proteins that occur in the ER.

Disulfide bond formation between cysteine residues

Carboxylation of glutamate

Hydroxylation of proline

Glycosylation of asparagine with N-acetylglucosamine

ovalbumin, the major protein of egg white, the signal sequence is located between residues 22 and 41 of a chain of 385 amino acids. Obviously, the passage of such a preprotein through the ER membrane is a more complex procedure than the situation for proteins with amino-terminal signals.

After the signal sequence has been cleaved, but still during translation of the protein, four possible **modifications of the protein can occur.** These are catalyzed by enzymes at the inner ER membrane or lumen and generally change the physical properties of the protein (Figure 6–7).

1. **Disulfide bond formation** between cysteine residues occurs on most proteins and is catalyzed by the luminal-resident, soluble enzyme (reticuloplasmin), protein disulfide isomerase. The resulting covalent linkages between different regions of the protein are important in its shape and, by implication, function.

2. **Carboxylation** of glutamate residues to produce the γ-glutamyl side chain gives the modified protein the ability to bind Ca^{++} ions. This is important in the blood-clotting protein prothrombin, where it binds tightly to clotting factors on cell surfaces. The enzyme that catalyzes the carboxylation is an integral protein of the ER membrane whose active site faces the lumen.

3. **Hydroxylation** of proline and lysine residues occurs on the animal extracellular matrix proteins collagen and elastin, where it acts to stabilize the interchain linkages of these fibrous proteins (see Chapter 13). Gene cloning and sequencing have shown that one of the polypeptide chains of peptidylproline hydroxylase is identical with protein disulfide isomerase. In plants, hydroxyproline is also formed (on cell wall proteins), but the hydroxylation occurs after translation has been completed.

4. **Glycosylation** of proteins in the ER of animal cells occurs via the addition of presynthesized oligosaccharide chains attached to the lipid carrier dolichol phosphate (Figure 1–19). These chains face the RER lumen and are composed of up to 60 residues of N-acetyl-glucosamine, mannose, and glucose. Attachment is to asparagine in the sequence -ser-x-asn- (where x is an amino acid) in the growing protein chain and only occurs after at least 45 residues past the acceptor site have been translated. This ensures that the target residues will be in the lumen in a proper three-dimensional configuration.

In addition to this N-linked addition of oligosaccharide chains, three other glycosylation events can occur within the ER. N-acetylglucosamine can be added to the proteins via the hydroxyl groups of the amino acids, serine and threonine; some glucose residues may be added to mannose units on the N-linked carbohydrate; and sugars are added to the glycosylphosphatidylinositol anchors of certain membrane proteins (Figure 1–14). Since many of the enzymes which catalyze these glycosylations face the inner surface of the ER membrane, there must be translocation mechanisms not only for the protein, but also for the individual sugars and oligosaccharide chains.

Following completion of the protein and its release into the RER lumen, the three glucose residues are removed by a glucosidase, and a single mannose is removed by a mannosidase. The reason for these deletions is not clear; it may constitute a signal that the protein is ready for the next step in the secretory process. The glycosylation events are summarized in Figure 6–8.

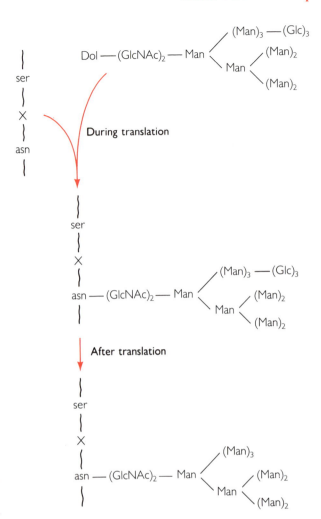

FIGURE 6–8

Addition and metabolism of an oligosaccharide to a protein in the lumen of the endoplasmic reticulum. (ser = serine; X = amino acid; asn = asparagine; GlcNAC = N-acetylglucosamine; Man = mannose; Glc = glucose; Dol = dolichol phosphate.)

Almost all proteins that pass through the ER are glycosylated. A notable exception is serum albumin, which is secreted unglycosylated from liver cells into the bloodstream. Why glycosylation occurs at all is a mystery, but some ideas have been proposed. For instance, carbohydrate might stabilize proteins from proteolytic attack, possibly by blocking accessibility to the active site of the protease. Or, carbohydrate residues may be essential for the targeting of proteins to organelles (e.g., phosphomannose to the lysosome). Or, for cell surface proteins, carbohydrate is important in recognition (see Chapter 14).

Posttranslational modifications in the ER differ in different cell types. For example, the pattern of N-glycosylation of the surface glycoprotein of influenza virus differs depending on the host cell in which the virus reproduces. This phenomenon means that the introduction of a cloned gene into a foreign cell may result in a different pattern of protein modifications than was present in the cell from which the gene was taken. This poses a serious problem for genetic engineers trying to get a cell such as yeast to correctly make an animal cell product (e.g., a virus coat for a vaccine), since the gene product made in the laboratory system may be quite different in its modification pattern (which is often important antigenically in vaccine generation, for example) from the naturally synthesized product.

Segregation and Transport

Once synthesized and processed, a secretory or storage protein inside the lumen of the ER is irreversibly **segregated**. Three factors prevent it from retracing its path and going through the membrane to the cytoplasm:

1. The removal of the signal sequence should stop the mature protein from interacting with the lipid bilayer, since the presence of the sequence is required for binding and translocation.

2. The protein is now in an aqueous environment and assumes its native globular shape. For example, in serum albumin, 11 of 17 disulfide bonds are formed by the time translation is completed, and the rest form shortly after.

3. Glycosylation with amino sugars renders the protein more hydrophilic. This makes interaction with the hydrophobic membrane core thermodynamically unlikely.

To these phenomena is added the process of **oligomerization,** the formation of multi-subunit proteins, which occurs soon after translation. Many proteins that pass through the ER are oligomeric. A major problem here is that the various subunits must not be allowed to aggregate nonspecifically by themselves. Its solution is the same as that in the plastids and mitochondria: A molecular chaperone binds to the subunits inside the ER lumen before they have a chance to fold and keeps them from folding inappropriately (Figure 6–9).

The best example of an ER chaperone is the first one identified— BiP, the binding protein for immunoglobulins (antibodies). These proteins are composed of two types of chains, light and heavy (for their respective molecular weights), and the oligomeric product has two of each kind of chain and is secreted from the cell that produces an antibody. Once a chain is made and inserted into the ER lumen, BiP "welcomes" it by binding to it, preventing it from aggregating. Once the other chains are made and are nearby, BiP is released and the oligomeric protein is assembled. Because BiP is a widespread resident of the ER lumen, its role is not restricted to immunoglobulins but probably is essential for the formation of most oligomers.

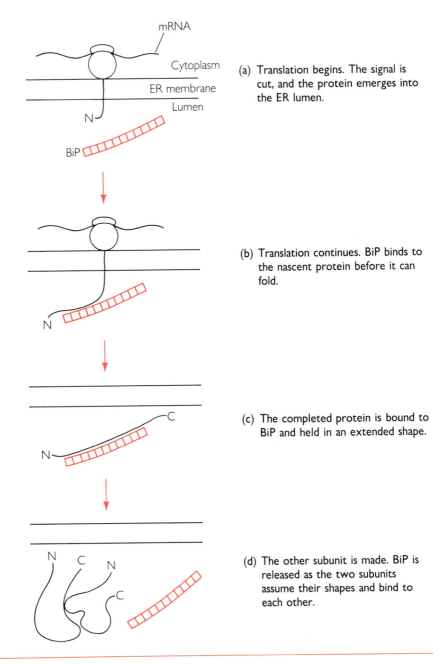

(a) Translation begins. The signal is cut, and the protein emerges into the ER lumen.

(b) Translation continues. BiP binds to the nascent protein before it can fold.

(c) The completed protein is bound to BiP and held in an extended shape.

(d) The other subunit is made. BiP is released as the two subunits assume their shapes and bind to each other.

FIGURE 6–9

The BiP cycle for the oligomerization of a protein within the ER lumen.

An elegant way to show the importance of correct oligomerization is to inject the mRNA for the heavy chain into a test cell, the frog oocyte. The oocyte goes ahead and makes the heavy chains, but they stay in the cell for days and are incorrectly assembled. If light chain mRNA is injected at the same time, however, light chains are also made; they associate with the heavy chains via BiP, and the mature immunoglobulin is secreted.

BiP has another role: that of quality control. If a protein in the ER is misfolded, misassembled into the wrong structure, or not assembled at all, BiP binds to it and prevents it from being secreted (recall the heavy chain example). How this occurs is not known, but it is clearly an important process. There is a specific protein-degrading activity in the ER that breaks down incorrectly folded proteins.

In addition to its role in "quality control," protein degradation within the ER appears to be an ongoing process for "quantity control." The enzyme hydroxymethyl-glutaryl CoA reductase (HMG CoA reductase) catalyzes a key early step in the synthesis of sterols. When sterols are abundant in the cell, this enzyme, normally resident in the ER, gets degraded. A second example involves apolipoprotein B100, a component of low-density lipoproteins (LDL), which carry cholesterol. When LDL levels are high, the apolipoprotein is degraded within the ER and does not continue on the secretory pathway. How external signals promote selective degradation within the ER is not known.

Reticuloplasmins are permanent residents of the ER lumen (Table 6–4). Remarkably, in mammals these proteins have a common C-terminal sequence composed of the amino acids lys-asp-glu-leu (KDEL, using the amino acid single-letter abbreviations). It is clear from chimeric protein experiments that this sequence, or something very similar to it, is a required "identity card" for an ER resident. For instance, deleting the sequence from a protein will cause it to be secreted, and if it is added to a protein normally targeted to another cell compartment, the protein now becomes an ER resident.

The latter experiments have revealed an interesting twist to the story. Cathepsin lacks the KDEL terminus and normally goes from the ER to the Golgi and then to its final destination, the lysosome. If KDEL is added to this protein, it ends up in the ER lumen, but it also has mannose-6-phosphate, a marker added in the Golgi (see Chapter 7). This means that ER proteins go first to the Golgi and then are recycled back to the ER! Mechanisms for this salvage pathway are not clear but apparently include a combination of shuttling vesicles from the Golgi and a binding signal at the ER.

This ER binding signal has been identified through yeast mutants termed ERD (ER retention defective). The ER protein coded for by the wild type genes for ERD has seven transmembrane domains, and several of them loop out into the lumen to form the binding site for the ER retention signal. A protein that binds KDEL has also been isolated from mammalian ER.

An interesting application of ER retention has been shown in the human disease α-1-antitrypsin deficiency. This protein normally protects the lining of the lungs from degradation by elastase, an enzyme that breaks down elastin (see Chapter 13). Genetic deficiency of the antitrypsin causes pulmonary emphysema. In one of the more common mutations, the deficiency is manifest by the antitrypsin protein aggregating in the ER, with a resulting low level in blood serum. The mechanism for this ER retention is not known but could be related to the signal-receptor interaction.

The rate of **transfer** of glycoproteins from the ER to the Golgi is quite variable. In human hepatoma cells, the half-times of transfer of various proteins range from 25 min (albumin) to 180 min (transferrin). Because these proteins leave the Golgi and are secreted at the same rate (20 min), there must be some mechanism for the differences in ER–Golgi movement.

Several mechanisms have been proposed for the transfer of proteins from the ER.

1. One way might be via small vesicles budding off from the ER. Such vesicles have been observed near the ER and Golgi in electron micrographs. For example, when animal cells are infected with vesicular stomatitis virus, its surface glycoprotein is synthesized and transported to the cell surface via the ER–Golgi route. If infection is carried out at a low temperature (15°C), the protein accumulates after leaving the ER and is detected in small vesicles. These then fuse with the Golgi if the block is released by a rise in temperature.

2. Proteins sequestered in the RER may move to their destinations by bulk flow of the organelle. If RER membranes are constantly broken down and resynthesized, there will be a flow of the organelle with the direction of its origin being the site of synthesis and its terminus, the site of degradation. Thus, if a flow of sequestered proteins to the Golgi is desired, ER membranes will break down near the Golgi and be synthesized near the nucleus. Membrane turnover rates in the microsomes of liver and immune cells have been calculated, and there is a rapid breakdown and resynthesis of much of the ER. But the polarity of this turnover remains to be determined.

3. Specific recognition markers on the proteins might "flag" them for their destination. These markers would interact with receptor proteins on the cisternal side of the RER. Receptor occupancy could lead to a pinching off of a vesicle or membrane alterations leading to a directed flow of the complex. The existence of a marker for lysosomal proteins in the form of phosphomannose supports this hypothesis, but this signal is actually completed in the Golgi. Therefore, sorting may occur in the latter organelle.

A promising system for the elucidation of ER transport is a series of temperature-sensitive mutant strains of yeast that fail to secrete enzymes such as invertase to the extracellular medium. These strains show their mutant phenotype at one temperature (restrictive) but are normal at other temperatures (permissive). In some of these mutants, invertase accumulates in the RER and does not pass to the Golgi, and, indeed, RER accumulates also at the mutant's restrictive temperature. At the permissive temperature, the invertase passes into the Golgi in an energy-requiring process and is then secreted. In others, small vesicles accumulate, indicating that these are the means by which proteins get to the Golgi (Figure 6–10).

The yeast mutants have been used to create a test-tube model for protein transport. Donor ER is from the mutant yeast and also lacks the ability to add mannose residues to the invertase; recipient Golgi is from another tissue, which has the mannose-adding enzyme. When the two are combined at the restrictive temperature, unmannosylated invertase accumulates. If the temperature is then made permissive, the addition of mannose provides a convenient measure of the rate of ER-to-Golgi transport.

Analyses of the secretion mutants have permitted a beginning to the definition of the molecules that are responsible for the process. A number of membrane proteins, some of them G-proteins, have been identified. Indeed, a nonhydrolyzable analogue of GTP blocks ER-to-Golgi transport in wild type cells. One of the proteins, called NSF because it is sensitive to N-ethyl maleimide, is essential not only for ER-to-Golgi transport but also for transport within the Golgi and receptor-mediated endocytosis. This points to a unitary mechanism for vesicle formation.

Whereas the yeast experiments point to a generalized mechanism for ER-to-Golgi transport, data from mammalian experiments point to some specificity. If liver cells (which make serum albumin) and fibroblasts (which make procollagen) are fused, the heterokaryon initially has ER and Golgi from both cell types. Remarkably, serum albumin is transferred only from liver ER to liver Golgi, and procollagen is transferred only from fibroblast ER to fibroblast Golgi! How this occurs is not known but could involve cell type-specific receptors.

Function: Electron Transport

Cytochrome P450

An electron transport system located in the microsomal membrane of plants and animals, in adrenal mitochondria, and in the cytoplasm of bacteria catalyzes the oxidation of a wide variety of compounds. This system is classified as a mixed function oxidase because it uses a reduced coenzyme along with molecular oxygen to form the oxidized product

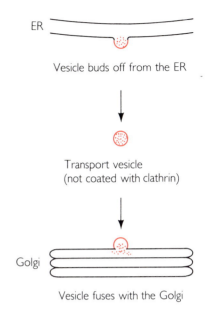

ER

Vesicle buds off from the ER

Transport vesicle
(not coated with clathrin)

Golgi

Vesicle fuses with the Golgi

FIGURE 6–10

A model for the transport of proteins from the endoplasmic reticulum to the Golgi complex.

and water. The reduced coenzyme is NADPH. Therefore, the overall reaction is:

$$RH + O_2 + NADPH + H^+ \rightarrow ROH + H_2O + NADP^+$$

where R is the substrate. The electron acceptor between NADPH and O_2 is cytochrome P450, so named because of its unusually high differential absorption at 450 nm when bound with carbon monoxide. The overall scheme is outlined in Figure 6–11a.

As noted above, these components are preferentially located in the SER outer surface, embedded into the membrane via short hydrophobic N-termini. Because they are ER membrane proteins, P450 and the reductase do not undergo cleavage of a signal when they are inserted into the membrane. Rather, the hydrophobic region serves as an anchor and a "stop-transfer" signal.

The importance of cytochrome P450 is readily apparent from the amazing variety of its substrates (Table 6–9). There are many forms of P450, and these seem to correlate with its many different functions. More than one P450 enzyme can act on a single compound, and the balance between these actions can be important in determining how the substance affects the organism.

DNA sequencing indicates that human P450 exists as a "super gene family," comprising at least nine separate gene families, with a total of about 50 genes. Within a gene family, the amino acid sequences of the proteins are at least 70% similar; between families, there is about a one-third similarity. Each P450 molecule is about 50,000 to 60,000 molecular weight.

FIGURE 6–11

The electron transport systems of liver microsomes: **(a)** cytochrome P450 and **(b)** fatty acid desaturase.

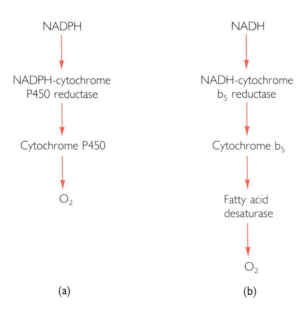

(a)   (b)

TABLE 6–9 Microsomal Cytochrome P450 Reactions

Substrate	Products	Organism/Tissue
Bile acid precursors	Bile acids	Liver
Cholesterol	Steroid hormone precursors	Liver
Fatty acids	OH-fatty acids	Kidney, yeast
Geraniol	OH-geraniol	Plants
Kaurene	Kaurenol	Plants
25-OH-vitamin D3	1,25-di-OH-vitamin D3	Liver
Steroids	21-OH-steroids	Adrenal
Xenobiotics	Alcohols, phenols, epoxides	Liver, yeast, insects

Nowhere is the essential and diverse role of P450 more evident than in the metabolism of xenobiotics such as drugs and pesticides. These compounds often enter the body through the digestive system. After absorption in the gut, they enter the circulation, and one of the first places they go is the portal system in the liver. Indeed, about 30% of the total blood output from the beating of the heart passes through this system. The liver contains abundant SER, and P450 is well situated to carry out biotransformations of these xenobiotics.

In most cases, these potentially toxic substances are inactivated by the P450 oxidation (Figure 6–12). There are notable exceptions: P450 converts aflatoxins (metabolites of certain fungi that contaminate foods) and nitrosamines (by-products of nitrites used to preserve meat products) into potent carcinogens. In the case of both detoxification and activation, P450 converts a lipid-soluble molecule into one that is more soluble in water and that can be excreted from the kidneys. In vertebrates, microsomal P450 is also found in the lungs, skin, gastrointestinal tract, and placenta, four other locations of entry of xenobiotics. Few, if any, of these substances escape the body unaltered by the P450 system.

About 1 in 10 Caucasians has a genetic disorder resulting in poor metabolism of such drugs as the antihypertensive debrisoquine; the anti-adrenergic bufuralol; and a variety of antidepressants and opioids. The defect appears to be in a certain member of the P450 superfamily (P450db1), specifically in the formation of a mature mRNA from this gene via correct RNA splicing. This error leads to a defective P450 and adverse reactions to a number of drugs.

Recognition of the potentially activating effect of cytochrome P450 has resulted in its inclusion in tests for potential carcinogens. The mutagenicity (and, possibly, carcinogenicity) of a substance is often simply tested by its effects on bacteria, fungi, or cultured mammalian cells.

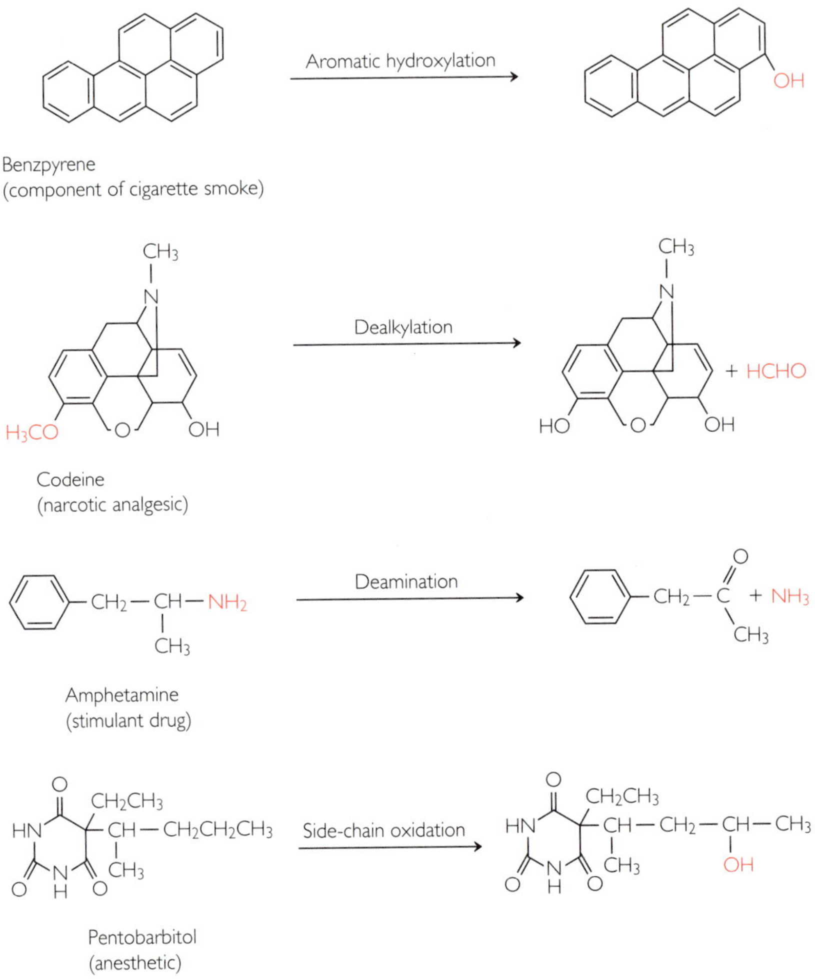

FIGURE 6–12

Some reactions catalyzed by liver microsomal cytochrome P450.

Prior to testing, the substance is incubated with liver microsomes or even intact hepatocytes, to simulate the situation in the intact organism. In many cases, this incubation is essential to activate a carcinogen.

Fatty Acid Desaturase

A second microsomal electron transport system uses a reduced coenzyme to oxidize carbon-carbon bonds in fatty acids, producing unsaturated chains. In mammals, the first reaction is:

$$\text{stearyl-CoA} + O_2 + \text{NADH} \rightarrow \text{oleyl-CoA} + H_2O + \text{NAD.}$$
$$\quad C_{18:0} \qquad\qquad\qquad\qquad\quad C_{18:1}$$

This reaction occurs in microsomes of adipose tissue, lung, and liver. In plants the initial desaturation is achieved by a cytoplasmic enzyme. In both animals and plants, further desaturations occur via the microsomal system.

The three proteins of the fatty acid desaturase complex (Figure 6–11b) are located in the SER membrane. In contrast to the P450 system, the desaturase system is not sensitive to carbon monoxide but is blocked by cyanide. It can be dissociated from the microsomal membrane, the three components can be individually isolated, and then they can be reconstituted with membrane vesicles. Both cytochrome b_5 and its reductase are inserted into the membrane by short hydrophobic regions, with much of the proteins protruding from the cytoplasmic surface. This makes functional sense, since the fatty acids are made by an enzyme complex in the cytoplasm and would be exposed to the protruding active site of the desaturase.

Induction of Endoplasmic Reticulum

A number of **chemical stimuli** induce the two microsomal electron transport systems. The desaturase complex is sensitive to diet: A high-carbohydrate, low-fat diet results in a proliferation of liver SER and increased desaturase activity, presumably to compensate for the reduction of dietary fatty acids. On the other hand, diabetes leads to a reduction in the desaturase; this can be reversed by insulin administration.

Several hundred drugs, environmental pollutants, and steroids can induce the cytochrome P450 system, and nearly all of these are substrates for the enzyme. For instance, within days of administration, the analgesic phenobarbital causes a massive increase in the SER and all of its components, especially cytochrome P450, its reductase, and cytochrome b_5. This even leads to a measurable increase in the weight of the liver! A different type of induction, such as that by methylcholanthrene, involves an increase in cytochrome P450 but not the dramatic SER proliferation. In these cases, the particular P450 form that is induced often differs from the forms present in uninduced cells.

Inducers can be additive in their effects, especially if they are of the two different types. This has important implications. If one drug has already increased the P450 activity, addition of a second drug will result in a further enhancement of drug metabolism. Thus, people are cautioned against combining the use of some recreational drugs or taking certain drugs along with alcohol.

A number of drugs, notably the diuretic tricrynaten, cause the body to make antibodies against cytochrome P450. This remarkable phenomenon appears to occur because the drug actually binds to the liver enzyme covalently and does not detach, and the bound complex acts as a target for the immune system. The autoimmunity ends up manifesting itself as hepatitis.

In both plants and animals, ER is induced naturally during development. As broad bean seeds germinate, the proportion of the cotyledon cell volume occupied by ER cisternae rises from 1% to 18%. In the mature cereal grain, hormone induction of the aleurone cells by gibberellic acid leads to an increase in RER, followed by the synthesis and secretion of hydrolytic enzymes. In mammalian liver, hepatocyte ER is arranged as random tubules prior to birth, but shortly after birth there is a considerable increase in RER and SER. This is accompanied by increases in the activities of the ER electron transport systems.

Function: Carbohydrate and Lipid Metabolism

Carbohydrates

When plant cells that are actively synthesizing cell wall polysaccharides are incubated in radioactive glucose, the label is associated with both the Golgi and the SER. A direct role for SER in cellulose and hemicellulose biosyntheses is unlikely, since the enzymes for these reactions are associated with the Golgi. Nevertheless, the presence of putative polysaccharides or precursors by radioactive labeling, as well as electron micrographs showing abundant SER near the cell surface when cell walls are being formed, indicates a functional role for the SER.

In mammalian liver and kidney, glycogen is deposited on the surface of the SER. This is especially true in fasted animals when resumption of feeding occurs and the polysaccharide accumulates. Hormonal (glucagon) stimulation causes a series of events (Figure 6–13):

1. Glycogen is broken down to glucose-1-phosphate.
2. The glucose-1-phosphate is converted into glucose-6-phosphate.
3. The enzyme glucose-6-phosphatase is present in the SER membrane, with the subunit containing its active site facing the lumen (Table 6–6). Hydrolysis releases glucose to the lumen.
4. The glucose is transported out of the cell and into the blood serum, in which it travels to such organs as brain and muscle.

The importance of this system was shown when a patient with Von Gierke's disease was found to have a deficiency in glucose-6-phosphatase activity. This disease is very rare (about 1 person in 100,000) and manifests itself with low blood sugar, liver enlargement, short stature, and delayed adolescence. It is now called glycogen storage disease, type 1a. In a related disorder, type 1b, a noncatalytic subunit of the enzyme is

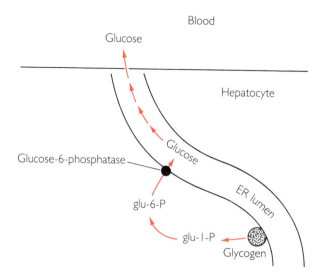

Blood

Glucose

Hepatocyte

Glucose-6-phosphatase

Glucose

ER lumen

glu-6-P

glu-1-P

Glycogen

FIGURE 6–13

Role of the endoplasmic reticulum in the release of glucose from glycogen. Hormonal stimulation leads to the hydrolysis of glycogen to glucose-1-phosphate. In the cytoplasm, this is converted to glucose-6-phosphate. An ER enzyme, glucose-6-phosphatase, removes the terminal phosphate, and the resulting free glucose goes from the ER lumen to the plasma membrane, where it is transported into the blood.

defective; however, this also causes the active site of the enzyme to be nonfunctional in vivo.

A final step in the metabolism of many xenobiotics, as well as bilirubin (the product of heme breakdown), is the addition of glucuronic acid. This is achieved via an enzyme in the ER lumen, glucuronic transferase. This event often renders the substance inert and more water-soluble for excretion. The heme breakdown product, bilirubin, is also eliminated by glucuronide formation.

Lipids

The ER has an important role in lipid anabolism, as it contains a number of enzymes involved (Table 6–3). The initial step in the incorporation of long-chain fatty acids to lipids, acylation with coenzyme A, occurs in the microsome. Fatty acyl-CoAs are then used as substrates for desaturation by the microsomal electron transport system or incorporation into phosphatides, also made in the ER. Glycolipids are synthesized by the SER, as are the other components of cellular membranes (see Chapter 1).

In plant and animal tissues, specialized lipids are often made in the SER. For example, a proliferation of the organelle occurs in plant cells synthesizing oils such as terpenes and the sticky fluid on the stigma of the flower that binds pollen. Animal cells involved in steroid syntheses, such as the adrenal gland, have a large amount of SER. The ER of liver is responsible for the formation of lipoproteins, which are composed of a protein moiety and cholesterol with other lipids. In addition, the ER in many tissues synthesizes prostaglandins from the C_{20} fatty acid, arachidonic acid.

Sarcoplasmic Reticulum

In 1902, E. Veratti observed a filamentous network running between muscle fibers under the light microscope. A half-century later, the structure of this network was revealed by electron microscopy. It was immediately realized that, morphologically, it is an ER and so was termed the sarcoplasmic reticulum (SR).

The SR is composed of three components (Figure 6–14):

1. A transverse component (T system) is connected to the plasma membrane of the muscle cell.

2. A longitudinal component resembling SER surrounds the muscle fibers.

3. A junctional complex connects the two.

The connection between the T system and longitudinal ER is indirect. This can be shown experimentally if muscle tissue is incubated in an electron-dense, high-molecular–weight marker such as the iron-containing protein ferritin. The protein will rapidly enter the T system but then stop at the junctional complex, although the membranes between the membrane systems are in contact. There appears to be a single large transmembrane protein in the junctional complex that mediates the connection between the two membrane systems.

Isolated SR vesicles contain three major membrane proteins. One of them is a **Ca^{++}-activated ATPase,** which accumulates this ion within the SR. Under the electron microscope, this protein appears as particles protruding 5 nm from the surface of the SR but not penetrating into the lumen. The amino acid sequence of this pump has been derived from its DNA sequence and describes a molecule with 10 transmembrane helices and 3 domains facing the cytoplasm, where the active site is located. The second protein is **calsequestrin,** which binds the accumulated Ca^{++}. It is located mostly within the SR that surrounds the muscle fibers. The Ca^{++} is released via a third protein, the **Ca^{++} channel,** also located at the muscle fibers.

During muscle contraction, a nerve impulse first arrives at the neuromuscular junction (Figure 2–24). Release of a neurotransmitter and the subsequent depolarization of the muscle plasma membrane sets off a series of events in which the SR plays a pivotal role (Figure 6–15):

1. The depolarization is carried down the T system to the junctional complex.

2. A membrane-spanning protein at the complex serves to bridge the T system with the longitudinal complex. In some unknown way, the signal is transduced to this complex.

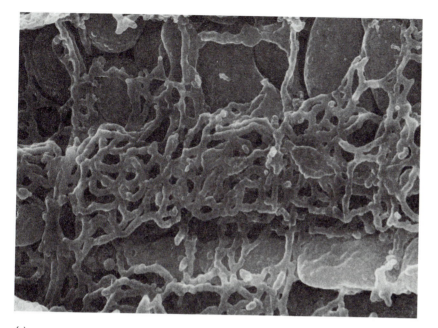

(a)

FIGURE 6–14

The sarcoplasmic reticulum (SR) in vertebrate muscle. **(a)** Scanning electron micrograph of SR in cardiac muscle. (Courtesy of Dr. T. Ogata.) **(b)** Diagram of the SR. The T system runs from the plasma membrane to a junctional complex, which connects it to a reticulum that surrounds the muscle fibers.

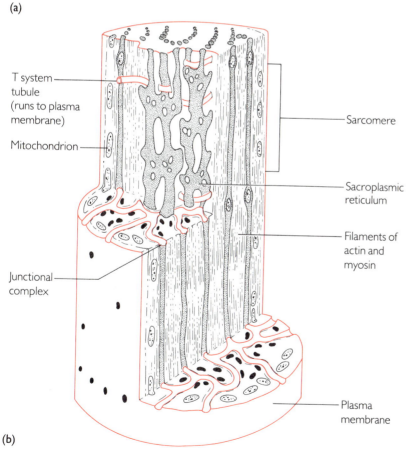

T system tubule (runs to plasma membrane)

Mitochondrion

Junctional complex

Sarcomere

Sacroplasmic reticulum

Filaments of actin and myosin

Plasma membrane

(b)

FIGURE 6–15

The role of the sarcoplasmic reticulum in muscle contraction. Depolarization at the plasma membrane also depolarizes the T tubule. Through a large membrane-spanning protein, the signal is transduced to the sarcoplasmic reticulum. This causes an inhibition of the Ca^{++} pump, a release of the Ca^{++} bound to calsequestrin, and its exit from the lumen via a channel. At the muscle fiber, the Ca^{++} binds to troponin. This alters the actin-myosin interaction and allows the two filaments to slide past one another for contraction.

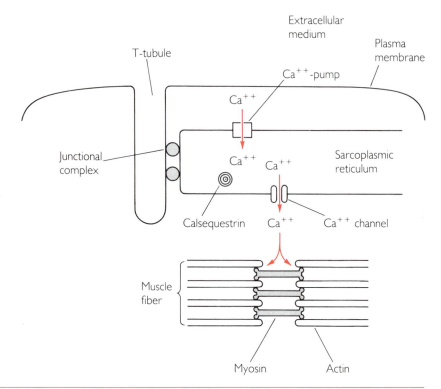

3. The Ca^{++} pump is inhibited.

4. Either via depolarization or via a second messenger (see Chapter 2), the gated Ca^{++} channel opens, releasing Ca^{++} to the muscle fibers.

5. The Ca^{++} binds to tropomyosin, a component of the muscle fiber. This allows myosin to bind to actin, and the two types of filaments slide past one another.

6. When the membrane is repolarized, the Ca^{++} pump is reactivated, Ca^{++} is removed from the muscle fibers, and relaxation occurs.

The SR is not the only Ca^{++}-binding structure in the cell. There is a strongly Ca^{++}-binding protein resembling calsequestrin in the lumen of practically all smooth ER. This protein has many binding sites for the cation, leading to an internal concentration of about 50 mg/mL within the ER. Because Ca^{++} levels can control a number of cellular activities including metabolism and the integrity of the cytoskeleton, release of the cation from the ER could be an important cell regulatory event.

Events at the plasma membrane can release the second messenger, inositol 1,4,5-triphosphate (Figure 2–21). Once inside the cell, this molecule interacts with its receptor, leading to the release of stored intracellular Ca^{++}. The receptor has been identified as an ER membrane protein, thus completing the regulatory loop.

Further Reading

Abeijon, C., and Hirschberg, C. Topography of glycosylation reactions in the ER. *Trends Biochem Sci* 17 (1992):32–36.

Blobel, G., and Dobberstein, B. Transfer of proteins across membranes. *J Cell Biol* 67 (1975):835–851.

Bonifacino, J., and Lippincott-Schwartz, J. Degradation of proteins within the endoplasmic reticulum. *Curr Op Cell Biol* 3 (1991):592–600.

Boobis, A., Caldwell, J., DeMatteis, F., and Davies, D. (Eds.). *Microsomes and drug oxidations.* London: Taylor and Francis, Ltd., 1985.

Burchell, B., and Burchell, A. Molecular pathologies of the hepatic endoplasmic reticulum. *Curr Op Cell Biol* 1 (1989):712–717.

Coon, M., Ding, X., Pernecky, S., and Vas, A. Cytochrome P450: progress and predictions. *FASEB J* 6 (1992):669–673.

Davey, J. Sorting out the secretory pathway. *BioEssays* 11 (1989):185–187.

Deschaies, R., Koch, B., Werner-Washburne, M., Craig, E., and Schekman, R. A subfamily of stress proteins facilitates translocation of secretory and mitochondrial precursor polypeptides. *Nature* 332 (1988):800–805.

Donaldson, R., and Luster, D. Multiple forms of plant cytochromes P450. *Plant Physiol* 96 (1991):669–674.

Entman, M., and Van Winkle, W. B. *Sarcoplasmic reticulum in muscle physiology.* Boca Raton, FL: CRC Press, 1986.

Freedman, R. Post translational modification and folding of secreted proteins. *Biochem Soc Trans* 17 (1989):331–335.

Garfield, S., and Cardell, R. Endoplasmic reticulum: Rough and smooth. *Int Rev Cytol Suppl* 17 (1987):255–275.

Gething, M., McCammon, K., and Sambrook, J. Protein folding and intracellular transport. *Meth Cell Biol* 32 (1989):185–206.

Gilmore, R. The protein translocation apparatus of the rough endoplasmic reticulum, its associated proteins and the mechanism of translocation. *Curr Op Cell Biol* 3 (1991):580–584.

Guengerich, F. P. Cytochromes P450. *Comp Biochem Biophys* 89C (1988):1–4.

Hicke, L., and Schekman, R. Molecular machinery required for protein translocation from the ER to the Golgi complex. *BioEssays* 12 (1990):253–258.

Hildago, C. Lipid-protein interactions and the function of the Ca^{++}-ATPase of the sarcoplasmic reticulum. *CRC Crit Rev Biochem* 21 (1987):319–396.

Hirschberg, C., and Snider, M. Topography of glycosylation in the rough endoplasmic reticulum and Golgi apparatus. *Ann Rev Biochem* 56 (1987):63–87.

Hortsch, M., and Meyer, D. I. Transfer of proteins through the membrane of the endoplasmic reticulum. *Int Rev Cytol* 102 (1986):215–242.

Hurtley, S., and Helenius, A. Protein oligomerization in the endoplasmic reticulum. *Ann Rev Cell Biol* 5 (1989):277–307.

Jones, A. Do we have the auxin receptor yet? *Physiol Plant* 80 (1990):154–158.

Kamura, S., Kanai, K., and Watanabe, J. Fine structure and function of hepatocytes during development. *J Electr Microscop Tech* 14 (1990):92–105.

Klausner, R., and Sitia, R. Protein degradation in the ER. *Cell* 62 (1990):611–614.

Koch, G. L. E. The endoplasmic reticulum and calcium storage. *BioEssays* 12 (1990):527–531.

Koch, G. L. E., Smith, M., Macer, D., Booth, C., and Wooding, F. Structure and assembly of the endoplasmic reticulum. *Biochem Soc Trans* 17 (1989):328–331.

Lingappa, V. More than just a channel: Provocative new features of protein traffic across the ER membrane. *Cell* 65 (1991):527–530.

Lodish, H. F. Transport of secretory and membrane glycoproteins from the rough endoplasmic reticulum to the Golgi. *J Biol Chem* 263 (1988):2107–2110.

Meyer, D. Protein translocation into the ER. *Trends Cell Biol* 1 (1991):154–159.

Miners, J. O., Birkerr, D. J., Drew, R., May, B., and McManus, M. (Eds.). *Microsomes and drug oxidations.* Philadelphia: Taylor and Francis, 1988.

Missaien, L., Wuytack, F., Raeymaekers, L., De Smedt, H., Droogmans, G., Declerck, I., and Casteels, R. Calcium ion extrusion across plasma membrane and calcium ion uptake by intracellular stores. *Pharmacol Ther* 50 (1991):191–232.

Mortonosi, A., Jona, I., Molnar, E., Seidler, N., Buchet, R., and Varga, S. Emerging views on the structure and dynamics of the Ca-ATPase in the sarcoplasmic reticulum. *FEBS Lett* 268 (1990):365–370.

Oritz de Montellano, P. (Ed.). Cytochrome P450. New York: Plenum Press, 1986.

Palade, G. Intracellular aspects of the process of protein synthesis. *Science* 189 (1975):347–358.

Pelham, H. The retention signal for soluble proteins of the endoplasmic reticulum. *Trends Biochem Sci* 15 (1990):483–486.

Pryme, I. F. Domains of rough endoplasmic reticulum. *Mol Cell Biochem* 87 (1989):93–103.

Rapoport, T. Protein transport across the ER membrane. *Trends Biochem Sci* 15 (1990):355–358.

Rapoport, T. A., Heinrich, R., Walter, P., and Schulmeister, T. Mathematical modeling of the effects of the signal recognition particle on translation and translocation of proteins across the endoplasmic reticulum membrane. *J Mol Biol* 195 (1987):621–636.

Robinson, A., and Austen, B. The role of topogenic sequences in the movement of proteins through membranes. *Biochem J* 246 (1987):249–261.

Rose, J. K., and Doms, R. Regulation of proteins export from the endoplasmic reticulum. *Ann Rev Cell Bio* 4 (1988):257–288.

Rothblatt, J., and Schekman, R. A hitchhiker's guide to analysis of the secretory pathway in yeast. *Meth Cell Biol* 32 (1989):3–36.

Sambrook, J. The involvement of calcium in transport of secretory proteins from the ER. *Cell* 61 (1990):197–199.

Siegel, V., and Walter, P. Functional dissection of the signal recognition particle. *Trends Biochem Sci* 13 (1988):314–317.

Sifers, R., Finegold, M., and Woo, S. Alpha-1-antitrypsin deficiency: Accumulation or degradation of mutant variants within the hepatic ER. *Am J Respir Cell Mol Biol* 1 (1989):341–345.

Sotaniemi, E. (Ed.). *Enzyme induction in man.* Philadelphia: Taylor and Francis, 1987.

Tavill, A. Intracellular pathway of protein synthesis and secretion in the hepatocyte. *Semin Liv Dis* 5 (1985):95–109.

Terasaki, M. Recent progress on structural interactions of the endoplasmic reticulum. *Cell Motil Cytoskel* 15 (1990):71–85.

Volpe, P. The unraveling architecture of the junctional sarcoplasmic reticulum. *J Bioenerget Biomembr* 21 (1989):215–225.

Von Heijne, G. Transcending the impenetrable: How proteins come to terms with membranes. *Biochim Biophys Acta* 947 (1988):307–333.

West, C. Current ideas on the significance of protein glycosylation. *Mol Cell Biochem* 72 (1986):3–20.

Whitlock, J. P., Jr. Regulation of cytochrome P450 gene expression. *Ann Rev Pharmac Toxicol* 26 (1986):333–369.

Wickner, W. Mechanisms of membrane assembly: General lessons from the study of M13 coat protein and *E. coli* leader peptidase. *Biochemistry* 27 (1988):1081–1085.

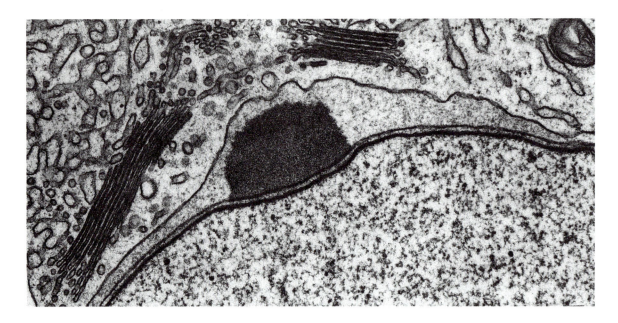

Golgi Complex

A Sorting and Packaging Center

With the major exception of those targeted for mitochondria and plastids (see Chapter 5), proteins destined for membrane-bound organelles and the extracellular environment are inserted into the lumen of the endoplasmic reticulum (ER). Each protein has an "address" or signal that is specific for its destination, but the actual "postal system" that sorts the molecules and sends them on their way is the Golgi complex.

This membrane-bound structure is also a cellular compartment with specific biochemical functions. One of these is the covalent modification of proteins, similar to what occurs in the ER. Some of these modifications are important for the ultimate function of the protein (e.g., the cutting of the precursor of insulin into two chains to make the hormone), and others are important for the proper targeting of the protein to an organelle (e.g., the synthesis of the phosphomannose marker for lysosomal proteins). In addition, the Golgi complex contains enzymes for the synthesis of certain polysaccharides that are secreted from the cell.

Structure

Microscopy

In 1898, Camillo Golgi applied silver salts to brain tissue secretions and observed precipitated metallic silver in a network surrounding the nuclei. He termed this the "apparate reticolare interno." He later won a Nobel Prize, but this did not stop others from doubting the existence of what came to be known as the Golgi complex. Indeed, over the next half century several thousand articles were published on the complex, many of which tried to show that it was an artifact.

Three phenomena caused this confusion. First, many structures, not just the Golgi, could reduce silver salts, and without a unique chemical marker it was hard to confirm the identity of the complex. Second, the structure was in different locations in different cells (e.g., near the cell membrane in secretory cells or near the center of the plant cell during division); this contrasted to the central nucleus, for example. Third, the harsh fixatives used to preserve the complex actually could cause its artifactual formation. It was only with the development of phase contrast and especially electron microscopy that the Golgi complex was established as a cell organelle.

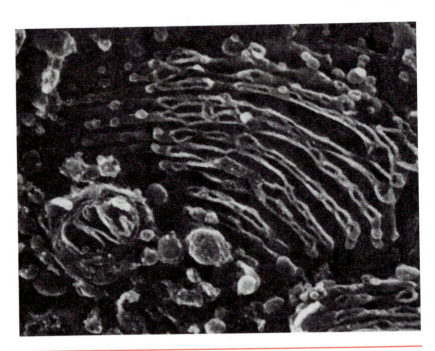

FIGURE 7–1

Scanning electron micrograph of a Golgi complex from a rat epididymal cell. (×97,000. Photograph courtesy of Dr. K. Tanaka.)

The Golgi complex under the electron microscope is composed of flattened sacs termed **cisternae** (Figure 7–1); each one is bounded by a smooth membrane and is 0.5–1.0 μm in diameter. Some cisternae have 60-nm–diameter tubules or flat perforations (fenestrae) extending from their periphery. These tubules are usually connected with the ER or adjacent cisternae. The perforated cisternae may coexist with the more flattened variety in a stack of, typically, five to eight parallel sacs, ranging up to 20 in some lower organisms.

Golgi cisternae are about 20 nm apart in the stack, and interconnections between them have been observed. But these connections may not be all that holds the complex together. Treatment of cells with nocodazole, a drug that induces microtubules to break down (see Chapter 9), causes the cisternae to separate from one another and become randomly distributed in the cytoplasm. This indicates that the cytoskeleton is involved in maintaining the integrity of the Golgi complex.

In many cells, the complex is usually located in a polar fashion, with the "cis" region facing the nucleus and endoplasmic reticulum, accompanied by small vesicles of 50 nm in diameter. The **"trans"** region, on the other side of the stack, faces the plasma membrane and has larger (up to 0.5 μm) vesicles associated with it. Between them are **"medial"** cisternae. Apparently, this orientation also is mediated by the cytoskeleton. Both coated and uncoated vesicles are commonly associated with the Golgi complex.

In individual cells, stacks may be interconnected structurally, but even if unconnected, they function in a coordinated manner. The term dictyosome ("net body") refers to individual stacks in plant cells, with Golgi apparatus being reserved for the entire network. In animal cells, the terminology is less precise, with Golgi complex referring to an individual stack of cisternae with associated vesicles. The number of stacks depends on cell type and physiological function. Certain cells in fungi have a single cisterna; hepatocytes have about 50 stacks; pollen tubes of higher plants have as many as 25,000 stacks.

Chemical Composition

In a cell stained with reactions specific for certain enzymes, light microscopy can be used to localize these enzymes to the cytoplasm or an organelle. This has been useful to find out which enzymes are present in the Golgi and to get an overall view of the distribution of the complex(es) in the cell. But electron microscope cytochemistry has shown that the enzymes are not uniformly distributed within the Golgi complex (Table 7–1). Different cisternae appear to have different compositions.

Osmium tetroxide is selectively reduced in some fashion by cellular membranes following prolonged incubation, but only the two cis cisternae reduce it. A number of stains are used for carbohydrate localization, including dyes such as ruthenium red for acidic groups and lectins, and there is a progression of increasingly intense staining from cis to trans cisternae, with the stain concentrated on the inner (lumen-facing) part of the membrane. This is consistent with a possible role for the Golgi in protein glycosylation and in forming vesicles to fuse with the plasma membrane. Following exocytosis, the formerly inner face of a vesicle becomes the outer face of the cell membrane, and this is the face that stains most intensively for carbohydrate.

Some Golgi enzymes are more concentrated in the cis or trans Golgi elements. This is a rough indication of some functional differen-

TABLE 7–1 **Cytochemistry of Golgi Stacks**

Cis Elements	Trans Elements
5'-nucleotidase	5'-nucleotidase
Adenylate cyclase	Adenylate cyclase
Weak carbohydrate staining	Strong carbohydrate staining
Osmium deposits	Thiamine pyrophosphatase
Galactosyltransferase	Inosine diphosphatase
α-mannosidase	Acid phosphatase
Sialyltransferase	

tiation across the organelle, and indeed biochemical studies have borne this out. As will be described below, some of the carbohydrate transferases act in sequence to modify proteins as they pass from the cis to the trans cisternae. Inosine diphosphatase has high activity in the Golgi and is used as a crude marker for the organelle.

Information on the precise chemical composition of the Golgi complex necessitates its isolation, but this has proved to be quite difficult because of its fragility. The Golgi cisternae detach easily, and when they fragment, the vesicles that form are hard to distinguish from smooth ER or plasma membrane. Three major technical advances have permitted the isolation of intact Golgi complexes:

1. The enzyme galactosyl transferase, which catalyzes the transfer of the sugar from uridine diphosphate (UDP)-galactose to N-acetylglucosamine, is a reliable specific marker for Golgi that is not present elsewhere in the cell.

2. The use of high salt concentrations or glutaraldehyde prevent disaggregation of the stacks.

3. Affinity chromatography using antibodies to marker enzymes can be used to selectively adsorb Golgi subfractions rich in those enzymes.

The **Golgi membrane** is composed of 60% lipid and 40% protein by weight. Its lipid composition differs from that of other cellular membranes but appears to be intermediate between those of the endoplasmic reticulum and plasma membrane (Table 7–2). Golgi membranes resemble those of the ER in the presence of sphingomyelin and a relatively low cholesterol content, but their phospholipid content is similar to that of the plasma membrane. The fatty acid profile of the Golgi is also in apparent transition between the ER and the plasma membrane, with intermediate levels in both the lengths of the chains and the degree of unsaturation.

These gradations can be seen even when the different cisternae of the Golgi are examined. For instance, there is increasing thickness from the nuclear envelope and ER, through individual cisternae, to trans vesicles, to the plasma membrane (Table 7–3). Taken together, these data suggest that the cis side of the Golgi resembles, and may come from, the ER. As the cisternae progress toward the trans face, they diverge from this composition, and the trans-associated vesicles resemble the plasma membrane with which they may fuse.

The actual biochemical differences between the cis and trans Golgi can be determined if these elements are separated from one another, and this has been done in three ways:

TABLE 7–2 Comparative Lipid Compositions of Rat Liver Membranes

Component	Percent Lipid Weight		
	Endoplasmic Reticulum	Golgi	Plasma Membrane
Cholesterol	6	8	21
Sphingolipid	6	10	
Phospholipid	81	69	70
Phosphatidylcholine	53	41	30
Phosphatidylethanolamine	20	18	15
Phosphatidylserine	3	2	5
Phosphatidylinositol	8	8	12
Fatty acids			
C_{16}	25	35	37
C_{18}	54	49	50
C_{20}	20	15	11
Unsaturated	49	41	30

TABLE 7–3 Gradations in Membrane Thickness

Membrane	Membrane Thickness (nm)	
	Rat Liver	Soybean Hypocotyl
Nuclear envelope	6.5	5.6
Endoplasmic reticulum	6.5	5.6
Golgi complex		
Cisterna 1 (cis)	6.5	5.6
Cisterna 2	6.8	5.8
Cisterna 3	7.2	6.1
Cisterna 4/5 (trans)	8.0	6.9
Secretory vesicle	8.3	7.8
Plasma membrane	8.5	8.8

FROM: Mollenhauer, H., and Morre, D. The Golgi apparatus. In Tolbert N. (Ed.). *The plant cell* (p. 469). New York: Academic Press, 1980.

1. Mild homogenization of tissues followed by centrifugation on a density gradient yields a separation of ER, cis Golgi, and trans Golgi, which are of decreasing density.

2. Affinity chromatography of isolated Golgi on columns with antibodies to microsomal enzymes such as NADPH-cytochrome c

reductase yields two fractions. One sticks to the antibody (and, therefore, has the microsomal enzyme), and the other does not stick. But microscopically, both are true Golgi fractions.

3. If intact Golgi is treated mechanically, the individual cisternae detach. These can be separated electrophoretically, since there is a gradation of charge from the cis to trans elements.

Biochemical analyses based on these separations have reinforced the idea that the cis regions are "ER-like." For instance, the cis elements contain such ER markers as NADPH-cytochrome P450 reductase, cytochrome P450, and NADH-cytochrome b_5 reductase. The trans elements, on the other hand, contain lysosomal (aryl sulfatase, glucosaminidase) and plasma membrane (5′-nucleotidase) markers. Moreover, the enzymes involved in glycosylation of proteins are distributed among the cisternae nonrandomly. As examples, the mannosidase that removes mannose added in the ER is associated with more cis cisternae, while galactosyltransferase is in the trans elements.

If the membrane proteins of the Golgi are examined by electrophoresis, five major and many minor species are observed. In a given organism and tissue, the protein profile is somewhat similar to both the ER and the plasma membrane, but there are differences. Also, secretory vesicle membrane proteins have a separate profile from those of the stacked membranes, and the patterns differ in different organisms. Most Golgi proteins are glycosylated. In animal cells, the primary sugars involved are sialic acid, glucosamine, and hexoses such as mannose and galactose. In plants, sialic acid is absent, but there are pentoses, such as arabinose.

Function: Protein Packaging

Experimental Approach

The first indication that the Golgi complex is involved in the packaging of secretory products came with the observation that cells that actively secrete molecules to the extracellular medium have extensive Golgi, in contrast to its rudimentary nature in nonsecretory tissues. In 1943, L. Worley showed that in mollusc embryos after gastrulation, yolk protein accumulated in the extensive Golgi region, and after the yolk was released from the cells, the Golgi got much smaller. In early electron micrographs, the Golgi was near the cell surface, with secretory granules apparently budding from the trans face of the organelle. The elaboration of the acrosome, a secretory organelle at the apex of the mammalian sperm cell, was also shown morphologically to be a Golgi function.

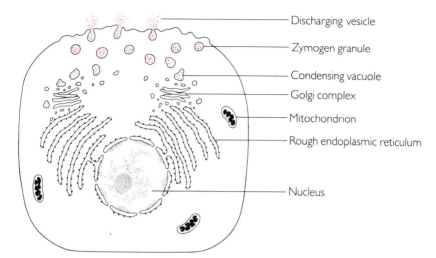

Lumen of pancreatic duct

Discharging vesicle
Zymogen granule
Condensing vacuole
Golgi complex
Mitochondrion
Rough endoplasmic reticulum

Nucleus

FIGURE 7–2
The arrangement of organelles in a pancreatic secretory cell. The cell secretes digestive enzyme precursors into the pancreatic duct, which connects to the small intestine. Note the polarized arrangement of the organelles, from which the pathway of the protein from its synthesis on rough ER to secretion into the lumen of the duct can be followed.

However, it was the development of electron microscope autoradiography and its application to secretory cells that provided experimental evidence for the involvement of the Golgi in cell secretion. The pancreatic secretory cell (Figure 7–2) has a polarized morphology suited to its role in the manufacture, storage, and secretion of digestive enzymes. The nucleus is usually at the base, the end of the cell opposite the lumen of the pancreatic duct. A highly proliferated rough endoplasmic reticulum (RER) surrounds the nucleus, and Golgi complexes lie distal to the ER. The Golgi complex has associated with its trans surface, large (about 1 μm), irregularly shaped condensing vacuoles, and just below the plasma membrane are the more compact zymogen granules.

If these cells are stained with antibodies specific for the enzymes, both condensing vacuoles and zymogen granules are shown to contain pancreatic hydrolases (Table 7–4). Slices of this exocrine tissue synthesize over 80% of all of their protein as secretory proteins. They are stored as inactive zymogens in granules prior to release in response to hormones (secretion, cholecystokinin) or neural stimulation. The actual conversion of the zymogens into active enzymes happens (fortunately) outside of the cell, in the lumen of the small intestine.

To follow the intracellular pathway of these proteins, G. Palade used a **pulse-chase** technique. Pancreas slices were incubated for a short time (three minutes) in a radioactive amino acid. Following this pulse, an excess of nonradioactive amino acid was added as a chase. This meant that any proteins synthesized during the pulse would be labeled with radioactivity, but those made during the chase would not be labeled.

TABLE 7–4 Hydrolases Synthesized In and Secreted by the Pancreatic Acinar Cell

Proteases

Carboxypeptidases A, B

Chymotrypsins A, B, C

Elastase

Trypsin

Lipases

Lipase

Phospholipase

Nucleases

Deoxyribonuclease

Ribonuclease

Amylases

α-amylase

If cells were examined after various periods of chase, the location of the radioactive protein could be determined in two ways. First, electron microscope autoradiography gave an indication of the fine structural location of the proteins. Second, this location was confirmed by cell fractionation—isolation of the various cell organelles—and determination of their relative contents of radioactive proteins. Two not unreasonable assumptions that underlie interpretations of this work are that the variety of secreted products (Table 7–4) are all made by the same pathway in the cell and that the cell was synthesizing mostly proteins destined for secretion.

Table 7–5 shows typical results for an autoradiographic pulse-chase experiment on pancreatic secretory proteins, and a clearly defined intracellular pathway is discernible. Proteins are synthesized on the RER and then sequestered within it (see Chapter 6). They are then transferred to the cis, then to the trans Golgi, and finally to the diffuse condensing vacuoles. This is followed by their appearance in zymogen granules. Later, they are released to the acinar lumen of the pancreas, which connects with the pancreatic duct.

A similar pathway has been found for secretory proteins in many other cell types. These include insulin in the pancreatic beta cells, dentin in odontoblasts, collagen in fibroblasts, immunoglobulins in lymphocytes, and the plant cell wall protein extensin in higher plant tissues. A major difference between these systems and the exocrine pancreas is that the pancreatic enzymes are stored in granules and secreted only

TABLE 7–5 The Secretory Pathway in Guinea Pig Pancreas

Cell Compartment	Pulse (3 min)	Chase (min)		
		+7	+37	+117
Rough endoplasmic reticulum	86.3	43.7	24.3	20.0
Golgi:peripheral vesicles	2.7	43.0	14.9	3.6
Golgi:condensing vesicles	1.0	3.8	48.5	7.5
Zymogen granules	3.0	4.6	11.3	58.6
Acinar lumen				7.1
Nuclei, mitochondria	7.0	4.6	1.1	3.2

Tissues were given a pulse of $^3H-$ leucine and then chased in nonradioactive leucine.

FROM: Palade, G. Intracellular aspects of the process of protein synthesis. *Science* 189 (1975):351.

when an appropriate stimulus reaches the cell. This regulated secretion contrasts with the constitutive secretion of the other systems in which the protein is secreted right after it has passed through the Golgi. For example, collagen is secreted by fibroblasts about 20 minutes after its arrival in the Golgi complex (see Chapter 13).

Transport from ER to Golgi

The overall pathway found in the pulse-chase studies implies a specific transport mechanism from one organelle compartment to the next. Because Golgi membranes are an obvious transition in composition between the ER and the plasma membrane (Tables 7–2 and 7–3), the idea arose that proteins move between the compartments in vesicles that bud off from one compartment and then fuse with the next. This is indeed the case (Figure 7–8).

Not unexpectedly, the vesicles are coated, as in the case of the clathrin-coated vesicles involved in endocytosis (see Chapter 2). But the vesicles budding off the ER are coated with a spike protein totally unrelated to clathrin. Once in the cytoplasm, these vesicles must somehow be targeted to their destination, the cis Golgi, and this may involve a "G" protein (see Chapter 2). Such a protein (called YPT1) has been isolated from yeast; when it is mutated, the yeast cells do not perform ER to Golgi transport, and instead the ER proliferates. Guanosine triphosphate (GTP) hydrolysis is required for the vesicle to arrive at the ER, where it loses its coating.

Modifications of Secretory Proteins

Secretory proteins are synthesized and processed by the RER, where glycosylation via asparagine residues also occurs (Figure 6–5). In the Golgi complex, a number of additional covalent modifications can occur (Figure 7–3). The most common of these is **glycosylation.** The carbohydrates are added in sequence, in separate compartments of the Golgi complex, as the protein passes through the cisternae. Indeed, if the enzymes are all made soluble and presented to the protein from the ER, glycosylation occurs inaccurately, if at all. Each reaction in the Golgi changes the protein to make it accessible to the enzyme for the next step (Figure 7–4 and Table 7–6).

FIGURE 7–3 Three covalent modifications of proteins that occur in the Golgi complex.

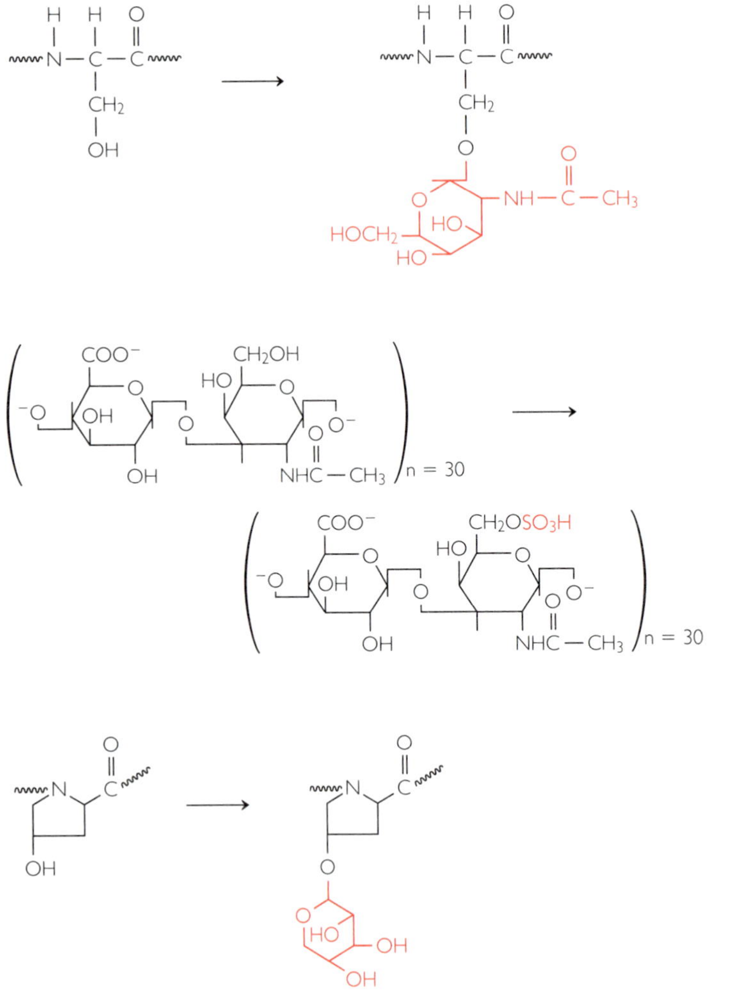

Glycosylation of serine with N-acetylgalactosamine, which occurs on many animal proteins

Sulfation of N-acetylgalactosamine (attached to glucuronic acid), which occurs in the production of chondroitin sulfate for the animal extracellular matrix

Glycosylation of hydroxyproline with arabinose, which occurs in the formation of plant cell wall proteins

1. When the glycoprotein arrives in the cis Golgi, a mannosidase removes some of the mannose residues in two stages: First, two mannose residues are removed; then, N-acetylglucosamine is added; then additional mannose residues are removed.

2. In the medial cisternae, "capping" sugars such as galactose (via the marker galactosyltransferase) and fucose are added.

3. In addition, at this stage certain serine residues are glycosylated by an O-glycosidic linkage to N-acetylgalactosamine.

4. Capping sugars (sialic acid, galactose) are also added sequentially to form this side chain.

In the assembly of the plant cell wall glycoprotein extensin, the rotein portion is made on RER and then glycosylated in the Golgi. Two ypes of glycoprotein linkages occur, and they both involve O-glycosidic

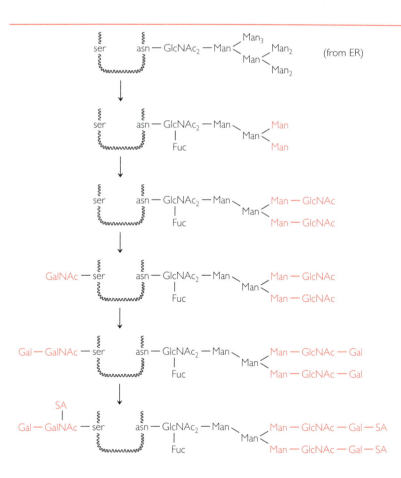

FIGURE 7–4

Modifications in the Golgi complex of a glycoprotein synthesized in the endoplasmic reticulum. (asn = asparagine; Man = mannose; Fuc = fucose; ser = serine; Gal = galactose; SA = sialic acid; GalNAc = N-acetylgalactosamine; GlcNAc = N-acetylglucosamine.)

TABLE 7–6 Compartmentation Within the Golgi Complex

Cis Golgi

Addition of phospho-N-acetylglucosamine

Removal of N-acetylglucosamine

Medial Golgi

Removal of mannose

Addition of N-acetylglucosamine

Addition of N-acetylgalactosamine

Addition of galactose

Trans Golgi

Addition of sialic acid

bonds. One is between hydroxyproline and arabinose, and the other is between serine and galactose, with the carbohydrate side chains usually less than 10 residues in length.

In plants, the formation of hydroxyproline, and its subsequent glycosylation, are essential for secretion of the glycoprotein to the cell wall. This is in contrast to the analogous situation in animal cells (see Chapter 13). Collagen (the major extracellular protein of fibrous tissues) that is not hydroxylated in the ER and therefore not glycosylated is secreted to the extracellular matrix, albeit at a lower rate, and is susceptible to degradation in the extracellular matrix. When people do not take in enough vitamin C (ascorbic acid, a cofactor needed for proline hydroxylation), they suffer from connective tissue defects because of the lack of collagen glycosylation.

The glycosyltransferases of the Golgi catalyze the transfer of sugars from a UDP donor to an acceptor glycoprotein:

$$UDPX + ACC \rightarrow UMP + ACC\text{-}X$$

These reactions apparently occur at the inner face of the trans cisterna. Galactosyltransferase and sialyltransferase are integral membrane proteins that have their active sites protruding into the Golgi lumen. The membrane contains a specific transport mechanism for the exchange of the nucleotides that serve as sugar donors, with carriers for facilitated transport.

A second Golgi modification of some secretory proteins is **sulfation,** which can be detected by incubation of tissues in radioactive sulfate. It is especially common in cells that form sulfated glucosamino-

TABLE 7–7 Occurrence of Sulfated Glucosaminoglycans in
Mammalian Tissues

Tissue	Chondroitin Sulfate	Keratosulfate	Heparin
Skin	+		
Cartilage	+	+	
Tendon, ligament	+		
Umbilical cord	+		
Heart valve	+		
Spinal disk	+	+	
Bone	+	+	
Cornea	+	+	
Liver			+
Lung			+
Arterial wall			+
Mast cells			+

glycans such as chondroitin sulfate, dermatan sulfate, keratosulfate, and heparin, which are important components of the extracellular matrix. These reactions occur in a wide variety of cells (e.g., skin, cartilage, tendons, cornea [see Table 7–7]).

Sulfation via a sulfotransferase occurs in the Golgi during glycosylation. The sulfate donor is phosphoadenosine phosphosulfate, which is formed from the reaction of inorganic sulfate with ATP. Sulfated proteoglycans usually contain long carbohydrate chains, with sulfated disaccharide repeating units. The anionic charge of these molecules gives them aggregation and water-retention properties that are important to their functions (see Chapter 13).

A third type of Golgi modification of some secretory proteins is **proteolysis.** The initial translation product of most secreted proteins is larger than the final product, and an amino-terminal signal sequence is cleaved from the precursor in the RER (see Chapter 6). For several peptide hormones such as insulin and parathyroid hormone, post-ER proteolytic processing occurs prior to secretion.

This process usually begins in the Golgi complex and is completed in secretory granules. In the case of insulin, the initial protein formed by pancreatic cells is preproinsulin, a chain of 103 amino acids, and in the RER, the amino terminal 23 residues are removed. The remaining chain is proinsulin, which is hormonally inactive. An internal 30-residue segment is excised from this chain in the Golgi and granules, leaving the 21-residue A chain and 30-residue B chain of active insulin (Figure 7–5).

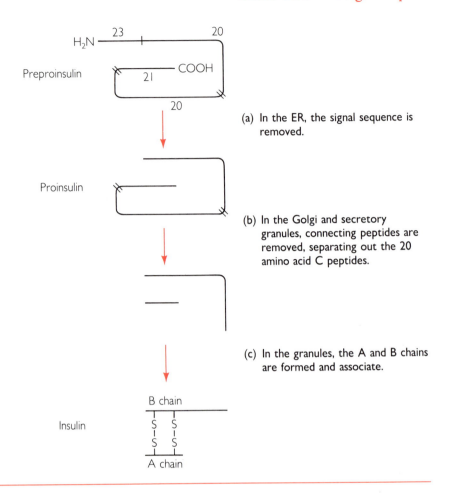

(a) In the ER, the signal sequence is removed.

(b) In the Golgi and secretory granules, connecting peptides are removed, separating out the 20 amino acid C peptides.

(c) In the granules, the A and B chains are formed and associate.

FIGURE 7–5

Proteolytic processing of insulin by pancreatic beta cells.

Movement Through the Golgi Complex

There are several proposals for the mechanism by which molecules move from cisterna to cisterna within the Golgi complex. One hypothesis involves the **migration of entire cisternae** from one face toward the other, accompanied by a loss of trans vesicles through exocytosis. Addition of more cis cisternae would be via vesicles from the ER and recycling of the plasma membrane by endocytosis. As the membrane "flowed" through the Golgi, its composition would change, in keeping with the biochemical, enzymatic, and cytochemical data.

Support for the concept of flow comes from observations of the alga *Pleurochrysis.* These cells have a single Golgi complex, with 30 cisternae. Synthesis of extracellular scales (see below) is sequential, beginning at the cis face and ending at the trans face, where a cisterna is lost with the completed scale (Figure 7–6). The entire process is visible under the light microscope in living cells and takes about 30 minutes. This

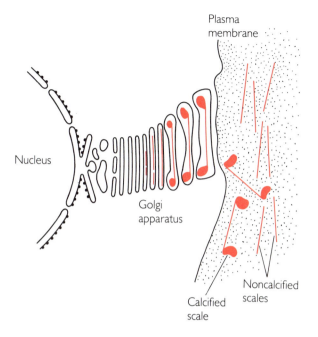

FIGURE 7–6
Diagrammatic representation of scale formation in Chrysophycean algae. Scales are formed within Golgi cisternae and are calcified prior to discharge to the extracellular matrix. The cisternae migrate through the Golgi. (After M. Brown.)

time is remarkably similar to that taken for the passage of secretory proteins through the Golgi complex in animal cells, as determined by pulse-chase experiments.

An alternate mechanism for movement through the Golgi is that it is a static structure. This is in keeping with the structural and functional differentiations that occur across the stack. Transfer of proteins would be through **interconnections between the cisternae,** in a vectorial fashion. Sometimes, electron micrographs show such connections, but they are difficult to see in most instances.

The third mechanism is a compromise between the other two models. It envisions that the cisternae bud off to form small **vesicles,** which shuttle from one cisterna to the next. Vesicles that fuse with the plasma membrane would essentially be returned to the Golgi via endocytosis. The visualization of such vesicles in electron micrographs favors this model, and additional evidence has come from elegant Golgi complementation experiments done by J. Rothman.

Two cell lines are used in such an experiment:

1. Line A is genetically deficient in a late step in glycosylation. Proteins enter its Golgi but stop before leaving it because of this deficiency.

2. Line B is genetically deficient in an early glycosylation step, and once again proteins enter but cannot leave it.

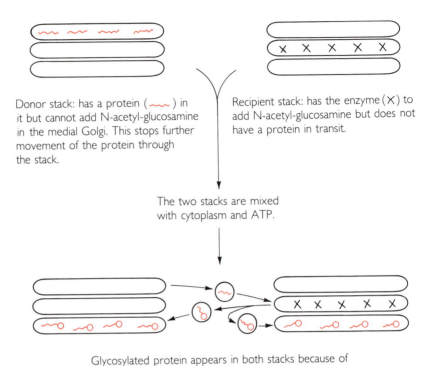

Donor stack: has a protein (～～) in it but cannot add N-acetyl-glucosamine in the medial Golgi. This stops further movement of the protein through the stack.

Recipient stack: has the enzyme (✕) to add N-acetyl-glucosamine but does not have a protein in transit.

The two stacks are mixed with cytoplasm and ATP.

Glycosylated protein appears in both stacks because of transport in vesicles between them.

FIGURE 7–7

Experimental demonstration of the transfer of proteins between two Golgi stacks in the test tube.

If these two cells are fused together, the proteins in the hybrid are correctly glycosylated and leave the Golgi. The only way this could happen is for the early step to occur in a Golgi stack derived from cell line A and the late step to occur in Golgi from line B. But the actual stacks remain intact during the experiment (they do not fuse or intermix), and the glycoprotein is never free in the cytoplasm. Electron microscopy shows that Golgi of line A (the "donor line") have vesicles budding from them, indicating that in this system the vesicles are the means of transport of the unfinished glycoprotein to the line B Golgi, where it can be completed.

This type of complementation experiment can even be carried out in the test tube, using isolated Golgi complexes from two different cell lines (Figure 7–7). Once again, the data are consistent with vesicular transport. But the test-tube experiments have an added benefit: They can be used to study the process of transport in isolation to determine its metabolic requirements and mechanism.

This has led to the actual isolation of the vesicles involved in intra-Golgi transport. As in ER-to-Golgi transport, spike-coated vesicles are involved (Figure 7–8). The coat proteins (called COPs) have been characterized, and one of them is similar to a protein called adaptin that

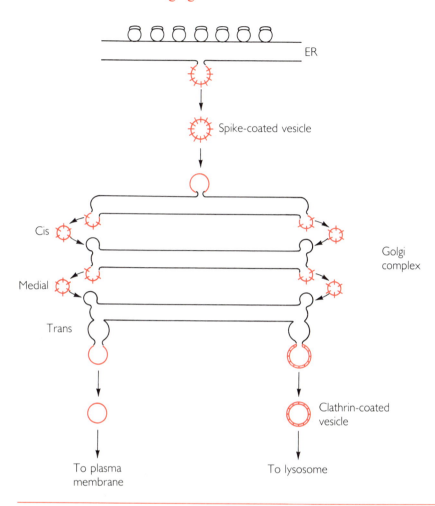

ER

Spike-coated vesicle

Cis

Medial

Trans

Golgi
complex

Clathrin-coated
vesicle

To plasma
membrane

To lysosome

FIGURE 7–8
The role of vesicles in the movement of proteins from the ER to
the Golgi, within the Golgi, and
out of the Golgi. Note the three
types of vesicles: smooth, spike-
coated (COP), and clathrin-
coated.

occurs on clathrin-coated vesicles where it apparently acts to bind the
vesicles to membranes. In the case of this particular COP, it has specificity for Golgi membranes. Indeed, brefeldin A, an antiviral agent isolated from a mold, inhibits transport between the ER and Golgi and
causes the Golgi stacks to disassemble, an indication that COPs are involved in membrane interactions in the Golgi.

Transport from one Golgi cisterna to another in the test tube requires both ATP and soluble factors from the cytoplasm. The cytoplasmic factor appears to be a protein, called NSF (because it is sensitive to
N-ethylmaleimide). NSF is homologous in sequence to a protein in yeast
(sec18), which, when mutated, blocks ER-to-Golgi transport. These proteins may act at some point in the fusion of the vesicle with the target
membrane. Before fusion, the vesicle loses its coat in a process requiring
the hydrolysis of GTP.

Packaging

Following transit through, and modifications by, the Golgi complex, secretory proteins often appear in amorphous condensing vacuoles and then in more concentrated secretory granules (Figure 7–2 and Table 7–5). Electron microscopy indicates that the granule membranes come from the Golgi membranes, but it is also possible that the granule membranes are part of a pool of vesicles that shuttle back and forth from the plasma membrane during an endocytosis-exocytosis cycle. In support of this idea, biochemical analyses of granule membranes (Table 7–2) indicate a lipid composition quite similar to that of the plasma membrane.

Concentration of the vesicle contents occurs in the presence of uncouplers of oxidative phosphorylation and so does not require the continued presence of ATP. This eliminates the possibility of an active transport pump. Instead, concentration may occur because of the formation of molecular aggregates in the condensing vacuoles. Electrostatic interactions between cationic and anionic secretion products (e.g., pancreatic enzymes and sulfated glucosaminoglycans) could form complexes that reduce the effective osmotic concentration in the vacuole. Water would then move out into the cytoplasm, thereby concentrating the vacuole to a granular appearance.

A similar mechanism appears to operate in the secretory granules of the adrenal medulla. These chromaffin granules concentrate not proteins but catecholamines, such as norepinephrine, and ATP. Although the sum of the individual components inside the granule gives an osmotic concentration of 0.8 molal (0.8 moles/kg solvent), the actual osmotic pressure is 0.3 molal. This is less than the surrounding cytoplasm, and thus the granule becomes concentrated as water leaves it.

Nuclear magnetic resonance (NMR) spectroscopy shows that the granule contents are in constant motion in a medium of relatively low viscosity. Because the catecholamine:ATP ratio is about 4:1, the cationic charge of the amines is neutralized by the anionic ATP, and this complex may be the agent that lowers the granule osmotic pressure. An interesting aspect of these granules is that they accumulate their amines by an electrogenic pump driven by a chemiosmotic pH gradient. This pumping mechanism is less likely to occur in the case of secretory proteins because of their much larger size and their concentration being independent of the presence of ATP.

Generally, the contents of secretory granules are a "mixed bag," with each granule having a random assortment of all of the molecules to be secreted. This can be shown by immunocytochemistry, where antibodies to numerous hydrolases stain a single granule. The same vesicle can contain proteins destined for secretion and ones destined to be part of the plasma membrane. In addition, autoradiographic experiments show that newly synthesized proteins and granules become distributed among the old.

Proteins That Are Not Secreted

In addition to proteins destined for secretion, the Golgi complex is responsible for modifications of nonsecreted proteins. These include ones that are bound for the plasma membrane, lysosomes, vacuoles, the ER, and proteins that remain in the Golgi.

A well-documented example of a membrane protein passing through the Golgi is the viral glycoprotein (viral G protein) of vesicular stomatitis virus. This virus infects animal cells and uses the host protein synthesis machinery to synthesize its five viral proteins. The viral G protein is an integral protein that spans the viral membrane. It is processed in a manner quite similar to that of secreted proteins in pancreatic cells.

The viral G protein is co-translationally glycosylated in the RER by the addition of glucosamine and polymannose (Figure 6–5). In the cis Golgi elements, mannose residues are removed, and in the trans Golgi, capping sugars are added (Figure 7–4). Direct evidence for this process comes from a host cell temperature-sensitive mutant that cannot carry out the "capping" glycosylations. At the mutant's restrictive temperature where capping does not occur, the protein in infected cells accumulates in the RER and, somewhat, in the cis Golgi elements. If the temperature is shifted to a permissive temperature where capping does occur, the protein spreads sequentially to the cis and trans elements.

Pulse-chase experiments of vesicular stomatitis virus-infected cells show an early wave of vesicles containing viral G protein with polymannose, the result of glycosylation in the RER. Later, a wave of hexose-capped glycoproteins is observed in spike-coated vesicles, this glycosylation having occurred in the Golgi complex.

Enzymes destined for the **lysosome** are mostly hydrolases (see Chapter 8) and so are readily seen by specific staining. The observation that in liver cells only the most trans ER cisterna and its associated vesicles stain for acid phosphatase activity led A. Novikoff to propose that this represents a specialized membrane system for segregating harmful lysosomal hydrolases from the rest of the cell. The term *GERL* (golgi-endoplasmic reticulum-lysosome) is used to describe this structure.

Secretory proteins are absent from GERL, but they are present near it in granules. This indicates that separation of lysosomal from secretory proteins occurs when the trans cisterna forms secretory granules, on the one hand, and vesicles containing hydrolases, on the other. Indeed, the lysosomal enzymes are packaged into clathrin-coated vesicles, which are not used for typically constitutively secreted proteins.

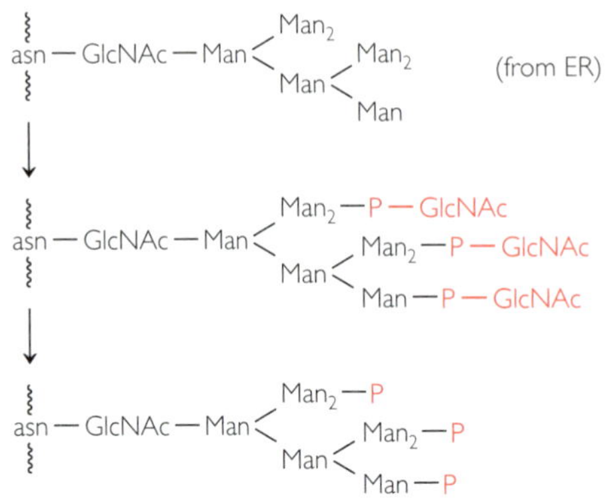

FIGURE 7–9

Golgi modifications of a lysosomal enzyme made in the endoplasmic reticulum. (asn = asparagine; GlcNAc = N-acetylglucosamine; P = phosphate; Man = mannose.)

The Golgi modifies lysosomal enzymes in a manner different from the secretory and membrane proteins (Figure 7–9):

1. The hydrolases are glycosylated in the "normal" (high-mannose) manner on the RER.

2. In the cis Golgi cisternae, a mannosidase removes some mannose residues.

3. In the medial Golgi, two unique reactions occur. First, N-acetylglucosamine phosphotransferase puts N-acetylglucosamine-1-phosphate on the mannose residues.

4. A glucosaminidase removes the N-acetylglucosamine. This leaves mannose-phosphate as the terminal sugar on lysosomal hydrolases.

Sorting

Concomitant with the packaging of lysosomal, membrane, and secretory proteins, there must be a mechanism to direct them to their respective destinations in the cell. The concept of a separate compartment for lysosomal hydrolases in the GERL is a structural solution to this sorting problem. However, although the GERL seems to exist in liver cells, it probably does not in other tissues. Instead, sorting is achieved by signals on the glycoproteins. The sorting marker on lysosomal enzymes has been identified through the study of mutant human cells.

Mucolipidosis II, or I-cell disease, is a recessively inherited human disease that was first described in 1966. Usually, patients are born with low birth weight, and the first six months of life are relatively normal. After this time, and until the 18th month, growth and neurological development slow down dramatically. The children usually die by age six. When fibroblasts of these children are grown in tissue culture, the cells' lysosomes contain little, if any, of most of the lysosomal hydrolases.

Thus, this is classified as a lysosomal storage disease (see Chapter 8). Remarkably, the missing intracellular hydrolases are found in the extracellular medium!

A number of lines of evidence indicate that the reason that I-cell fibroblasts secrete massive amounts of lysosomal enzymes is that they lack the enzyme N-acetylglucosaminylphosphotransferase, which is involved in the production of a recognition marker for these enzymes in protein sorting.

1. Actual enzyme analyses show that fibroblasts from normal people have this transferase enzyme activity, whereas those from the patients are severely deficient.

2. If hydrolases from normal cells are put into the medium surrounding I-cells, they are taken up by the cells and inserted into lysosomes. This shows that the mutant cells are not deficient in recognition of the enzymes.

3. If the normal hydrolases are first incubated in endoglucosaminidase H, which cleaves high-mannose oligosaccharide chains, they are not taken up by mutant or normal cells. This indicates that the sugar chain has the recognition marker.

4. Uptake of untreated, normal enzymes is competitively inhibited by phosphomannose, and it is a much stronger inhibitor than mannose.

Thus the recognition factor on the extracellular lysosomal hydrolase for uptake by fibroblasts is phosphomannose (Figure 7–9), a sugar not usually found on nonlysosomal enzymes. However, in normal cells the bulk of lysosomal hydrolases are not first secreted and then taken up by the cells for delivery to the lysosome. Instead, sorting occurs in the trans Golgi elements. If secretion-reuptake were the normal process for lysosomal hydrolase sorting, the presence of phosphomannose in the medium would be expected to severely reduce the intracellular content of these enzymes in normal cells. But this does not happen. The two enzymes involved in phosphomannose production (Figure 7–9) are located in the cis and medial elements of the Golgi complex (Table 7–6), and the modified enzyme then moves to the trans region (or GERL), where sorting occurs.

The actual sorting mechanism is a **receptor** for the phosphomannose-coated protein. This receptor is a membrane protein located in the trans Golgi elements and binds the Golgi-modified proteins via their mannose phosphate groups, probably by attraction to basic amino acids. Binding is best at slightly acidic pH; if the pH falls below 5.5, the ligand and receptor dissociate.

The membrane with its receptor attached to the lysosomal enzyme then buds off the Golgi as a clathrin-coated vesicle. A proton pump, which will be essential to the functioning of the lysosome (see Chapter 8), gradually acidifies the vesicle, and this causes the enzyme to dissociate from the receptor. The enzyme goes on to be part of the lysosome, while the receptor is recycled back to the trans Golgi, where it can pick up another lysosome-targeted enzyme. Table 7–8 summarizes the roles of different organelles in the synthesis and targeting of lysosomal hydrolases by this pathway.

The phosphomannose recognition system apparently is not operative in all tissues. While connective tissue cells, certain kidney cells, and Schwann cells contain the enzymes for production of the marker, other cells do not. Despite the absence of phosphomannose, lysosomal hydrolases are properly sorted by the Golgi. For instance, the hepatocytes of I-cell patients have a full complement of hydrolases; thus some other recognition scheme must be used to direct the hydrolases in this case. In addition, several hydrolases (e.g., acid phosphatase and alpha-glucosidase) are properly directed by the Golgi even in I-cell fibroblasts. Their sorting also must not involve phosphomannose. A hint as to the signals comes from studies on yeast, where a 4 amino acid region is used to direct hydrolases to the lysosomal vacuole.

Sorting of proteins to some other hydrolytic cell compartments also occurs in the Golgi, but the details are not as well understood as the lysosome-mannose phosphate pathway. In the case of the **peroxisome** (see Chapter 8), the signal for localization is about 15 to 25 amino acids at the carboxy-terminus of the protein, with a required tripeptide of ser-lys-leu. Peroxisomal proteins without this signal are transported from the Golgi to the cytosol.

TABLE 7–8   **Roles of Different Organelles in the Synthesis and Targeting of Lysosomal Hydrolases**

Organelle	Role
Ribosomes	Synthesis of protein
ER	Addition of N-linked oligosaccharides
Cis and medial Golgi	Addition of P-GlcNAc
	Removal of GlcNAc
Trans-Golgi	Binding of glycoprotein to receptor
Vesicle	Acidification
	Debinding of receptor from glycoprotein
	Receptor recycled to Golgi
Lysosome	Glycoprotein targeted

In plant cells, the **vacuole** is the organelle analogous to the animal cell lysosome (see Chapter 8). Proteins are targeted for this compartment in the Golgi, via a region of the protein separate from the hydrophobic signal for ER insertion. A similar mechanism appears to exist for sending proteins to the plant **protein bodies,** which are major storage organelles (see Chapter 8).

Another sorting function of the Golgi is to target the proteins that are to be resident in the **ER.** As described in Chapter 6, these proteins, which usually have the KDEL terminal amino acid signal, actually go to the cis elements of the Golgi and are then recycled back to the ER in spike-coated vesicles.

Finally, there must be a signal for the retention of **Golgi-resident** proteins, such as the enzymes catalyzing protein modifications. It appears that some part of the protein causes it to remain in the Golgi, but there may not be a specific sequence of amino acids required. For example, the protein El coded by infectious bronchitis virus is localized in the cis Golgi and has three membrane-spanning sequences. The first of these is apparently essential for Golgi retention: If it is missing, the protein is secreted; if it is fused to another protein normally destined for the plasma membrane (e.g. the G protein of vesicular stomatitis virus), that fusion protein remains in the Golgi. In most other cases examined, there is a similar need for a transmembrane domain to retain a protein in the Golgi. It is unclear what the information content of these domains is. Nevertheless, it is not surprising that Golgi-resident proteins are all membrane proteins and there appear to be no soluble proteins that reside in the Golgi.

It seems that there are no signals for the targeting of proteins from the Golgi to the plasma membrane or the extracellular medium. Instead, **secretion is the default pathway,** which occurs when other signals are not detected at the Golgi. Several experiments point to this conclusion:

1. The antimalarial drug chloroquine prevents vesicle acidification. In treated cells, lysosomal enzymes are not targeted to that organelle but are secreted, presumably because they cannot bind to the phosphomannose receptor.

2. In yeast, an N-terminal region of a protein (e.g., carboxypeptidase) targets it to the vacuole, and deletion of this signal and insertion of the altered gene into yeast cells leads to the secretion of the protein.

3. Construction of a chimeric gene for a plant protein containing the ER insertion signal of a protein normally inserted to the lysosome-like protein body (see Chapter 8) and fused to a cytoplasmic protein leads to the insertion of the chimera into the

ER and transport to the Golgi, from which it is secreted. This indicates that the chimera, lacking a targeting signal but arriving at the Golgi, is directed via the default pathway to be secreted.

The vesicles that carry secreted proteins to the cell surface are not coated (Figure 7–8). This is in sharp contrast to the clathrin-coated vesicles that carry lysosomal enzymes. Apparently, there are proteins (in addition to the clathrin) in the latter that have the mechanism for carrying specific cargo. In the secretory vesicles, these recognition proteins are absent, so the vesicle contains a mixture of proteins. But there is an additional complication: In some cells (see below), secretory proteins are stored in vesicles prior to secretion. This regulated pathway must somehow be separated from the constitutive pathway in which proteins are secreted immediately after passing through the Golgi. There may be receptors in the trans Golgi to achieve this separation.

Directed secretion of proteins occurs in polarized epithelial cells, such as those lining the intestine (Figure 7–10). Here, two types of asymmetry are seen: First, digestive enzymes (e.g., acid phosphatase) are secreted from the apical surface only. Second, there is a distinct asymmetry of membrane proteins (see Chapter 1). A model for how these events occur is as follows:

1. Proteins destined for secretion are packaged by the Golgi into vesicles and sent to the apical plasma membrane, where there is a receptor for them. This receptor is absent from the basal and lateral plasma membranes. These proteins usually contain a glycosphingolipid anchor for membrane insertion.

FIGURE 7–10

A model for the sorting of proteins in polarized epithelial cells.

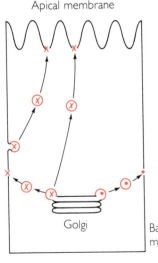

Apical membrane

A protein destined for the apical surface (or secretion) is either directed by a vesicle-receptor system or first inserted in the basolateral membrane, then endocytosed, and finally reinserted in the apical membrane.

A protein on the basolateral surface is inserted by a default pathway.

Golgi

Basolateral membrane

2. Proteins destined for the lateral and basal membranes are packaged by the Golgi into vesicles and sent by a default pathway to these membranes. A 17 amino-acid segment on the cytoplasmic domain is required for the protein to be directed basolaterally.

3. Some secretory proteins may be in the vesicles that fuse with the lateral membranes. In this case, the membrane containing the proteins is endocytosed, and the resulting vesicle goes to its proper destination, the apical plasma membrane. This phenomenon is called transcytosis.

Clearly, these mechanisms require several recognition events. The nature of these is under investigation, and the model appears to be confirmed.

Function: Polysaccharide Synthesis

Plant cells are enclosed within an often-extensive cell wall (see Chapter 13). This structure is composed of polysaccharides such as cellulose and hemicelluloses, a structural glycoprotein, and various enzymes. Considerable morphological and biochemical evidence implicates the Golgi complex in the formation of the wall. Both the glycoprotein and a number of the wall-bound enzymes are packaged by and secreted from the Golgi in a manner similar to the process in animal cells. However, some wall-degrading enzymes apparently bypass this organelle and are secreted directly from the ER.

Electron micrographs of plant cells active in cell wall synthesis show abundant Golgi near the plasma membrane (Figure 7–11). Large secretory vesicles, budding from the cisternae, are often observed to contain fibrillar material similar morphologically to components in the cell wall. Rapidly growing root tips or pollen tubes, which must assemble both cell membrane and wall, are especially rich in Golgi complexes. If the Golgi vesicles are used for new plasma membrane, the rate of growth of a lily pollen tube is such that over 1000 vesicles per minute must be exported from the Golgi. Another place where Golgi complexes are prominent is in the formation of the new cell wall and membrane at cytokinesis when cells divide.

Autoradiographic experiments, similar to those done for secretory proteins, have been used to follow polysaccharide synthesis. Pulse-chase studies using radioactive glucose show that labeled polysaccharides first appear in the Golgi complex and then sequentially move to secretory vesicles, the plasma membrane, and the cell wall (Table 7–9). Extraction of the labeled material indicates that most of the radioactive glucose has been converted to galactose and is present in hemicelluloses and pectins.

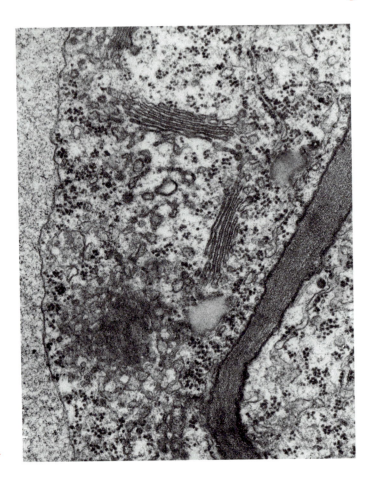

FIGURE 7–11

Part of an outer root cap cell of maize, showing a dictyosome and secretory vesicles. (×38,000. Photograph courtesy of Dr. E. H. Newcomb.)

Plant Golgi complexes contain many glycosyl transferase complexes in medial cisternae, which catalyze the transfer of sugars from nucleotide donors to oligosaccharide acceptors. There is evidence that some cell wall polysaccharides are built up in the Golgi while part of glycoproteins. If this is the case, the organelle must contain an enzyme that cleaves the protein-sugar linkage, because secretory vesicles contain free cell wall polysaccharides. The biosynthesis of cellulose apparently does not occur in the Golgi complex but instead occurs at the plasma membrane (see Chapter 13).

With the many roles of plant Golgi in synthesis and secretion, the question arises as to whether there is some cisternal specificity. Staining with antibodies to cell wall components, followed by electron microscopy, indicates that such specificity can occur. In root tip cortical cells, antibodies to rhamnogalacturonan stain only cis and medial cisternae, and this polysaccharide leaves from the medial compartment. In contrast, antibodies to xyloglucan stain only the trans cisternae, from which this carbohydrate exits the organelle. The cell wall glycoprotein, extensin, goes through all cisternae before exiting the trans region. (See Chap-

TABLE 7–9 **Autoradiography of Wheat Root Cap Cells Incubated in 3H-Glucose**

Cell Region	Degree of Labeling				
	Pulse 5 min	Pulse 10 min	10 Pulse + Chase, 10′	10 Pulse + Chase, 20′	10 Pulse + Chase, 30′
Golgi cisternae	High	High	Low	Low	Low
Golgi vesicles	Low	High	High	Low	Low
Plasma membrane	Low	Low	High	High	Low
Cell wall	Low	Low	Medium	High	High
Rest of cell*	Low	Low	Low	Low	Low

*Except chloroplasts

ADAPTED FROM: Northcote, D., and Pickett-Heaps, J. A function of the Golgi apparatus in polysaccharide synthesis and transport in root cap cells of wheat. *Biochem J* 98 (1966):159–167.

ter 13 for a description of these cell wall components.) The mechanism for this spatial organization within the Golgi is not known.

Exocytosis and Membrane Fusion

Exocytosis

Secretory products packaged by the Golgi complex can be released to the extracellular medium immediately via exocytosis of secretory vesicles. This occurs with plant cell wall components, where the transit time from synthesis to secretion is approximately 30 minutes. But in some animal cells, secretory proteins are stored, and instead, exocytosis is initiated by an appropriate hormonal, neural, or ionic stimulus. This phenomenon is termed regulated secretion (or stimulus-secretion coupling), in contrast to the continuous constitutive pathway.

A well-studied regulated secretory system is the release of vesicles of neurotransmitters from synapses (see Chapter 2). When the nerve impulse arrives at the synapse, it causes the opening of a voltage-gated Ca^{++} channel. Because this ion is in much higher concentration outside of the cell, it diffuses in, quickly raising the intracellular Ca^{++} from 10^{-8}M to 10^{-6}M. This then induces exocytosis of the neurotransmitter. Movement of the secretory vesicle to the plasma membrane is apparently mediated by the cytoskeleton, with a network of microtubules along which vesicles can be seen to move under video-enhanced microscopy.

In a number of physiological situations, development of the microtubule network parallels the development of the ability to move vesicles from the Golgi to the plasma membrane. Notable examples are the lymphocyte classes that kill other cells. When they bind to a target cell, these lymphocytes undergo a profound cytological reorganization. The Golgi reorients to face the target, as does the microtubular network, resulting in a directed exocytosis of a protein from the killer cell to the target cell, where it forms pores in the target cell plasma membrane. This lyses the target cell. A similar reorientation occurs when lymphocytes termed T helper cells bind to and stimulate T cells in the immune system.

A microtubule-associated ATPase, kinesin, has been proposed to be the "motor" that propels the vesicles to their destination (see Chapter 9). Agents that disrupt microtubules prevent secretory vesicles from leaving the Golgi region.

As is the case with ER-to-Golgi movement as well as transport within the Golgi, "G" proteins, which bind GTP to cause a cellular function (see Chapter 2), have been implicated in vesicle transport from the Golgi. A temperature-sensitive mutation in yeast causes the accumulation of secretory vesicles derived from the Golgi but unable to go to the plasma membrane and perform exocytosis. The defective protein in this mutant has been isolated and is a GTP-binding and GTP-hydrolyzing protein. The relationship between this protein, which is essential for secretion, and GTP-proteins needed for microtubule elongation has not been elucidated.

Membrane Fusion

The exocytotic event involves membrane fusion of the secretory vesicle with the plasma membrane, implying some reorganization of the lipid bilayers (Figure 7–12). There have been many studies of this phenomenon using phospholipid vesicles. These vesicles are surrounded by a layer of water (**hydrated**) because of its attraction with the polar "heads" on the lipids, and this water inhibits fusion. The phospholipid content of the outer lipid leaflet can influence hydration: Phosphatidylcholine is heavily hydrated, whereas phosphatidylethanolamine is less hydrated. Thus a change to more of the latter phospholipid, perhaps by transversion (see Chapter 1), could enhance fusion. Removal of the water of hydration by a solvent such as polyethylene glycol also promotes membrane fusion, and this substance is commonly used to fuse animal cells in culture.

Ca^{++} **ions** are also necessary for membrane fusion. This cation neutralizes the negative charges on the membrane surface due to the phospholipids, thereby lessening the membrane's affinity for water. The localized region of nonpolar surface is then primed for fusion with an adjacent nonpolar region on another membrane. Apparently, Ca^{++} forms cross-bridges between the two membranes.

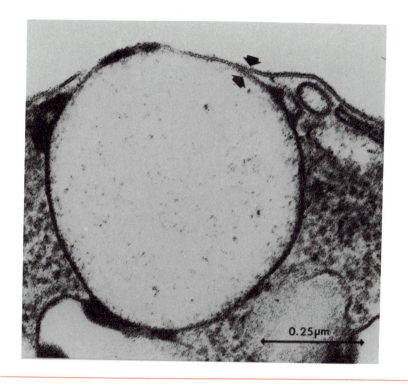

FIGURE 7–12
Fusion of a granule membrane with the plasma membrane. Note the single bilayer where the two membranes have fused and the more dense fusion zone where the two bilayers are apparently still present. (×120,000. Photograph courtesy of Dr. P. Pinto da Silva.)

A third factor required for membrane fusion is the absence (or presence) of **proteins.** Treatment of membrane vesicles with proteases promotes polyethylene-glycol mediated fusion, and electron microscopy of membranes so treated often shows an absence of membrane particles in the fusing region. This zone of particle absence is also seen when vesicles fuse to plasma membranes and when the mammalian sperm fuses with the egg. It implies that it is the lipids that do the fusing of two membranes and that proteins merely "get in the way."

On the other hand, very fast freezing of fusing membranes, followed by freeze-fracturing (see Chapter 4), has revealed proteins in the fusing region. In fact, a specific protein, termed synexin, has been implicated as being required for the fusion of chromaffin granules in the adrenal medulla. In neurons, synaptophysin, a Ca^{++}-binding protein is needed for fusion of the synaptic vesicle with the terminal region of the axonal plasma membrane.

As is the case with transport within the Golgi, analysis in yeast has been especially useful in gaining insight into the role of proteins in membrane fusion. Both biochemical analyses (e.g., inhibition of fusion by the sulfhydryl group modifying agent N-ethyl maleimide) and mutations (e.g., the strain sec17 accumulates unfused transport vesicles) have led to the description of four protein components necessary for a membrane to fuse with another:

1. An intrinsic protein receptor that resides permanently in the membrane;

2. A soluble protein that binds to the receptor;

3. An N-ethyl maleimide sensitive protein that binds to the soluble protein reversibly, hydrolyzing ATP in the process;

4. A GTP binding protein.

Unraveling the complexities of the interactions of these components is the subject of ongoing research.

A biological system that illustrates the three requirements for membrane fusion is the degranulation of mast cells in humans. These cells are closely associated with blood capillaries. On their surfaces, mast cells have receptors for immunoglobulin E (IgE), an antibody molecule. When a foreign substance (e.g., a protein on the surface of a pollen grain) enters the organism, it binds to IgE on the mast cell surface. This triggers granular vesicles inside of the cells to fuse with the plasma membrane and release their contents, histamine and serotonin, to the bloodstream. This event, termed degranulation, has great effects on physiology, as histamine promotes a series of reactions called the allergic response (e.g., constriction of the smooth muscles lining the respiratory tract). Hay-fever sufferers are quite familiar with this response.

A key event in degranulation is the fusion of the vesicle and plasma membranes, and this is under intensive study in the hope of designing specific antiallergy drugs. Just after the antigen binds to IgE on the mast cell surface, plasma membrane permeability to Ca^{++} increases, allowing this ion to enter the cells to promote fusion. Intermembrane particles then appear to leave a small region on the adjacent vesicle and cell membranes, and a small channel, 10 nm in diameter, forms. This channel later enlarges as the two membranes fuse over most of their areas.

Two models have been proposed to account for membrane behavior during exocytotic fusion (Figure 7–13):

1. In the first model, the region of contact between vesicle and cell membrane becomes structurally continuous. A bilayer composed of the outer lipid layer of the cell membrane and inner layer of the vesicle is formed by dissolution of the other two lipid layers. This bilayer then disrupts to form a diaphragm, and the granule's contents are released.

2. In the second model, a diaphragm does not form. Rather, apposition of the two membranes leads to the formation of lipid micelles. These are unstable and rearrange to produce a fusion of the cell membrane and vesicle.

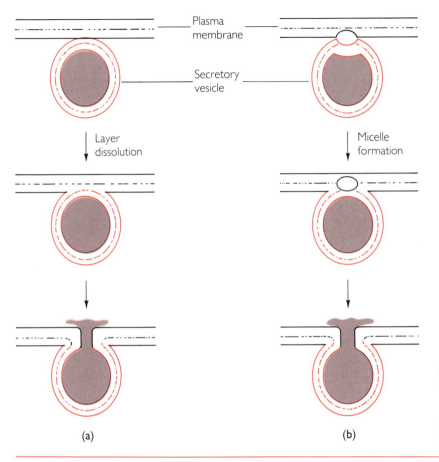

Plasma membrane

Secretory vesicle

Layer dissolution

Micelle formation

(a)   (b)

FIGURE 7–13

Two models for exocytosis in secretory cells: **(a)** structural continuity of apposed membranes; **(b)** micelle formation.

Exocytosis results in addition of material to the plasma membrane, which in growing plant cells is an important part of the process of cell expansion. However, studies of cultured cells have shown that the cell wall and turgor act to restrain membrane expansion in nonexpanding plant cells. This makes the situation analogous to that in animal cells, where cell size remains constant in the face of the addition of membrane via fusion of vesicles. This implies that membrane must be removed at the same rate as it is added.

Information on the fate of the secretory vesicle membrane comes from labeling of the membrane with electron-dense tracers. Both uncharged dextrans and ferritins of various charges can be used to follow the vesicle membrane, and the results of such experiments on a number of types of secretory cells indicate considerable membrane reuse. Following exocytosis, the former vesicle membrane component is endocytosed, and most of it fuses with the dilated edges of the Golgi cisternae, which ultimately will form condensing and secretory granules.

FIGURE 7–14

Exocytosis, endocytosis, and then exocytosis in intestinal epithelial cells. 1. Glycoprotein assembly in the ER. 2. Modification in the Golgi and insertion into the plasma membrane. 3. Diffusion to the basal membrane facing the blood. 4. Binding of IgA to the membrane protein. 5. Endocytosis of the IgA-protein complex. 6. Intracellular transport to the apical surface. 7. Exocytosis of the IgA-protein complex. (From: Targan, S. R.: Immunologic mechanisms in intestinal diseases. *Ann Int Med* 106 (1987):857.)

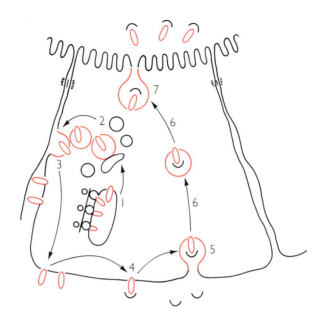

A striking example of the relationship of endocytosis and exocytosis is provided by human epithelial cells (Figure 7–14). Such cells as intestinal mucosa secrete to their extracellular matrix a complex mixture, containing the immunoglobulin IgA, which is associated with a glycoprotein termed SC (secretory component). The IgA acts as a barrier to such external agents as bacteria and viruses, preventing their entry into the tissue. This first line of defense stops these substances from eliciting an immune response, thereby "saving" the immune system for other roles.

SC is synthesized and inserted into the ER, modified by the Golgi, and inserted in a directed fashion into the plasma membrane at the basal surfaces of the epithelial cell. It then acts as a receptor for IgA from the extracellular environment, and the IgA-SC complex is internalized via coated pits. The resulting intracellular vesicle is then transported in a directed fashion to the apical surface, where it fuses with the membrane in an exocytic event, resulting in the secretion of the complex. Because both the receptor and the ligand are secreted, this system has been termed a "sacrificial" one, to differentiate it from the more typical situation where receptors are recycled.

Further Reading

Almers, W. Exocytosis. *Ann Rev Physiol* 52 (1990):607–624.

Balch, W. Biochemistry of interorganelle transport. *J Biol Chem* 264 (1989):16965–16968.

Beaudoin, A. R., and Grondin, G. Secretory pathways in animal cells with emphasis on pancreatic acinal cells. *J Elec Micros Tech* 17 (1991):51–69.

Bermann, J., Tokuyasu, J., and Singer, S. J. Passage of an integral protein, the VSV virus glycoprotein, through the Golgi apparatus enroute to the plasma membrane. *Proceed National Acad Sci* 78 (1981):1746–1750.

Brandli, M. Mammalian glycosylation mutants as tools for the analysis and reconstitution of protein transport. *Biochem J* 276 (1991):1–12.

Breitfield, P., Casanova, J., Simster, N., Ross, S., McKinnon, W., and Mostov, K. Sorting signals. *Curr Op Cell Biol* 1 (1989):617–623.

Brumley, L., and Marchase, R. Receptor synthesis and routing to the plasma membrane. *Am J Med Sci* 302 (1991):238–243.

Burger, K., and Verkleij, A. Membrane fusion. *Experientia* 46 (1990):631–643.

Burgess, T. L., and Kelly, R. B. Constitutive and regulated secretion of proteins. *Ann Rev Cell Biol* 3 (1987):243–293.

Burgoyne, R. Secretory vesicle associated proteins and their role in exocytosis. *Ann Rev Physiol* 52 (1990):647–659.

Burgoyne, R. Control of exocytosis in adrenal chromaffin cells. *Biochim Biophys Acta* 1071 (1991):174–192.

Castle, J. D. Sorting and secretory pathways in exocrine cells. *Am J Respir Cell Mol Biol* 2 (1990):119–126.

Chrispeels, M. J., and Tague, B. W. Protein sorting in the secretory system of plant cells. *Int Rev Cytol* 125 (1991):1–43.

Domozych, D. Golgi apparatus and membrane trafficking in green algae. *Int Rev Cytol* 131 (1991):213–253.

Dunphy, W. G., and Rothman, J. E. Compartmental organization of the Golgi stack. *Cell* 42 (1984):13–21.

Gruenberg, J., and Howell, J. E. Membrane traffic in endocytosis: Insights from cell-free assays. *Ann Rev Cell Biol* 5 (1989):453–481.

Jamieson, J., and Palade, G. Production of secretory proteins in animal cells. In Brinkley, B., and Porter, K. (Eds.), *International cell biology* (pp. 308–317). New York: Rockefeller University Press, 1977.

Jones, R., and Jacobsen, J. Regulation of synthesis and transport of secreted proteins in cereal aleurone. *Int Rev Cytol* 126 (1991):49–87.

Kornfeld, S. Trafficking of lysosomal enzymes. *FASEB J* 1 (1987):462–468.

Kreis, T. Role of microtubules in the organization of the Golgi apparatus. *Cell Motil Cytoskel* 15 (1990):67–70.

Kuhn, L., and Kraehenbuhl, J. The sacrificial receptor translocation of polymeric IgA across epithelia. *Trends Biochem Sci* 7 (1982):299–302.

Lasic, D. The mechanism of vesicle formation. *Biochem J* 256 (1988):1–11.

Mellman, I., and Simons, K. The Golgi complex: In vitro veritas? *Cell* 68 (1992):829–840.

Mollenhauer, H., and Morre, D. J. The Golgi apparatus. In Tolbert, N. (Ed.), *The plant cell* (pp. 438–488). New York: Academic Press, 1980.

Morre, D. J. The Golgi apparatus. *Int Rev Cytol Suppl* 17 (1987):211–258.

Niemann, H., Mayer, T., and Tamura, T. Signals for membrane-associated transport in eukaryotic cells. *Subcell Biochem* 15 (1990):307–365.

Northcote, D. Involvement of the Golgi apparatus in the biosynthesis and secretion of glycoproteins and polysaccharides. In Mason, L. (Ed.), *Biomembranes,* Vol. 10 (pp. 51–75). New York: Plenum, 1979.

Pagano, R. The Golgi apparatus: Insights from lipid biochemistry. *Biochem Soc Trans* 18 (1990):361–366.

Paulson, J., and Colley, K. Glycosyltransferases. *J Biol Chem* 264 (1989):17615–17618.

Pavelka, M. Functional morphology of the Golgi apparatus. *Adv Anat Embryol Cell Biol* 106 (1987):1–94.

Pfeffer, S. Mannose-6-phosphate receptors and their role in targeting proteins to lysosomes. *J Membr Biol* 103 (1989):7–16.

Pfeffer, S. GTP binding proteins in intracellular transport. *Trends Cell Biol* 2 (1992): 41–45.

Pinto da Silva, P. Geometric topology of membrane fusion. In Ohki, S., Doyle, D., Flanagan, T., Hui, S., and Mayhew, E. (Eds.), *Molecular mechanisms of membrane fusion* (pp. 521–529). New York: Plenum, 1988.

Roth, J. Localization of glycosylation sites in the Golgi apparatus using immunolabeling and cytochemistry. *J Electr Microscop Tech* 17 (1991):121–131.

Rothman, J. E. Molecular dissection of vesicular transport. *Chemtracts* 1 (1990):89–102.

Rothman, J. E., Miller, R. L., and Urbani, L. Intercompartmental transport in the Golgi is a dissociative process: Facile transport of a membrane protein between two Golgi populations. *J Cell Biol* 99 (1984):260–271.

Schekman, R. Protein localization in yeast. *Ann Rev Cell Biol* 1 (1986):115–143.

Schwartz, A. Cell biology of intracellular protein trafficking. *Ann Rev Immunol* 8 (1990):195–229.

Simons, K., and Wandinger-Ness, A. Polarized sorting in epithelia. *Cell* 62 (1990):207–210.

Staehelin, L. A., and Chapman, R. L. Secretion and membrane recycling in plant cells: Novel intermediary structures visualized in ultrarapidly frozen sycamore and carrot suspension-culture cells. *Planta* 171 (1987):43–57.

Steer, M. W. The role of calcium in exocytosis and endocytosis in plant cells. *Physiol Plant* 72 (1988):213–220.

Von Figura, K., and Hasilik, A. Lysosomal enzymes and their receptors. *Ann Rev Biochem* 55 (1986):167–193.

Whaley, W. G. *The Golgi apparatus.* New York: Springer-Verlag, 1975.

Wilschut, J. Intracellular membrane fusion. *Curr Op Cell Biol* 1 (1989):639–647.

Wilson, D., Whitehead, S., Orci, L., and Rothman, J. E. Intracellular membrane fusion. *Trends Biochem Sci* 16 (1991):334–337.

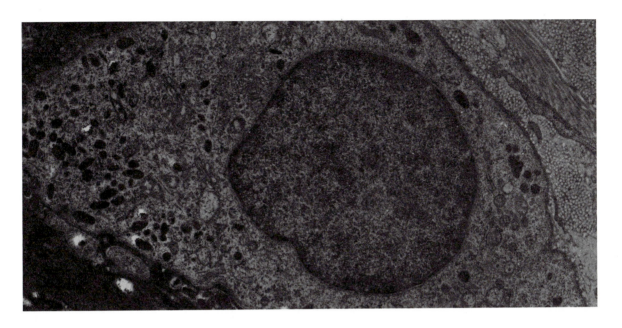

Lysosomes, Vacuoles, and Microbodies

A Hydrolytic Compartment

In 1949, C. de Duve began a series of studies designed to determine how insulin affects the liver. Using the recently developed techniques of gentle tissue homogenization and cell fractionation, he localized glucose-6-phosphatase, an important enzyme in glycogen mobilization, in the microsome fraction of the cells (see Chapter 7). As a control, he also assayed acid phosphatase, and this led to two puzzling observations:

1. If the tissue was homogenized gently, the activity of this enzyme was much lower than if more violent means of breaking open the cells were used.

2. If he stored the gently homogenized extract in the refrigerator for several days and then assayed for acid phosphatase, its activity was much higher than it was before storage. In fact, it was just as high as the activity he found when the tissues were disrupted in a blender.

These observations were explained by the presence of acid phosphatase in a membrane-bound organelle. The delicate membrane that separated the substrate from the enzyme could be disrupted by the blender or could break down on refrigeration. Subcellular localization studies initially placed acid phosphatase in the microsomal fraction and then in the mitochondria. But more refined fractionation showed that the enzyme was in a slowly sedimenting "mitochondrial" fraction, intermediate in density between organelles containing cytochrome oxidase (mitochondria) and glucose-6-phosphatase (microsomes). This fraction also contained other hydrolases such as nucleases and proteases and was called the lysosome ("lytic body").

The lysosome is part of the cell's endomembrane system, which also contains the endoplasmic reticulum (ER), Golgi, and associated coated and uncoated vesicles. But the lysosome is more than a derivative of these organelles; it has its own distinguishing feature—it is the ultimate destination of soluble proteins taken in by the cell via endocytosis. Within the lytic compartment, enzymes catalyze the hydrolyses of the complex endocytosed proteins to their building blocks, which can then be used by the cell for energy or anabolism. In addition, the organelle has a similar role in the recycling of cellular components. Other specialized hydrolytic functions are carried out in microbodies (peroxisomes and glyoxysomes).

Lysosomes

Structure

Microscopically, lysosomes are defined in terms of their acid phosphatase content. Although this enzyme does occur elsewhere in the cell, it is most concentrated in lysosomes. The Gomori reaction visualizes acid phosphatase cytochemically (Figure 8–1). Phosphate liberated from the hydrolysis of β-glycerophosphate is precipitated as lead phosphate, which can be converted to lead sulfide for visualization under the light or electron microscopes. Other substrates can be used to assay this, as well as other hydrolases.

Lysosomes are relatively spherical in shape (Figure 8–2) and range in diameter from 0.2 to 0.8 μm, with the liver organelles averaging 0.4 μm. They are enclosed within a single membrane, 10 nm in thickness, with a "halo" of electron-lucent material often just below the membrane. The internal material is quite heterogeneous, ranging from a dense matrix to granular to flaky, and these structural variations correlate with the functioning of the organelle. Primary lysosomes are ho-

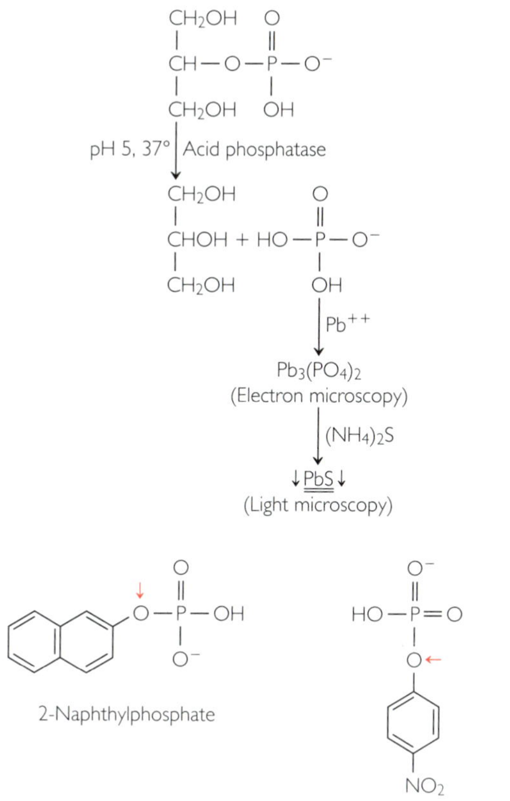

FIGURE 8–1

How acid phosphatase activity is visualized so that lysosomes can be stained for microscopy. Two other substrates for the enzyme are shown at bottom.

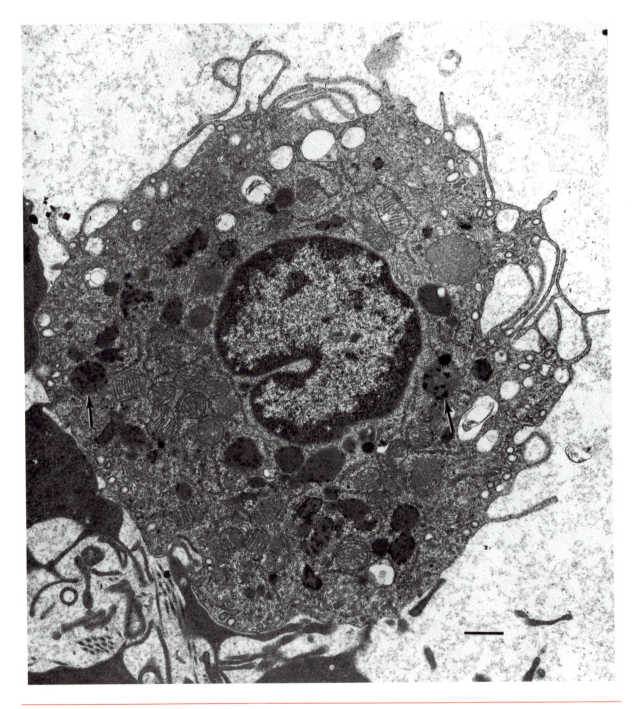

FIGURE 8–2   Mouse macrophage with numerous (arrow) darkly stained lysosomes. These
cells actively endocytose and digest macromolecules and cells in the bloodstream. (From: J. Bozzola and L. Russell: *Electron Microscopy: Principles and
Techniques for Biologists*. Jones and Bartlett, 1992, p. 257.)

mogeneous and dense, whereas secondary lysosomes and residual bodies (see below) are more heterogeneous.

Almost all phyla of multicellular animals have lysosomes. But there is some doubt about whether they exist as such in plants, where their lytic function is served by vacuoles. Lysosomes are most numerous in cells that take up macromolecules or larger substances from their environment. For example, the intestinal epithelial cells and phagocytic reticuloendothelial cells can have up to several hundred lysosomes. On the other hand, cells without an absorptive role, such as secretory cells of the pancreas, contain few lysosomes. Often, lysosomes are preferentially located in a certain part of the cell. For example, in the proximal kidney tubule epithelium they are located near the tubular lumen, and in the hepatocyte, they are located near the bile canaliculus.

Chemical Composition

Lysosomes are distinctive in their composition of **hydrolytic enzymes.** Over 60 of these enzymes have been associated with the organelle (Table 8–1). In general, they catalyze the reaction:

$$A\text{-}B + H_2O \rightarrow AH + BOH$$

Obviously, the enclosure of these enzymes within an organelle compartment is essential to the survival of the cell, lest cellular components be

TABLE 8–I   **Lysosomal Enzymes**

Type of Bond Hydrolyzed	Number of Enzymes	Examples
Acid anhydride	3	Nucleoside triphosphatase, adenylsulfatase
Amide	2	Ceramidase, aspartylglucosaminidase
Carboxylic ester	5	Arylesterase, phospholipases, cholesterol esterase
Glycoside	24	Lysozyme, α-glucosidase, gangliosidase, glucocerebrosidase, iduronidase, mannosidase
Nitrogen sulfur	I	Heparin sulfamidase
Peptide	17	Cathepsins, carboxypeptidases, elastase, collagenase
Phosphoric diester	5	DNase II, sphingomyelinase
Phosphoric monoester	3	Acid phosphatase, phosphoprotein phosphatase
Sulfur ester	6	Arylsulfatases, cerebroside sulfatase

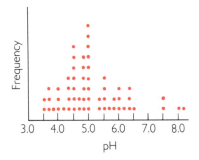

FIGURE 8–3

pH optima of lysosomal enzymes. Each point represents a single hydrolase. Note the clustering at pH 4–6.

inappropriately broken down. Indeed, lysosomal leakage has been implicated in cancer, aging, and a host of degenerative diseases.

The presence of hydrolases inside the lysosome poses two problems for the organelle:

1. Most lysosomal enzymes have pH optima in the acidic range (Figure 8–3). But the surrounding cytoplasm has a pH of 6.5 to 7.5. This means that there must be some mechanism to lower intralysosomal pH.

2. Hydrolyses inside the lysosome result in osmotic alterations. If a protein containing 100 amino acids is hydrolyzed within the organelle, the osmotic concentration of the organelle would rise, and in theory, the lysosome would burst because of the osmotic entry of water. There must be a mechanism to allow the low molecular weight compounds that are hydrolytic products to pass out into the cytoplasm.

The pH inside the lysosome can be measured in two ways. First, cells can be induced to endocytose and put into lysosomes particle-bound indicator dyes that change color in solutions of different pH. The second method relies on the fact that the distribution of weak acids or bases between isolated lysosomes and the surrounding medium is dependent on pH. These small molecules can be made radioactive and the distribution of radioactivity can be measured, or a marker attached to the weak acid or base can be a target for an antibody-based stain.

Both of these methods reveal an acidic intralysosomal pH of about 4.5 to 6.0. Although this is above the pH optimum of some lysosomal enzymes, most have a rather broad range of highest activity, ranging up to one pH unit above or below their optimum. These methods also show that the lysosome is not the only acidic compartment of the cell. Vesicles associated with the trans Golgi region, endosomes associated with endocytosis, and clathrin-coated vesicles also have a lower pH than the surrounding cytoplasm.

The maintenance of the cytoplasm-lysosome pH difference is a property of the lysosomal membrane. One way to achieve this difference would be through a membrane-bound, ATPase-driven **proton pump.** This pump would have to be electrically neutral, with a positive charge being released (or negative one pumped in) for each proton entering the organelle. A number of studies show that ATP stimulates intralysosomal degradation of macromolecules in vitro, and the stimulation is inhibited by proton ionophores, suggesting that the ATP effect is via a proton gradient. Indeed, when ATP is added to lysosomes in the test tube, the intralysosomal pH quickly drops.

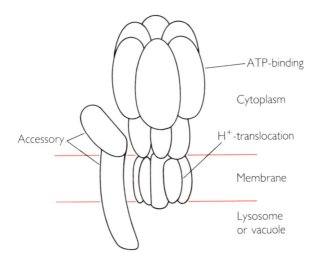

ATP-binding

Cytoplasm

Accessory

H$^+$-translocation

Membrane

Lysosome
or vacuole

F I G U R E   8–4
A model for the lysosomal-vacuo-
lar proton pump. The ATP-binding "lollipop" protrudes into the
cytoplasm. The H$^+$ channel is
embedded in the lipid bilayer.

As predicted, a proton-pumping ATPase activity has been isolated
from the membranes of both lysosomes and plant vacuoles (see below).
This protein pumps two H$^+$ into the organelle per ATP hydrolyzed, and,
as predicted, a Cl$^-$ transporter acts along with the pump to import
anions and keep the situation electrogenically neutral.

Remarkably, the lysosomal proton pump is widespread in the cell,
occurring also in other organelles that become acidified—clathrin-
coated vesicles, endosomes, and chromaffin granules. In addition to its
lysosomal role in activating hydrolases, the pump in the vesicles is involved in the dissociation of receptor-ligand complexes.

This proton pump is quite different from the F$_1$F$_0$-ATPase of the
mitochondria and plastids (see Chapters 3 and 4): The lysosomal pump
has different inhibitors (e.g., N-ethylmaleimide) than the F$_1$F$_0$ pump
(e.g., oligomycin). In addition, the lysosomal pump only hydrolyzes ATP
to pump protons, while the other one is reversible. Finally, the polypeptide structures of the two types are somewhat different, although
both have a catalytic site on the membrane surface and a proton channel
in the membrane (Figure 8–4).

The lysosomal pump has a "lollipop" that protrudes into the cytoplasm and is composed of three copies of two large subunits (which
bind ATP) and a copy each of three others that attach it hydrophobically
to the membrane. Five polypeptides (four copies of one subunit and one
copy on another) make up the membrane H$^+$ channel. The roles of the
two accessory subunits are not known.

Gene sequencing of the two types of ATPases shows 25 to 50%
homology, depending on the subunits and species examined. They differ
much less from each other than from a third class of proton pumps,

exemplified by the H^+-ATPase at the epithelium of the stomach (see Chapter 2). The finding of a lysosomal-type proton pump in the most primitive bacteria (the *Archaebacteria*) indicates that the ability to acidify the cell or a compartment was an important event in cellular evolution.

The second problem for lysosomes, that of the **membrane permeability** to hydrolysis products, has been investigated in a manner similar to Overton's investigations of cell permeability, with two modifications: Instead of erythrocytes, isolated lysosomes are used. Instead of osmotic swelling and then hemolysis due to entry of water following the small molecule, a lysosomal enzyme's accessibility to its substrate (a function often blocked by an intact membrane) is measured.

Tests on hundreds of molecules have shown that small molecules (molecular weight less than 200) cross the membrane much more readily than do large molecules. This means that the products of lysosomal hydrolysis can diffuse out of the organelle, to be used by the cell for synthesis and catabolism.

The **membrane composition** of the lysosome is rather similar to that of the plasma membrane, especially in its protein:lipid ratio (Table 1–2) and high cholesterol content (Table 1–3). However, the high content of sphingomyelin in the lysosome more resembles the Golgi membrane. These facts may be a reflection of the multiple origins of organelles defined as lysosomes (Figure 8–5). Some may arise via fusion of plasma membrane-derived endocytotic vesicles (see Chapter 1), while others are derived from the trans cisternae of the Golgi complex or GERL (see Chapter 7).

Qualitative analyses of lysosomal membrane proteins by gel electrophoresis show at least 40 polypeptides. The major species compose a closely related family termed the lysosomal glycoproteins (lgp). Their overall structure has three domains: a short (10 to 11 amino acids) cytoplasmic tail, a single hydrophobic region that spans the lipid bilayer, and a long (about 400 amino acids) region that protrudes into the lysosome. This region is heavily glycosylated, and it is this property that may make these proteins resistant to the actions of proteases that are inside the organelle. (It is not clear how the enzymes within the organelle are resistant.)

Acid phosphatase is a lysosomal membrane protein while on its way to the interior of the organelle. This enzyme is made as a membrane-bound precursor, which has a hydrophobic tail to insert itself into the bilayer. This tail is then cut off by a membrane-bound peptidase, and the enzyme is released to the lumen of the lysosome. Other lysosomal enzymes do not follow this pathway: They arrive at the organelle inside clathrin-coated vesicles via the mannose-6-phosphate receptor at the Golgi. The fact that acid phosphatase does not have the mannose-6-

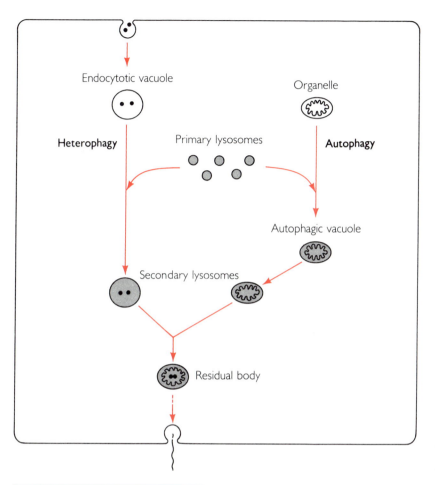

Endocytotic vacuole

Organelle

Heterophagy   Primary lysosomes   **Autophagy**

Autophagic vacuole

Secondary lysosomes

Residual body

FIGURE 8–5

The lysosomal cycle. Primary lysosomes contain hydrolases and may fuse with either an endocytotic vesicle (heterophagy) or an autophagic vacuole (autophagy). The resulting secondary lysosomes are the sites of digestion.

phosphate marker leads to its rather unusual mode of transport to the lysosome.

Function: Heterophagy

The role of lysosomes in the intracellular digestion of extracellular macromolecules was predicted from the work of E. Metchnikoff in 1893. In the context of his discovery of phagocytosis, he showed that the interior of food vacuoles in invertebrates and unicellular organisms is acidic because it turned ingested litmus paper from blue to red and that it contains digestive enzymes.

The modern concept of heterophagy is shown on the left half of Figure 8–5. Macromolecules or aggregates enter cells by endocytosis, which may be receptor-mediated as in the uptake of low-density lipoproteins by fibroblasts. Alternatively, it may be phagocytic, as happens when macrophages digest entire cells (Figure 8–2). In either case, an endocytotic vacuole forms, which is derived from the plasma membrane. This vacuole gradually becomes acidified, dissociating a ligand from its receptor, and then fuses with a lysosome or lysosomes. In many instances, the vacuole may be much larger than the lysosomes.

The fusion event essentially makes the vacuole and lysosome(s) continuous, and the resulting organelle is termed a **secondary lysosome.** Hydrolysis of the macromolecules present in the vacuole now occurs, and the products of hydrolysis are released to the cytoplasm for reuse in biosynthesis or energy metabolism. Indigestible material remains, and the residual body may eliminate it by exocytosis.

An example of these phenomena is the uptake and digestion of hemoglobin by mammalian hepatocytes as studied by A. Novikoff. This is a physiologically important process, since an estimated 9 million erythrocytes are destroyed each hour in a human, and recycling of some of the components of hemoglobin would be biochemically economical. That bilirubin, a heme breakdown product, is formed in the liver implicates hepatocytes in this process. These cells are situated between the blood sinuses and bile duct in the liver. Because both hemoglobin (by the presence of a peroxidase activity in its heme iron) and lysosomes (by the presence of acid phosphatase) can be visualized cytochemically, the intracellular events can be followed by microscopy (Figure 8–6).

When rats are injected intravenously with large doses of hemoglobin, it is immediately observed in large (up to 15 µm) endocytotic invaginations at the hepatocyte plasma membrane. Several minutes later, the hemoglobin is inside the cell within vacuoles. At this point, several lysosomes fuse with each vacuole, and the number of free primary lysosomes in the cell is reduced. By 30 minutes after injection, all of the hemoglobin vacuoles contain lysosomal enzymes and are now secondary lysosomes. Sixteen hours later, there is no stainable hemoglobin remaining in the hepatocyte; it has all been broken down, and normal cell morphology returns.

Less massive doses of injected hemoglobin disappear from the hepatocyte within one hour. Very low doses are not taken up by these cells but rather by Kupffer cells, which are reticuloendothelial cells lining the blood sinus in the liver. Thus, the hepatocyte takes up and digests excess hemoglobin, which occurs in hemolytic anemias. In these patients, the liver becomes loaded with iron deposits, which are in residual bodies.

The return of normal hepatocyte morphology after the intracellular digestion of hemoglobin has two implications. First, more lysosomes are produced; this occurs via the Golgi complex. Second, endocytotic vacuole membranes are recycled to the plasma membrane. Evidence for this comes from experiments using labeled membranes, in which the labeled vacuole membranes end up at the cell surface. This shuttling system is

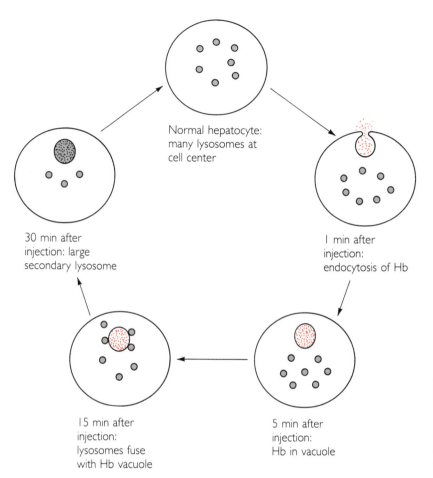

Normal hepatocyte: many lysosomes at cell center

1 min after injection: endocytosis of Hb

30 min after injection: large secondary lysosome

15 min after injection: lysosomes fuse with Hb vacuole

5 min after injection: Hb in vacuole

FIGURE 8–6

Diagram of hemoglobin uptake and processing in a hepatocyte after intravenous injection of hemoglobin (Hb). Only organelles involved are shown.

similar to that involving Golgi vesicles (Chapter 7) and indeed may be part of it.

A process in some ways the reverse of heterophagy involves the extracellular release of lysosomal hydrolases. The epithelial cells of the prostate gland secrete acid phosphatase, which is a component of prostatic fluid. This secretion involves lysosomal vesicles, which exocytose their contents by fusing with the plasma membrane. In the sperm cell, a coalescence of Golgi vesicles forms the acrosome at the anterior tip. This structure contains lysosomal enzymes that, when released on contact with the outer coat of the egg, digest the coat and allow the sperm to reach the plasma membrane of the egg.

Function: Autophagy

Autophagy ("self-feeding") is the intracellular degradation of organelles or macromolecules. The pathway for this phenomenon is shown on the

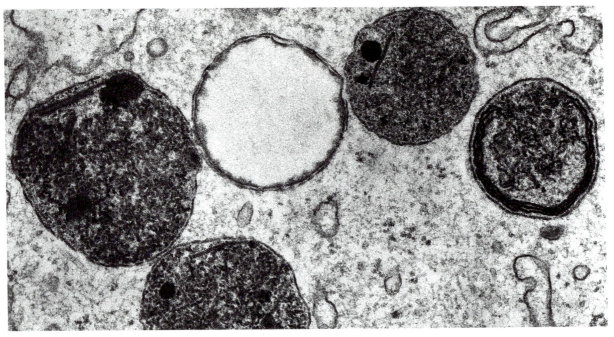

FIGURE 8–7   Secondary lysosomes in a mouse macrophage. Each organelle is surrounded by a single membrane. These lysosomes contain granular, filamentous and membranous materials. (×60,000. From: Bozzola, J. and Russell, L.: *Electron Microscopy: Principles and Techniques for Biologists.* Jones and Bartlett, 1992, p 461.)

right side of Figure 8–5. Initially, cytoplasmic components are sequestered within a membrane-bound structure, forming an **autophagic vacuole** (Figure 8–7). The source of this membrane may be new synthesis, or it may come from some other organelle such as the endoplasmic reticulum (ER) or Golgi complex. Following vacuole formation, fusion with lysosomes occurs, and the resulting secondary lysosome is the site of breakdown of the cellular macromolecules.

Autophagy is an ongoing cellular activity, with proteins (and organelles) constantly being degraded and resynthesized ("turned over") (Table 8–2). There structures appear to have finite lifetimes, after which they begin to break down spontaneously. For example, amino acids begin to change their structure (racemize). There is no selection for older or damaged structures, with the autophagic vacuole a random trap, surrounding recently synthesized, normal-appearing organelles or macromolecules as well as older components. Presumably, this ensures that molecules will be recycled before they become nonfunctional.

This breakdown of normal macromolecules also occurs when the lysosomal membrane invaginates around an adjacent area of cytoplasm

TABLE 8–2 Turnover Rates of Total Proteins of Liver Cell Fractions

Fraction	Half-Life (Days)
Homogenate	3.3
Nuclear	5.1
Mitochondrial	6.8
Microsomal	3.0
Cytosolic	5.1
Plasma membrane	1.8

FROM: Glaumann, H., Ericsson, J., and Marzella, L. Mechanisms of intralysosomal degradation with special reference to autophagocytosis and heterophagocytosis of cell organelles. *Internat Rev Cytol* 73 (1981):152.

and essentially "endocytoses" it. This slow process is termed **"micro-autophagy"** and presumably provides the healthy cell with a continuous pool of free amino acids and degrades proteins before they become nonfunctional.

A third type of autophagy is observed when cells in culture are deprived of serum growth factors. In this case, the cell is being slowly starved of amino acids. Normally, the other two types of autophagy would help to keep up the supply of amino acids. But after a long time, random autophagy might deplete the cell of essential enzymes. Instead, specific proteins are taken up by the lysosome and degraded. These proteins all have a marker peptide (KFERQ — lys-phe-glu-arg-gln) and may be ones that are most "dispensable" to the cell in an emergency.

The steady-state level of a cellular substance is a result of the opposing actions of synthesis and degradation. When liver smooth endoplasmic reticulum (SER) proliferates after phenobarbital administration (see Chapter 6), lysosomal activities remain constant. This implies that the increase in ER is due to increased synthesis rather than to decreased degradation. But when the drug is no longer present, the ER levels return to their preinduction status, a process that correlates with an increase in the number of both lysosomes and their enzymes. Thus, degradation is important in this case.

Another instance of degradation being rate limiting occurs in regenerating liver. If half of a mammalian liver is excised, the remaining portion will regenerate what was lost, primarily through decreased autophagy. This has implications in liver transplants: Generally, only part of the liver is surgically grafted from donor to recipient (especially in children), and the rest of it soon regenerates.

A number of developmental processes involve considerable tissue breakdown. For example, the neonatal human womb at birth weighs over 2 kg, but nine days later it weighs only 50 g; this decrease is caused in part by more autophagic vacuoles. Lysosomes are also responsible for the degradation of embryonic Müllerian ducts (which would form oviducts) in the developing male. The involution of the mammary gland when lactation ceases and of the prostate following castration are both partly autophagic events, as is the resorption of the tail by a tadpole as it metamorphoses into a frog. Most of these events are hormonally controlled, and it appears that the increased level of autophagy is due to increased synthesis of lysosomal enzymes.

Lysosomes and Disease

The importance of the lytic actions of lysosomes is indicated by a number of **inborn errors** of human metabolism in which the activity of one or more of the hydrolases is absent or very low. Some of these genetic

TABLE 8–3 Some Lysosomal Storage Diseases

Disease Name	Enzyme Deficiency	Substrate Accumulated
Glycogenosis		
Pompe	α-glucosidase	Glycogen
Mucopolysaccharidoses		
Hunter	α-iduronosulfate sulfatase	Heparan and dermatan sulfates
Hurler	α-iduronidase	Heparan and dermatan sulfates
Maroteaux-Lamy	Arylsulfatase B	Determatan sulfate
Morquio	N-acetylhexosamine-6-sulfate sulfatase	Keratan and chondroitin sulfates
Sanfilippo A	Heparan sulfate sulfatase	Heparan sulfate
Sanfilippo B	N-acetyl-α-glucosaminidase	Heparan sulfate
Sphingolipidoses		
Fabry	α-galactosidase	Ceramide hexoside
Farber	Ceramidase	Ceramide
Gangliosidosis GM$_1$	β-galactosidase	GM$_1$ ganglioside
Gaucher	β-glucosidase	Glucocerebroside
Metachromatic leukodystrophy	Arylsulfatase A	Galactosylsulfate ceramide
Niemann-Pick	Sphinogomyelinase	Sphinogomyelin, cholesterol
Tay-Sachs	N-acetylhexosaminidase-A	GM$_2$ ganglioside

diseases are listed in Table 8–3. In all cases, the substrate of the missing enzyme accumulates in the cell and is microscopically detectable in vacuoles. Although they share a common mechanism, these diseases have diverse clinical presentations. But few are benign. Most lead to neurological problems and many to early death.

The extreme case of a lysosomal disease is I-cell disease, in which many hydrolases are properly targeted to lysosomes of connective and nervous tissues. In these patients, secondary lysosomes with breakdown products rarely form, and vacuoles filled with undigested materials fill the cells (Figure 8–8). Presumably, these vacuoles are formed from the fusion of endosomes with "ghosts" of primary lysosomes.

A problem in the pathology of these diseases is to determine whether the damage is caused by the accumulated substrate or the absence of metabolically important breakdown products. This is a question common to the study of all such congenital defects. For example, several of the mucopolysaccharidoses result in accumulations of sulfated glucosaminoglycans. These are normally broken down to sulfate, which is excreted, and glucosamine and uronides, which enter the glycolysis and pentose phosphate pathways, respectively. These compounds would seem to be of little importance in the provision of energy to the cell, but the fact that if they accumulate a serious disease results belies this idea. So it must be proposed that there is a developmental phase during which they are an essential energy source. Alternatively, the accumulated vacuoles in the cells of these patients could be detrimental to cell function.

Several **intracellular parasites** can take advantage of the heterophagy pathway. These organisms enter eukaryotic cells by endocytosis, but instead of being hydrolyzed in secondary lysosomes, they thrive there and reproduce, ultimately killing the host cell.

An example of this phenomenon is the rickettsia parasite *Coxiella burnetii,* which causes Q fever. This is an influenza-like respiratory infection that can be fatal and that affects cattle, sheep, and goats with considerable frequency (up to 50% in the United States). It can be transmitted via unpasteurized milk or airborne particles to humans, where it causes a milder pneumonitis. Studies of *C. burnetii* show that its metabolite transport systems and energy-yielding biochemical pathways function best at the low intralysosomal pH. In contrast, other rickettsias require a neutral pH. Moreover, the high proton concentration within the lysosome may be used by the parasite for chemiosmotic energy generation, as occurs in the microbe leishmania.

Mycobacteria, such as those that cause leprosy *(Mycobacterium leprae)* and tuberculosis *(M. tuberculosis)* escape degradation by the host in a different way. In this case, the agent is ingested into white blood cells via endocytosis, but fusion of the endosome with the lysosome is inhibited.

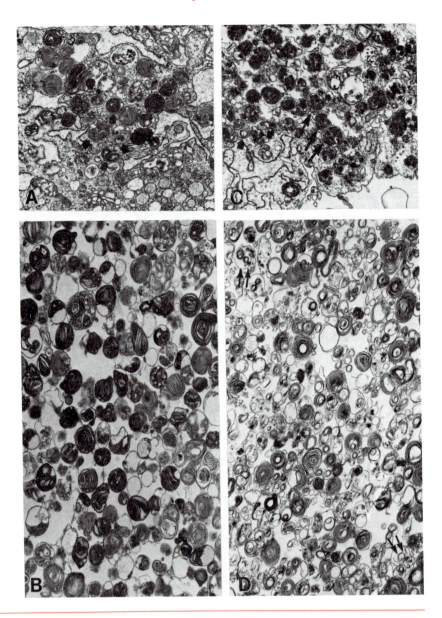

FIGURE 8–8

Lysosomes from normal (A, B) and I-cell (C, D) fibroblasts. (Photographs courtesy of Dr. A. Miller.)

Thus the means (lysosomal hydrolases) that the cell has to destroy the invader cannot get to the ingested bacterium. Apparently, the bacterial cell surface is essential for this inhibition, because if the surface is coated with antibodies, endosome-lysosome fusion occurs normally. Host protection to these invaders has been thought to be via cell-mediated immunity, where T cells of the immune system bind to the macrophages. Because of the effects of antibodies on indirectly promoting the destruction of the bacterium, they may be useful in treating people already infected with the bacteria.

The lysosomal membrane protects the cell from hydrolytic damage due to the hydrolases within the organelle. Clearly, leakiness of the membrane can lead to a generalized autophagy and cell death. When miners inhale silica or asbestos dust, the particles are endocytosed by lung macrophages and end up in secondary lysosomes, and chemical interactions between silicic acid and components of the organelle membrane render it leaky. The cells die, releasing more particles, which leads to further deaths. Fibroblasts are then stimulated to deposit collagen fibers, which decrease the elasticity of the lung. Theories along this line have been proposed for the etiology of syndromes ranging from aging and arthritis to cancer, but these cases have yet to be proved.

A number of parasitic organisms survive by phagocytically ingesting cellular constituents. These include the agents causing malaria and amoebiasis. The drug chloroquine (Figure 8–9) is effective in treating these diseases. It is a weak base and is protonated as soon as it enters the lysosome. The source of these protons is the lysosomal contents, and so the pH inside of the organelle rises. The accumulation of chloroquine is often so great that it raises the intralysosomal osmotic concentration and water enters the organelle and makes it swell. The expanded volume also contributes to the elevation of pH. This pH change inactivates the lysosomal hydrolases, and as a result, the pathogenic agent dies by intracellular starvation.

But there is a disadvantage to chloroquine treatment: Agents that thrive in an elevated lysosomal pH will now be able to grow and reproduce. Thus, people treated with chloroquine are susceptible to infections with *Toxoplasma, Legionella, Chlamydia,* and *Nocardia.*

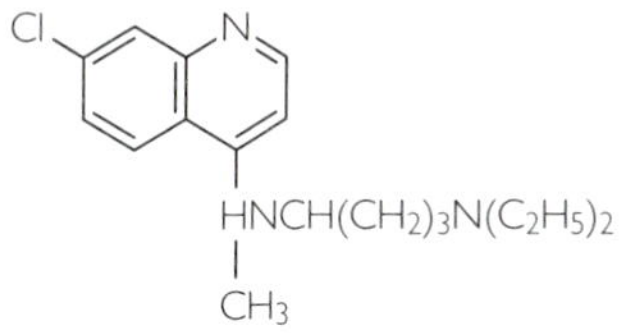

FIGURE 8–9

Chloroquine, a drug that raises the intralysosomal pH. The amine groups act as proton acceptors, and this lowers the H^+ concentration in the organelle. This drug is used to combat the malaria parasite.

Vacuoles and Protein Bodies

Structure

The membrane-bound components of the lysosomal cycle in animal cells form what is termed an exoplasmic space, a cellular compartment that houses catabolic and some storage functions. In plant cells, the exoplasmic space is typically housed within a large central **vacuole.**

In contrast to lysosomes, the origin of the large plant vacuole is not clear. Two scenarios have been proposed (and both may be correct):

1. In young dividing cells, vacuoles are absent and electron microscopy reveals an extensive, interconnected tubular SER. As cells expand and differentiate, this ER pinches off small vesicles or may become dilated to form large ones. These then appear to coalesce to form the mature vacuole (Figure 8–10).

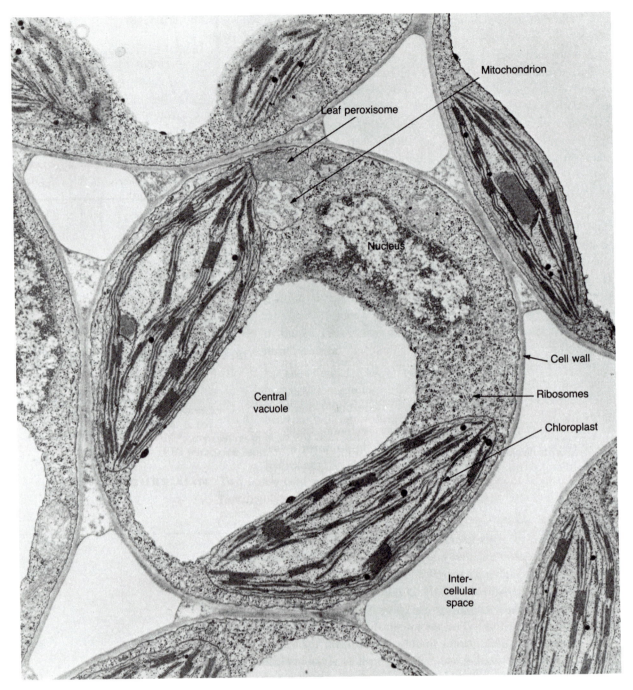

FIGURE 8–10   Electron micrograph of a cell in a leaf of *Coleus blumei*. The principal structures visible are the cell wall, chloroplasts, central vacuole, nucleus, mitochondrion, a peroxisome, and ribosomes. (Micrograph by W. P. Wergin, courtesy of Eldon H. Newcomb.)

2. Another scenario has the small vacuole precursors arising much the same as lysosomes, from the trans Golgi. But instead of the multiple vesicles fusing with an endocytotic vacuole, in plants the vesicles fuse directly with each other.

Most plant cells have a single vacuole, and it is certainly the largest organelle, occupying up to 80% of the volume of the cell. The membrane that surrounds this organelle is called the **tonoplast.** Chemical analyses of the tonoplast from sugar beets show that it is 40% lipid and 60% protein, although these percentages are reversed in other plants and tissues. The major lipids are phosphatidylcholine, phosphatidylethanolamine, and glycolipids. Electrophoresis reveals 5 to 15 major polypeptides and a similar number of minor polypeptides in the membrane. α-Mannosidase, present only in the vacuolar sap, is a marker enzyme for the vacuole.

Many plants store proteins in seeds, and these so-called reserve proteins are later used by the developing embryo. As these seeds are formed, major changes occur in the protein-storing cells, as each single vacuole becomes divided into many smaller ones. This occurs concomitantly with the synthesis (on rough endoplasmic reticulum [RER]) and deposition of the storage proteins into the small vacuoles, which when filled are termed **protein bodies.** In developing cotyledons of the pea plant, the single vacuole in each cell forms over 175,000 protein bodies, each 1 to 2 μm in diameter and bound by a single membrane!

Late in seed development, protein bodies are formed from the budding off of portions of RER. The transport of storage proteins from RER to the protein bodies occurs via the Golgi complex, where additional glycosylation occurs. Newly synthesized storage proteins appear first in Golgi-derived vesicles, which then fuse with the vacuolar membrane.

A protein destined for the vacuole must have targeting information, and this appears to reside in sequences of relatively few amino acids. In barley lectin, these amino acids are in the carboxy-terminal region of the protein. Deletion of this region results in the lectin being secreted via the default pathway, while fusion of it to a normally secreted protein results in this protein ending up in a vacuole. An interesting case is the enzyme, chitinase in tobacco. This protein is present in the vacuole as well as in the cell wall, where it is important in disease resistance. Not surprisingly, only the vacuolar enzyme has an extra C-terminus, containing the amino acids glu-leu-leu-val-asp-thr-met.

Some vacuolar proteins have their targeting information in the amino- (or N-) terminal region. For instance, sporamin, a storage protein from sweet potato, has the sequence asn-pro-ile-arg-leu-pro, near its N-terminus and this is required for vacuolar localization. Still other plant proteins have the information in the middle of the protein, where it is

TABLE 8–4 Hydrolases in Plant Vacuoles

Acetylglucosaminidase

Acid invertase

Acid phosphatase

Acid protease

Carboxypeptidase

Deoxyribonuclease

α-galactosidase

β-glucosidase

α-mannosidase

Phosphodiesterase

Phytase

Ribonuclease

Data compiled from 12 different species.

presumably exposed when the protein assumes its final three-dimensional configuration.

Finding the signals as well as the rules for packaging these storage proteins is no mere academic exercise. Because about 80% of the human diet comes directly from plants, and because plant proteins often are deficient in one or another of the essential dietary amino acids, a major goal of agriculture is to improve the nutritional quality of the plant proteins people eat. This could be done by manipulating a food plant to produce a different protein through genetic engineering. But this new (and improved) protein will be of no use if it cannot be properly packaged into the protein bodies of the seed. This is where knowledge of the cell biology of protein targeting and packaging is essential.

Function

Vacuoles contain a variety of **hydrolytic enzymes** (Table 8–4), which are detectable cytochemically and through careful isolation of intact vacuoles. As is the case in lysosomes, the hydrolases have acidic pH optima and exist in a vacuolar sap whose pH is maintained at an estimated 5.0. A proton pump similar in structure and function to the lysosomal H^+-ATPase is largely responsible for the intravacuolar pH. In some plants, vacuolar pH is very low: Begonia and oxalis plants accumulate oxalic acid in their vacuoles, resulting in a pH of 1.2.

Electron micrographs showing organelles or organelle fragments within the vacuole indicate that it has an autophagic function. The tonoplast appears to invaginate around a structure to be digested. The heterophagic pathway involving endocytosis is difficult to envision for plant cells, as they are surrounded by a cell wall, but it apparently does occur to a limited extent in some cells.

In addition to hydrolases, plant vacuoles may contain inhibitors of serine endopeptidases. These enzymes are not important components of plant cells but are present in animal lysosomes. When an insect eats plant tissue, it is endocytosed into the insect cells and ends up in secondary lysosomes, where digestion occurs. However, the presence of the protease inhibitor blocks this part of the digestive process. This interaction of plant and animal hydrolytic compartments can be used as a defense mechanism for the plant.

The second major function of plant vacuoles is **solute accumulation.** These solutes include ions (Cl^-, K^+, and Na^+), amino acids, sugars (sucrose), organic acids (malate and citrate), and secondary metabolites (pigments and alkaloids). Some of these apparently are synthesized at least in part within the vacuole, but most are imported from the cytoplasm. The mechanism by which these substances are concentrated across the tonoplast is not clear, but a number of studies indicate that metabolite transport is coupled to H^+ transport. For example, metabolite transport in isolated vacuoles and tonoplast vesicles is inhibited by proton ionophores.

Solute accumulation is important in the maintenance of turgor, the osmotic force that maintains the "hydrostatic skeleton" of the plant. Plant cells actively pump out protons across their plasma membranes, creating an electrical gradient that drives the inward flow of K^+. The pH change in the cytoplasm due to proton extrusion is compensated for by the synthesis of organic acid anions such as malate or the import of Cl^-. However, cytoplasmic accumulations of these two anions feed back to block their further synthesis or transport, respectively. Therefore, they are removed from the cytoplasm by being accumulated in the vacuole. This continued accumulation creates an osmotic pressure so that water enters the cells and they become turgid.

The function of protein bodies in the mature seed is to **store** proteins that will ultimately be broken down for use by the developing embryo (or an animal that eats the seed). When the seed germinates, proteases are synthesized on RER and deposited in the protein bodies, where the food reserves are broken down.

This process of protein degradation can be followed microscopically, as the dense internal material of the protein body appears to dissolve. After protein digestion in a storage tissue such as the cotyledon, autophagy occurs and the cells die, a process mediated by protein bodies acting as lysosomes. Hydrolytic enzymes such as acid phosphatase, as well as organelle membrane fragments, are observed within the protein bodies after their proteolytic function is completed (Figure 8–11).

Microbodies

Structure and Composition

First described by J. Rhodin in 1954, microbodies are ubiquitous components of virtually all eukaryotic cells, except mature mammalian erythrocytes. These organelles are generally spherical, with diameters ranging from 0.3 to 1.5 μm. They are enclosed by a 7-nm membrane and, by electron microscopy, have a granular matrix (Figure 8–12). Within the matrix, there is an electron-dense inner core that varies from species to species in structure from a crystalline to a tubular array. The presence of the core is correlated with the presence of the enzyme urate oxidase in many animal cells and catalase in plants. In humans and birds, where urate oxidase is lacking, the microbodies lack core material.

Microbodies are cytochemically defined by their **catalase** activity. This enzyme protects the cell from a potentially dangerous accumulation of hydrogen peroxide, a highly reactive reducing agent formed by flavin oxidase. Thus:

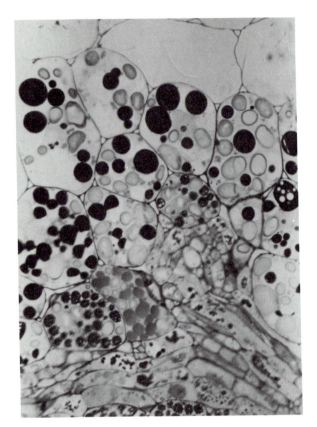 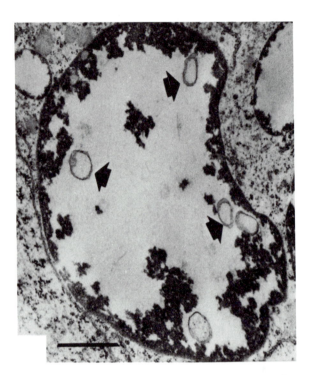

FIGURE 8–11 Protein bodies in the storage material of a mung bean cotyledon. *Left:* Light micrograph, showing the densely stained protein bodies. The cells at the top (away from the vascular tissue) are ones in which the protein has been degraded. *Right:* Electron micrograph showing a protein body breaking down its stored material. The arrows indicate membrane fragments. (Courtesy Dr. Maarten Chrispeels.)

$$AH_2 + O_2 \rightarrow A + H_2O_2$$

is the flavin oxidase reaction, where A is a flavin-containing molecule such as urate oxidase. The catalase reaction is:

$$2H_2O_2 \rightarrow 2H_2O + O_2.$$

This enzyme, when supplied with the proper substrate, can act as a **peroxidase.** For cytochemical studies, diaminobenzidine (DAB), which turns brown when oxidized, is used:

$$H_2O_2 + DAB_{reduced} \rightarrow 2H_2O + DAB_{oxidized}$$

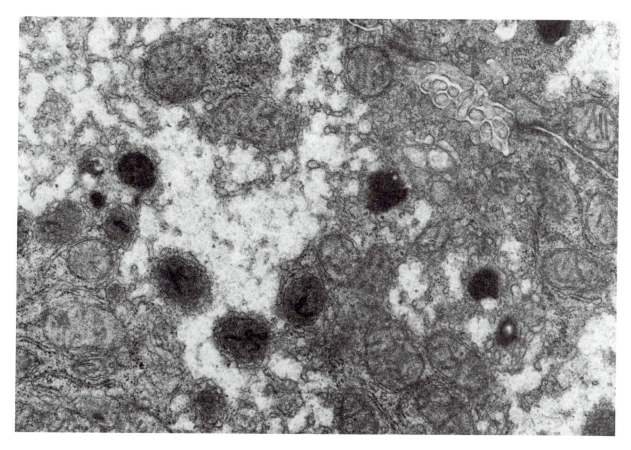

F I G U R E 8–12   Peroxisomes in a liver cell, stained darkly for catalase. Note the nearby ER and mitochondria. (×32,000. Photograph courtesy Dr. J. Watanabe.)

It is difficult to isolate microbodies by centrifugation because their membrane is fragile. But because they are richer in protein than other cell organelles of similar size (e.g., mitochondria, with internal membranes), microbodies are somewhat denser. Their catalase activity (the enzyme makes up about one-fourth of all the protein in the organelle) makes identification straightforward.

Isolated microbodies are generally permeable to the diffusion of small molecules and ions. The pH optimum of the flavin oxidase is about 8.5, and those of catalase and other microbody enzymes are usually greater than 7. This implies that the intramicrobody pH is slightly basic.

Although catalase and flavin oxidase are a universal presence in microbodies, specialized enzyme compositions occur in different tissues. Three main classes of this organelle are liver peroxisomes, leaf peroxisomes, and seed glyoxysomes. These organelles appear to arise from pre-

TABLE 8–5 **Enzyme Content of Liver Peroxisomes**

Catalase

Oxidases
  Glycolate
  Fatty acyl-CoA
  Urate (not in primates)
  Amino acid
  Polyamine

Dehydrogenases
  NAD: hydroxy fatty acyl-CoA
  NAD: glycerol-P
  NADP: isocitrate
  Xanthine

Fatty acid β-oxidation enzymes

Acyl transferases
  Carnitine acyl-CoA
  Acyl-CoA: dihydroxyacetone-P

Aminotransferases
  Alanine-glyoxylate
  Serine-pyruvate
  Leucine-glyoxylate

Other
  NADH-cytochrome c reductase
  Lipase
  Allantoinase
  Allantoicase

existing microbodies, possibly by vesicles that are then filled with microbody products from the cytoplasm.

Liver Peroxisomes

A typical hepatocyte contains approximately 1000 microbodies termed peroxisomes. This number is even larger during fetal and postnatal growth and during liver regeneration. Peroxisome proliferation can be induced by such drugs as aspirin and clofibrate, a hypolipidemic agent (see below).

Peroxisomes contain a wide variety of enzymes (Table 8–5), but the most important of them quantitatively is the flavin oxidase-catalase system, which accounts for up to 20% of the oxygen consumption of liver tissue. This electron transport chain is fundamentally different from the one in mitochondria in that the free energy lost on oxidation is not coupled to ATP synthesis. Instead, it is released as heat, a phenomenon that may have importance in thermogenesis. When an animal is exposed to cold temperatures, brown fat cells proliferate peroxisomes, whose metabolism could provide some of the heat needed by the animal.

Several flavin-containing enzymes produce hydrogen peroxide in the peroxisome. An α-hydroxyacid oxidase (glycolate oxidase) acts on glycolate or lactate.

$$\text{Glycolate} + O_2 \rightarrow \text{glyoxylate} + H_2O_2$$

$$\text{Lactate} + O_2 \rightarrow \text{pyruvate} + H_2O_2$$

The latter reaction can be coupled with the reconversion of pyruvate to lactate in the cytoplasm:

$$\text{Pyruvate} + \text{NADH} \rightarrow \text{lactate} + \text{NAD}$$

This may act as a "shuttle" system to reoxidize cytoplasmic NADH.

A second peroxide-generating oxidase occurs in the β-oxidation scheme for fatty acids. This series of steps breaks down medium-chain (C_{14} to C_{18}) fatty acids to acetyl-CoA. An initial step is catalyzed by fatty acyl-CoA oxidase:

$$\text{Fatty acyl-CoA} + \text{FAD} \rightarrow \text{unsaturated fatty acyl-CoA} + \text{FADH}_2$$

In the mitochondrion, the reduced FAD passes its electrons via a transfer protein to cytochrome b (Figure 3–7). In the microbody, the electrons are passed to flavin oxidase:

$$\text{FADH}_2 + O_2 \rightarrow \text{FAD} + H_2O_2$$

Most animals use peroxisomal oxidase to convert uric acid, a product of the breakdown of nucleotides (purines), into allantoin. But primates lack this enzyme activity and so accumulate uric acid. If too much is accumulated, uric acid concentrations exceed its solubility in body fluids, and it forms crystals. This occurs in gout, which can be thought of as a genetic disease of primates that is due to the loss of a peroxisomal protein.

Several other important metabolic functions occur in the peroxisome, including:

1. The initial steps in synthesizing plasmalogens, ether-linked glycolipids that are especially important in myelin, occur there.

2. Fatty acids with long (C_{22} to C_{26}) chains are metabolized by the β-oxidation system in the peroxisome. For catabolism of shorter chains, the analogous system in mitochondria is preferentially used. However, plants and fungi lack the mitochondrial system, so that their peroxisomal β-oxidation is essential to generate energy from fatty acid breakdown. The importance of this pathway is shown by the X-linked genetic disease adrenoleukodystrophy: These people have a defect in the catabolism of long-chain fatty acids. Children develop normally until age four to eight, when progressive and severe neurological problems begin, leading to death by age 10.

3. The peroxisome contains the enzymes for the α-oxidation of fatty acids, especially necessary in the catabolism of chlorophyll. The long phytol tail of chlorophyll is a fatty acid with methyl groups attached. The standard β-oxidation enzymes do not act on such a substituted chain. But after the terminal carbon is removed by α-oxidation, the β-oxidation system can take over.

4. Some synthesis of cholesterol occurs in this organelle, and once again this pathway is redundant to one in the ER. Although the peroxisome accounts for less than 30% of the cholesterol a liver cell makes, its pathway may be important at certain times. In rat liver, the microsomal pathway is most active during the night, while the peroxisomal pathway is mostly active during the day.

A rare autosomal recessively inherited disease, **Zellweger syndrome,** casts some light on the biogenesis of peroxisomes. This disorder is characterized by severe neurological and metabolic impairment in early life, and death in the first year. A. Goldfischer found that these patients lack recognizable peroxisomes. Some proteins normally present in the peroxisomal matrix, such as catalase, are present in normal amounts but are free in the cytoplasm. Others are made but rapidly degraded, leading to deficiencies in such metabolic pathways as lipid synthesis and breakdown.

At first, it was thought that the defect in these patients was in the assembly of the peroxisomal membrane and that this is what led to enzymes not being packaged. But careful examination of the patients' cells showed that membrane-bound peroxisomal "ghosts" are present. Apparently, the membrane is made, but the machinery for importing proteins into the organelle is defective.

A clue as to the nature of the defect comes from a yeast mutant that, like Zellweger patients, fails to target peroxisomal proteins to the organelle but instead leaves them in the cytoplasm. The gene product of the wild type allele for this yeast mutant is a protein with ATPase activity, which has sequence similarity to proteins involved in such diverse biological processes as vesicle-mediated transport and the control of cell division. The precise role of this protein in peroxisome biogenesis, and its relationship to Zellweger syndrome, are under investigation.

Direct sequencing of proteins imported into peroxisomes has been used to search for directional **signals** in these proteins. In many cases, the N-terminal amino acids are serine-lysine-leucine (SKL) or serine-histidine-leucine (SHL). Addition of this region to nonperoxisomal proteins causes them to end up in that organelle. On the other hand, elimination of the signal causes a peroxisomal protein to be secreted (the default pathway [see Chapter 7]). But this is a short signal, and many cytoplasmic proteins have it yet do not go to the peroxisome. This means that there must be some aspect of the regions around the signal that is also essential for import, a situation rather different from the signals for plastids, mitochondria, and lysosomes, where specificity of sequence appears to be the rule.

This signal is apparently recognized by a receptor on the peroxisome membrane. Nonperoxisomal proteins do not bind to the receptor, so it is specific. Little is known about the mechanism of translocation across the membrane and into the organelle, except that ATP is required and there is an ATPase activity closely linked to the signal receptor protein.

Like plastids and mitochondria, peroxisomes come from the fission of preexisting peroxisomes. Electron micrographs show peroxisomes connected via narrow tubules, probably a late stage in the fission process. This organelle does not contain DNA or the molecular apparatus for protein synthesis. What controls peroxisome division is not known.

A special form of the peroxisome in the kinetoplastid trypanosomes (see Chapter 5) also is useful in the elucidation of peroxisome biogenesis. These protozoans have peroxisomes that contain catalase as well as some other typical microbody enzymes. But they also contain a unique set of glycolytic enzymes that convert glucose and glycerol to phosphoglycerate. For this reason, these organelles are called **glycosomes.**

The presence of a set of enzymes inside a peroxisome that is the same as a cytoplasmic set has allowed a comparison of the two sets to determine if there are special signals for import on the microbody enzymes. Most glycosomal glycolytic enzymes are somewhat larger than their cytoplasmic counterparts. The extra amino acids are internal rather than N-terminal as in secreted proteins, and they are mostly positively charged. Apparently these amino acids are involved in the import of these proteins into the glycosome.

Acatalasemia is an example of a genetic disease affecting a single peroxisomal enzyme. Patients in this case have very low catalase activity, and many of them develop ulcers in their mouths, especially around the teeth. This seems to occur because of their inability to metabolize hydrogen peroxide produced by bacteria living in the buccal region. Among other effects, the peroxide converts hemoglobin to methemoglobin, a form much less prone to release its oxygen. The tissues become oxygen-deprived and die as a result. Although the peroxisome contains a β-oxidation system for fatty acids, it is redundant to the one in mitochondria, and therefore, these patients can break down fats.

In addition to genetic diseases, peroxisomes can be involved in cancer. A diverse set of chemicals such as certain drugs used to lower blood lipids, herbicides, analgesics, and plasticizers cause the proliferation of liver peroxisomes (Figure 8–13). How this occurs is not known, but one consequence of prolonged exposure to these agents is the induction of liver tumors. When the DNA of these tumors is examined, oxidative damage to the bases is commonly seen, but the agents themselves do not damage DNA directly. Rather, it is the induction of more peroxisomal oxidation, with its accompanying production of H_2O_2 that overwhelms the ability of catalase to metabolize the peroxide. The latter then acts as the carcinogen by the production of free radicals, which damage DNA.

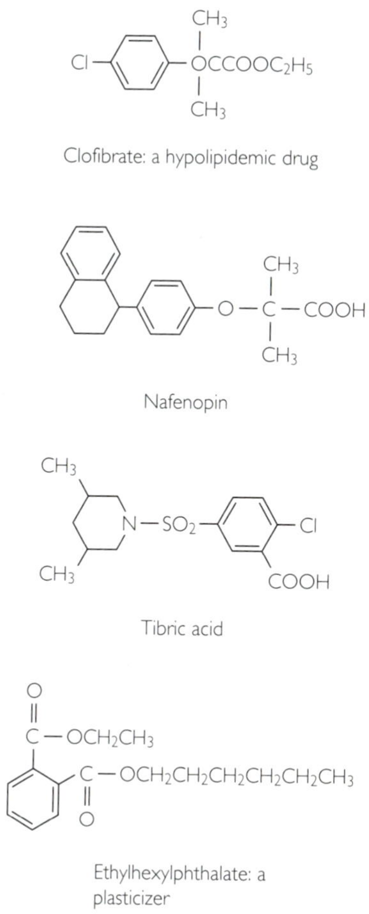

Leaf Peroxisomes

Peroxisomes are present in the photosynthetic cells of green plants, especially in palisade cells of C3 leaves and bundle sheath cells of C4 leaves (see Chapter 4). Estimates place the ratio of microbodies to mitochondria to chloroplasts in these cells at 1:2:3. Leaf peroxisomes develop at the same time as plastids, and indeed the two organelles often are near each other in the cell (Figure 8–14).

This relationship is also a biochemical one, as the peroxisome plays an important role in photorespiration. In the chloroplasts of a C3 plant, RubisCO (see Chapter 4) can catalyze either the reductive fixation of carbon dioxide or, acting as an oxygenase, the production of glycolic

Clofibrate: a hypolipidemic drug

Nafenopin

Tibric acid

Ethylhexylphthalate: a plasticizer

FIGURE 8–13

Four compounds that induce proliferation of hepatic peroxisomes and that can indirectly cause cancer.

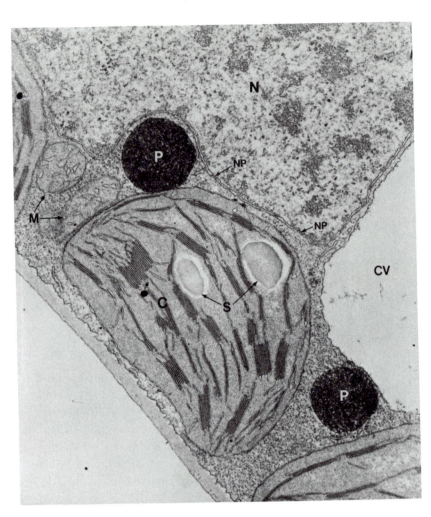

FIGURE 8–14

Peroxisomes in a tobacco leaf, staining darkly for catalase. Note their close association with the plastid. (Photograph courtesy of Dr. E. H. Newcomb.)

TABLE 8–6 Enzyme Content of Leaf Peroxisomes

Catalase

Oxidases: Glycolate
          Urate

Dehydrogenases:
    NAD: malate
    NAD: glycerate

Aminotransferases:
    Glutamate: glyoxylate
    Serine: glyoxylate
    Glutamate: oxaloacetate

Other: NADH-cytochrome c reductase

acid. On a sunny, hot day, the stomatas of the leaf may close to conserve water. Photosynthesis continues, but gas exchange with the atmosphere is reduced, raising the ratio of $O_2:CO_2$ and favoring the oxygenase activity.

The problem in the leaf, then, is how to dispose of the glycolate. This is solved by the peroxisome, which has the enzymatic machinery (Table 8–6) to metabolize the glycolate produced by the chloroplast and return it to that organelle as glycerate, usable in the production of hexose sugar. This pathway is outlined in Figure 8–15.

Phosphoglycolate is dephosphorylated in the chloroplast, and the resulting glycolate diffuses out of this organelle and into the peroxisome. A flavin oxidase, glycolate oxidase, converts the glycolate into glyoxylate.

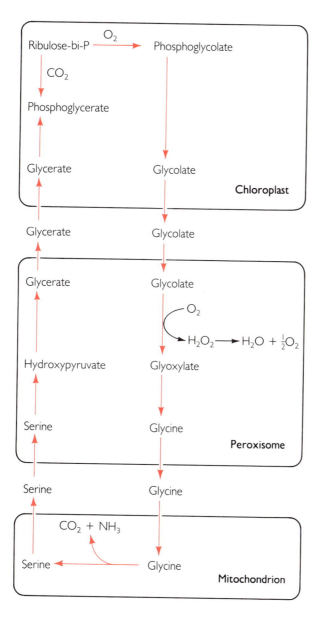

FIGURE 8–15

Role of the leaf peroxisome in photorespiration. Note the cooperation of three organelles—the plastid, the peroxisome and the mitochondrion—in the metabolism of phosphoglycolate.

In the process, H_2O_2 is produced (see above), and this is oxidized by catalase.

Meanwhile, an enzyme uses glutamate as the amino donor to convert glyoxylate to glycine, which leaves the peroxisome, enters the mitochondrion via a translocase, and is converted to serine. This reaction, which releases CO_2, is a major factor in the loss of fixed carbon in photorespiration. Serine is then converted in the peroxisome to glycerate, which can be used by the chloroplast to produce hexose.

Seed Glyoxysomes

Glyoxysomes are microbodies that occur for a brief time in the lives of plants such as certain beans and nuts, which store fats as an energy reserve in their seeds. The mature, dry seed does not contain these organelles. However, in the first few days after germination they appear in endosperm (storage) cells in association with lipid bodies (Figure 8–16), an event preceded by a proliferation of ER. After the storage fats are broken down, these organelles disappear.

The appearance of glyoxysomes correlates with a massive conversion of stored fat into hexose sugar. The glyoxysome contains the enzymes of fatty acid oxidation, as well as those of a unique pathway, the glyoxylate cycle (Table 8–7). Although a number of the enzymes of this pathway occur in other microbodies and organelles (e.g., mitochondria), two—isocitrate lyase and malate synthetase—are present only in glyoxysomes. The metabolic role of the glyoxysome is outlined in Figure 8–17.

FIGURE 8–16   Glyoxysomes (dark) associating with lipid bodies in a tomato seedling cotyledon cell. (×29,000. Photograph courtesy of Dr. E. H. Newcomb.)

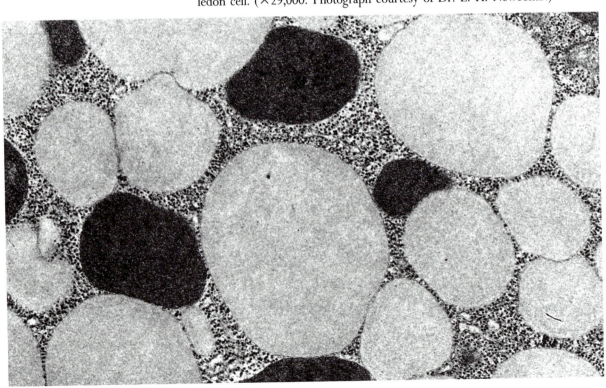

Storage lipids in fat bodies are initially hydrolyzed to glycerol and fatty acids. The latter enter the glyoxysome and, via β-oxidation, are converted into acetyl-CoA. Thus far, the situation is similar to that occurring in mitochondria in animal and other plant tissues. However, in the latter cases, the acetyl-CoA enters the Krebs cycle and is oxidized to CO_2:

$$\text{Acetyl-CoA} \rightarrow 2CO_2 + \text{CoA-SH}$$

In the glyoxysome, an alternate pathway, the glyoxylate cycle, converts acetyl-CoA into C4 acids:

$$2 \text{ acetyl-CoA} \rightarrow C_4 \text{ acid} + 2\text{CoA-SH}$$

TABLE 8–7 Enzyme Content of Bean Cotyledon Glyoxysomes

Catalase
Oxidases:
    Glycolate
    Fatty acyl-CoA
    Urate
Dehydrogenases:
    NAD: hydroxy fatty acyl-CoA
    NAD: malate
    NAD: glycerate
Fatty acid β-oxidation enzymes
Aminotransferases:
    Glutamate: gyloxylate
    Serine: glyoxylate
Glyoxylate cycle enzymes
Other:
    NADH-cytochrome c reductase
    Lipase

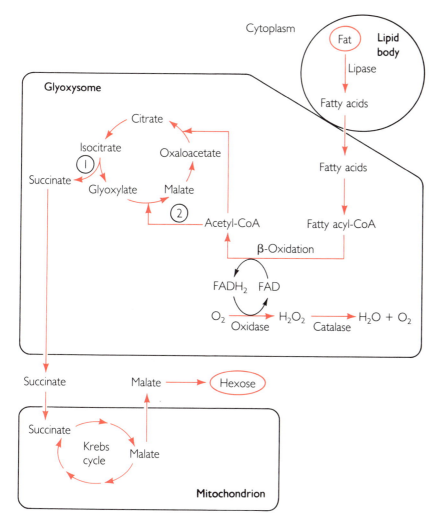

FIGURE 8–17

Interaction of the glyoxysome and mitochondrion in the conversion of fats to carbohydrates in the germinating bean seed.

The two C4 acids commonly formed are succinate and malate, and these diffuse out of the glyoxysome. In the case of succinate, the mitochondrial Krebs cycle is used to convert it to malate, and in the cytoplasm, malate is then converted to glucose.

The net conversion of fat to carbohydrate, a property of the glyoxysome, does not occur in mammalian (including human) cells.

Melanosomes

Human skin color results from the visual interaction of melanins (black-brown), carotenoids (yellow), oxidized hemoglobin (red), and reduced hemoglobin (blue), all present in the skin. Most of the variation among humans is due to the degree of melanization. Melanin synthesis occurs in specialized cells, melanocytes, and in specialized organelles, melanosomes.

These oval-shaped, membrane-bound bodies, 0.3 to 1.3 μm in length (Figure 8–18), are identified cytochemically by the presence of **tyrosinase,** the enzyme that catalyzes the multistep conversion of tyrosine to melanin. The actual structure of melanin has not been clearly

FIGURE 8–18

Melanocyte from the human epidermis, showing the dark melanosomes. (×20,000. Photograph courtesy of Dr. E. Flynn.)

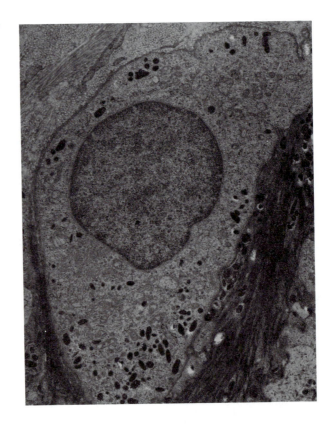

determined because it is very insoluble and therefore resistant to most methods of chemical analysis. But the overall reactions catalyzed by tyrosinase are:

$$\text{tyrosine} \rightarrow \text{dihydroxyphenylalanine} \rightarrow \text{melanin}$$
$$(\text{``dopa''})$$

Melanosomes appear to arise from both the ER and the Golgi. Electron microscopy shows fine filaments within the ER cisternae and in vesicles budding from the ER, and these vesicles soon appear with the filaments lined up parallel to each other. Biochemical and cytochemical evidence indicates that tyrosinase is synthesized on RER, packaged and glycosylated in the Golgi, and released in small vesicles. These fuse with the ER-derived premelanosome and form the mature melanosome. Following the import of tyrosine from the cytoplasm, possibly via a translocase requiring ATP (Table 8–8), melanin is synthesized and deposited on the filaments until the organelle is uniformly dark. These events are summarized in Figure 8–19.

A number of factors can influence melanosome development. **Melanocyte-stimulating hormone** (MSH) induces hyperpigmentation,

TABLE 8–8 **Enzyme Content of Melanosomes**

Tyrosinase
ATPase
N-Acetylglucosaminidase
β-Galactosidase
β-Glucuronidase
Cathepsin
Peroxidase
Aryl sulfatase
Acid phosphatase

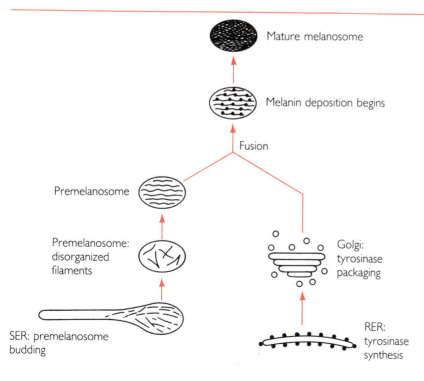

FIGURE 8–19
The formation of a melanosome from the ER and Golgi.

accompanied by an increased tyrosinase activity and proliferation of melanosomes. The initial event appears to be an activation of existing tyrosinase; later, synthesis of the enzyme is stimulated. The early response is mediated by a cell membrane receptor for MSH, which, when bound to the hormone, stimulates adenylate cyclase to synthesize cAMP. In addition, this hormone is internalized by receptor-mediated endocytosis and may travel to the Golgi, where its endocytotic vesicle fuses with a tyrosinase vesicle.

A second factor that induces melanosomes is **ultraviolet light.** The familiar tanning response that is a protective mechanism against the mutagenic effects of ultraviolet light involves increased tyrosinase and melanosomes. The ultimate degradation of melanosomes may be catalyzed by the hydrolases present in the organelle (Table 8–8). Otherwise, the function of these lysosomal enzymes in this organelle is not known.

Mature, pigmented melanosomes do not remain in the melanocyte but migrate to fingerlike "dendritic" processes. Each melanocyte is associated with 36 keratinocytes, which are the major cells of the skin. These cells take up the melanosomes by pinching off dendritic processes and endocytosing them. Once in the new cells, the melanosomes remain within the endocytotic vesicles. In Caucasians, the vesicles often contain several melanosomes and may fuse with lysosomes, leading to degradation of the melanosomes. On the other hand, in black individuals, there is only one melanosome per vesicle and no lysosomal fusion. Thus the autophagic degradation of melanosomes plays a role in skin pigmentation.

Many physiological abnormalities involve the melanocyte. When the adrenal glands are not producing adequate amounts of the cortisol, a feedback loop to the pituitary gland in the brain causes the overproduction of ACTH (adrenocorticotropic hormone), which is an attempt to overcompensate for the sluggish adrenals. However, ACTH and MSH are derived from the same precursor protein, and so MSH is also overproduced. This stimulates the melanocyte to produce more melanosomes with more melanin, and the skin soon darkens. The darkening over pressure points in the skin (e.g., backbone, knees, knuckles, and elbows) is a hallmark of primary adrenal insufficiency (Addison's disease).

At the other end of the spectrum is **albinism,** in which mature melanosomes do not form, leading to very white skin and hair. There are several distinct forms of albinism, each inherited as a genetic recessive. The most common type is due to a lack of activity of tyrosinase, even though antityrosinase antibodies usually can detect the presence of the protein. In melanocytes of these people, the premelanosome buds off the ER, but tyrosinase vesicles are not formed, and so the melanosome remains in its rudimentary state. Albinos have decreased visual acuity as well as great sensitivity to sunlight, especially ultraviolet light. As a result, skin cancers are common by the third decade of life.

About 1 to 3% of all cancers are **malignant melanoma,** which is due to an inappropriate proliferation of altered melanocytes. These usually pigmented lesions of the skin can be distinguished by a clinician from the benign moles on the basis of several pathologic criteria. If not excised at an early stage, this tumor can spread to other organs of the body and be lethal. The precise factors that cause a melanocyte to become a melanoma are not known; however, some melanomas appear to arise from overexposure to the sun.

Further Reading

Anderson, R., and Orci, L. A view of acidic intracellular compartments. *J Cell Biol* 106 (1988):539–543.

Barbier, H., Renaudin, J., and Guern, J. The vacuolar membrane of plant cells. *Biochimie* 68 (1986):417–425.

Blackwell, R., Murray, A., Lea, P., Kendall, A., Hall, N., Turner, J., and Wallsgrove, R. The value of mutants unable to carry out photorespiration. *Photos Resyn* 16 (1988):155–176.

Boller, T., and Wiemken, A. Dynamics of vacuolar compartmentation. *Ann Rev Plant Physiol* 37 (1986):137–164.

Cattley, R., Marsman, D., and Popp, A. Cell proliferation and promotion of the hepatocarcinogenicity of peroxisome proliferating chemicals. *Prog Clin Biol Res* 340D (1990):123–132.

Chekedel, M., and Zeise, L. Sunlight, melanogenesis and radicals in the skin. *Lipids* 23 (1988):587–591.

Del Rio, L., Sandalio, L., and Palma, J. A new cellular function for peroxisomes related to oxygen free radicals? *Experientia* 46 (1990):989–992.

Forster, S., and Lloyd, J. B. Solute translocation across the mammalian lysosome membrane. *Biochim Biophys Acta* 947 (1988):465–491.

Gahl, W. Lysosomal membrane transport in cellular nutrition. *Ann Rev Nutr* 9 (1989):39–61.

Goldfischer, S., Novikoff, A., Albala, A., and Biempica, L. Hemoglobin uptake by rat hepatocytes and its breakdown within lysosomes. *J Cell Biol* 44 (1970):513–529.

Hearing, V., and Jiminez, M. Mammalian tyrosinase: The critical regulatory control point in melanocyte pigmentation. *Int J Biochem* 19 (1987):1141–1147.

Holtzman, E. Lysosomes. New York: Plenum, 1989.

Jimbow, K., Salopek, T., Dixon, W., Searles, G., and Yamada, K. The epidermal melanin unit in the pathophysiology of malignant melanoma. *Am J Dermatol* 13 (1991):179–188.

Kindl, H. Lysosomes and peroxisomes. *Int Rev Cytol Suppl* 17 (1987):325–356.

Kornfeld, S. Lysosomal enzyme targeting. *Biochem Soc Trans* 18 (1990):367–374.

Krogstad, D., and Schlesinger, P. H. Acid-vesicle function: Intracellular pathogens and the action of chloroquine against plasmodium falciparum. *New Engl J Med* 317 (1987):542–548.

Lazarow, P. B. Peroxisome biogenesis. *Curr Op Cell Biol* 1 (1989):630–634.

Lock, E., Mitchell, A., and Elcombe, C. Biochemical mechanisms of hepatic peroxisome proliferation. *Ann Rev Pharmacol Toxicol* 29 (1989):145–163.

Mannaerts, G., and Veldhoven, P. The peroxisome: Functional properties in health and disease. *Biochem Soc Trans* 18 (1990):87–93.

Marin, B. P. (Eds.). *Biochemistry and function of vacuolar ATPase in fungi and plants.* Berlin: Springer-Verlag, 1986.

Matile, Ph. Biochemistry and function of vacuoles. *Ann Rev Plant Physiol* 29 (1978):193–213.

Monnens, L., and Heymans, H. Peroxisomal disorders: Clinical characterization. *J Inherit Metab Dis Suppl* 1 (1987):23–32.

Moser, H. Peroxisomal diseases. *Ave Pediatr* 36 (1989):1–38.

Nelson, N., and Taiz, L. The evolution of H^+ ATPases. *Trends Biochem Sci* 14 (1989):113–117.

Osmundsen, H., Bremer, J., and Pederson, J. Metabolic aspects of peroxisomal beta oxidation. *Biochem Biophys Acta* 1085 (1991):141–158.

Osumi, T., and Fujiki, Y. Topogenesis of peroxisomal proteins. *BioEssays* 12 (1990):217–222.

Pisoni, R., and Thoene, J. The transport systems of mammalian lysosomes. *Biochim Biophys Acta* 1071 (1991):351–374.

Rao, M. S., and Reddy, J. K. The relevance of peroxisome proliferation in peroxisome proliferator-induced hepatocarcinogenesis. *Drug Metab Rev* 21 (1989):103–110.

Rea, P., and Sanders, D. Tonoplast energization: Two H^+-pumps, one membrane. *Physiol Plant* 71 (1987):131–141.

Rudnick, G. ATP-driven H^+-pumping into intracellular organelles. *Ann Rev Physiol* 48 (1986):403–415.

Small, G., and Lewin, A. Peroxisomal proliferation: Mechanisms and biological consequences. *Biochem Soc Trans* 18 (1990):85–94.

Stone, D., Crider, B., Sudhof, T., and Xie, X.-S. Vacuolar proton pumps. *J Bioener Biomembr* 21 (1989):605–621.

Storrie, B. Assembly of lysosomes: Perspectives from comparative molecular biology. *Int Rev Cytol* 111 (1988):53–101.

Szabo, G., Hirobe, T., Flynn, E., and Garcia, R. Biology of the melanocyte. *Adv Pigm Cell Res* 1 (1988):463–474.

Vamecq, J., and Draye, J. P. Pathophysiology of peroxisomal beta oxidation. *Essays Biochem* 24 (1990):115–201.

Vitale, A., and Chrispeels, M. Sorting of proteins to the vacuoles of plant cells. *Bioessays* 14 (1992):152–158.

Von Figura, K. Molecular recognition and targeting of lysosomal proteins. *Curr Op Cell Biol* 3 (1991):642–646.

Wilson, G., Holmes, R., and Hajra, A. Peroxisomal disorders: Clinical commentary and future prospects. *Am J Med Genet* 30 (1988):771–792.

The Cytoskeleton

A Network of Internal Filaments

The picture of a eukaryotic cell as a static structure in which complex processes occur within different compartments is misleading. Far from being immovable bricks in the building of a larger organism, the cells of plants and animals can move and change their shapes. During embryonic development, this happens as cells migrate before differentiating. Some specialized cells are in constant motion throughout their lifetimes, as in certain epithelia and blood cells in animals and pollen tubes and root hairs in plants. Even within most cells, cytoplasmic streaming commonly occurs.

These movements are mediated by an internal cytoskeleton that is based on three major types of structures — microtubules, microfilaments, and intermediate filaments. But as its name implies, the cytoskeleton also has a structural role in determining cell shape, as was seen in the submembrane skeleton of the erythrocyte (see Chapter 2). Because these filamentous structures are polymers, the possibility of cycles of polymerization and depolymerization allows the cell some measure of reversible control over its shape.

The third and least-defined role of the cytoskeleton is in organizing the cytoplasm. The endoplasmic reticulum (ER), Golgi, vesicles, and other membrane-bound organelles are usually associated with cytoskeletal components, and these associations are probably not coincidental. Even the "soluble" cytoplasm may be organized, with the various enzymes bound up in the cytoskeletal lattice so that substrates can be channeled from one enzyme to the next in a biochemical pathway.

The main cytoskeletal structures were first defined by light and electron microscopy. Then they were isolated, and the molecules that make them up were characterized, as well as the ways that they polymerize. The discovery that other proteins attach to these filaments to regulate their polymerization and structure led to the concept that the filaments can in some cases be a "workbench" on which other processes occur. Finally, the function of the cytoskeleton in both plant and animal cells has been a subject of ongoing intensive investigations. These studies have been aided by the availability of mutations for the genes coding for the skeletal components, as well as drugs that specifically bind to them.

Microtubules

Structure

Since the 1800s, light microscopists have reported that cells and their extensions often contain a fibrous component. This was especially noticeable in cilia and flagella, which appeared to have parallel arrays of fibers, and this was confirmed by their birefringence in the polarizing microscope. When the electron microscope was first used to investigate cilia, their fibrous nature was again confirmed, but such fibers were not seen in the cytoplasm.

Unfortunately, two of the techniques that the electron microscopists were using—fixation in osmium tetroxide and cold temperature—tended to destroy cellular filaments. In 1963, D. Slautterback fixed tissues from the invertebrate *Hydra* in glutaraldehyde at room temperature and observed in the electron microscope numerous rods that he termed microtubules. Over the next year, these organelles were found in many tissues in which they had previously been missed.

Microtubules are found in virtually all eukaryotic cells. In animal cells, they usually radiate from a centrosome located near the nucleus, while in plant cells they are more near the plasma membrane. As seen with the electron microscope, they appear in longitudinal sections as long, thin rods (Figure 9–1) or in cross sections as hollow tubes ringed by 13 subunits.

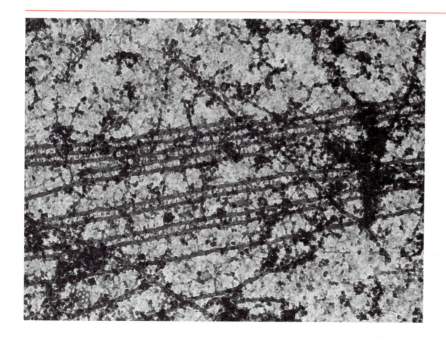

FIGURE 9–1

Microtubules in a protoplast from the plant *Nicotiana plumbaginifolia*. Note the bundling, which is probably caused by associated proteins. (×86,400. Courtesy of Drs. H. Kengen and J. Derksen.)

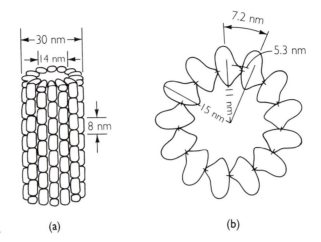

(a)   (b)

FIGURE 9-2

FIGURE 9-2

Diagrammatic longitudinal (a) and cross sections (b) of a microtubule. Each protofilament unit is a tubulin dimer. (After Dr. L. Amos.)

Both microscopy and X-ray diffraction data show that the outer and inner diameters of these tubes are 30 nm and 14 nm, respectively. Therefore, the wall thickness is 8 nm. The lumen of the microtubule is generally electron-lucent and often appears to be "empty," although this term is certainly inaccurate. A clear, unstained zone, 5 to 20 nm in width, often surrounds the outer wall of the microtubule.

The lengths of microtubules vary between cell types and species. For example, in plants, lengths range from 0.3 μm in dividing corn cells to 3.5 μm in root tip cells to an amazing 1 cm in certain mosses. The tubules often appear in highly organized bundles with regularly spaced bridges between the individual units, as in certain protozoan cellular extensions. In other cases, bundles are less regular with fewer cross-bridges, as in the mitotic spindle.

The wall of the microtubule is made up of 13 identical **protofilaments** (Figure 9–2), although exceptions exist, such as those in the crayfish nerve cord, which has 12. These longitudinally arranged structures are in turn made up of subunits composed of dimers of the protein tubulin. These dimers are held together more tightly than the tubule. This can be shown by the fact that under carefully controlled conditions of depolymerization, the protofilaments first dissociate from each other, and then their subunits dissociate.

Chemical Composition

Biochemical analyses of microtubules have been simplified by use of the drug colchicine (Figure 9–3). This alkaloid, from the meadow saffron, has been prescribed by physicians for many years for the treatment of joint pain in gout. In 1889, B. Pernice noted that the drug induced a proliferation of mitotic cells in the dog intestine, and later work showed that the actual effect was a destruction of the mitotic spindle, preventing chromosome separation. Electron microscopy showed that colchicine caused microtubules in the mitotic spindle and other cells to depolymerize.

The availability of radioactive colchicine in the 1960s soon led to the realization that it binds to a single component of the mitotic spindle and that this component is also present in other cell types. The colchicine-binding component was named tubulin, and it occurs in different cells in different amounts (Table 9–1).

Tubulin exists as a dimer, of molecular weight 110,000. It has been purified from a wide variety of plants, animals, and protists. Separation methods include affinity chromatography (using tubulin-binding drugs as ligands), centrifugation, and repeated cycles of polymerization and depolymerization. In the latter technique, tissue is homogenized in the cold to disaggregate the microtubules. Following the removal of cellular debris and cold-stable tubulin by centrifugation, the extract is warmed and the tubulin polymerizes. The microtubules thus formed are then purified by centrifugation and put through a second cold-warm cycle, resulting in organelles that are essentially pure.

The tubulin dimer is composed of alternating monomers, each 55,000 molecular weight, termed α and β, and the protofilament is constructed so that one type of monomer alternates with the other. Dissociation into α- and β-tubulin can be achieved by the use of detergents or similar agents that disrupt noncovalent, hydrophobic interactions between molecules, an indication that the subunits are not held together by covalent bonds.

The two tubulins are rather similar in their amino acid sequences but not identical, with about half of their sequences in common. In fact, an antibody raised against rat brain tubulin will bind to tubulin from carrots, indicating that the basic structure of this molecule and the microtubule itself are the same across the kingdoms of organisms. But the picture is complicated by the occurrence of heterogeneity both within

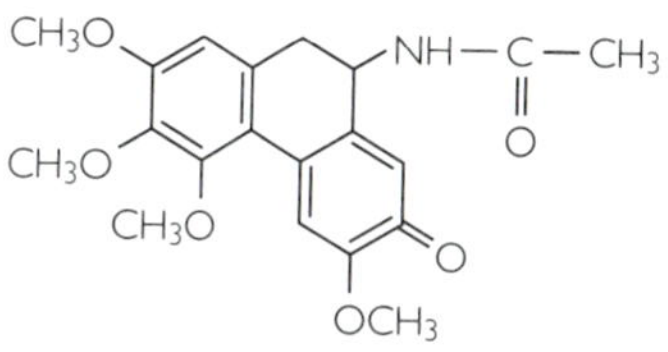

Colchicine

(a)

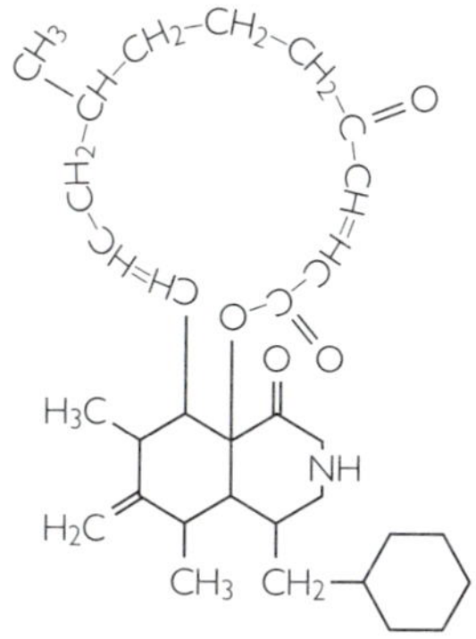

Cytochalasin B

(b)

FIGURE 9–3

Two drugs that affect the integrity of the cytoskelton: **(a)** The microtubule inhibitor colchicine and **(b)** the microfilament inhibitor cytochalasin B.

TABLE 9–1 **Tubulin Contents of Cells**

Cell or Tissue, Organism	Tubulin, % of Total Protein
Lymphocyte, human	0.3
Hepatocyte, rat	0.8
Seedling, cowpea	1.0
Egg, sea urchin	4.0
Platelets, human	12.0
Brain, pig	27.7

and between species. For instance, the parasite *Crithidia fasciculata* has distinct subpopulations of alpha tubulin in its flagella and cytoplasm, indicating some specialization of the monomer.

Microheterogeneity also occurs in most organisms, partly due to the fact that there are multiple genes for the two chains (typically five genes for α and six for β). This heterogeneity is primarily observed at the carboxy-terminal 15 residues of the 450 amino acid polymer. Also, when certain cells are transformed to become cancerous, the primary tubulin in the cellular pool switches from α to β.

Following its synthesis, tubulin may undergo several **covalent modifications:**

1. Tyrosylation is the most unusual modification. An ATP-requiring enzyme, tubulin-tyrosine ligase, adds a single tyrosine residue to the carboxy-terminal ends of most tubulin after it is synthesized but while it is still free. However, another enzyme acts in reverse to detyrosylate the protein once it is at the end of a tubule; on the average, this latter reaction occurs on a fraction of the tubulin molecules.

2. Phosphorylation of serine and threonine residues occurs preferentially in the α subunit.

3. Glycosylation of tubulin via a glucosaminyltransferase occurs in test-tube systems, but its significance for the intact cell is doubtful, since purified tubulin does not contain carbohydrate.

4. Acetylation occurs on lysine residues, especially in ciliary microtubules. Like detyrosylation, acetylation is correlated with increased microtubule stability.

The functions of these covalent modifications are not clear, but there are some correlations with development. For example, tyrosylation rises in developing chick brain between days 9 and 16 and then declines, and tyrosylated microtubules predominate in cells that are rapidly dividing. In the test tube, and in some cells, increased detyrosylation is correlated with increased tubulin polymerization. However, the exact relationships are not known. The same can be said of enzyme activities associated with purified tubulin. Protein kinase, GTPase, ATPase, and protein phosphatase are often detected in "pure" tubulin preparations, and they may have roles in tubulin polymerization.

Several exogenous and endogenous molecules exhibit **tubulin binding,** and all of their binding sites appear to be distinct:

1. In animal cells, there is one colchicine-binding site per dimer, while in plants binding activity is less than one site per 10 dimers.

2. Vinblastine, an anticancer and antimicrotubule drug (see below) binds to two sites per dimer.

3. Calcium ions, which promote microtubule disassembly, bind to 1 high-affinity site and 16 low-affinity sites.

4. Magnesium ions, which promote assembly, bind to 48 sites.

5. Glycerol can bind to tubulin, displacing water and thereby stabilizing assembled microtubules. This may be important in poikilotherms, whose body temperature fluctuates with the environmental temperature. Without glycerol or a similar substance, their microtubules would depolymerize at cold temperatures.

6. The nucleotide GTP, a cofactor for assembly, binds to two sites on a tubulin dimer. This binding domain is very different from that of the signal transducing G proteins (see Chapter 2), which have a common binding sequence.

Microtubules purified by cycles of polymerization-depolymerization do not contain only tubulin. Other proteins are consistently associated with the monomers, often in precise proportions. These proteins, as well as others found bound to tubulin isolated from different tissues, are termed **microtubule associated proteins** (MAPs), and several have been identified (Table 9–2).

MAPs 1 and **2** have been characterized from brain tissue as the HMW (high molecular weight) MAPs. MAP1 is actually composed of three different polypeptides, not surprisingly designated MAPs 1A, 1B, and 1C, while MAP2 has A and B subunits. Sequence data show that there is some homology between these two MAPs, possibly related to a common role in the cell. The tubulin-binding domain of these proteins lies in the carboxy-terminal region.

TABLE 9–2 Some Microtubule-Associated Proteins

Protein	Major Source	Mol. Wt. (Kdaltons)
MAP1	Brain: white matter; cell bodies, axons	350
MAP2	Brain: gray matter; mostly dendrites	270
MAP3	Brain: glia; myelinated axons	180
MAP4	Brain: glia; cultured cells	200
Tau	Brain: axons	60
Kinesin	Squid axon	110
Dynein	Cilia, flagella	>400
Dynein (cytoplasmic)	Brain: axons	>350
Dynamin	Brain	100

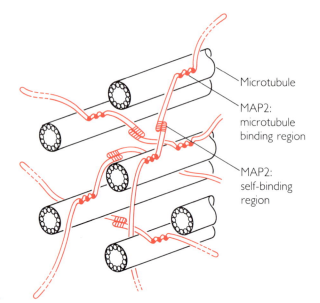

Microtubule

MAP2:
microtubule
binding region

MAP2:
self-binding
region

FIGURE 9–4

Cross-linking of microtubules by MAP2. There are separate regions for the binding of the MAPs to the tubule and to each other. (After N. Cowan.)

When tubulin is polymerized in the test tube with the MAPs, the resulting microtubule appears to be decorated with regularly spaced, thin projections, which are absent from MAP-free microtubules. This is not a mere in vitro phenomenon, since antibodies to brain MAPs bind to regions near microtubules in brain tissue sections. Immunofluorescence has been used to show that the various MAPs have unique tissue and cell distributions (Table 9–2).

The brain MAPs seem to be involved in the three-dimensional interactions of microtubules in the cytoplasm. Injection of MAP2 or tau, another brain MAP, into nonneuronal cells causes a bundling of cellular microtubules similar to that which occurs in neurons. This bundling appears to be actual cross-linking of adjacent microtubules via the MAP, and electron micrographs clearly show this. The MAPs actually have domains that allow them to bind to both microtubules and to each other (Figure 9–4). Clearly, this MAP is responsible for the tight bundles of microtubules seen in axons.

This bundling has morphological consequences. If tau synthesis is suppressed by blocking its mRNA with a complementary RNA (antisense RNA), cultured neurons will elaborate processes (neurites), but they are shorter and less polarized than they would be in cells expressing tau. The indications are that the MAP bundling is not essential for initiation of process extension but rather for stabilization once they are formed.

A second role for the two large MAPs may be to link microtubules with other structures in the cell. For example, MAP1 can bind to neu-

rofilaments, a class of intermediate filament, and MAP2 can bind to actin, the monomer of microfilaments (see below). Tau has similar binding
properties and has been implicated in Alzheimer's disease. In neurons of
people with this disease, dense tangles of fibers are present in areas of
the brain involved with memory. The tangles apparently lead to a disruption of neuronal functioning, and memory loss ensues. Among these
fibers is tau, probably cross-linking them. The structure of tau seems
well suited to this purpose, as it is elastic and can extend to threefold its
normal length, presumably to cover many tubules.

Another MAP with a known function is **dynein,** which binds to
the microtubules of cilia and flagella. This protein is a Mg^{++}-ATPase
and generates the energy for the bending of the cilium or flagellum (see
below). A recently characterized MAP, **kinesin,** is also an ATPase, and
it appears to act in the movement of other substances along microtubules. In tissue culture cells, MAPs often associate with both microtubules and with other filaments, such as microfilaments, thereby linking
different components of the cytoskeleton. In addition, tubulin has been
found in cellular membranes and nuclear chromatin. It appears, then,
that the microtubule can act as a structural frame on which different,
functional MAPs can be attached.

Assembly

Microtubule assembly involves the addition of subunits noncovalently to
lengthen the polymer. It is an impressive process in terms of chemistry,
because the subunits are essentially self-assembling, and in terms of rapidity, which can be up to micrometers per minute. Assembly (and disassembly) are carefully regulated by the cell and occur at specific times
during cell differentiation and the cell cycle.

In terms of the origin of the tubulin subunits, there are two
possibilities:

1. The tubulin dimers are newly synthesized for the purpose of
   lengthening the polymer, or
2. The polymer uses tubulin from a preexisting cytoplasmic pool.

Addition of new dimers occurs in lower organisms during the formation of new cilia and flagella, which contain microtubules (see below).
An intermediate mechanism, involving both synthesis and a pool, occurs
when cilia or flagella are regenerated to replace ones that are lost or
resorbed. In these cases, measurements of the tubulin pool indicate that
there is enough to assemble only about half of the necessary microtubules, and the remaining tubulin is synthesized de novo.

But most cytoplasmic microtubules are made almost entirely from
tubulin pools. When microtubules essential to cells form in protozoa,

cultured mammalian fibroblasts or neurons are depolymerized in situ, they reform from the subunits already present. For example, measurements of the total cellular tubulin in rat liver cells indicate that only about 15% of it occurs in microtubules; the rest is in the soluble pool. The mitotic apparatus, a structure based on microtubules (see Chapter 12), is also formed in this way. In the mature sea urchin egg, the soluble tubulin pool is about 5% of the total egg soluble protein. During the first cleavage, about 5% of the tubulin is used to assemble the mitotic apparatus. As further cell divisions occur, the pool is further depleted until, later in development, new tubulin must be synthesized.

The **polymerization** of tubulin is conveniently studied in the test tube. If adequate concentrations of tubulin, cofactors, and, possibly MAPs, are present, purified dimers will self-assemble into microtubules. This increases the viscosity of the solution, providing a measure of polymerization. It does not occur, and the tubules break down if the temperature is too low or colchicine is present. Given that tubulin has one high-affinity binding site for the drug, it might be expected that 50% inhibition of polymerization would occur when 50% of the free tubulin is bound with colchicine, but such is not the case. Half-maximal inhibition is caused by binding to only 2% of the tubulin in solution.

The substoichiometric colchicine effect is explained by the facts that only free tubulin binds the drug, that the tubulin-colchicine complex can bind to a growing tubule, and that its addition blocks the addition of further tubulin, drug-bound or free.

But this does not explain how colchicine induces microtubules to depolymerize. This phenomenon is caused by an equilibrium that exists between free and bound tubulin:

free tubulin in soluble pool $\rightleftarrows$ bound tubulin in microtubules

Obviously, when a microtubule is growing, the equilibrium is shifted to the right. When colchicine-tubulin dimers bind irreversibly to the microtubule, the equilibrium is shifted to the left.

Experiments using chemically marked microtubules (e.g., one end deriving from a flagellum) show that colchicine-tubulin binds to only one end of the microtubule. At this end, assembly is favored over disassembly. In contrast, the reverse occurs at the other end of the growing organelle, and disassembly is favored. These relationships are shown in Figure 9–5.

Under steady-state conditions, net assembly occurs at the growing end, which equals net disassembly at the other end, and thus the microtubule length remains constant. Data on microtubules from flagella of the alga *Chlamydomonas* indicate that the growing end has a net growth of 1 dimer because of addition of 11 and loss of 10, and the subtracting end loses 1 dimer because of the addition of 3 and loss of 4. Of course,

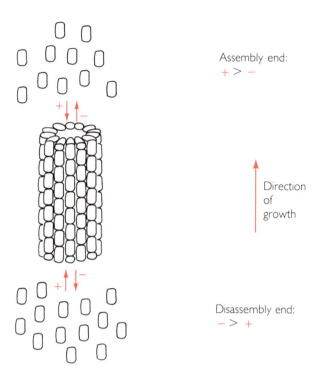

Assembly end:
+ > −

Direction
of
growth

Disassembly end:
− > +

FIGURE 9–5
A model for the growth of a microtubule. For net growth, the rate of assembly is greater than the rate of disassembly. At steady state, the two rates are equal, and there is a constant microtubule length.

if the microtubule is to get longer, net assembly should be greater than disassembly, and it is.

Tubulin subunits added at one end are lost at the other end, a process termed **treadmilling,** which can be demonstrated in the test tube by labeling experiments using GTP. This molecule binds irreversibly to beta-tubulin at the growing end of the microtubule. If microtubules in steady state are given a brief pulse of radioactive GTP, some of the labeled tubulin is lost to the medium, indicating the equilibrium set up at the growing end. However, if the pulse is followed by a chase in unlabeled GTP, little of the labeled tubulin is lost until it has "migrated" to the other end of the microtubule.

This treadmilling occurs at a rate of about 1 μm per hour in vitro. It is maintained by the hydrolysis of tubulin-bound GTP to GDP at the growing end of the microtubule. The less energetic GDP-tubulin is then depolymerized at the other end. But there has not been a convincing demonstration of this process in vivo, and so it remains hypothetical.

An alternative model for microtubule growth attempts to explain the **dynamic instability** of microtubules in vivo. The organelles are able to grow and disappear rapidly, much more so than could be explained by simple addition at one end and subtraction at the other (Figure 9–6). An example of this rapidity is the appearance and disappearance of the mitotic spindle (see Chapter 12). Another example is the

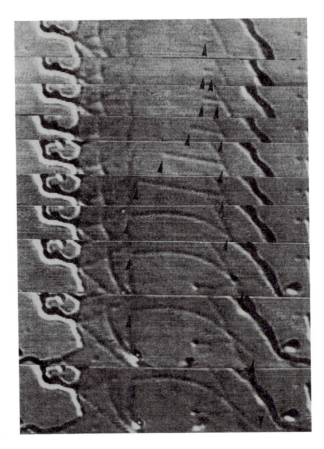

FIGURE 9–6

Rapid microtubule assembly in the giant amoeba *Reticulomyxa*. The numbers refer to seconds. At time 0, the assembly process began, and the growth of the tubule is indicated by the arrows. The bar represents 1 μm. The images are made by electronically enhanced interference contrast optics. (×3300. Courtesy of Dr. M. Schliwa.)

rapid, oriented elongation of microtubules in polymorphonuclear leukocytes toward the source of a stimulating chemical. These changes in length are in micrometers.

Instability can be estimated by the use of fluorescently tagged tubulin. If this is allowed to add to microtubules in vivo, much of it is transferred from the microtubule to the soluble pool within 15 minutes, indicating a very rapid turnover time for the organelle. Observations of the lengths of such tagged microtubules with the fluorescence microscope bear this out: There are frequent elongation and shortening cycles.

T. Mitchison and M. Kirschner have proposed that microtubules whose terminal tubulin has bound GTP will not decrease in length. If the GTP is hydrolyzed to GDP, rapid shrinkage can occur, and the rate of this shrinkage is dependent on a very high dissociation constant for tubulin-GDP as compared with tubulin-GTP. A typical microtubule might have a core of tubulin-GDP, which is "protected" from depolymerization by a tubulin-GTP cap. When the core is exposed (by treadmilling or hydrolysis of GTP), the microtubule collapses. "Rescue" of the collapsing tubule can be achieved rapidly if a tubulin-GTP cap is added (Figure 9–7).

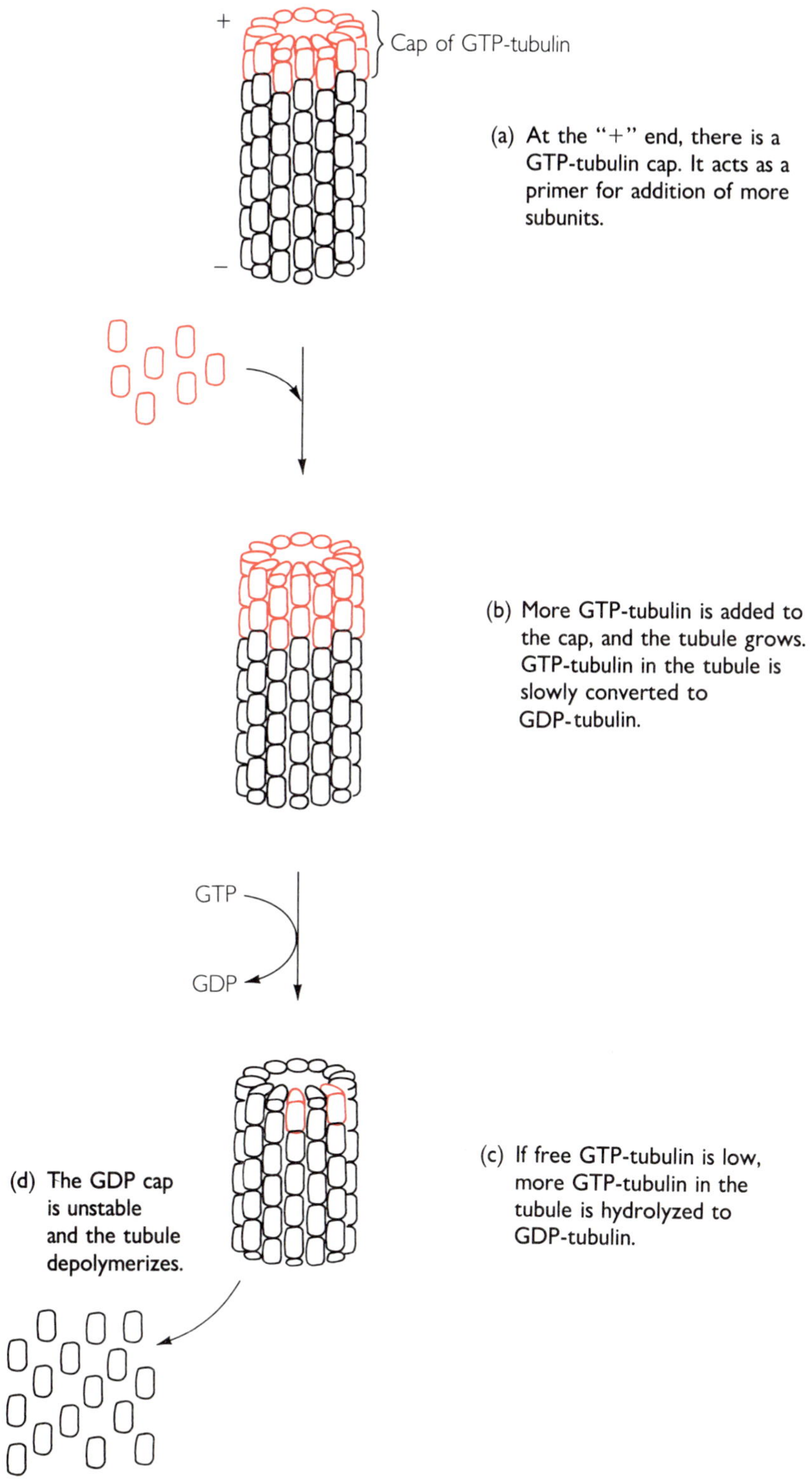

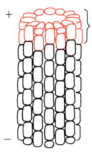

Cap of GTP-tubulin

(a) At the "+" end, there is a GTP-tubulin cap. It acts as a primer for addition of more subunits.

FIGURE 9–7

A model for the dynamic instability of microtubules.

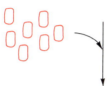

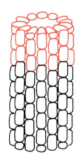

(b) More GTP-tubulin is added to the cap, and the tubule grows. GTP-tubulin in the tubule is slowly converted to GDP-tubulin.

GTP

GDP

(d) The GDP cap is unstable and the tubule depolymerizes.

(c) If free GTP-tubulin is low, more GTP-tubulin in the tubule is hydrolyzed to GDP-tubulin.

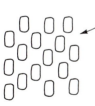

The factors that control instability are not known, but one could be covalent modification of the tubulin itself. Genetic evidence indicates that an important event in the progression of cells from interphase to metaphase during cell division is the action of a protein kinase (see Chapter 12). One of the targets of this kinase is tubulin, and the phosphorylation of tubulin appears to be necessary for its rapid polymerization to form the mitotic spindle.

The initial polymerization of tubulin often occurs at certain regions of the cell termed **microtubule organizing centers** (MTOCs), and the location and orientation of these regions determine the pattern of future growth of the organelles. MTOCs have several forms:

1. Basal bodies form the microtubular doublets of cilia and flagella and have a defined structure.

2. Centrioles, which form the mitotic spindle in animal cells, have a more amorphous structure.

3. Kinetochores are regions on chromosomes from which spindle microtubules are also elaborated.

4. The MTOCs that form the spindle in plant cells appear in the electron microscope as dense regions, often associated with the plasma membrane.

5. In interphase cells, the major MTOC is termed the centrosome. In cells with centrioles, it contains them and often is located near the nucleus.

An MTOC can promote microtubule assembly in the test tube. For example, isolated centrioles form asterlike structures when supplied with tubulin, and kinetochores from human chromosomes form spindlelike bundles when supplied with chick brain tubulin.

Structural polarity can be shown if microtubules are polymerized in unusual buffer conditions, including dimethylsulfoxide. Hooked protofilament appendages, which appear clockwise when viewed in cross section from a point distal to the MTOC, form in these conditions. These appendages, therefore, are markers of tubule structural polarity, and microtubules elongated from MTOCs in the test tube have the same polarity.

Polarity studies have shown that the MTOC appears to regulate microtubule growth by rendering the depolymerizing ("minus") end biochemically inert. Because tubulin is no longer lost from the minus end (perhaps because of treadmilling), any addition of tubulin at the plus end results in net growth of the microtubule. Microtubules radiating from an MTOC are polarized such that the "minus" ends are near the MTOC, and the "plus" ends are distal to the MTOC.

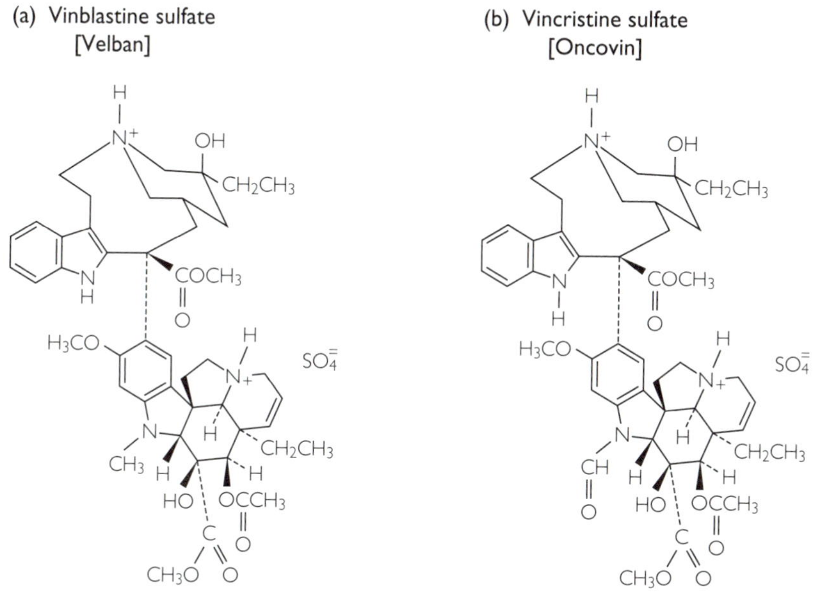

(a) Vinblastine sulfate
[Velban]

(b) Vincristine sulfate
[Oncovin]

FIGURE 9–8

Two anticancer agents from the periwinkle plant *Vinca*. They bind to tubulin and prevent polymerization.

As noted above, other **exogenous agents** besides colchicine can bind to microtubules, and many of them affect assembly. The vinca alkaloids vinblastine and vincristine have been purified from the periwinkle plant, *Vinca rosea,* and are widely used to treat rapidly growing cancers, such as lymphomas, lung and breast cancers, and leukemias (Figure 9–8). These drugs act to block cell division by binding to tubulin dimers at a site distinct from the colchicine-binding site and then forming crystalline arrays of tubulin. The equilibrium of free:bound tubulin (see above) is shifted to the free monomer, resulting in loose arrays of tubulin crystals and multinucleate cells.

Unfortunately, these alkaloids also inhibit division of other rapidly proliferating tissues, such as the intestinal epithelium and bone marrow, and so hair loss, anemia, and leukopenia (depressed numbers of white blood cells) are side effects of treatment. In addition, because microtubules are involved in axonal transport (see below), neurotoxic effects such as loss of reflexes, headache, and convulsions can occur. Nevertheless, these drugs are widely used in cancer chemotherapy with considerable success.

Several **endogenous agents** affect microtubule assembly. If cells are treated with the ionophore A23187, which increases intracellular Ca^{++}, or if cells are microinjected with Ca^{++}, microtubule assembly is blocked and disassembly is promoted. These observations are consistent with Ca^{++} inhibition of tubulin polymerization in the test tube. Because the cytoplasmic concentration of Ca^{++} can be finely regulated by intracellular pumping and sequestration into mitochondria or the ER (see Chapters 3 and 5), the idea of using this cation to control tubulin polymerization is attractive. But if and exactly how this occurs in the cell are not known.

Also unclear is the possible role of MAPs in assembly. Although MAPs 1 and 2 and tau lower the effective tubulin concentration needed for polymerization in the test tube, several other substances, both physiological and nonphysiological, have the same effect. High concentrations of Mg^{++}, along with low Ca^{++}, as well as polycations such as polylysine and histone, promote microtubule formation.

Functions: Cell Structure and Motility

The role of microtubules in maintaining **cell structure** can be inferred from electron micrographs from a wide variety of tissues. For instance, in the neuronal axon, bundles of microtubules lie parallel to the longitudinal axis, and in growing plant cells, cortical microtubules lie parallel to the axis of cell elongation. But these images are correlative and do not prove that microtubules cause the structure observed. One method to show cause and effect is to disaggregate the microtubules with a specific inhibitor and then determine if the cell shape has changed. Fortunately, a molecule somewhat specific for microtubule depolymerization, colchicine, is available. This drug has been applied to a wide variety of cells in many different situations.

Echinosphaerium is a large heliozoan amoeba. Projecting from the cell surface are long (400 μm), thin (10 μm) axopods that are used to capture food that initially adheres to them. The axopods then "melt" to the cell surface, thereby delivering the food to an endocytotic vacuole. These axopods have filaments running along their length that are clearly seen under polarizing microscopy. The electron microscope reveals these to contain two interlocking coils of up to 500 microtubules, each one traversing most of the length of the axopod.

If these amoebae are exposed to colchicine, the axopods collapse (Figure 9–9). There is no birefringence in the polarizing microscope, and no microtubules are seen in the electron microscope. Similar results are obtained with low temperature. In this case, raising the temperature leads to a reformation of the microtubules and concomitant regrowth of the axopods. These experiments clearly implicate microtubules in the formation of these cellular structures.

In plants, tracheary elements are part of the xylem, the tissue that conducts water throughout the plant. These elements are derived from undifferentiated cells that first elongate and then form elaborate secondary cell walls. This whole process can be studied in culture, and if an

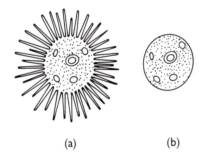

(a)   (b)

FIGURE 9–9

Cells of the protozoan *Echinosphaerium* before **(a)** and after **(b)** treatment with colchicine disaggregate microtubules. The extending axopods have bundles of microtubules running parallel along their axis.

undifferentiated cell is given the proper hormones, it will form a xylem element within four days.

Early on, the undifferentiated cells' microtubules appear in a random orientation in the cytoplasm. Then, new tubules are made, and they become oriented parallel to the longitudinal axis of the cell. Finally, there is massive tubulin synthesis, and the new and existing tubules become oriented perpendicular to the cell's long axis. This latter orientation is exactly like the orientation of the cellulose fibrils in the mature element, which then differentiates (Figure 9–10). If colchicine is applied to these cells, they fail to form the elaborate wall structure. Although the exact role of microtubules in wall formation is unclear (see Chapter 13), they do exert a profound effect on wall fibril orientation.

The application of colchicine to *Echinosphaerium* or plant cells inhibits **cytoplasmic motion** in the cells. This observation, made during the 1960s, led to similar experiments on cellular and intracellular motility in other systems.

In neurons, material is constantly flowing from the cell body down the long axon to the terminal region, where synaptic transmission to the next cell occurs. There are three general classes of transport:

1. Fast (2–5 μm/sec): moves glycoproteins and acetylcholinesterase, usually in granules and vesicles

2. Medium (0.2–0.6 μm/sec): moves mitochondria and their proteins, as well as some filaments

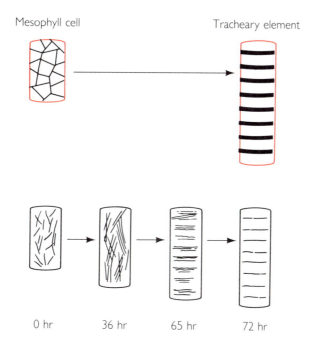

FIGURE 9–10

Reorientation of microtubules in Zinnia mesophyll cells in culture as they differentiate to form xylem tracheary elements. Initially, the tubules are randomly oriented (see bottom series), then they successively become oriented along, and finally they are perpendicular to the long axis of the cell. In the latter orientation, they presage the orientation of the cell wall of the mature xylem cell (see top series).

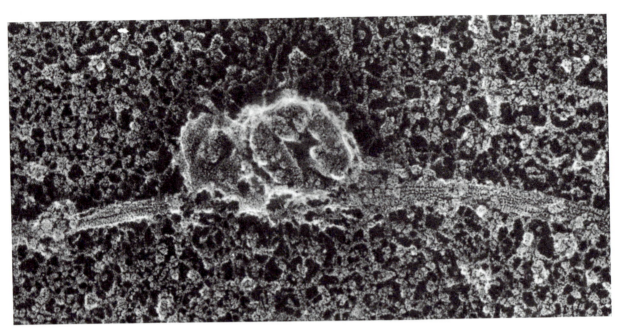

FIGURE 9–11   An organelle attached to microtubules in the axon of a crayfish neuron. (×100,000. Courtesy of Dr. R. Vale.)

3. Slow (0.002–0.01 μm/sec): moves actin, tubulin, microfilaments, and microtubules

Both fast and slow transport appear to move substances away from the cell body (anterograde transport, as opposed to the retrograde transport toward the cell body). While anterograde transport carries materials to the synapse at the periphery of the neuron, retrograde transport carries substances from the synapse back to the cell, such as nerve growth factor, tetanus toxin, and certain viruses such as rabies.

Because the axon is full of bundles of microtubules, they have been implicated in transport (Figure 9–11). Experiments with colchicine confirm this. For instance, fast transport is inhibited if colchicine is applied to the neuron.

Teleost fish can change color quickly to camouflage themselves when predators are nearby, and these color changes are mediated by microtubules. The pigments involved are in granules called chromatophores. When they are in the cell center, the fish has a light color; when they disperse to the cell periphery, the fish turns darker.

Dispersal occurs along processes that contain bundles of microtubules. Unfortunately, they are rather insensitive to colchicine, but cold temperature, which disaggregates cytoplasmic microtubules, inhibits pigment granule migration. Microtubules are even involved in the movement of large cell organelles. In the alga *Closterium* the microtubules form a trail along which the newly formed nucleus is drawn following cell division. A distance of 20 μm is traveled in seven minutes.

There are two possibilities for the mechanism of movement of organelles along microtubules:

1. The organelles have their own propulsive machinery, and microtubules form channels along which the organelles move.

2. The microtubule has the motor that propels the organelle along. This motor would be expected to attach to both microtubules and organelles and have ATP binding and hydrolysis activities.

These organelle movements have been investigated by a combination of microscopic and biochemical methods. R. Allen developed a method to electronically enhance the image produced by interference contrast (Nomarski) optics. This allows the visualization of individual microtubules in the cell and in vitro (Figure 9–6). It is possible to remove the cytoplasm from cells such as the neuron (axon), amoeba, and fibroblast and show that isolated microtubules support the movement of vesicles. This movement is stimulated by the addition of ATP, suggesting that a motor that requires energy is involved.

Simply put, the evidence favors the existence of a microtubular motor. In fact, it has been isolated and called **kinesin** (from the Greek word *kinein,* to move). When extracted from cells, this protein has the ability to cross-link between microtubules and latex beads. Addition of ATP causes the beads to move on the microtubules at a rapid rate similar to the movement of organelles and vesicles on microtubules in vivo. These movements can be quantified by electronically enhanced microscopy.

Kinesin can be purified from cells because it binds tightly to AMP-PNP, a nonhydrolyzable analogue of ATP. This binding induces very stable cross-links through kinesin between organelles and microtubules. When these complexes are isolated and ATP is added, the analogue is unbound, and tubulin-induced ATP hydrolysis induces unidirectional movement of the organelles along the microtubule toward the "plus" (net growing) end.

Kinesin is a large molecule (Table 9–2), with at least two each of subunits of molecular weights 120K and 65K. The heavier chains contain sites for both the tubulin and ATP binding. Under the electron microscope, kinesin appears as an elongated structure over 100 nm in length, with a stalk that attaches to the microtubule and a globular head that attaches to the organelle. The head region has ATPase activity (Figure 9–12).

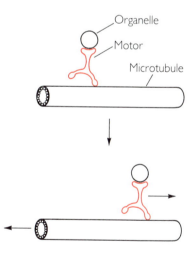

FIGURE 9–12

Movement of an organelle along a microtubule via a microtubule motor (e.g., kinesin). The motor is attached to the organelle by its stem. The flexible head binds to tubulin, then detaches, moves ahead, and reattaches. The cycle is catalyzed by an ATPase in the head that is activated by tubulin.

Kinesin moves along a microtubule in a rather straight line, as it attaches to a protofilament of a microtubule then detaches and reattaches to the next protofilament down the wall. All the while the motor remains firmly attached to the organelle, so the latter appears to move along the microtubule.

The functional importance of kinesin is indicated by a mutation, unc-104 in the worm, *Caenorhabditis elegans.* These worms show uncoordinated movements (hence the abbreviation *unc*) due to defective interactions at the nerve-muscle synapses. Instead of neurotransmitter-filled vesicles accumulating at the presynaptic region, they stay in the cell body. This implies some defect in intracellular transport along the axon, and indeed molecular analysis of the wild type gene for the unc-104 mutation shows that it codes for the heavy chain of kinesin.

In axons, kinesin appears to be the motor only for anterograde transport. Antibodies to kinesin injected into neurons inhibit anterograde transport but have no effect on transport toward the cell center. The motor in this case is MAP1C, which has been shown to cause organelle movement along microtubules in vitro in a direction opposite to that induced by kinesin. This protein has been called **cytoplasmic dynein,** since like the protein in cilia (see below) it uses energy to move toward the "minus" end of the microtubule. Possible roles for this motor in nonneuronal cells are in the transport of organelles toward the cell center and in endocytosis.

A third microtubule motor in nonciliated cells, **dynamin,** acts in a way quite different from the other two. In this case, both ends of the motor are attached to adjacent microtubules, so that the net effect is cross-linking. This motor binds and hydrolyzes GTP and so can be classified as a "G protein." Hydrolysis of the GTP is implicated in causing the microtubules to slide past one another.

Function: Cilia and Flagella

Cilia and flagella are motile cell projections in which microtubules and MAPs are important components. Cilia are short (5–10 μm) and numerous (hundreds or thousands per cell). They occur in Protozoa, such as *Paramecium,* where they propel the organism through its aqueous environment and serve a similar function in invertebrates such as echinoderm larvae. Ciliated epithelia line the airway and oviduct of mammals, where their action moves liquid and mucus along and helps eliminate foreign solids (Figure 9–13). Flagella are long (usually greater than 150 μm), and there are one or two per cell. They occur as propulsive organelles on cells in an aqueous environment, such as flagellated protozoa and mammalian sperm.

The microtubules of cilia and flagella come from specialized MTOCs, the **basal bodies.** These structures, 0.2 μm in diameter and

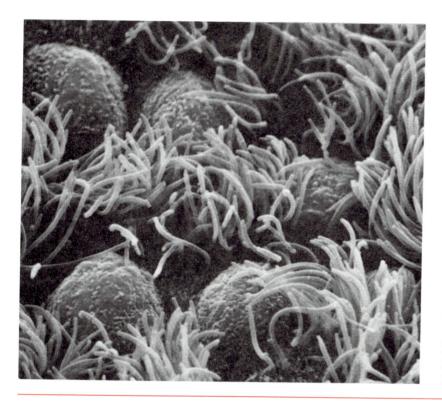

FIGURE 9–13
Cilia lining the epithelium of the
human bronchus. (Copyright
Fawcett/Gehr, Science Source/
Photo Researchers.)

0.5 μm in length, are cylinders open at both ends except in the case of a cilium, where a ciliary plate separates the tubules from the MTOC. Basal bodies are often identical to the centrioles and have their own proteins in addition to tubulin. There is even some evidence that they have their own DNA, although probably not their own machinery for synthesizing proteins, like the plastids and mitochondria. Nevertheless, basal bodies should be included in the serial endosymbiosis theory (see Chapter 5).

Arising from the basal body are nine microtubule triplets plus a central singlet. Two of the subtubules of each of the nine triplets give rise to doublet microtubules in the cilium or flagellum. The third one terminates at the cell boundary except in insects, where it forms an additional ciliary microtubule. This growing projection is enclosed within an extension of the plasma membrane. The basal body remains as the cellular point of attachment for the microtubules of the mature flagellum or cilium (see Figure 9–14).

In cross section, a cilium (and flagellum) has a "9 + 2" configuration, with nine microtubule doublets (A and B subfibers) around a central pair (Figure 9–15). In addition to the tubules, there are several prominent MAPs that appear to cross-link them. The A subfibers have

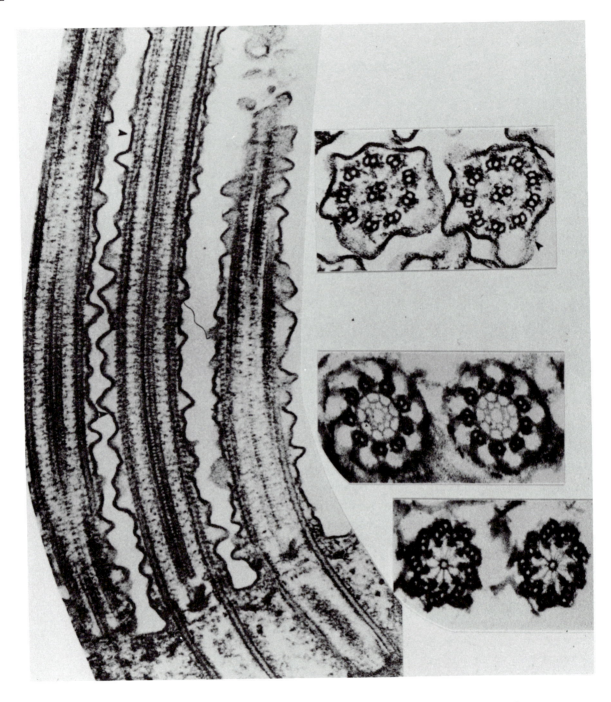

FIGURE 9–14 Longitudinal *(left)* and transverse *(right)* sections of cilia from the protozoan *Tetrahymena* showing the microtubular arrangements in the basal body *(bottom)* and up through the cilia *(top)*. Note the triplets of microtubules in the basal body and doublets in the cilia. (Photographs courtesy of Dr. W. M. Dentler.)

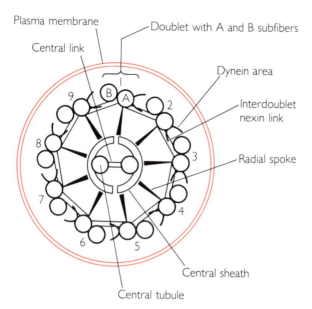

Plasma membrane
Central link
Doublet with A and B subfibers
Dynein area
Interdoublet nexin link
Radial spoke
Central sheath
Central tubule

FIGURE 9–15

Diagrammatic cross section of a cilium.

microscopically distinct structures (MAPs) attached to them, the most obvious being the paired dynein arms and radial spokes. Longitudinally, the dynein arms occur every 24 nm along the microtubule, and the central "sheath" is projected from each central tubule at 16-nm intervals. The nexin links, which connect the A subfibers of the doublets, are spaced down the tubules at 96-nm intervals.

Variations in the "9 + 2" pattern are widespread. One striking variation is the occurrence of nine additional fibers, situated outside the doublets and apparently not connected to them. This arrangement is seen in some mammalian and insect sperm tails, where these extra fibers appear to be involved in stiffening the flagellum.

Chemical analysis of ciliary proteins is possible if the membrane is digested away with a detergent. Whereas the structural diagrams show only a few major proteins, two-dimensional SDS-acrylamide gel electrophoresis reveals about 130 polypeptides, and only some of them have been characterized.

Not surprisingly, over 70% of the total protein is **tubulin,** the major structural building block of the microtubules. Both α and β tubulins are present in equimolar amounts, and their properties are similar to the tubulins of cytoplasmic microtubules, except that ciliary alpha-tubulin is acetylated.

The best-characterized of the flagellar MAPs is **dynein** (dyne: a unit of force), which composes about 15% of the total protein in the organelle. Dialysis of demembraned cilia against the chelator EDTA at low ionic strength removes the dynein arms (as well as the other MAPs) from the A subfibers. If ions such as Mg^{++} are added back in adequate amounts, the arms will reattach, indicating that they are held on the tubules ionically.

The solubilized dynein arm has strong Ca^{++}-Mg^{++} activated ATPase activity. It is a large molecule, with two to three individual chains of molecular weight about 400,000. Intermediate (two to three chains of molecular weight about 100,000) and small (four to six chains of molecular weight about 15,000) subunits appear to be responsible for attachment of the dynein onto the A microtubule. The large complex has the overall structure of other microtubule motors (Figure 9–12), with a fixed stem and flexible, globular arms that have ATPase activity.

If time-lapse cinematography is used to follow the movements of a flagellum, it is clearly shown that the beat occurs as a series of circular arcs, with the arcs moving toward the tip. A wave motion is generated (Figure 9–16a). Individual cilia are more difficult to visualize because of their small size and large numbers, but their motion also involves bending. In this case, there is a rigid effective stroke and a bending recovery stroke (Figure 9–16b). The propagation of waves in cilia and flagella provides energy to propel the organism and/or move the liquid environment. But because energy is released for this movement of the extracellular medium, the wave propagation must be continuously replenished in an energy-requiring process.

The mechanism for generating bending is proposed by B. Afzelius and others to be **sliding of the doublet microtubules.** Microtubules cannot contract and do not bend in the flagellum or cilium. But in a cilium, for example, the doublet at the leading edge of the effective or recovery stroke slides toward the tip relative to the doublets at the opposite side of the cilium (Figure 9–17). Sliding is an energy-requiring step involving the breaking and reformation of dynein cross-bridges between the A subfiber of one doublet and the B subfiber of the adjacent doublet. Hydrolysis of ATP by dynein ATPase provides energy for this process.

The movement of dynein along the microtubule toward the "minus" end indicates that it is a motor, in this case very similar to dynamin, that acts to move microtubules past each other in nonciliated cells (see above). Indeed, dynein can act just like dynamin in the test tube, since either will translocate microtubules if attached to a glass surface and supplied with ATP. But by itself, sliding cannot cause the cilium to bend. In fact, the microtubules will just slide past each other and the cilium will fall apart if the process continues. Instead, the radial spokes and nexin links remain intact during sliding and this shear resistance generates bending.

There is considerable evidence consistent with the sliding microtubule model:

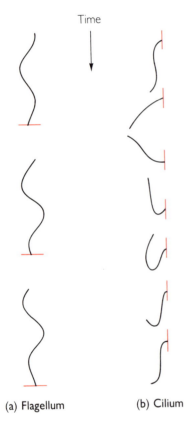

Time

(a) Flagellum   (b) Cilium

FIGURE 9–16

Movements of a flagellum **(a)** and cilium **(b).** In both cases, movement of the cell is down the plane of the page. Note the propagation of the wave in the flagellum and the bending of the cilium.

1. Electron micrographs of cross sections of cilia tips during various phases of the beat cycle show that doublets at the leading edge do indeed project into the tip more than doublets at the opposite side.

2. Flagella can be isolated and, following membrane removal, they are not active. But if ATP and Mg^{++} are supplied, the flagella beat regularly until the nucleotide is used up.

3. Microinjection of an ATP pulse at a localized region of a demembraned flagellum causes the flagellum to bend only at that region.

4. Dynein must be present for the flagellum to bend. If it is removed and ATP is added, bending does not occur.

5. Strong evidence for the sliding model comes from experiments on isolated flagella. If they are briefly incubated with the protease trypsin prior to ATP addition, the nexin and radial spokes are broken down, and bending does not occur. But when ATP is then added, the flagellum progressively disintegrates. Seen under the electron microscope, the microtubule doublets have "telescoped," that is, slid relative to one another toward the flagellar tip (Figure 9–18). The explanation for this phenomenon is that before ATP is added, the dynein arms are attached to both the A subfiber of their "home" doublet and the B subfiber of the

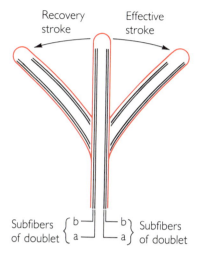

FIGURE 9–17

Model for sliding microtubule doublets during ciliary motion. The doublet lengths remain the same, but they slide relative to one another, generating local bending because they are held to the central region and other doublets by the radial spokes and nexin links (see Figure 9–15). (After P. Satir.)

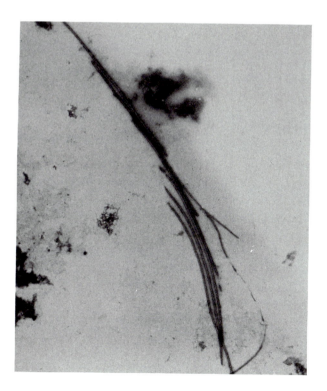

FIGURE 9–18

Telescoping microtubule doublets from cilia of the rabbit trachea. The cilia have been digested with trypsin, and ATP has been added. This caused the dynein to move along the doublets, but there were no radial spokes or nexin links to hold them in place. (×20,000. Photograph courtesy of Dr. E. R. Dirksen.)

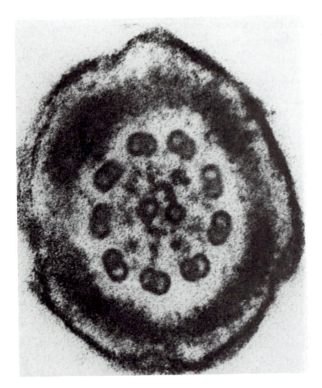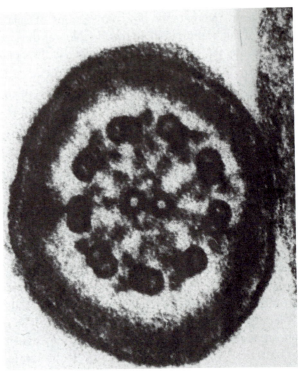

FIGURE 9–19 Cross sections of human sperm tails from a normal man *(right)* and a man with immotile-cilia syndrome *(left)*. Note the absence of dynein arms in the latter. (Photographs courtesy of Dr. B. Afzelius.)

adjacent doublet. When ATP is added, detachment of the arm from the B fibers occurs, leaving the doublets free to slide relative to each other.

6. Genetic evidence indicates the importance of individual ciliary MAPs in the bending process. In humans, about 1 person in every 20,000 born has the immotile cilia syndrome. This is a recessively inherited genetic disease in which ciliary and flagellar functions are severely limited or absent. Chronic bronchitis, sinusitis, and pneumonitis, as well as male sterility due to immotile sperm, occur in these patients. Electron microscopy shows missing MAPs, and an absence of dynein (Figure 9–19), nexin, and radial spokes has been reported in different individuals.

Similar mutations have been studied in the unicellular, flagellated alga *Chlamydomonas*. In the case of one such strain, flagella are immotile despite the fact that microtubule doublets slide when reactivated with ATP. This mutant lacks functional radial spokes, thus showing their importance in the bending mechanism. Several genes have been characterized that suppress the radial spoke defect. This indicates that a number of the unidentified flagellar polypeptides are involved in regulating the function of the organelle.

Microfilaments

Composition and Structure

In addition to containing microtubules, cells have many smaller microfilaments. By conventional electron microscopy, these are shorter ($1-2$ μm) and thinner ($5-7$ nm vs. 30 nm) than the microtubules. While there had been many correlations of these filaments with cell structure and function, studies of them were greatly facilitated when it became clear that they are composed of the protein actin.

The discovery of actin filaments in nonmuscle cells was made in 1952 by A. Loewy, when he added ATP to a cytoplasmic extract of a slime mold, *Physarum*. The result was an immediate drop in viscosity of the extract, coupled with the release of phosphate, showing that an ATPase was present. Because no membrane-bound organelles (such as mitochondria) were present, he proposed that some sort of contractile protein, perhaps similar to muscle actomyosin, was responsible for his observations.

But the demonstration that microfilaments of higher eukaryotes are composed of actin came from two techniques developed in the 1970s. The first is **immunofluorescence.** The principle of this technique is to use a fluorescent antibody to a muscle protein (or, more commonly, a fluorescent antibody to the antimuscle protein antibody) to localize the protein in a nonmuscle cell. The results of this procedure can be quite beautiful (Figure 9–20), with the microfilaments often seen as parallel arrays. In tissue culture cells, these are termed stress fibers, implying a role in cell structure and motility. But in other cells, the arrangement is more diffuse.

The second technique involves **decoration with heavy meromyosin,** the portion of the muscle myosin molecule that interacts specifically with muscle actin by forming a cross-bridge. Actin treated with heavy meromyosin not only binds the fragment but does so in a specific fashion (Figure 9–21). Arrowheads "decorate" the actin filaments, all of them pointing in the same direction. In muscle cells, the arrowheads point in the direction of sliding when actin slides past myosin during muscle contraction, and the same is believed to occur with microfilaments in nonmuscle cells. This means that heavy meromyosin decoration not only indicates the presence of actin in microfilaments but also gives information on their polarity.

Actin is one of the most abundant cellular proteins, accounting for up to 25% of the total protein in some nonmuscle cells. Although these cells contain myosin, the ratio of actin to myosin is much higher than in muscle. For example, rabbit skeletal muscle has an actin:myosin ratio of 6, whereas in the human platelet this ratio is 110. This implies that microfilaments are largely actin rather than typical muscle actomyosin. Amino acid sequences reveal striking (over 95%) similarities between muscle and nonmuscle actins, including the presence in both of the unusual amino acid N-methylhistidine.

Actin is a polymer (F-actin) of individual globular monomers (G-actin). **Polymerization** starts with a rate-limiting step where three monomers come together with ATP and Ca^{++}, forming a nucleus similar to that formed in microtubule polymerization. But there is no analogue of an MTOC to give the growing filament polarity; instead, growth of the polymer occurs at a rapid rate and at both ends but is tenfold more rapid at the "nonpointed" end of the filament (if it is decorated with heavy meromyosin).

Each monomer of G-actin has a bound ATP, which is hydrolyzed to ADP when the monomer is added to the growing chain. Actin filaments form a double helix, 6 nm wide, with globular units 5.5 nm long

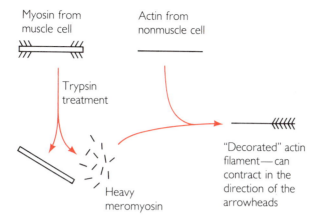

FIGURE 9–21

Determination of the polarity of
an actin microfilament.

and a 37-nm helix repeat. Like muscle actin, the nonmuscle filaments can reversibly bind myosin in the test tube and activate its ATPase activity in the presence of Mg^{++}. Presumably, these reactions occur in the cell as well.

There are a number of **microfilament binding proteins,** and the best-characterized of these is myosin. Nonmuscle myosin is similar to that of muscle cells in that it has two heavy and four light chains and forms "heads" with ATPase and actin-binding activity. It differs from muscle myosin in that it forms smaller filaments (10 nm $\times$ 0.3 μm vs. 18 nm $\times$ 1.5 μm). Calmodulin (similar to muscle troponin), the Ca^{++}-binding protein, affects cytoplasmic actin-myosin interactions. It acts as a regulatory subunit of a protein kinase that phosphorylates myosin light chains, thereby rendering myosin ATPase insensitive to actin activation. Other components of the muscle regulatory system, tropomyosin and α-actinin, are also present in nonmuscle cells. But because it is in such great excess, most nonmuscle actin does not interact with myosin but instead acts as a structural framework. The high association constant of actin monomers ensures that virtually all of the protein is in polymers.

Regulation of polymerization occurs via specialized actin-binding proteins, some of which are outlined in Table 9–3. These can be classified into several types, depending on their actions:

1. Proteins that break actin filaments. **Gelsolin** breaks microfilaments by inserting between two adjacent actin subunits and capping the ends of the growing filaments, preventing polymerization. Sequence data show that the region of gelsolin that binds to actin is at least 60% similar to the site on G-actin involved in actin-actin interactions. This leads to the proposal that gelsolin acts by mimicking an actin monomer, binding to the end of a growing filament and blocking further elongation.

TABLE 9–3 Some Microfilament-Regulating Proteins

Name of Protein	Effect
α-Actinin	Bundles actin filaments; binds them to membranes
Capping protein	Inhibits polymerization at barbed end
Filamin	Cross-links actin filaments into a gel
Fimbrin	Bundles actin filaments
Gelsolin	Breaks actin filaments
Profilin	Binds to actin monomers to prevent polymerization
Spectrin	Cross-links actin filaments
Villin	Cross-links and bundles actin filaments
Vinculin	Anchors actin filaments to membranes

However, gelsolin also binds actin monomers to form a nucleus, thus promoting assembly. Which of these opposing activities predominates seems to depend on the cell Ca^{++} concentration. If the latter is <1 μM actin polymerization is favored, while at >1 μM, depolymerization readily occurs.

2. Proteins that cross-link actin monomers in filaments. **α-actinin** has more than one binding site for actin and so can bind to more than one actin subunit in a microfilament. The actin-binding domain of this protein is similar in structure to those of gelsolin and the other binding proteins. A minor muscle protein, dystrophin, also has this actin-binding domain. This protein is the one that is mutated in the hereditary disease Duchenne muscular dystrophy. α-actinin (and dystrophin) also has a domain that binds to an integral membrane protein, a property that allows it to act as the link between the cytoskeleton and the plasma membrane. This is a similar situation to the structural role of band 3 protein in erythrocytes (see Chapter 2).

Function: Cell Motility

There is strong evidence for actin filament involvement in **localized cell membrane movements.** A good example of this phenomenon is the brush border of the intestinal epithelium. This is the cell layer that lines the lumen of the gut, and each cell has a plasma membrane that is folded at its apex into microvilli to increase its absorptive surface area (Figure 9–22).

These microvilli contain parallel arrays of 20 to 30 microfilaments that run along the longitudinal axis and are bundled by cross-bridges made up of actin-binding proteins such as fimbrin and villin (Table 9–3). The "plus" ends of the actin filaments point toward the tip, and they are

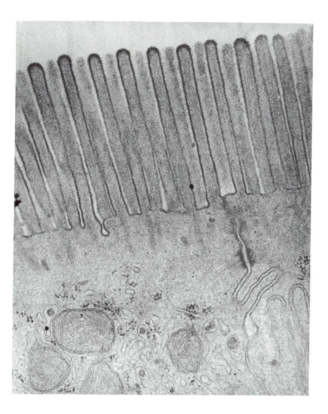

FIGURE 9-22
Isolated brush border from the mammalian intestine showing the bundles of filaments in the microvilli. (Photograph from Dr. K. Porter.)

connected to the adjacent plasma membrane via protein cross-bridges. At the tip of the microvillus is an electron-dense region that may serve as the growing point for actin polymerization. At the base of the microvillus is a second set of microfilaments, which run perpendicular to the projections and appear to interact with myosin and filaments of a fibrous protein similar to erythrocyte spectrin (see Chapter 2), which anchors the structure (Figure 9–23).

Microvilli are in constant motion, apparently to circulate the fluid around them and increase absorption. This movement is probably mediated by the sliding of actin filaments toward the base of the microvilli, a sliding that could be generated by the myosin present, cross-bridges being broken by a regulating protein if necessary. If brush borders are isolated, they contract if ATP is supplied, an observation consistent with the involvement of actin and myosin.

Another example of actin involvement in cell membrane shape occurs when the sperm of an animal such as a sea urchin binds to the egg. At the very tip of the sperm is the acrosome, a Golgi-derived vesicle containing numerous hydrolases. Just below the tip is a pool of unpolymerized actin, largely so because it is bound to profilin. When the sperm comes in contact with the outer jelly coat of the egg, Ca^{++} enters

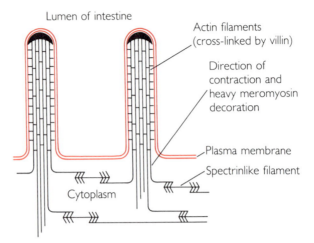

Lumen of intestine

Actin filaments
(cross-linked by villin)

Direction of
contraction and
heavy meromyosin
decoration

Plasma membrane

Spectrinlike filament

Cytoplasm

FIGURE 9–23

Diagram showing the types of filaments in the brush border (Figure 9–22). (After Dr. J. Weatherbee.)

the sperm, causing the acrosome membrane to fuse with the plasma membrane. At the same time, the intracellular pH rises, causing profilin to be released from actin, which is now free to polymerize into microfilaments. The sudden growth of actin filaments provides a propulsive force for the release of the acrosomal hydrolases into the jelly coat, which is then dissolved so that the sperm plasma membrane can fuse with that of the egg.

Cytoplasmic streaming occurs to some extent in all cells but is most striking in Characean algae, where it was first seen over 200 years ago. These cells are several centimeters long and over 500 μm in diameter and have distinct cytoplasmic zones with respect to intracellular motions. The outer cortical zone is stationary and contains the chloroplasts and microtubules, while the inner endoplasm zone surrounds the central vacuole and streams at speeds up to 100 μm/sec.

Actin filaments in bundles of 100 are located at the boundary between these two regions. If ATP is locally depleted from the cell, streaming stops along the microfilament bundles and resumes when ATP is reintroduced. This suggests an involvement of actin and myosin in streaming. Ca^{++}, which at the right concentration can cause actin filaments to depolymerize, inhibits streaming when injected into these cells. Because the intracellular levels of this cation are regulated by a pump at the ER membrane (see Chapter 6), it has been proposed that streaming is modulated by the ER Ca^{++} pump.

A third involvement of microfilaments is in the **movements of whole cells.** Their role here can be indicated by the use of the cytochalasins (*chalasis*: relaxation). These molecules (Figure 9–3b) from the mold *Helminthosporium* bind to actin filaments and reduce their growth and association into larger filaments. They also strongly inhibit crosslinking between filaments.

Both cultured cell migrations and ameboid movement are blocked by cytochalasin. Ameboid motion has been intensively studied, since large protozoan cells are easily manipulated in the laboratory. The driving force for the extension of the cell and its movement appears to be an increased viscosity (or contraction) at the leading edge of the cell. This causes cytoplasm, which exhibits viscoelastic properties, to flow to the leading edge. Microfilaments form a network in this region and may contract to provide the propulsive force.

Tissue culture cells move in a manner similar to that of amoebae, forming filamentous pseudopodia called filopodia, as well as lamellipodia, which may detach from the substratum as ruffles. Although these cells contain microfilaments (Figure 9–20), their role is not clearly defined. Stress fibers found near points of attachment to the substratum can be shown by immunofluorescence to contain actin, myosin, filamin, calmodulin, vinculin, and tropomyosin, while ruffles and filopodia contain largely actin. Stress fibers can contract if supplied with ATP, and it is possible that they pull the rear of the cell forward during locomotion.

The formation of filopodia appears to involve a conversion from a cross-linked meshwork of actin filaments to more organized bundles. This could occur by the action of regulating proteins such as gelsolin and villin. Similar changes occur in blood platelets. When activated for clotting, these cells change shape in seconds from disks to spheres with numerous filopodia. These changes are accompanied by microfilament changes from a cross-linked granular arrangement to bundles.

Intermediate Filaments

Intermediate filaments have a diameter (7–11 nm) that lies between those of microtubules and microfilaments (Figure 9–24). Unlike the other two types, which have a rather homogeneous structure and composition, intermediate filaments show wide variations in antigenicity, composition, and solubility. Immunological methods have been important in studies of these filaments since it was found that they are targets for autoantibodies made by both humans and rabbits. These antibodies have been used to localize intermediate filaments in cells (by immunofluorescence) and to purify the proteins making up the filaments (by chromatography).

Intermediate filaments have been found in many, but not all, cell types in both vertebrates and invertebrates, as well as plants. There are several major classes of intermediate filaments (Table 9–4), and these classes are similar across species lines. For example, an antibody to desmin from human skeletal muscle will react with desmin from virtually all other vertebrates. This antigenic conservation (and amino acid sequence conservation—within a class the sequences are 50–70% identical) implies a conservation of function. The nuclear lamins, referred to

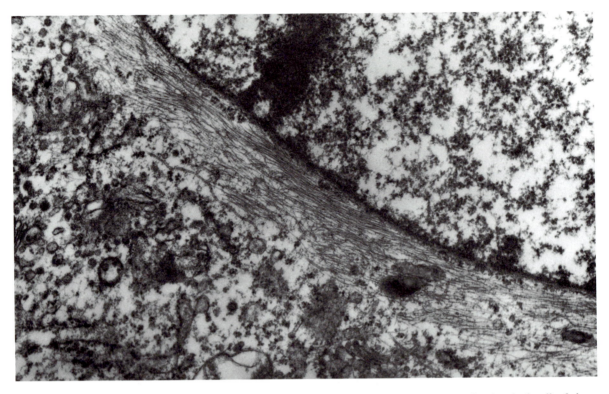

FIGURE 9–24   Intermediate filaments near the nuclear envelope of a decidual cell of the mouse endometrium. (×28,000. Photograph courtesy Dr. P. Abrahamsohn.)

as intermediate filaments on the basis of their structure, will be discussed in Chapter 10.

The various intermediate filaments are different yet are remarkably homologous in overall structure. All of them have a 310 amino acid central, rodlike helical region, which has a repeating pattern of seven amino acids with the first and fourth as hydrophobic. This pattern allows the hydrophobic residues of adjacent chains to interact, causing the chains to coil about each other. The central region is surrounded by amino- and carboxyl-termini of varying lengths and sequences, which do not have interchain interactions (Figure 9–25).

The basic unit of the intermediate filament is the two-chain coil, which forms from helix interactions. In the cases of all types except cytokeratins, the two chains are identical (homopolymer). On the other hand, cytokeratins are heterodimers, with one chain of one type and the second of a different type. Once the dimer has formed, it interacts with similar forces with a second dimer to form a tetramer, still in parallel array. Two of these units then associate in antiparallel fashion to form a protofilament, an octamer of the basic chain. Four protofilaments make

TABLE 9–4   **Classes of Intermediate Filaments**

	Filament Class	Subunit Mol. Wt.	Tissue Distribution
I:	Acidic keratins	40,000–70,000	Epithelia
II:	Basic, neutral keratins	40,000–70,000	Epithelia
III:	Desmin	52,000	Muscle, others
	Glial fibrillary	45,000	Glia
	Peripherin	57,000	Nervous
	Vimentin	53,000	Mesenchyme, others
IV:	Neurofilaments	68,000–200,000	Neurons
V:	Nuclear lamins	65,000–75,000	Nuclei of many cells

up the 10-nm–diameter filament seen in cells. In all, about 25,000 individual proteins associate to form a 40 μm filament.

There is often a linkage between intermediate filaments and the nuclear envelope, seen in electron micrographs as well as when the nuclei are isolated. This leads to the suggestion that the nuclear envelope is somehow involved in filament polymerization. In fact, if filament proteins are injected into cells, they associate with the envelope and polymerize. But the question of an IFOC (intermediate filament organizing center) remains unresolved.

The assembly and disassembly of these filaments apparently involve the amino-terminal extension. At the onset of mitosis, there is phosphorylation of the amino-terminal "head" of the filaments, and they disaggregate; after division, there is dephosphorylation, and they reaggregate. In addition, proteolysis of the "head" region, either in the test tube or in the cell, prevents the monomeric filament from polymerizing. The exact mechanism by which the "head" region promotes polymerization is not clear.

As is the case with microtubules and microfilaments, several proteins appear to be specifically associated with intermediate filaments. These IFAPs include paramenin and symenin, which cross-link the filaments into networks. Tighter cross-linking is achieved by small molecular weight proteins that bind up the keratins of hair; filagrin, which

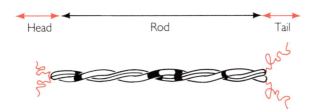

Head   Rod   Tail

FIGURE 9–25

Basic structure of an intermediate filament. The dark bands on the chains are nonhelical regions. The amino-terminal "head" and carboxy-terminal "tail" are shown.

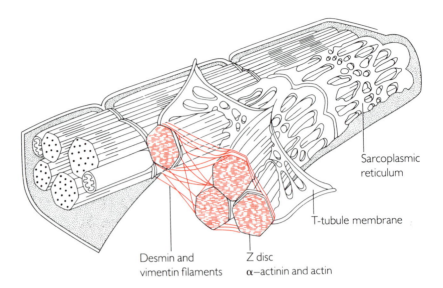

Sarcoplasmic
reticulum

T-tubule membrane

Desmin and
vimentin filaments

Z disc
α—actinin and actin

FIGURE 9–26

The role of intermediate filaments at the "Z"-line in muscle.
(From Dr. E. Lazarides.)

acts in epidermal cells; and plectin, which acts in fibroblasts. Plectin, a widespread binding protein, also acts at junctions to bind the filaments at their central rod regions. It also binds to MAPs 1 and 2 and so can cross-link microfilaments, microtubules, and intermediate filaments.

Desmin is found primarily in cardiac, skeletal, and smooth muscle. In the latter, it forms a cytoplasmic connection with membrane-bound plaques at the ends of the cells, helping the cells to integrate into a tissue. In skeletal muscle, desmin is present at the "Z"-line, which is at the end of each sarcomere at the terminus of the actin filaments. The Z-disc is thought to maintain the actin and sarcomeres in parallel registry. E. Lazarides has shown that as myogenesis proceeds, randomly oriented intermediate filaments move to the Z-disc and become an organized network. This may be the mechanism for lining up the sarcomeres to integrate muscle function (Figure 9–26).

Neurofilaments move with slow axoplasmic transport, a process that probably occurs not by motors but by diffusion. Neurotoxins, such as aluminum salts, induce a degeneration of neurofilaments in the axon but an increase in the mass of them in the cell body. A similar situation occurs in patients with Alzheimer's disease (senility). This situation appears to be a secondary effect due to a disruption in the cytoskeleton, which disturbs normal axoplasmic flow. The filaments in the axon are important in maintaining its cross-sectional area, and indeed the diameter of neurofilament bundles correlates well with axonal diameter.

Peripherin was named because it was first found in the sympathetic and parasympathetic (peripheral) nervous system. However, it is also found in several regions of the brain, where it is expressed along with neurofilaments in the same cells, in many cases. But it is distinct in amino acid sequence from the neurofilament class. Its role is not known.

Vimentin filaments are present in mesenchymal-derived cells such as adipocytes, the endothelium of blood vessels, and fibroblasts. They often surround other cell structures, especially the nucleus and fat droplets, and this has led to speculation that their role is to hold the organelles in place and prevent them from aggregating with other structures.

Cytokeratins are a class of intermediate filament that occurs in epithelia. They are different from the more familiar keratins, which are smaller proteins that make up feathers and scales. Over 30 types of cytokeratins have been identified, and they fall into either acidic or basic classes. A filament dimer is composed of two nonidentical chains, one from the acidic class and one from the basic class. The cytokeratin filaments are typically organized into bundles (tonofilaments), which can be attached to cell junctions (see Chapter 14) called desmosomes.

Different cytokeratin chains are expressed in different tissues and at different times. For example, in the epidermis, basal cells synthesize two types (molecular weight 50,000 [K5] and 56,000 [K14]), but once in the upper layers, they make two entirely different types of keratins (molecular weight 57,000 [K10] and 65,000 [K1]). When epidermal tissue is wounded, the healing cells, which previously made K1 and K10, switch to two different forms, molecular weight 47,000 (K16) and 56,000 (K6). Vitamin A, which has long been known to affect the differentiation of epithelia, induces cells to change their pattern of keratin gene expression, apparently via binding to specific intracellular receptors.

While these differences provide the experimenter with useful systems to study cell differentiation, they also provide the clinician with **cell markers.** Specific monoclonal antibodies can be raised to the different cytokeratins and other intermediate filaments, and when used in immunofluorescence, these antibodies provide a simple test for the presence of a type of filament. This can be used to determine the identity of a tumor.

For example, adenocarcinomas (which come from epithelial cells) contain cytokeratins, but lymph nodes and bone marrow cells do not. Breast cancer adenocarcinomas may become metastatic, with even single cells being carried through the blood system and invading other tissues. Immunofluorescence using anticytokeratin stain can be used to detect such metastatic cells in tissues such as lymph nodes that do not contain cytokeratin.

In some cases, distinguishing between similar-looking tumor cells is quite difficult. For example, thymoma and lymphoma cells can look quite alike. Because the former comes from an epithelium and the latter does not, the diagnosis can be made by applying cytokeratin immuno-fluorescence to the cells, since it will only stain the epithelium-derived cells.

The Cytoskeleton

Microtubules, microfilaments, and intermediate filaments interact in the cell to form an elaborate cytoskeleton. Just how elaborate this is can be inferred if the networks of the three components are visualized in a single cell. A triple stain has been developed that uses antibodies against microtubules and vimentin, and a direct chemical stain for actin. The results of such staining (Figure 9–27) show considerable overlaps on the extensive networks of each of the three components.

Connections between them can be detected with cytochemical markers. For example, tubulin and neurofilament proteins migrate at the same rate in the axon, indicating some association. Injection of antibodies to tubulin not only disrupts the microtubules in these cells but also disrupts the neurofilaments. Immunofluorescence staining shows MAP2 between the two types of structures, indicating that it or plectin may be the cross-linker.

If tissue culture cells are gently lysed in the detergent Triton X-100 in the presence of a Ca^{++}-chelating agent, the plasma membrane dissolves in 5 seconds, and organelle membranes dissolve 10 seconds later. The cytoskeleton remains intact. Conventional electron microscopy has inadequate voltages to examine thick specimens. But information in depth can be obtained from the high-voltage electron microscope, an instrument that employs voltages 10 times higher than those used in the conventional microscopes. This technique reveals a fourth cellular filament, 2 to 3 nm in diameter and 30 to 300 nm long.

The four components of the cytoskeleton are extensively inter-connected. Actin filaments abut the other types and also form side-by-side contacts. In addition, microfilaments branch extensively, in a manner similar to that induced in the test tube by regulating proteins. Remarkably, microtubules appear to bend, a remarkable observation since, in the test tube, they polymerize in a straight line. Bends may occur in the cell because of tensions imposed by direct interactions with other elements in the cytoskeleton. The 2- to 3-nm filaments form lateral contacts with all of the other elements, acting as cross-linkers.

Superimposed on the network of filaments appears to be a **microtrabecular lattice,** of strands 3 to 6 nm thick. The lattice can be seen when whole cells are prepared for high-voltage electron microscopy. Apparently, it is extracted by the detergent procedures used to isolate the cytoskeleton.

FIGURE 9–27
Triple immunofluorescence visualization of the cytoskeleton of a single human fibroblast. (Photographs courtesy of Dr. J. V. Small.)

(a) Coumarin-phalloidin staining for actin.

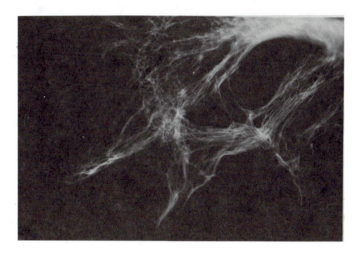

(b) Rhodamine staining for tubulin.

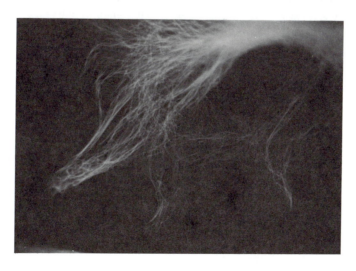

(c) Fluorescein staining for vimentin.

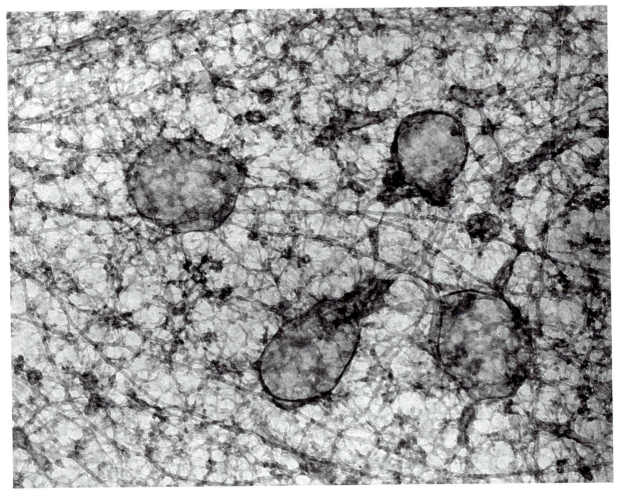

FIGURE 9–28   High-voltage electron micrograph of rat kidney cell cytoplasm. The vesicles are smooth ER; the long tubes are microtubules. (Photograph courtesy of Dr. K. Porter.)

The trabeculae extend throughout the cytoplasm and interconnect with the filaments as well as membrane-bound organelles (Figure 9–28). They appear not only in the cytoplasm but also in its extensions, such as cilia and microvilli. Examination of pigment granule migration in chromatophores (see above) shows that the pigments are associated with the trabeculae. Indeed, the lattice may impose considerable spatial organization on what was once thought of as amorphous cytoplasm. The fact that cytoplasm has a viscosity two to six times greater than that of pure water may be partly due to the lattice.

The cytoskeleton certainly interacts with cell organelles, but its complexity makes it difficult to elucidate these relationships. There is considerable microscopic and biochemical information on cytoskeletal interactions with the proteins of the plasma membrane. In the erythrocyte (see Chapter 2), cytoplasmic actin filaments complex with spectrin and band 4.1 protein. Because a microfilament band runs below the red cell membrane, it is reasonable to assume that its interaction with the membrane skeleton is important in determining cell shape. The same is true of intestinal microvilli (see above).

Much of the evidence linking the cytoskeleton and membrane proteins is indirect. For example, cell surface receptors to the lectin concanavalin A are randomly distributed over the membrane in cells treated with the ligand. However, if colchicine is present, the lectin induces patch and cap formation (Figure 1–15). This experiment implies that intact microtubules are needed for the random distribution of the receptors, but other components of the cytoskeleton could be involved, since microtubules are connected to them. Indeed, membranes can be carefully isolated and often show a cytoskeleton associated with membrane proteins that includes the three main types of cytoskeletal components.

Cancer cells undergo cell surface alterations that allow them to divide and form large masses and, in some cases, grow in inappropriate locations (e.g., metastasis—see Chapter 14). The cytoskeleton of cancer cells is changed from that of the normal cell. For example, normal tissue culture cells often have their microfilaments arranged as stress fibers (Figure 9–20), while in cancer cells in culture, the arrangement is diffuse. In many cases, a single gene is responsible for the transformation from a normal cell into a tumor cell. This gene, which differs for different types of cancer, codes for a product that either directly or indirectly affects the cytoskeleton. These alterations can lead to pathological changes in the cell.

Further Reading

Afzelius, B. Genetical and ultrastructural aspects of the immotile-cilia syndrome. *Am J Hum Genet* 33 (1981):852–864.

Avila, J. Microtubule functions. *Life Sci* 50 (1992):327–334.

Battifora, H. Clinical applications of the immunohistochemistry of filamentous proteins. *Am J Surg Pathol* 12 (1988):24–42.

Bayley, P. M. What makes microtubules. *J Cell Sci* 95 (1990):329–334.

Bearer, E. Cytoskeleton in development. *Curr Top Dev Biol 26 (1992).*

Bershadsky, A., and Vasiliev, J. *The cytoskeleton.* New York: Plenum, 1988.

Bignold, L. P. Ameboid movement: A review and proposal of a membrane ratchet model. *Experientia* 43 (1987):860–867.

Blanchard, A., Ohanian, V., and Critchley, D. The structure and function of alpha actinin. *J Musc Res Cell Motil* 10 (1989):280–289.

Bloom, G. Motor proteins for cytoplasmic microtubules. *Curr Op Cell Biol* 4 (1992): 66–73.

Bourguignon, L., and Bourguignon, G. Capping and the cytoskeleton. *Int Rev Cytol* 87 (1984):195–223.

Bray, D. Cell movements. New York: Garland, 1991.

Bretscher, A. Microfilament structure and function in the cortical cytoskeleton. *Ann Rev Cell Biol* 7 (1992):337–374.

Caplow, M. Microtubule dynamics. *Curr Op Cell Biol* 4 (1992): 58–65.

Carraway, K. Membranes and microfilaments: Interactions and role in cellular dynamics. *BioEssays* 12 (1990):90–92.

Cassimeris, L., Walker, R., Pryer, N. K., and Salmon, E. D. Dynamic instability of microtubules. *BioEssays* 7 (1988):149–154.

Cleveland, D. The multitubulin hypothesis revisited. *J Cell Biol* 104 (1988):381–383.

Cohn, S. A. The mechanochemistry of kinesin. *Mol Chem Neuropathol* 12 (1990):83–94.

Cooper, J. A. The role of actin polymerization in cell motility. *Ann Rev Physiol* 53 (1991):585–605.

Cross, R. A., and Kendrick-Jones, J. (Eds.). Motor proteins. *J Cell Sci Suppl* 14 (1991).

Dentler, W. L. Cilia and flagella. *Int Rev Cytol Suppl* 17 (1987):391–458.

Dubreuil, R. Structure and evolution of the actin crosslinking proteins. *BioEssays* 13 (1991):219–225.

Dustin, P. *Microtubules,* 2nd edition. New York: Springer-Verlag, 1984.

Dutcher, S., and Lux, F. Genetic interactions of mutations affecting flagella and basal bodies in *Chlamydomonas. Cell Motil Cytoskel* 14 (1989):104–117.

Elson, E. Cellular mechanics as an indicator of cytoskeletal structure and function. *Ann Rev Biophys Biophys Chem* 17 (1988):397–430.

Foisner, R., and Wiche, G. Intermediate filament associated proteins. *Curr Op Cell Biol* 3 (1991):75–81.

Fukuda, H., and Kobayashi, H. Dynamic organization of the cytoskeleton during tracheary element differentiation. *Develop Growth Differ* 31 (1989):9–16.

Fulton, A. B. *The cytoskeleton: Cellular architecture and choreography.* New York: Chapman-Hall, 1984.

Gaertner, A., Ruhnau, K., Schroer, E., Selve, N., Wagner, M., and Wegner, A. Probing nucleation, cutting, and capping of actin filaments. *J Musc Res Cell Motil* 10 (1989):1–10.

Gelfand, V., and Bershadsky, A. Microtubule dynamics: mechanism, regulation and function. *Ann Rev Cell Biol* 7 (1991):93–116.

Gibbons, I. R. Dynein ATPases as microtubule motors. *J Biol Chem* 263 (1988):15837–15840.

Griffin, J., and Watson, D. F. Axonal transport in neurological disease. *Ann Neurol* 23 (1988):3–13.

Hartwig, J., and Kwiatkowski, D. Actin-binding proteins. *Curr Op Cell Biol* 3 (1991): 87–97.

Kamiya, N. Physical and chemical basis of cytoplasmic streaming. *Ann Rev Plant Physiol* 32 (1981):205–236.

Kartenbeck, J. Intermediate filament proteins: Diagnostic markers in tumor pathology. *Interdis Sci Rev* 14 (1989):278–283.

Klymkowsky, M., Bachant, J., and Domingo, A. Functions of intermediate filaments. *Cell Motil Cytoskel* 14 (1989):309–331.

Kohno, T., and Shimmen, T. Calcium induced fragmentation of actin filaments in pollen tubes. *Protoplasma* 141 (1987):177–179.

Kreis, T. Role of microtubules in the organisation of the Golgi apparatus. *Cell Motil Cytoskel* 15 (1990):67–70.

Kuroda, K. Cytoplasmic streaming in plant cells. *Int Rev Cytol* 121 (1990):267–307.

Lane, E., and Alexander, C. The use of keratin antibodies in tumor diagnosis. *Semin Cancer Biol* 1 (1991):165–179.

Levy, M., Spino, M., and Read, S. Colchicine: state of the art review. *Pharmacotherapy* 11 (1991):196–211.

Lloyd, C. (Ed.). The cytoskeleton: Cell function and organization. *J Cell Sci Suppl* 5 (1986).

Luby-Phelps, K., Lanni, F., and Taylor, D. L. The submicroscopic properties of cytoplasm as a determinant of cellular function. *Ann Rev Biophys Biophys Chem* 17 (1988):369–396.

Mandelkow, E., and Mandelkow, E.-M. Microtubular structure and tubulin polymerization. *Curr Op Cell Biol* 1 (1989):5–9.

Margolis, R., and Wilson, L. Microtubule treadmills—possible molecular machinery. *Nature* 293 (1981):705–711.

McIntosh, J. R., and Porter, M. Enzymes for microtubule-dependent motility. *J Biol Chem* 264 (1989):6001–6004.

Mitchison, T. J., and Kirschner, M. Dynamic instability of microtubule growth. *Nature* 312 (1984):237–242.

Murray, J. Structure of flagellar microtubules. *Int Rev Cytol* 125 (1991):47–95.

Nagle, R. B. Intermediate filaments: Basic biology. *Am J Surg Pathol* 12 (1988):4–16.

Niggli, V., and Burger, M. Interaction of the cytoskeleton with the plasma membrane. *J Membr Biol* 100 (1987):97–121.

Olmsted, J. B. Microtubule-associated proteins. *Ann Rev Cell Biol* 2 (1986):421–457.

Omoto, C. Mechanochemical coupling in cilia. *Int Rev Cytol* 131 (1991):255–295.

Otto, J. Vinculin. *Cell Motil Cytoskel* 16 (1990):1–6.

Pollard, T. Assembly and dynamics of the actin filament system in nonmuscle cells. *J Cell Biochem* 31 (1987):87–95.

Porter, M. Dynein structure and function. *Ann Rev Cell Biol* 5 (1989):19–51.

Rao, K., and Cohen, H. Actin cytoskeleton in aging and cancer. *Mutat Res* 256 (1991):139–148.

Sawin, K., and Scholey, J. Motor proteins in cell division. *Trends Cell Biol* 1 (1991):122–129..

Schroeder, T. E. The contractile ring and furrowing in dividing cells. *Ann NY Acad Sci* 582 (1990):78–87.

Schroer, T., and Sheetz, M. P. Functions of microtubule based motors. *Ann Rev Physiol* 53 (1991):629–652.

Shay, J. (Ed.). *The cytoskeleton.* New York: Plenum, 1987.

Sheetz, M. P. What are the functions of kinesin? *BioEssays* 7 (1988):165–168.

Small, J. V., Zobeley, S., Rinnerthaler, G., and Faulstich, H. Coumarin-phalloidin: A new actin probe permitting triple immunofluorescence microscopy of the cytoskeleton. *J Cell Sci* 89 (1988):21–24.

Staiger, C., and Lloyd, C. The plant cytoskeleton. *Curr Op Cell Biol* 3 (1991):33–42.

Stebbings, H. How is microtubule-based organelle translocation regulated? *J Cell Sci* 95 (1990):5–7.

Stewart, M. Intermediate filaments: Structure, assembly and molecular interactions. *Curr Op Cell Biol* 2 (1990):91–100.

Tash, J. Protein phosphorylation: The second messenger signal transducer of flagellar motility. *Cell Motil Cytoskel* 14 (1989):332–339.

Tellam, R., Morton, D., and Clarke, F. A common theme in the amino acid sequences of actin and many actin-binding proteins. *Trends Biochem Sci* 14 (1989):130–133.

Vale, R. D. Intracellular transport using microtubule-based motors. *Ann Rev Cell Biol* 3 (1987):347–378.

Vallee, R. B. Molecular characterization of high molecular weight microtubule associated proteins: Some answers, many questions. *Cell Motil Cytoskel* 15 (1990): 204–209.

Vallee, R. B., and Shpetner, H. S. Motor proteins of cytoplasmic microtubules. *Ann Rev Biochem* 59 (1990):909–932.

Vandekerchove, J. Structural principles of actin binding proteins. *Curr Op Cell Biol* 1 (1989):15–22.

Vorobjev, I. A., and Nadezhdina, E. The centrosome and its role in the organization of microtubules. *Int Rev Cytol* 106 (1987):227–288.

Warner, F. D., and McIntosh, J. R. *Cell movement II: Kinesin, dynein and microtubule dynamics.* New York: Alan R. Liss, 1989.

Warner, F. D., Satir, P., and Gibbons, I. R. *Cell movement I: The dynein ATPases.* New York: Alan R. Liss, 1989.

Watterson, J. The role of water in cell architecture. *Mol Cell Biochem* 79 (1988): 101–105.

Wiche, G., Oberkanins, C., and Himmler, A. Molecular structure and function of microtubule-associated proteins. *Int Rev Cytol* 124 (1991):217–270.

CHAPTER 10

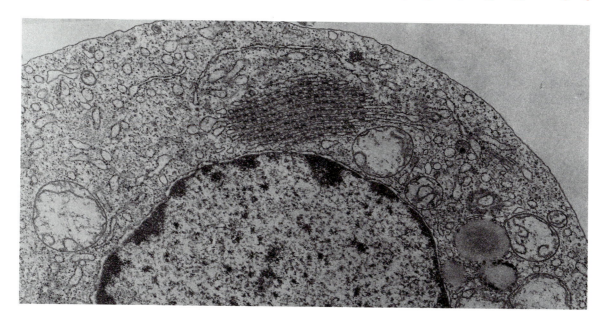

Nucleus and Cytoplasm

Compartmentation of Genetic Material

The nucleus is certainly the most prominent cell organelle, and early cytologists assumed that it had a similarly important role in cell function. All eukaryotic cells that are metabolically active have a nucleus except for a brief period when they are dividing, a notable exception being the mature erythrocyte in mammals. But even this cell has a nucleus when it is formed in the bone marrow and synthesizes its hemoglobin. Phloem cells that conduct sugar in plants are also anucleate, but they depend on adjacent nucleated cells for their survival.

More dramatic evidence for the need for a nucleus comes from experiments in which the nucleus is surgically removed from a large cell, such as an ameoba or an oocyte. In these cases, the cell survives for a while, but it cannot divide and eventually dies. In the single-celled plant Acetabularia, the nucleus is located at the base of the cell, and if this base is removed and cultured, it will program the regeneration of a new cell. Moreover, this regeneration is species specific: The nucleus of one species will regenerate the cell morphology of only that species (Figure 10—1). These experiments, done by A. Haemmerling over 50 years ago, clearly show that the nucleus determines not only cell survival but the activities of the cytoplasm that result in cell specificity.

It is now well established that the nucleus exerts this control by virtue of its DNA. As is the case with other membrane-bound organelles, an advantage of having the genetic material inside of an organelle is that access to it and the export of its products become controllable. The surrounding cytoplasm has many ions that could alter DNA and chromosome structure; for instance, the binding of proteins to DNA is largely ionic, and alterations in the nuclear environment could change this. Also, the cytoplasm has some proteins that could bind to DNA and even covalently change it; for example, there are cytoplasmic basic proteins that bind avidly to DNA in the test tube.

In addition to the protection it affords, the nuclear envelope controls the exit of the products of nuclear metabolism, such as RNA. In prokaryotes, translation of mRNA into protein begins before transcription is completed (that is, the two processes are coupled in space and time). But in nucleated cells, transcription occurs in a compartment (the nucleus) separated from translation (the ribosome in the cytoplasm). This affords a measure of control over when an mRNA will prime the production of a protein, and one place where this could occur is at the nuclear envelope.

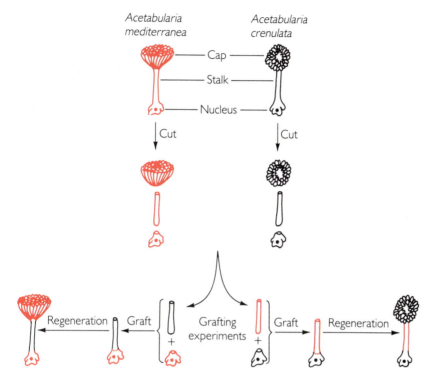

FIGURE 10–1   Experiments with the single-celled alga *Acetabularia*, which showed that the nucleus is responsible for cell specificity. The cell is about 1 cm in length and adheres to rocks in the intertidal zone. The nucleus is located at the base of the cell.

The nucleus has several prominent features (Figure 10–2). The most obvious is one or more nucleoli; these will be the subject of the next chapter. The rest of the material inside the nucleus is DNA complexed with proteins (chromatin), the RNA transcribed from DNA, and many other proteins unattached to DNA in the nucleoplasm. The organelle boundary is a double membrane, the nuclear envelope.

Two key events occur within the nucleus, and both involve DNA. The first is DNA replication, which occurs during S phase of the cell cycle; this will be discussed in Chapter 12. The second is transcription of DNA into RNA. This is a central subject of molecular biology, and its relationship to nuclear structure will be emphasized here. The two major foci of the chapter will be on nuclear structure and on the role of the nuclear envelope in regulating nucleocytoplasmic interactions.

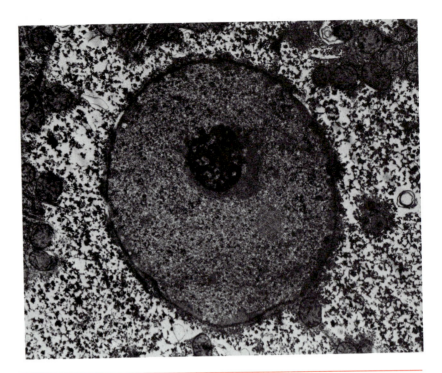

FIGURE 10–2

Electron micrograph of a rat hepatocyte, showing the nucleus with a prominent nucleolus. (×15,000. Courtesy of Dr. K. Brasch.)

Nuclear Envelope

Structure

The nuclear envelope is composed of two membranes, each 6.5 nm thick, with a 10- to 30-nm intermembrane space. On its nucleoplasmic surface, much of the inner membrane is coated with a fibrous layer (lamina) of varying thickness, up to 100 nm, which is associated with nuclear chromatin. On its cytoplasmic surface, the outer membrane is often studded with 20-nm–diameter ribosomes, which can be active in protein synthesis. For example, the outer membrane is the location of synthesis of much of the G protein (the major protein of the membrane) of vesicular stomatitis virus; this protein is then cleaved by a signal peptidase (see Chapters 6 and 7) at the membrane and put into the intermembrane space. Often, the intermembrane space is continuous with the lumen of the rough endoplasmic reticulum (RER).

A distinctive feature of the envelope is the presence of microscopically visible **pores** (Figure 10–3), first observed by E. Hertwig in 1876 using light microscopy. Electron microscopy shows the pores to be membrane-lined channels in which the outer and inner membranes of the envelope are fused. The diameter from membrane to membrane in a pore is 60 to 90 nm.

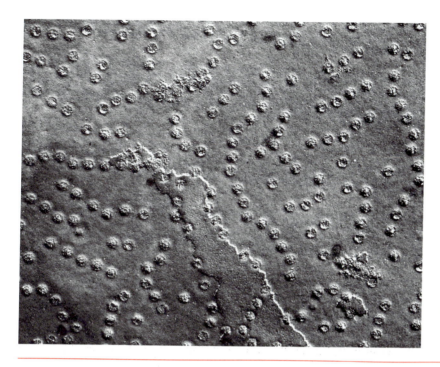

FIGURE 10–3
Nuclear envelope from a barley seed, showing numerous pores. (×44,000. Photograph courtesy of Drs. K. Platt-Aloia and W. Thompson.)

The number of pores varies between cell types and species from 1 to 100 per square micrometer of nuclear surface. Although on typical plant and animal nuclei the proportion of the surface area occupied by pores is small (less than 5%), it can be quite large, as in amphibian oocytes where it approaches 30% (50 million per nucleus).

These numbers change during developmental cycles. For example, in the mold *Physarum,* the pore frequency rises from $14/\mu m^2$ nuclear surface during S phase of the cell cycle to $22/\mu m^2$ just prior to mitosis. When rat seminal vesicle epithelium is exposed to the sex hormone testosterone, the nuclear pore density also increases. Why these changes occur is not clear, but because the pores are involved in the transport of RNA out of the nucleus, one hypothesis is that the number of pores should be positively correlated with the extent of RNA synthesis by the nucleus.

Stacks of flattened vesicles with pores occur in the cytoplasm of certain animal embryonic cells, tumor cells, red algae, spermatocytes, and oocytes (Figure 10–4). The density of pores on these stacks (called **annulate lamellae**) is usually higher than that on the nuclear envelope of the same cell. Up to half of the surface of these annulate lamellae may be occupied by the pores, which are identical to those on the nucleus. The lamellae often begin as nonpored extensions of the nuclear enve-

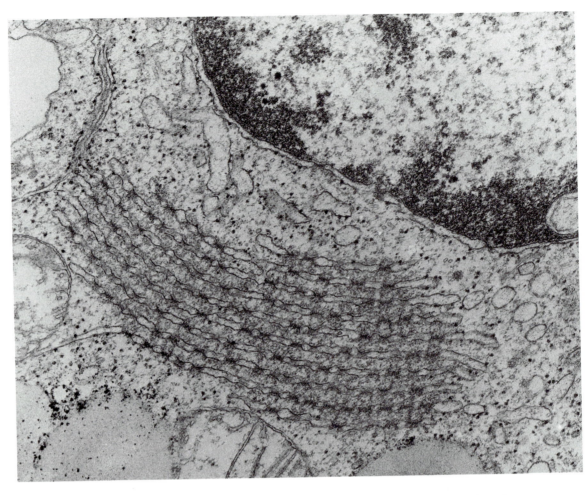

FIGURE 10-4 Annulate lamellae adjacent to the nucleus of a Burkitt's lymphoma tumor cell in culture. The darker areas are nuclear pore complexes. (×80,000. Courtesy of Dr. E. D. Allen.)

lope, which pinch off to form the ER-like sacs, and then the pores are inserted.

Extracts of frog eggs, when suitably incubated in the test tube in the absence of DNA or chromatin, will form annulate lamellae. However, when either of these two components is present, lamellae formation is reduced, and if the chromatin concentration is high, lamellae do not form. This indicates that the nuclear envelope will form onto chromatin if it is present but remains as annulate lamellae if there is insufficient chromatin onto which the envelope material can attach. Thus it is not surprising that annulate lamellae occur in rapidly dividing cells, where rapid nuclear envelope formation is required.

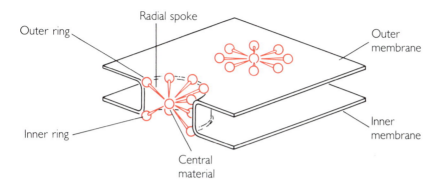

FIGURE 10-5
A model for the nuclear pore complex. (After Dr. W. Franke.)

Extensive observations of pores using the electron microscope reveal a common structure in diverse organisms, with both radial and bilateral symmetry (Figure 10–5). There are three principal components:

1. On both the nucleoplasmic and cytoplasmic surfaces are rings of eight 10- to 25-nm–diameter granules that form an annulus.

2. Projecting from each granule of these rings is a radial spoke, and these eight spokes converge as the lining of the actual opening, which is about 9 nm in diameter.

3. A granular region is often in the center of the pore and might be material on its way through the pore.

Composition

The nuclear envelope and its associated pores are quite stable. If rat liver nuclei are treated with sodium bicarbonate, the envelope becomes leaky, and **nuclear ghosts,** consisting of only the envelope, can then be isolated by differential centrifugation. The purity and polarity of the ghosts can be monitored because the nuclear pores remain intact, and the associated material (ribosomes on the outside, lamina on the inside) remain attached. These morphological markers must be used because there is no enzymatic marker available to monitor envelope purification. Indeed, many enzymes believed to be specific to other organelles are often associated with "pure" envelope preparations.

The envelope is about 25% lipid (Table 1–2), and this is virtually all of the nuclear lipid. In several respects, the lipid composition of the envelope resembles that of the endoplasmic reticulum (ER): There are low amounts of cholesterol and sphingomyelin (Table 10–1). This is in contrast with the compositions of the Golgi and plasma membrane, where these two components are more prominent. But the similarity to ER is not so pronounced when membrane structural proteins are examined. Gel electrophoresis in detergent reveals little similarity between

TABLE 10–1 Lipid Composition of the Nuclear Envelope

% Lipid Dry Weight	Rat Liver	Onion Root Tip
Phospholipids		
Phosphatidylcholine	55	25
Phosphatidylethanolamine	22	17
Phosphatidylserine	2	2
Phosphatidylinositol	6	7
Phosphatidic acid		34
Sphingolipids	5	
Cardiolipids		2
Cholesterol	10	

DATA FROM: Harris, J. Biochemistry and structure of the nuclear envelope. *Biochim Biophys Acta* 515 (1978):55–104.

the envelope and other cellular membranes. A few major proteins, about 60,000 molecular weight and termed lamins, account for about one-quarter of the proteins; the rest are heterogeneous.

If nuclear ghosts are extracted with detergents in the presence of high ionic strength salts, the membrane dissolves away, leaving the pores and lamina. This pore-lamina complex is almost entirely composed of protein, and gel electrophoresis reveals only a few polypeptides. Two of these, designated gp210 and p62 by their molecular weights (in kilodaltons), are the **pore proteins.**

Immunofluorescence using antibodies to gp210 shows that it is the major integral pore protein, is present in 25 copies per pore, and has a role in anchoring the pore complex to the membranes. The smaller protein, p62, binds the lectin, wheat germ agglutinin (see Chapter 2), via an unusual carbohydrate linkage, an O-linked N-acetylglucosamine. This protein is also located at the pore but is extrinsic at the cytoplasmic face. Because lectin binding blocks nucleocytoplasmic transport, the smaller protein is thought to be essential for transport through the pores.

The nuclear lamina is composed of one to four different polypeptides termed **lamins.** These molecules are 60,000–70,000 daltons, and in mammals three types, A, B, and C, have been characterized. Lamins A and C have identical sequences of 566 amino acids, except for the presence of an extra carboxy-terminal 98 amino acids in lamin A. Apparently, lamins A and C are generated by differential RNA splicing after transcription.

The lamins are intermediate filaments, as indicated by their amino acid sequences and their ability to form long, α-helical fibers (see Chapter 9). Electron microscopy of the lamina confirms this, showing a mesh-like network of fibers, just like the cytoplasmic intermediate filaments (Figure 9–23). But the other types of intermediate filaments stay in the cytoplasm because they lack the nuclear targeting signal present on the lamins.

The lamins may be the "glue" that sticks chromatin to the inner nuclear envelope. During mitosis, the lamina (and nuclear envelope) disaggregate, possibly because of phosphorylation; however, lamin B still adheres to envelope vesicles throughout mitosis and appears to act as a "nucleator" to which the rest of the lamina attaches when the envelope reforms after mitosis. A receptor protein for lamin B is an integral protein in envelope membranes and probably provides the binding site for the lamina to the envelope. In addition, lamin B has a carboxyl-terminal "CaaX box," where C = cysteine, a = aliphatic amino acid, X = any amino acid. This motif is modified by the attachment of hydrophobic isoprenoid groups, which facilitate insertion of the protein into the membrane. Because lamins A and C bind both to lamin B and to chromatin, the proposed relationships of the lamina to the envelope and chromatin are:

envelope receptor ↔ lamin B ↔ lamins A/C ↔ chromatin.

This multiple lamin model may be suitable for cells such as Chinese hamster ovary tissue culture cells, which have several lamins, but the model cannot apply to mouse embryos (lamin B only) or frog embryos (lamin C only). In fact, these two lamins in these cells do not associate with nuclear envelopes reforming around chromatin during mitosis. This implies a need for some other intermediary between the envelope and chromatin. A valuable experimental tool in analyzing this process is the egg of the frog *Xenopus.*

A frog egg contains enough stored nuclear envelope components and chromosomal proteins to form nuclei after many cell divisions. This storage is important because the developing embryo does not synthesize its own proteins until gastrulation (after several dozen cell divisions). If this stored envelope material is removed from the egg, it can be used as vesicles to make "nuclei" in the test tube if DNA is supplied. Remarkably, any DNA will do the job, even the DNA of a bacteriophage that does not have a nucleus! This experiment strongly indicates that the specificity of nuclear formation resides with the proteins and not with the DNA.

But if the envelope vesicles are first treated with the protease trypsin before testing for reassembly, they do not bind to chromatin and surround it. This implies that there is a protein sticking out from the envelope surface that acts as a receptor for chromatin. But in this case, the bridging protein is not a lamin. Thus, in these nuclei the cross-linking is:

envelope receptor ↔ cross-linking protein ↔ chromatin.

In addition to these structural proteins, the nuclear envelope contains a number of **enzymes,** which can be detected by cytochemistry of fixed cells and biochemical assays of isolated envelopes (Table 10–2). Several of these are also in the ER membrane (compare Table 6–3), such as the two ER electron transport systems. These systems may be a source of energy for the nucleus, which could use them for oxidative phosphorylation. The presence of an envelope ATPase adds to this speculation.

Function: Nucleocytoplasmic Exchange

In a typical mitotic cell (see Chapter 12), the nuclear envelope breaks down, and organelles such as mitochondria, ER, and vesicles are near the chromosomes. But in interphase, the presence of the nuclear envelope clearly excludes these organelles from the vicinity of the chromatin. This familiar observation is the starting point for discussion of the nuclear envelope as a barrier between the nucleus and cytoplasm, but it indicates little about molecular traffic across the envelope.

TABLE 10–2   **Enzymes in the Nuclear Envelope**

Acetyl-CoA carboxylase

Acid phosphatase

Adenyl cyclase

Aryl sulfatase

Alkaline phosphatase

ATPase

Cyclic AMP phosphodiesterase

Cytochrome b_5

Cytochrome oxidase

Cytochrome P450

Glucose-6-phosphatase

NADH, NADPH-cytochrome c reductases

Nucleotidase, 5′

Protein kinase

DATA FROM: Harris, J. The biochemistry and structure of the nuclear envelope. *Biochim Biophys Acta* 515 (1978):55–104.

Many substances must cross the nuclear envelope to get from the cytoplasm to the nucleus or vice versa. These include ions and small molecules such as nucleotides, macromolecules such as nuclear proteins, and aggregates such as ribonucleoprotein particles. Most of these substances probably cross via the pores in the envelope, but this may not always be the case. Exchange could occur directly through the bilayers, via the continuity with the ER cisterna or by vesicles pinching off or fusing with the membrane.

Intracellular measurements show that K^+ is more concentrated in the nucleus than in the cytoplasm, whereas the converse is true for Na^+, which has a greater cytoplasmic concentration. These imbalances are not maintained by active transport but rather by the binding of the ions to DNA and chromosomal proteins. Large cells (e.g., *Drosophila* salivary gland), when impaled with a microelectrode, show a negative intranuclear potential relative to the cytoplasm.

The permeability of the nuclear envelope to molecules has been studied in large cells, such as amoebae and oocytes. The strategy of these experiments involves microinjection of a molecule into the cytoplasm of a living cell. The molecule can be labeled with radioactivity, in which case its passage to the nucleus is monitored autoradiographically. Or, the label can be a fluorescent compound, and the fluorescence microscope can be used to follow nuclear uptake quantitatively (Figure 10–6).

The results of an experiment using the fluorescence method are shown in Table 10–3. Two trends are apparent from these data:

1. Small molecules enter the nucleus more readily than do large ones. A consistent finding in this type of experiment is that the size cutoff appears to be around 10 nm. This is about the size of the central granule

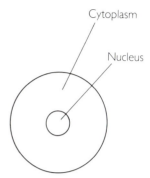

(a) A single large cell.

(b) A micropipet is used for injection of a fluorescently labeled compound such as a protein.

(c) The distribution of fluorescence is measured after a time.

FIGURE 10–6

Experiment using microinjection of a fluorescently labeled compound to demonstrate transport from the cytoplasm into the nucleus.

TABLE 10–3 Cytoplasm to Nucleus Passage of Proteins Inserted into Oocytes

Protein	Mol. Wt. (×1000)	Mol. Dimensions (nm)	Nuc/cyt Ratio	Time (hr)
Cytochrome c	12.4	3.4 × 3.4 × 3.0	1.30	5
Lysozyme	14.6	4.5 × 3.0 × 3.0	2.42	1
Myoglobin	17.8	4.3 × 3.5 × 2.3	2.15	2
Ovalbumin	44.3	6.3 × 2.3	0.37	2
Bovine serum albumin	67.0	7.0	0.08	5
Ferritin	465.0	9.4 × 9.4	0	5

DATA FROM: Paine, R., and Feldherr, C. Nucleocytoplasmic exchange of macromolecules. *Exp Cell Res* 74:91–98.

of the nuclear pore complex and is circumstantial evidence for passage through this region.

2. Some molecules appear to be preferentially taken up by the nucleus. If the molecule is freely diffusing, it should be equally concentrated in the nucleus and cytoplasm. But this is not always the case (see Table 10–3, lysozyme and myoglobin), as some proteins preferentially accumulate in the nucleus. This would be expected for nuclear constituents such as histones. When amphibian oocytes are injected with labeled histones or other nuclear proteins in their cytoplasm, they are concentrated over 100-fold in the nucleus.

As is the case with other membranous organelles, the selective uptake of nuclear macromolecules is mediated by **signal sequences.** Unlike the signals for membrane and ER proteins, nuclear signals are not cleaved off at the membrane boundary. Instead, they are a permanent fixture of the proteins. This is fortunate, since nuclear contents are dispersed during cell division and must reassemble at the end of cell division, and a protein without the signal might be lost at this time.

The nuclear-targeting signal on the sea urchin protein nucleoplasmin was among the first to be identified. In this case, the signal resides in the carboxy-terminal region of 50 amino acids out of a protein with over 1000 residues. If this "tail" is removed, the rest of the protein remains cytoplasmic, but the tail itself enters the nucleus! Construction of chimeric proteins by genetic engineering shows that the tail can target any protein to the nucleus, including those normally confined to the cytoplasm.

Small size, basicity, and a nearby proline residue are attributes of most nuclear signals (Table 10–4). They can occur at any location in the protein, as long as they are exposed on the surface. The **signal receptor** probably has an acidic region that would interact electrostatically with the basic signal. Evidence supporting this idea comes from an experiment in which an antibody is made to the acidic peptide asp-asp-glu-asp. When applied to nuclei, this antibody blocks uptake of exogenous proteins.

Moreover, in such experiments the antibody binds to a region on the nuclear envelope, indicating that it is here rather than within the nucleus that reception of the transported proteins initially occurs. This observation confirms other experimental data: If nuclear proteins are injected into the nucleoplasm of a frog oocyte whose nuclear envelope has been removed, the proteins rapidly leave the nuclear region and diffuse to the cytoplasm. This makes an intranuclear receptor unlikely.

Nuclear uptake selectivity can change in different physiological situations. For example, in the frog oocyte, small ribonucleoproteins are cytoplasmic, where they remain until well after fertilization. But during

TABLE 10-4 Nuclear-Targeting Signal Sequences

Protein	Signal Sequence	First Amino Acid Position of Signal
Histone H1	pro-arg-arg-lys-ala-lys-arg	30
Histone H2A	lys-ala-arg-ala-lys-ala-lys	9
Histone 2B	lys-lys-arg-lys-arg-ser-arg-lys	10
Histone H3	lys-lys-pro-his-arg-tyr-arg	36
Histone 4	ala-lys-arg-his-arg-lys-val	16
HMG14	pro-lys-arg-arg-ser-ala-arg	16
Lamins A and C	lys-ala-arg-asn-thr-lys-lys	137
SV40 T antigen	pro-lys-lys-lys-arg-lys-val	126

the 12th cell division of the zygote, these complexes suddenly begin to migrate into the nucleus. In another instance, a protein kinase is cytoplasmic until it is bound with cyclic AMP, at which time it enters the nucleus. Presumably in these cases, cryptic targeting signals are being exposed after some three-dimensional change in the protein.

A more striking example of changing the localization of a nuclear protein is the transformation of the c-abl oncogene product into a protein that causes leukemia in mice. This protein is a tyrosine protein kinase and is made by normal cells. But it differs from the typical membrane-bound tyrosine kinases in that it has the C-terminal nuclear localization signal. When the normal white blood cell becomes leukemic, the c-abl protein suddenly becomes cytoplasmic. The one event that causes this change is deletion of part of the c-abl gene, which leaves the nuclear signal sequence intact but in a masked three-dimensional configuration. Once in the cytoplasm the kinase phosphorylates proteins that cause rapid cell proliferation.

Microscopic evidence indicates that macromolecular aggregates such as ribonucleoproteins leave and enter the nucleus through the pores in the nuclear envelope. Electron micrographs show this dense material in and around the pores and not elsewhere around the envelope. In an especially informative study, colloidal gold particles coated with nucleoplasmin were traced into the nucleus after injection into frog oocyte cytoplasm. The particles were easily observable under the electron microscope and clearly seen in the nuclear pores at the location of the central granule.

A parallel study has been made of gold particles coated with RNA and injected into the oocyte nucleus (Table 10–5). In this case, both natural (tRNA and 5S rRNA) and artificial (poly A)-coated aggregates were observed at the pores and translocated (Figure 10–7). Particles

TABLE 10–5 Nucleus to Cytoplasm Passage of Coated Gold Particles Injected Into Frog Oocyte Nuclei

Particle	Number of Particles in Region		Particles/Pore
	Pore	Adjacent Cytoplasm	
tRNA-gold	437	414	4.3
5S RNA-gold	458	425	4.4
Poly A-gold	309	479	3.9
Albumin-gold	5	3	0.04
Ovalbumin-gold	2	9	0.05

DATA FROM: Dworetzky, S., and Feldherr, C. M. Translocation of RNA-coated gold particles through the nuclear pores of oocytes. *J Cell Biol* 106 (1988):575–584.

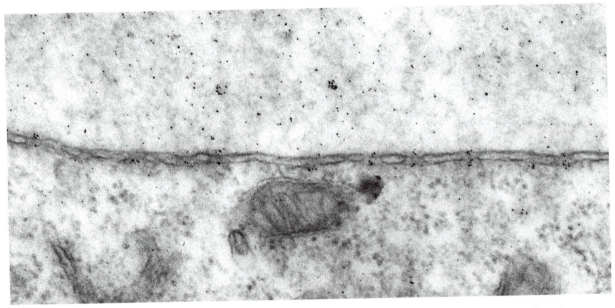

FIGURE 10–7 Nucleus (top) and cytoplasm (bottom) of a frog oocyte 1 hr after the nucleus was injected with 5S RNA-gold particles. The particles are seen in the nuclear envelope pores, as well as in the adjacent cytoplasm. (×100,000. Photograph courtesy of Dr. C. M. Feldherr.)

coated with proteins (e.g., ovalbumin) were not translocated to the cytoplasm, indicating that the translocation machinery is specific. As in the case of the protein import studies, particles much larger (23 nm) than the pore size were observed at the pores, indicating some specialized transport mechanism. The fact that these particles enter the nucleus at rates much greater than their diffusion rates also is evidence for some sort of mediated transport.

That the gold particles go through the pore intact indicates that transported substrates do not need to be unfolded. This is in sharp contrast to the situation in other membrane-bound organelles, where a special class of molecular chaperones helps unfold proteins so that they can cross the hydrophobic membrane. On the other hand, larger aggregates entering the nucleus often appear in the pores in a constricted "dumbbell" configuration, showing that some molecular "squeezing" can occur.

Transport across the nuclear envelope occurs in two steps, similar to the translocation across many other cellular membranes:

1. **Binding:** The transported molecule binds to proteins at the nuclear pore complex. This requires the nuclear signal sequence and can be blocked if the small glycosylated proteins at the pore are not present or if they are bound up with the lectin wheat germ agglutinin. The signal binding is a chemical affinity process and so can be saturated. It is of interest that some proteins have more than one copy of the signal, and in these cases the proteins are usually transported at a greater rate than ones with only a single signal.

2. **Translocation:** This step requires the continued presence of ATP, suggesting that some active transport and envelope Mg^{++}-ATPase may be involved. In one case (fruit fly embryos) a protein immunologically related to myosin, with ATPase activity, has been found in a pore-lamina preparation. When this ATPase activity rises, as in certain developmental situations, the rate of RNA efflux from the nucleus rises; when the ATPase is inhibited, a parallel inhibition of transport occurs.

Vesicles prepared from nuclei and containing mRNA will allow the export of the RNA only if ATP is hydrolyzed, presumably by the membrane ATPase. For translocation of most mRNAs through nuclear envelope, the poly A "tail" on the RNA must be present. In fact, poly A, either alone or as part of RNA, stimulates the envelope ATPase. The exact role of ATP hydrolysis is not known, but several possibilities are that it is required for the release of the transported molecule from the envelope receptor, that it is needed to physically expand the pore like a diaphragm so that material can pass through it, and/or that it is actually required for material to move through the pore.

Chromatin Chemistry

DNA Complexity

The nuclear interior contains DNA, protein, and RNA, in the typical (rat liver) mass ratio of 1.0:3.0:0.5, respectively. Most nuclear RNA is ribosomal (see Chapter 11) and pre-mRNA, but some is small nuclear RNA (snRNA, involved in RNA splicing). Nuclear DNA is bound up with proteins to form chromatin, so named by W. Flemming in 1882 because it becomes colored with certain cytochemical stains.

The nuclear DNA content varies considerably between organisms (Table 10–6), and this variability even extends to species in the same genus. On an intuitive level, it seems to make sense that a human cell has 150 times the amount of DNA as a yeast cell, but such an anthropocentric approach breaks down when one observes that a lily cell has 17 times more DNA than a human cell.

Another problem with DNA amounts is mutational load: There is a spontaneous rate of mutation, much of which occurs during DNA replication, and some of these mutations can lead to inactivation of a protein with harmful effects. If such a mutation occurs during the formation of a sex cell, it can be passed on to the offspring. All organisms (including people) carry a small number of harmful mutations, in almost all cases in recessive form masked by a dominant (normal) gene. Obviously, the number of these harmful mutations must be small (some estimates for humans put it at three to five), or the species might not survive. This essentially puts an upper limit on the number of mutable genes in an organism. Geneticists have used such reasoning to propose that only a small fraction (less than 10%) of the human genome is actually used for essential functions of the organism.

Such arguments lead to a question: A lot of DNA is there, but what is most of it there for? Analyses of genome organization by molecular

TABLE 10–6 Haploid DNA Contents of Various Organisms

Organism	DNA Content: Picograms/Cell	Base Pairs
Yeast (Saccharomyces)	0.02	20,000,000
Fruit fly (Drosophila)	0.1	100,000,000
Grab (Plagusia)	2.0	2,000,000,000
Mouse (Mus)	2.5	2,500,000,000
Human (Homo)	3.1	3,100,000,000
Corn (Zea)	7.5	7,500,000,000
Newt (Triturus)	45.0	45,000,000,000
Lily (Lilium)	53.0	53,000,000,000

techniques such as nucleic acid hybridization and DNA sequencing have revealed some common patterns in eukaryotes. Simply put, most of the "excess" DNA can be explained by repetitive DNA and intervening sequences.

The overall sequence complexity of nuclear DNA can be analyzed by hybridization kinetics (Figure 10–8). This method involves several steps:

FIGURE 10–8

Analysis of DNA by reassociation kinetics. Each letter represents a base sequence of several hundred nucleotides. A and A' are complementary.

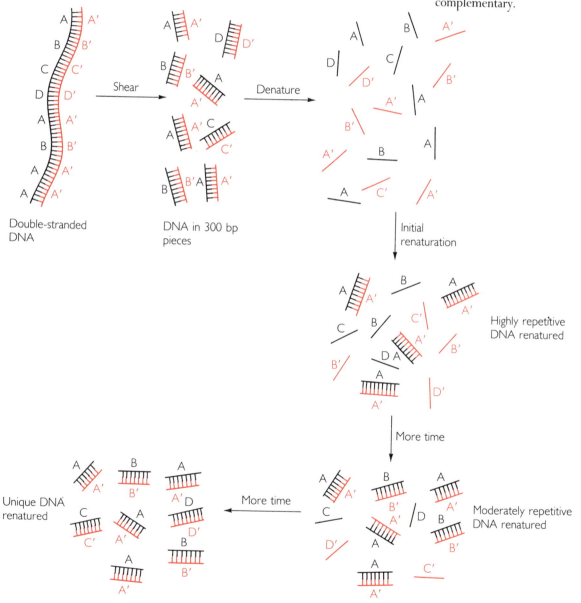

1. DNA is cut to a uniform size (typically about 300 base pairs).

2. The fragments are denatured by heat or high pH so that the hydrogen bonds holding the two strands together are broken and the two strands of the molecule separate.

3. The mixture is cooled slowly so that the single strands can renature by hydrogen bonding.

4. Some of the fragments reassociate quickly: These must be able to find their complementary strand in the solution because there are many copies of it in the genome; this is termed highly repetitive DNA.

5. Other fragments reassociate less quickly and are moderately repetitive.

6. Still other fragments take a very long time to find their complement because there is only one copy present; this is termed unique DNA.

Typical reassociation curves for prokaryotic *(E.coli)* and eukaryotic (calf) DNAs show great differences (Figure 10–9). For the prokaryote, all of the DNA reassociates slowly, as if each sequence is present only once or at most a few times. But the calf DNA has a triphasic behavior:

1. A small part of the DNA (about 10%) reassociates very quickly and is highly repetitive.

2. About 40% of the DNA renatures quickly but somewhat more slowly than the highly repetitive component; it is moderately repetitive.

3. About half of the DNA renatures quite slowly; it is essentially nonrepetitive and is termed unique.

These classes of nuclear DNA are summarized in Table 10–7; analyses of them is a major task of molecular genetics.

Highly repetitive DNAs are reasonably well characterized. They are often short stretches (as few as a dozen base pairs) that may have a base composition different from the bulk of the genome. For example, in many cases their % G + C (guanine + cytosine) is different, and so the DNA can be separated on the basis of its different density because GC base pairs (held together by three hydrogen bonds) are denser than AT base pairs (held together by two H-bonds).

Two examples of highly repetitive sequences are:

AATAACATAG . . . present in fruit flies

ACACAGCGGG . . . present in Kangaroo rats

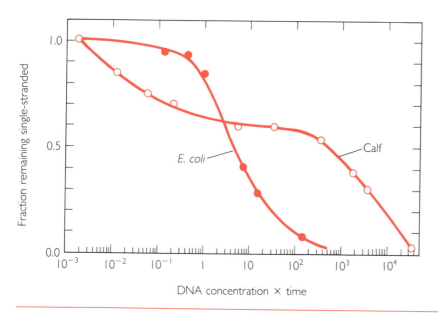

FIGURE 10-9

Reassociation kinetics of denatured DNA from a prokaryote *(E. coli)* and a eukaryote *(calf)*. The prokaryotic DNA shows a single component, which reassociates very slowly. The eukaryotic DNA has three components: About 10% reassociates very quickly; about 40% reassociates moderately quickly; and about 50% reassociates very slowly.

TABLE 10-7 Classes of Nuclear DNA

Organism	Percent of Genome as		
	Highly Repetitive	Moderately Repetitive	Unique
Mouse *(Mus musculus)*	10	20	70
Fly *(Drosophila melanogaster)*	15	15	70
Sea urchin *(Strongyolcentrus)*	19	29	52
Pea *(Pisum sativum)*	50	17	33

If these DNAs are made radioactive, they can be hybridized to denatured chromosomal DNA on a slide. Microscopic techniques can then be used to visualize the locations of these sequences, with an interesting result (Figure 10–10): The sequences often occur only at a few locations, often at the centromeres. At these locations, these simple sequences are repeated up to 100,000 times right beside each other (a tandem repeat).

The role of these tandem repeats is not known. One thing for certain is that they do not code for protein, as most are never transcribed into RNA. One possibility is that they may be involved in recognition of chromosomes at meiosis, when homologous chromosomes line up in close association (see Chapter 12). In some mammals, highly repeated sequences differ in species-specific patterns, but this is by no means a

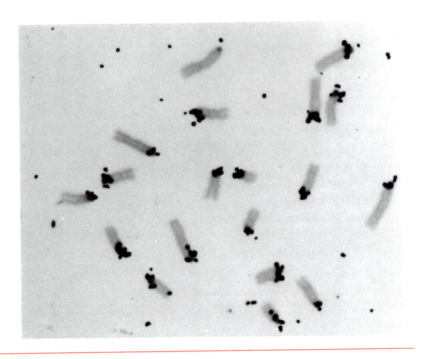

FIGURE 10–10

The localization of highly repetitive DNA on mouse chromosomes. The DNA was labeled with radioactivity (^{3}H), denatured, and applied to denatured chromosomes. The black dots indicate radioactivity. These chromosomes have their centromeres at the ends, and so the labeled DNA is at the centromeres. (Photograph courtesy of Dr. M. L. Pardue.)

common phenomenon. Another possibility, investigated in plants, is that these sequences are important in the separation of chromosomes during meiosis.

In humans, some of the highly repetitive sequences are not in large tandem arrays, but instead are in smaller blocks that have 20 to 50 copies each. The lengths of these blocks vary and are easily detectable by molecular techniques. These variations can be used to make a genetic "fingerprint" of an individual and are coming into widespread use for identification purposes.

There are two major types of **moderately repetitive** sequences:

1. The first type of these sequences code for RNAs that the cell needs in large amounts. For instance, ribosomal RNA accounts for about 80% of the RNA in a typical cell. To make these prodigious amounts, the DNA coding for these structural RNAs is moderately repeated. For example, a human cell has about 5 million ribosomes, and each one of them has a copy of four ribosomal RNAs. If the cell is dividing once a day, it must essentially make 5 million of these RNA molecules per day to supply the two cells. It is not surprising that there are 250 copies of the DNA coding for rRNA in the human genome.

Once again these moderately repetitive copies are not scattered throughout the genome but are tandemly repeated, in this case at a specialized region called the nucleolar organizer (see Chapter 11). In a similar fashion, each DNA sequence coding for a tRNA is present in about 30 copies in the human genome.

Some genes that code for proteins are moderately repeated, and these also are in clusters. Histone proteins are present in the cell in large amounts (see below), and so there is a need to have their genes repeated. In sea urchins, there are 200 to 1000 copies of each of the genes coding for the five histones, arranged in 6500 base pair stretches of the genome in an invariant order: H1-H4-H2B-H3-H2A. In the fruit fly, the order of the cluster is H2B-H2A-H4-H3-H1. In many organisms, these repeated clusters are in tandem at a single location, but in others the clusters are scattered.

2. The second type of moderately repetitive sequences do not code for structural RNA or proteins. Typically present in 10,000 to 300,000 copies per genome, these DNA sequences may be up to 300 base pairs in length. But in contrast to most other moderately repetitive sequences, these do not occur in tandemly arranged blocks. Instead, they are interspersed amongst unique genes throughout the genome, in an arrangement such as:

. . . moderately repet . . . unique . . . moderately repet . . . unique . . .

A well-studied example of such a sequence is the so-called Alu repeat in humans. This DNA sequence is present about 300,000 times in the human genome and is scattered throughout it. The name *Alu* is used because this DNA has the recognition sequence for a restriction endonuclease, Alu I. It is of some interest that a part of the Alu sequence is almost identical to part of the RNA of the signal recognition particle involved in protein synthesis on rough ER (see Chapter 6). Alu sequences are also striking in that they have a 10 to 15 base pair identical sequence at each end (a direct repeat). This apparently allows them to move out of their fixed location and to different ones in the genome. But how often this happens, or what the role of this repeated sequence is in the cell, is not known.

Unique DNA codes for most of the proteins in the cell. There are two ways to determine this:

1. A DNA sequence that codes for a protein can be isolated, denatured, and then hybridized to the total denatured DNA of the organism. If the kinetics of hybridization are monitored, this sequence takes a long time to renature with its complement on bulk DNA, indicating that it is present in a very few copies.

2. The total mRNA of a cell can be isolated and made into a DNA copy, which is then denatured and hybridized to total denatured DNA of the organism. Again, the slow kinetics indicates only a few copies of each sequence. An interesting piece of additional data from such an experiment is that only a small portion of the unique DNA is used to make mRNA. In the sea urchin embryo, for instance, this figure is 2.6% of the total unique DNA that is transcribed. This small percentage might be predicted from genetic load arguments.

Given the three types of DNA sequence repetition, the next problem is to determine how they are physically linked. In chloroplasts and mitochondria, each genome consists of a single DNA molecule (see Chapter 5), but such is not the case in the nucleus. Nevertheless, the nuclear genome does exist as a relatively small number of very large DNA molecules.

This can be demonstrated by gentle lysis of cells and measurement of the size of the largest intact DNA molecules. DNA molecules can be stretched by stirring them in solution, and if the stirring is stopped abruptly, the macromolecules will recoil back into a random coil configuration. Because larger molecules take longer to recoil, the time for the solution to come to a stop is proportional to the size of the largest molecule in that solution.

For both *Drosophila* (Table 10–8) and yeast, experiments using this technique have shown that the largest DNA molecule in the cell is the same size as the DNA content of the largest chromosome. A second technique, pulsed-field gel electrophoresis, has likewise shown giant single molecules of DNA extracted from the nucleus.

Thus it appears that the eukaryotic **chromosome** is composed of a limited number of very large DNA molecules. In humans, the total

---

TABLE 10–8 **Chromosome-Sized DNA Molecules in *Drosophila***

Genetic Strain	Morphology of Largest Chromosome	DNA Content of Largest Chromosome	Largest DNA Extracted
		(mol. wt.)	
Wild type		42×10^9	41×10^9
Inversion		42×10^9	41×10^9
Translocation		58×10^9	58×10^9
Deletion		24×10^9	24×10^9

DATA FROM: Kavenoff, R., and Zimm, B. Chromosome-sized DNA molecules from *Drosophila*. *Chromosoma* 44 (1973):1–13.

loid DNA content (Table 10−6) accounts for about 3×10^9 nucleotide pairs. There are 24 different chromosomes in the human species (22 autosomes, an X chromosome, and a Y chromosome), and an average chromosome has a DNA molecule with about 125 million base pairs.

Each chromosome has three defined functional types of DNA sequences. In fact, these are both necessary and sufficient for a piece of eukaryotic DNA to be called a chromosome (Figure 10−11):

1. There must be an **origin of DNA replication,** where DNA polymerase, the enzyme that catalyzes the duplication of the molecule, can attach and initiate the process. In both prokaryotes and eukaryotes these sequences are usually about 100 to 200 base pairs long and have an affinity for proteins that bind to DNA to initiate replication. A typical human chromosome has several hundred of these origins, but all chromosomes need at least one because an unreplicated piece of DNA is not passed on to the progeny cells during division and is ultimately lost.

2. There must be specific sequences at the ends of the chromosome, or **telomeres.** These are commonly variants of the sequence in humans:

. . . AGGGTT . . .

and are usually moderately or highly repeated. Their role is to act as a signal for the terminus in DNA replication. Another highly repetitive sequence is often adjacent to them.

3. There must be a specific sequence at the **centromere.** In both yeast and higher eukaryotes this sequence has 100 to 200 base pairs and some specificity in that it has required nucleotides in certain locations (especially at the ends) in all chromosomes of the organism. This sequence is needed to bind proteins of the kinetochore, a structure involved in attachment and elaboration of the mitotic spindle (see Chapter 12). Failure to have this sequence leads to loss of the chromosome during cell division.

Gene Structure

Many nonrepeated, protein-coding DNA regions have now been sequenced, and these sequences are usually represented as **antisense DNA,** reading from the $5'$ to the $3'$ end. The DNA molecule has two antiparallel strands, each with a free $5'$ phosphate end and a free $3'$ hydroxyl end:

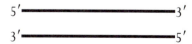

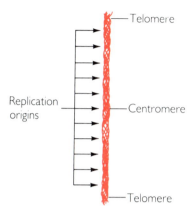

FIGURE 10−11

DNA structure of a typical eukaryotic chromosome. There is a single DNA molecule, with two telomeres, a single centromere, and many origins of DNA replication.

For a given region of DNA, only one of the two strands gets transcribed into mRNA. This is termed the sense strand of DNA, and the nontranscribed region is called the antisense strand:

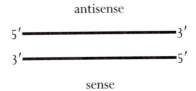

antisense

5′ ━━━━━━━━━━━━ 3′
3′ ━━━━━━━━━━━━ 5′

sense

Because the antisense strand is a direct reading of the RNA sequence transcribed from the sense strand (recall complementary base pairing during transcription—see Appendix 2), the antisense strand is the one usually written in DNA sequence studies, reading from 5′ to 3′ in the left-to-right direction:

5′ ━━━━━━━━━━━━ 3′

To illustrate, the two strands of DNA might have the base sequence:

5′ ━━━━ATGCCATAAACT━━━━3′
3′ ━━━━TACGGTATTTGA━━━━5′

If the mRNA copied from this sequence is:

5′ ━━━━AUGCCAUAAACU━━━━3′

the sense strand must be the bottom strand of DNA, and the antisense strand written in sequences would be the top strand of DNA.

A sequenced gene that illustrates many of the generalities that apply to other genes is the DNA coding for human β-**globin.** This polypeptide is part of hemoglobin, the oxygen-binding component of red blood cells. Two β-chains associate with two somewhat similar α-chains and the porphyrin pigment, heme, to make up hemoglobin.

If human β-globin DNA is denatured and then hybridized to denatured total human DNA under strict hybridization conditions that do not allow mismatched base pairing, the rate of hybridization is very slow, characteristic of a unique sequence. This means that β-globin is present only once in the haploid genome. But if the hybridization conditions are a bit looser, allowing for some small degree of mismatching to occur, the rate is somewhat faster. Calculations show that there are six sequences in the human genome quite similar to β-globin. These six genes compose the human β-globin **gene family.**

FIGURE 10-12 The arrangement of members of the β-globin gene family on human chromosome 11. The total length of this region of DNA is 60,000 base pairs. The blocks indicate globin genes, and the lines are spacer DNA.

It is important to emphasize that these are not moderately repetitive genes that have the same sequences. Rather, a gene family has members that are different from yet very similar to each other. In the case of the β-globins (referred to as globins in this discussion), the six members are clustered at a specific region of human chromosome 11, each member of the family being separated by **spacer DNA,** whose role is not known (Figure 10–12).

Each member of the beta family codes for a polypeptide chain 146 amino acids in length, but they differ in their sequences and therefore in the kind of hemoglobin formed when they are expressed and combined with the α-chains and heme:

- Epsilon chains are the first made by the developing human embryo. They are synthesized in the yolk sac and make a hemoglobin that binds oxygen tightly.

- Gamma chains are made by fetal liver and spleen from 8 to 38 weeks' of fetal gestation (full term is about 40 weeks). Again, the hemoglobin made (called fetal hemoglobin) tightly binds oxygen from the mother's circulation.

- Beta (98%) and delta (2%) chains are made in bone marrow cells and are the typical adult hemoglobin components.

This reinforces the concept of **differential expression** of members of the gene family.

A more detailed examination of the sequences of each globin gene reveals that although their protein-coding regions are somewhat different, they have many features in common (Figure 10–13). Each gene has the following:

1. The coding region has 441 base pairs (triplets coding for 146 amino acids plus a triplet stop codon).

2. About 30 base pairs before (5′ to) the coding region is a TATAAAT sequence of DNA, and another 50 base pairs back is

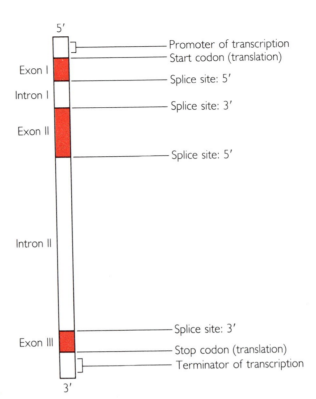

FIGURE 10–13

The structure of a member of the β-globin gene family. The entire DNA region is 1500 base pairs. The protein-coding part is in color.

the sequence ACACCC. These two sequences compose the **promoter,** a region responsible for binding RNA polymerase, the enzyme that catalyzes transcription (see below).

3. About 30 base pairs after (3′) the coding region of each gene is the sequence AATAA, which acts as a transcription **terminator** and causes RNA polymerase to cease transcription.

4. The coding region is interrupted by two sequences that are in the DNA but do not end up in the mature RNA. These intervening sequences (or **introns**) occur between codons 30 and 31 (130 base pair intron) and 104 and 105 (850 base pair intron). This means that the coding part of the gene is divided into three **exons.**

5. The **intron-exon boundaries** have specific sequences:

      . . . exon . . . G|GTAAGT . . . intron . . . CAG|CG . . .

                                    exon . . .

These are essential for the process of removing the intron after transcription and then splicing the two exons together prior to export of the mRNA to the cytoplasm.

A final characteristic of the globin family is the presence of a **pseudogene.** Like the other members of the family, this gene has a sequence quite similar to that of the prototype β-globin but with an essential difference: It is nonfunctional. Instead of coding for a complete β-chain, it has a base alteration that leads to a stop codon early in the mRNA, so that no functional globin is made. Pseudogenes are common in gene families, they may be there as a result of an insertion of a gene made from cytoplasmic mRNA, or they may simply be mutant genes that are nonlethal because of multiple copies of the parent gene.

It is important to note all the sequences of DNA that are non-coding in this 60,000 base pair portion of a human chromosome: promoters, terminators, introns, pseudogenes, and spacers. Indeed, the coding DNA accounts for only about 3% of the base pairs in this region!

Chromosomal DNA should not be thought of only in terms of clustered families of unique genes interspersed with repeated sequences. Many families are not clustered at all but are dispersed through the genome, with members on several different pieces of DNA (chromosomes). A good example of this is the family that codes for tubulins, the components of microtubules (see Chapter 9). In humans, there are over 20 tubulin genes scattered over at least a half-dozen different chromosomes. Nevertheless, the β-globins provide a good representation of the structure of typical protein-coding regions of eukaryotic nuclear DNA.

Structural Proteins

Nuclear DNA is associated with a specific group of proteins, the **histones,** which occur in chromatin in an aggregate mass about equal to that of the DNA. Virtually all eukaryotes have the same five classes of histones (Table 10–9), which are distinctive in their positive charge at physiological pH. This charge is due to chemical groups on the basic amino acids lysine and arginine, which are particularly abundant in these proteins.

Histones can be dissociated from DNA by high salt concentrations or low pH, indicating that they are bound to DNA electrostatically via

---

T A B L E 10–9 **The Histones**

Type	Molecular Weight	Composition
H1	21,000	Lysine-rich
H2a	14,000	Moderately lysine-rich
H2b	13,800	Moderately lysine-rich
H3	15,300	Arginine-rich
H4	11,300	Arginine-rich

the negative phosphate groups on the DNA. The basic amino acids are generally concentrated in the N-terminal third of the molecule as well as the C-terminal sixth. The middle half has a more or less typical amino acid content and a globular shape.

Several of the histones are remarkably similar in amino acid sequence when different organisms are compared. For example, H4 from peas and cows differs in but 2 amino acids of 102. Based on the hypothesized time of evolutionary divergence of different organisms, estimates of evolutionary amino acid changes in the histones can be made. For H3 and H4, there is estimated to have been only one change per 100 residues per billion years; for H2A and H2B, the rate is one change per 100 million years. These are among the lowest rates of protein-coding gene evolution known. The conservation of exact sequences indicates the essential role of the histones in DNA packaging (see below). The exception to the rule is histone H1, which has more sequence diversity than the other histones and even has variability among tissues of a given organism.

Histone amino acids can be posttranslationally modified, and these modifications can change the charge on the proteins, affecting their interaction with DNA. For instance, acetylation of lysine amino groups reduce their charge; phosphorylation of serine residues adds negative charges, which reduces the net positive charge. These modifications can cause the histones to detach from DNA and may be important for DNA transcription or replication.

Nonhistone proteins account for about two-thirds of the total mass of nuclear protein and occur in two types and quantities. The quantitatively major proteins are probably involved in packaging DNA. In many organisms, fewer than a dozen nonhistones account for one-third to one-half of this class of nuclear proteins. In a number of tissues, one of these major nonhistones has been identified as actin, a contractile protein whose presence suggests a role for it in the condensation of chromatin. Indeed, if an antibody against actin is injected into the oocyte nucleus, chromosome condensation is inhibited.

Enzymes and Transcription

In addition to the histones and nonhistones that are structurally important, there are many **DNA-binding proteins** that have specific roles in transcription of DNA to form mRNA. These proteins have specific structures that allow them to bind to the double-stranded DNA helix. Two of these three-dimensional structures are most common (Figure 10–14):

1. In the **helix-turn-helix** motif, the protein is a dimer, with two chains running in antiparallel directions. Protruding from the chains are

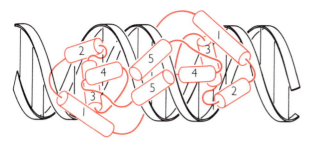 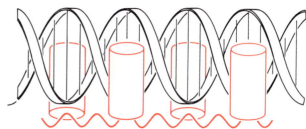

(a) Helix-turn-helix: Regular α-helices insert into the wide groove of DNA.

(b) Zinc finger: Amino acids coordinated to Zn$^+$ form protruding regions that insert into the wide groove of DNA.

FIGURE 10–14   Two types of DNA-binding proteins.

α-helical regions of just the right size and shape to fit in the grooves of DNA. These helices can recognize specific DNA base sequences on the two strands because the different bases within the helix stick out slightly when they pair to form hydrogen bonds.

2. In proteins with **zinc fingers,** coordination of two cysteine and two histidine residues with a Zn$^+$ ion produces a region that sticks out of the protein like a finger. This region can insert itself into the major (wider) groove of the DNA helix, and again specificity is possible because of the combination of the structure of the finger and the base pairs in the DNA.

Most DNA-binding proteins can bind to any DNA sequence, and indeed they can usually be found associated with DNA in the nucleus. But each binds even more tightly to certain DNA sequences, their "target DNA." The identification of this target can be made by "footprinting" analysis, in which the protein is mixed with DNA, and those DNA sequences that are unbound are digested away with nucleases, leaving the target sequence for analysis. Two general classes of such proteins are the RNA polymerases and proteins that regulate its function, transcription.

RNA polymerase is the nuclear protein that catalyzes the transcription of DNA to RNA:

$$\text{DNA template} + \text{NTPs} \rightarrow \text{RNA}$$

where NTPs are the four nucleoside triphosphates, ATP, GTP, CTP, and UTP.

Footprinting shows that in prokaryotes, RNA polymerase binds directly to the promoter, which is 5′ to the start of the coding region of a gene (see above). As indicated for the globin genes, the promoter determines which strand is the "sense strand" of DNA at a given point, because the particular promoter sequence will be on only one of the two strands. It is made up of two components:

1. A recognition region, at about 70 base pairs from the start codon, where that actual initial binding of the polymerase to DNA occurs

2. An AT-rich region, at about 25 base pairs from the start, where the two strands of DNA begin to unwind so that the template bases can be accessible to the enzyme

If the DNA sequence at the promoter is altered, RNA polymerase will not bind to it. This is strikingly shown in the β-globin genes of certain people who carry a promoter mutation. The recognition region is normally ACACCC, but in these people it has been mutated to ACATCC. The result is a total lack of transcription of the β gene, and no β-chains are made. Instead of normal hemoglobin, which has two α- and two β-chains, these people make a protein composed of only α-chains. This protein has aberrant oxygen-carrying capacity, and red blood cells carrying the pigment are readily destroyed. The severe anemia that results is called **thalassemia.**

In eukaryotes, there are commonly three different RNA polymerases, which act at different promoters to transcribe the various kinds of RNA in the cell:

1. RNA polymerase I occurs at the nucleolus and synthesizes rRNA.

2. RNA polymerase II occurs in the nucleoplasm and synthesizes mRNA.

3. RNA polymerase III also occurs in the nucleoplasm and synthesizes tRNA and some small nuclear RNAs.

The rate at which transcription occurs is often determined by the presence of molecules on DNA in addition to RNA polymerase. These are the **transcriptional regulatory proteins.** Often, they bind to DNA regions near RNA polymerase and either enhance or suppress its interaction with DNA.

An example of these factors is involved in the transcription of RNA polymerase I, which makes rRNA (Figure 10–15). The gene for this RNA has a promoter that is absolutely required for RNA polymerase to initiate transcription. But 5′ to the promoter is a second DNA region, at about 100 to 200 base pairs from the start codon. This region has an

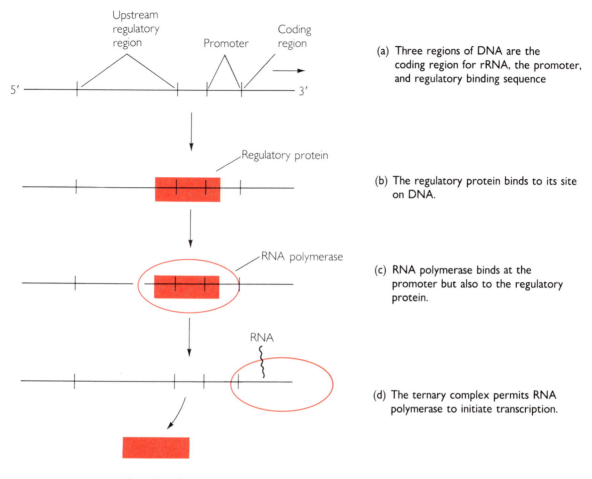

(a) Three regions of DNA are the coding region for rRNA, the promoter, and regulatory binding sequence

(b) The regulatory protein binds to its site on DNA.

(c) RNA polymerase binds at the promoter but also to the regulatory protein.

(d) The ternary complex permits RNA polymerase to initiate transcription.

FIGURE 10–15   The role of a DNA-binding protein in regulating transcription by RNA polymerase.

affinity for a regulatory protein called B and its associated subunit, S, and once these are bound to the DNA, the RNA polymerase binds and transcription can begin. B is a positive regulator, because if it is absent, transcription will still occur but at only about 5% of the normal rate. Similar initiation factors occur for RNA polymerases II and III in a number of cases. Transcription factors (TF's) are required for RNA polymerase II binding to the promoter. In fact, three of them (TFIID, TFIIA, TFIIB) bind sequentially to the promoter region prior to the polymerase, and a fourth (TFIIE) must bind later for transcription to begin.

Another type of nonpromoter DNA sequence that acts with a protein factor to stimulate transcription is **enhancers.** These sequences, usually about 150 base pairs in length, strongly stimulate transcription but in a way that is very different from the promoter and its 5′ sequences. Enhancers stimulate transcription at any nearby location, up to several thousand base pairs from the gene, 5′, 3′, or even within a gene's coding region! How they do this is less clear than the sites with more fixed locations, but it is known that enhancers bind nuclear regulatory proteins that often interact with RNA polymerase bound to the promoter. One way this could happen at a distance is for the enhancer-DNA complex to loop over to the RNA polymerase-promoter complex (Figure 10–16).

FIGURE 10–16   How an upstream enhancer element can stimulate transcription by RNA polymerase II.

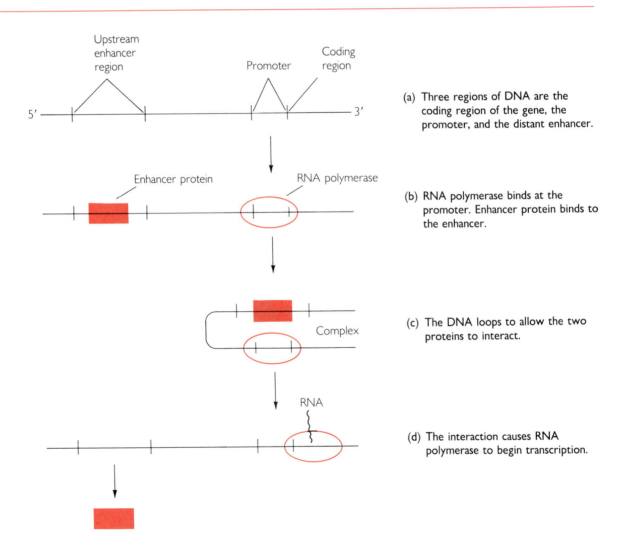

(a) Three regions of DNA are the coding region of the gene, the promoter, and the distant enhancer.

(b) RNA polymerase binds at the promoter. Enhancer protein binds to the enhancer.

(c) The DNA loops to allow the two proteins to interact.

(d) The interaction causes RNA polymerase to begin transcription.

Hormone receptors can also be transcriptional regulatory proteins. For example a steroid hormone is small and hydrophobic and can easily diffuse across the plasma membrane. Once within the cell, it binds to a specific receptor protein, altering that protein's shape. This altered protein then binds to specific DNA sequences, the steroid-responsive elements. This stimulates RNA synthesis, which leads to new proteins being made in the cell in response to the hormone.

An example of a steroid receptor is the protein that binds to the male sex hormone testosterone. In this case, DNA binding leads to enhanced transcription of proteins involved in the so-called secondary sex response (deepening of the voice, more muscle mass, more body hair, and so on). In the genetic disease testicular feminization syndrome, a defect in the DNA coding for the testosterone receptor leads to a lack of the transcriptional stimulation. Although testosterone is present in these males in adequate amounts, it cannot act to alter mRNA synthesis, and these men do not develop male secondary sex characteristics.

Following its synthesis, a eukaryotic RNA undergoes a series of **posttranscriptional modifications.** These include both the addition of nucleotides and the loss of some nucleotides. All of these reactions are catalyzed by nuclear proteins.

For mRNA, there are additions at both ends of the molecule (Figure 10–17):

1. At the 5' end (the first end made by RNA polymerase), a special nucleotide, 7-methyl guanosine is added as a "cap" and forms a new 3' end; this means that the mRNA now has two 3' ends. Because there are ribonucleases that attack RNAs at their 5' ends, this modification makes the resulting mRNA resistant to digestion. The methyl group apparently has a role in the export of the mRNA from the nucleus, since nonmethylated, or even overly methylated, caps result in the mRNA staying within the nucleus. Finally, the cap is involved in binding mRNA to the ribosome.

2. At the 3' end of the primary transcript, a sequence of RNA nucleotides is recognized by a nuclease, which cuts the RNA to make a new 3' end. It is important to note that this cut occurs in an RNA region, the trailer, which does not include any amino acid coding sequences. Then, a second enzyme adds up to 200 adenosine moieties, to make a "poly A tail." This tail is necessary for the transport of most mRNAs through the nuclear pore (see above) and seems also to be important for the longevity of the mRNA in the cytoplasm. Yet histone mRNA is made without a poly A tail and lasts the normal period of time.

The poly A tail serves a useful purpose also for the cell biologist: Poly T can be covalently attached to chromatographic beads, and when

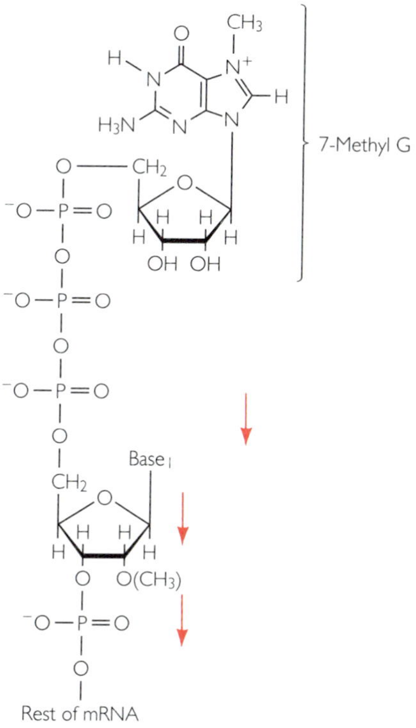

(a)

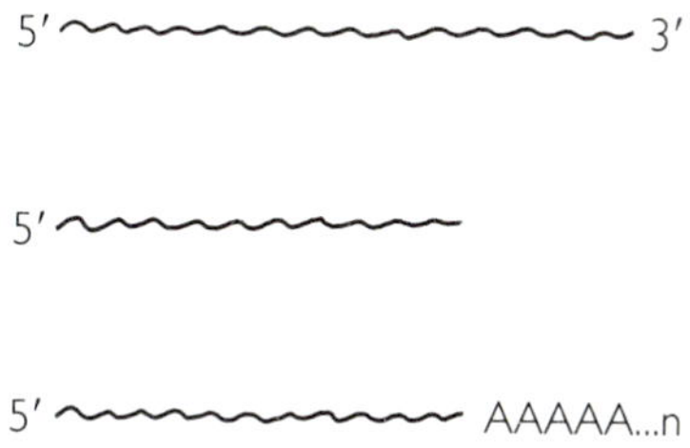

(b) At the 3' end, a cut is made and a poly A "tail" is added.

FIGURE 10–17

Posttranscriptional modifications of the 5' and 3' ends of eukaryotic mRNA.

total cell RNA is poured over a column of these beads, only poly A containing RNA (mRNA) sticks to the beads by base pairing. This provides a convenient way to purify mRNA.

Perhaps the most remarkable posttranscriptional change of mRNA is the excison of introns and subsequent joining of exons, a process called **RNA splicing.** As noted for the globin genes, the junctions between exons and introns have rather specific sequences, and these are recognized by an RNA-protein complex appropriately called the spliceosome. The RNA component of this particle is a small nuclear RNA, part of whose nucleotide sequence is complementary to the splice junction, especially at the 5′ end. It is this sequence recognition, this time by an RNA, that leads the complex to its target site at the exon-intron boundary.

Recognition failures at the splice sites have adverse consequences. Once again mutations of the β-globin gene are informative. The first two bases in the first intron of β gene are GT (actually GU in the mRNA). In some people, there is a mutation such that the bases are not AT. As a result, the splicing machinery does not recognize the site, the first intron is not spliced out of the pre-mRNA, and a larger mRNA is made. As a result, there are many more amino acids than in normal β-globin. The consequence is β-thalassemia.

Once the splicing complex lands at the splice site, the six proteins of the spliceosome take over and catalyze the splicing reactions (Figure 10–18):

1. The RNA is cut at the 5′ splice site.

2. The intron loops over and is covalently bound internally to form a "lariat" structure.

3. The free 3′ end of the exon (formed after the initial cut) then attacks the 3′ splice site and covalently binds to the second exon at this site.

4. The looped intron is made linear by a cut and released.

While it is clear that introns must be removed from the mRNA precursor in the nucleus, it is also apparent that introns serve some role in mRNA production. If a gene that normally has introns has them removed and then is inserted into a cell, it will not prime the synthesis of functional mRNA. The reason for this is not clear and intuitively makes no sense, since introns are regions that do not code for amino acids.

The evolutionary importance of introns has been a subject of much speculation. They often separate regions of DNA that code for distinct functional domains of a polypeptide. For instance, the middle exon of the β-globin gene (Figure 10–13) codes for part of the protein that

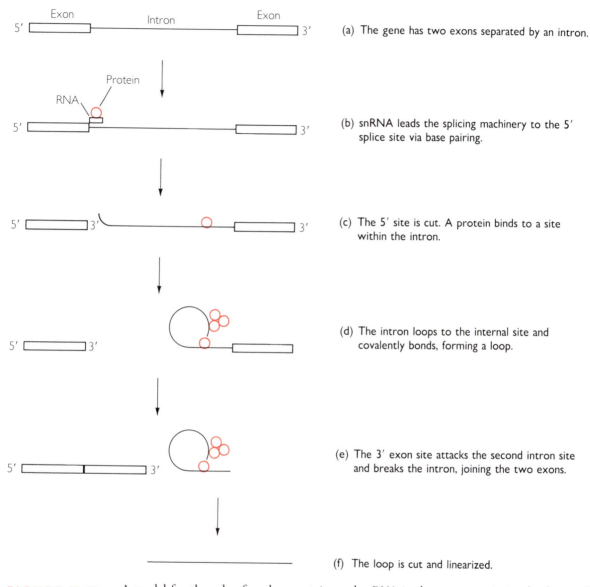

(a) The gene has two exons separated by an intron.

(b) snRNA leads the splicing machinery to the 5' splice site via base pairing.

(c) The 5' site is cut. A protein binds to a site within the intron.

(d) The intron loops to the internal site and covalently bonds, forming a loop.

(e) The 3' exon site attacks the second intron site and breaks the intron, joining the two exons.

(f) The loop is cut and linearized.

FIGURE 10-18   A model for the role of nuclear proteins and snRNA in the posttranscriptional splicing of pre-mRNA.

binds heme, and the third exon codes for the region responsible for binding to the other chains in the oligomeric protein. It is possible that these regions once were separate proteins in the genome, and a genetic recombination event brought them closer together ("exon shuffling").

Examples of single domains in multiple proteins have already been encountered in earlier chapters. Some of these are:

- Sequences that target proteins to organelles such as plastids, mitochondria, and ER
- Receptor signaling sequences, such as tyrosine kinase
- GTP binding sequences on G proteins

A variation of this exon shuffling happens in extant organisms. In a number of instances, the cell "selects" which introns to splice out and therefore which exons end up in mRNA. This alternate splicing is a way of producing different proteins from the same DNA sequence.

The nonhistones not mentioned so far are very heterogeneous and occur in individually smaller quantities. Over 400 different polypeptides have been identified as bands or spots on gel electropherograms. Some of these have been assigned functions, and many are enzymes. Chromatin contains many enzymes of nucleic acid metabolism such as ligases, various nucleases, and nucleotide modifying enzymes. In addition there are enzymes for the metabolism of nuclear proteins, such as kinases and acetyltransferases.

Chromatin Structure

Nucleosomes

A human somatic cell nucleus contains almost 2 meters of DNA. If one assumes that there are 46 chromosome-sized molecules, this means that the average human DNA molecule is over 4 cm long, and since the nucleus is only several micrometers in diameter, the packaging problem is indeed prodigious.

Thin-section electron microscopy of nuclei reveals little about chromatin structure, except a jumble of fibrils and granules. But when chromatin is isolated and put first into a hypotonic solution and then an aqueous medium, whole mounts show extended individual fibrils. These contain 10-nm−diameter beads, separated by a thin fiber of variable length (Figure 10−19).

These beads are similar in size and shape to ones produced when chromatin is digested with micrococcal nuclease, which hydrolyzes exposed DNA. They are also formed when naked DNA is incubated with equimolar amounts of the four histones (omitting H1). The beads, which represent the first order of packing of nuclear DNA, are termed **nucleosomes.**

Extended incubation of chromatin with nuclease leaves a core particle, containing 146 base pairs of DNA and two molecules each of H2A, H2B, H3, and H4. These four pairs of histones interact noncovalently to form a globular unit that has dimensions of 11 × 11 × 5.7 nm and that has a dyad symmetry. Each half of the core is identical, containing H3

(a)

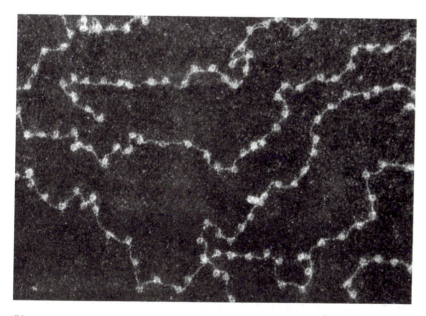

(b)

FIGURE 10–19

(a) Chromatin from a mouse tissue culture cell, showing nucleosomes. ($\times$20,000. Photograph courtesy of Dr. B. Hamkalo.) **(b)** Closer view of the nucleosomes ($\times$100,000).

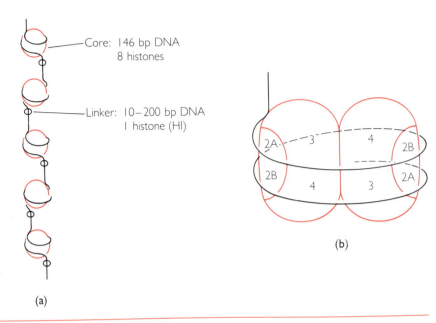

Core: 146 bp DNA
8 histones

Linker: 10–200 bp DNA
1 histone (H1)

FIGURE 10–20

Overall structure of nucleosomes in chromatin **(a)** and model for an individual nucleosome **(b).**

(a)

(b)

and H4 on the inside and H2A and H2B at the periphery of the disc (Figure 10–20). The specificity of interactions of the closely packed histones may be the reason that they have been conserved throughout evolution.

The DNA of the nucleosome winds around the outside of the histone core. With 146 base pairs, there is enough DNA to wind almost twice around the core, and this essentially means that about 50 nm of linear DNA is packed into a 10-nm particle.

But nuclear DNA is in the form of long, continuous molecules rather than 146 base pair units, so there must be DNA that connects the nucleosomes to each other. This "linker DNA" varies in length from 10 to over 200 base pairs. Histone H1 binds to the linker DNA, with binding being strongest near the nucleosome. It stabilizes nucleosome assembly in the test tube by preventing disaggregation in solutions of low ionic strength, and it is hypothesized to perform the same function in vivo.

Not all nuclear DNA is in regularly spaced nucleosomes and linkers. Some of it is nucleosome-free, and these are usually sequences that bind to other proteins, especially those that regulate transcription. In addition, the nucleosome must clearly come apart for an instant at places where DNA or RNA polymerases act, because these enzymes are large proteins that require a denatured DNA template.

When they were first observed, it was thought that **nucleosomes prevent transcription.** This has been clearly shown in many instances, a good example being the *Xenopus* (frog) 5S rRNA gene. Transcription of this gene requires a nonhistone factor, TF IIIA, in addition to RNA polymerase. If nucleosomes are assembled onto this gene in the test tube, transcription is inhibited because of the physical blockage of the particles. But if TF IIIA is present, it displaces the histones, and transcription can proceed; thus, TF IIIA indirectly activates transcription. However, if histone H1 is present, it stops the action of TF IIIA, and the nucleosomes stay on the gene; thus, H1 in this case indirectly inhibits transcription.

The absence of nucleosomes in actively transcribing regions of nuclear DNA allows exogenous proteins to bind, and this can be used as a marker for examining the structure of chromatin in these regions. One of the proteins widely used as an exogenous marker is pancreatic DNase I. This nuclease will not bind to, and cut, DNA if there are nucleosomes present. But if they are absent, the region is **DNase hypersensitive (DH).** Such regions are typically about 200 base pairs long.

When mapped on transcriptionally active genes, DH sites are often in regulatory regions such as promoters and enhancers, and when a specific gene is induced to become active, these sites become DH just before transcription begins. For instance, during the developmental switch from γ-globin to β-globin, which occurs late in fetal gestation (see above), the β gene develops DH sites 5′ to its coding region.

An interesting example of altered transcription patterns and DH sites occurs when cultured fibroblasts are transformed into tumor cells. In the normal cells, many of the DH sites lie near the nuclear envelope, but when the cell is transformed (by a carcinogen, for instance), the DH sites are more diffusely distributed in the nucleus. The peripheral distribution in normal cells apparently requires an intact cytoskeleton, which becomes disaggregated in transformed cells.

DH sites are locations where specific transcriptional activators can bind. A good example of this phenomenon is the heat-shock genes of the fruit fly, *Drosophila.* When flies (and other organisms) are exposed to 40° C, the syntheses of several proteins are increased. The chromatin structure at one of these genes coding for the heat-shock protein, hsp26, has been studied in detail (Figure 10–21).

The regulatory region 5′ to hsp26 DNA has two DH sites, separated by over 200 base pairs. Both of them contain sites for the binding of a nonhistone protein, HSF, which is activated only during heat shock. The one nearest the gene itself also has a binding site for a protein that acts to inhibit transcription by binding to RNA polymerase. So the situation prior to heat shock is that the inhibitory protein is on the DNA, and RNA polymerase is blocked from transcribing the hsp26 sequence.

On heat shock, HSF now binds to its two sites in the DH region. Binding to both sites results in an interaction with the transcriptional inhibitor, which now allows the RNA polymerase to transcribe hsp26. But the two HSF sites are over 200 base pairs apart, and the bound

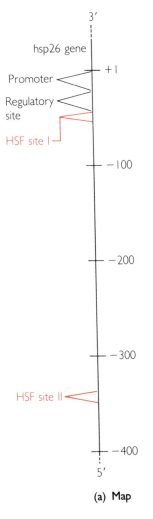

(a) Map

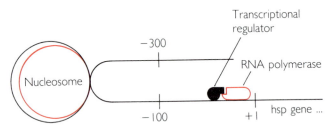

(b) Without heat shock: The transcriptional regulator blocks RNA polymerase from transcribing.

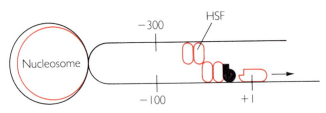

(c) In heat shock, proteins (HSF) bind to sites aligned because of a nucleosome. This alters the regulatory protein, so it no longer inhibits RNA polymerase. hsp26 is transcribed.

FIGURE 10–21   Chromatin structure and transcriptional regulation in the heat-shock gene hsp26 in the fruit fly. The overall map of the region of 400 base pairs (at left) shows the promoter and two regulatory regions separated by 200 base pairs. (After S. Elgin.)

proteins would not be expected to interact at such a distance. This problem is solved by the presence of a nucleosome between the two sites, which accommodates the 200 base pairs and brings the two HSF sites near each other. This is an excellent example of the nucleosome being essential to gene regulation because of its structural properties.

In addition to transcription, nucleosomes must come apart during **DNA replication.** This is a transient phenomenon, as the histone octamers detach from the DNA just as the DNA polymerase passes and then reassemble onto one of the newly synthesized double helices immediately after it is made. New histone octamers are assembled onto the other daughter helix.

Fibers

Nucleosomes are a fivefold packing of the linear DNA molecule. But the actual packing ratio of DNA into chromatin is 100-fold greater. Electron microscopy of chromatin frequently shows loops that are 20 to 30 nm in diameter. These loops consist of rows of two to three nucleosomes packed on one another in an often helical array and will progressively dissociate into chains of nucleosomes when exposed to buffers of low ionic strength. The DNA packing ratios of these ionically held aggregations are 40- to 50-fold.

Histone-depleted chromosomes can be prepared by incubating them with polyanions, such as dextran sulfate, which compete with histones for sites on the DNA. Such preparations show a central core, or **scaffold,** from which loops of DNA, 60,000–100,000 base pairs long, radiate (Figure 10–22).

FIGURE 10–22

A histone-depleted chromosome, showing the loops of DNA emanating from the chromosome scaffold. (Photograph courtesy of Dr. U. K. Laemmli.)

The looped structure that is made visible through chemical treatment of isolated chromatin has a naturally occurring counterpart. During the diplotene stage of meiosis in most mammals, the chromosomes decondense, and looped structures are seen. These lampbrush chromosomes can be isolated by dissection from the nucleus, and microscopy reveals many DNA loops extending from a protein core, with each loop containing 5000 to 300,000 base pairs of DNA (Figure 10–23).

But how does the packing occur? A hint at the answer may come from the identification of the scaffold proteins, of which there are two major species. One of them is an enzyme, DNA topoisomerase II, which, along with its counteracting analogue, DNA topoisomerase I, controls the supercoiling of looped DNA:

$$\text{supercoiled DNA} \underset{\text{topo II}}{\overset{\text{topo I}}{\rightleftarrows}} \text{relaxed DNA}$$

Supercoiling can be likened to holding a rubber band at its end and then twisting it until it collapses upon itself. To produce supercoiling, topo II must produce double-strand breaks in DNA, since this molecule

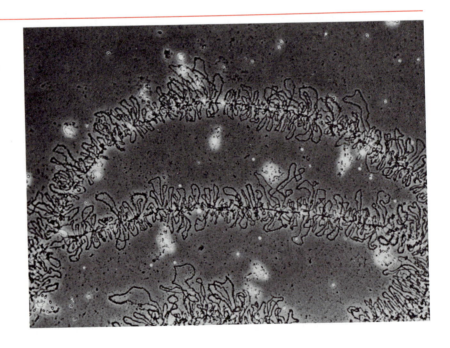

FIGURE 10–23

Lampbrush chromosomes in an amphibian oocyte. There are two homologous chromosomes with loops of DNA radiating from them. (Courtesy of Dr. J. Gall.)

is not elastic. The activity of scaffold topo II can be envisioned to cause the further looping of an already looped section of DNA. Indeed, inhibitors of topo II activity also inhibit nuclear assembly in the test tube. During DNA replication, topo II would be useful in untangling the intertwined loops of DNA. The sequences of DNA that are associated with the scaffold include the binding site for topo II, and antibodies to topo II bind to the scaffold region rather than to the loops radiating from it.

Topo I has also been identified as a nonhistone protein in the nucleus, thereby completing the regulatory network for DNA packaging. It is interesting to note that topo I is involved as an antigenic target in the human autoimmune disease scleroderma. This may mean that this nonhistone protein may control the overproduction of collagen in these patients by changing the looping, and availability for transcription, of collagen DNA.

A number of observations indicate that different chromosomes have a specific distribution within the nucleus. For example, grass hybrids can be made by crossing two species. If a probe is made to DNA sequences of one of the species, hybridization should reveal their location in the nucleus. Such experiments often show that the two species' chromosomes occupy different parts of the nucleus.

How this segregation could occur is not clear, but one point of attachment of the chromosomes might be the nuclear envelope. In 1885, C. Rabl showed that specific chromosomes appeared to attach themselves to the nuclear membrane as it reformed late in meiosis. These attachments have been confirmed and are at the telomeres (ends). On the other hand, a careful study of large polytene chromosomes of fruit flies showed that they are arranged with their centromeres attached to one end (pole) of the nuclear membrane and telomeres at the other end. These observations, coupled with the emerging ideas on a nuclear skeleton, offer fascinating possibilities for an overall structural and functional architecture of the nuclear interior.

Nuclear Matrix

If nuclei are treated with moderately elevated temperatures (37°C), a detergent to solubilize the nuclear envelope, a high salt concentration to solubilize chromatin, and nucleases to remove residual nucleic acids, a network of fibrils retaining the shape of the nucleus remains. Remarkably, this network looks somewhat like the untreated nucleus, and is called the nuclear matrix (Figure 10–24). It has three consistent components:

- A residual envelope with pore complexes
- A residual nucleolus

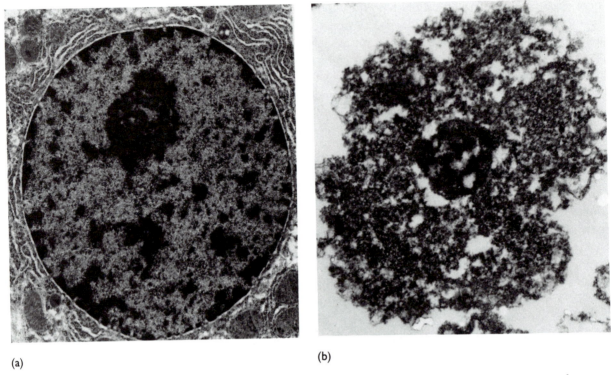

(a)

(b)

F I G U R E   10–24   **(a)** Rat liver nucleus (×9000) and **(b)** nuclear matrix (×13,000). (Photographs courtesy of Dr. M. Berezney.)

- An internal matrix that resembles the nucleoplasm of the intact nucleus

This structural framework has been isolated from a wide variety of organisms, ranging from ciliated protozoa to mammals.

The protein matrix has about 20% of the total nuclear protein but no histones (or nucleosomes). Rather, qualitative analysis of matrix proteins by gel electrophoresis reveals three major acidic polypeptides with molecular weights in the 65,000 dalton range, one of which may be DNA topo II. During cell division, the matrix disperses at prophase and reassembles at telophase.

The nuclear matrix apparently has both structural and functional aspects. If it is isolated in conditions that preserve nucleic acids, over half of the nuclear DNA remains attached to a fibrillar component. This DNA is in loops of over 100,000 base pairs and is often the DNA that is active in **transcription.** For instance, the oviduct of the chicken synthesizes the egg protein, ovalbumin, but liver cells do not. When nuclear matrices are isolated from these two tissues, the ovalbumin gene is attached to the matrix from oviduct but not from liver.

In support of the idea of matrix associations with transcription, most newly synthesized RNA of the nucleus is initially bound to the matrix. While there, much of the processing of RNA (e.g., the capping and splicing of mRNA) occurs, and then the mature mRNA is released in an ATP-requiring reaction. Inhibitors of DNA topoisomerases also inhibit RNA release, implicating the scaffold proteins in the process.

The initiation of DNA replication occurs on a component of the matrix. If cells are given a brief pulse with radioactive thymidine, radioactivity (in DNA) can be observed by electron microscope autoradiography to be in the matrixlike nuclear fibrillar zone. This newly synthesized DNA can be isolated along with the matrix, and the matrix has an active DNA polymerase activity, which can synthesize DNA in the test tube.

The multiple origins of DNA **replication** in the large eukaryotic chromosome appear to be permanently attached to the matrix. Pulse-chase experiments show that initially the replicated DNA is in small pieces (less than 1000 base pairs), but later it ends up in the much larger loops. These loops approximate the size of a replicating unit of DNA (replicon). Thus the nuclear matrix may be responsible for organizing the replication of the genome.

Regulation of transcription and replication may also occur on the nuclear matrix. For example, it has the receptors for androgenic and estrogenic steroids (see above). These receptors are tissue specific: In a rooster, the liver nuclear matrix contains less than 100 molecules of the receptor for estrogen. However, in the egg-laying hen there are high levels of these hormones and a great increase (to over 600) in nuclear receptors. The liver cells then synthesize vitellogenin, the precursor to the major egg protein. In addition, administration of androgens to castrated rats leads to a rapid accumulation of receptors for these hormones in the nuclear matrix of prostate cells.

Although the nuclear matrix and scaffold represent an attractive framework on which to build a basic nuclear structure, care should be taken in inferring their existence in situ. High salt concentrations cause nuclear proteins and nucleic acids to aggregate nonspecifically, and single-stranded (newly replicated) DNA has a strong affinity for such aggregates. The fact that bacteriophage DNA can be injected into oocyte cytoplasm and induce the formation of a "nucleus" out of nuclear proteins and envelope components implies that DNA sequence specificity is not necessary for nuclear structure. Nevertheless, the consistent appearance of a small number of proteins in matrix preparations, as well as certain enzymes and transcriptionally active DNA, strongly implies that the matrix and scaffold are natural nuclear structures.

Heterochromatin

The highest level of packaging of nuclear material can be visualized with the light microscope or low magnifications of the electron microscope. During the first quarter of the 20th century, cytologists observed that the interphase nucleus contains both condensed and diffuse regions. In 1929, E. Heitz used the terms *heterochromatin* and *euchromatin* to describe these respective regions. He noted two types of heterochromatin: Some remained condensed throughout interphase; the other went through condensation-decondensation cycles. Microscopists now recognize at least three classes of condensed chromatin in the nucleus (Table 10–10).

Constitutive heterochromatin is always condensed. It occurs in mitotic cells near centromeres or the ends of chromosomes and stains preferentially with the cytochemical dye Giemsa, especially when DNA is denatured. This procedure is known as C-banding because the constitutive heterochromatin shows up as discrete, reproducible bands on the chromosomes or clumps in the interphase nucleus.

The presence of some heterochromatic regions can affect genes adjacent to them. In fruit flies, translocation of a heterochromatic chromosomal segment to a site adjacent to the wild type eye color gene suppresses the latter, and white-eyed flies result. Several genes distal to the eye color gene can also be inhibited from expression. In maize, a gene, "Dissociator," associated with heterochromatin, can move from one place to another on the genome, suppressing the adjacent genes wherever it inserts. Such mobile elements have also been found in fungi and mammals.

The DNA of constitutive heterochromatin is highly repetitive satellite DNA. Indeed, most highly repetitive sequences in the genome are localized in this type of chromatin. These sequences are not transcribed and generally are replicated at a unique time during genome replication, usually late in the process.

A different kind of constitutive heterochromatin is observed in the chromosomes of dipteran (e.g., fruit fly) salivary gland cells. Here, the noncentromere DNA has undergone 10 rounds of replication without mitotic separation of the chromosomes. The highly replicated strands lie just beside each other, and when complexed with proteins dense bands

TABLE 10–10 **Classes of Condensed Chromatin**

Class	Comments
Constitutive heterochromatin	Always present
Facultative heterochromatin	Sex chromatin
Inactivated euchromatin	Reversible

are seen, separated by less dense regions (Figure 10−25). In this case, DNA within the bands does code for proteins and is transcribed, so the role of heterochromatin is unclear.

Facultative heterochromatin is condensed for only part of the life of the cell. The best example of this class is the mammalian sex chromatin. At about the 16th day of embryonic development in human females, one of the two X chromosomes in each cell is heterochromatized. This X chromosome remains heterochromatic for the lifetime of the cell and its descendants, and it can be detected in somatic cells as a dense clump at the nuclear periphery (Barr body). It is random as to which X (the one deriving from the father or the one from the mother) is affected.

The heterochromatic X is largely transcriptionally inactive. Therefore, on the average, half of a female mammal's cells have an active paternal and inactive maternal X chromosome; in the other half, the reverse is true. This means that if the two X chromosomes differ genetically, the female can be a phenotypic mosaic. For example, in cats the X chromosome carries the gene for coat color. A female with one X having the gene for black color and one for orange color has patches of both on her coat due to random facultative heterochromatization. This is termed a calico cat.

Inactivated euchromatin is essentially a type of heterochromatin in which the DNA sequences are heterochromatic for a period, then they are euchromatic, and then they can revert to heterochromatic again, and

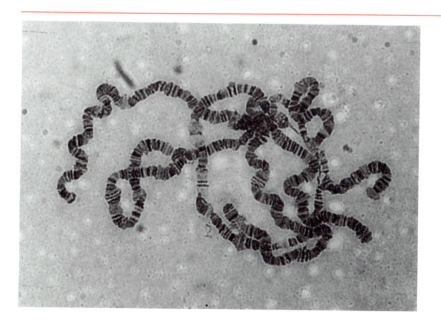

FIGURE 10−25

Polytene chromosomes from the salivary gland of the fruit fly, *Drosophila*. The dark staining bands are heterochromatin. (Photograph courtesy Dr. S. Celniker.)

so on. As the name implies, this heterochromatin is transcriptionally suppressed (although not all suppressed genes are heterochromatic). For example, in mature mammalian sperm cells, the entire nucleus is heterochromatic, and these cells do not synthesize RNA and protein. A similar situation holds for the terminally differentiated avian erythrocyte.

Cycles of inactivation and reactivation of euchromatin occur throughout the lifetimes of different tissues. These cycles do not involve changes in the amount of DNA, as heterochromatic nucleated erythrocytes contain as much DNA as their more euchromatic reticulocyte precursors.

An approach to determining the mechanism of these cyclic changes has been devised by N. Ringertz (Figure 10–26). Chicken erythrocytes,

FIGURE 10–26

Cell fusion experiment showing the reactivation of the heterochromatic chick erythrocyte nucleus.

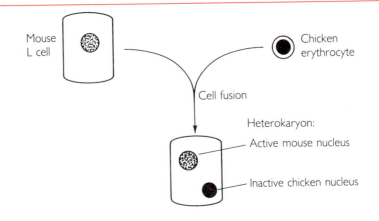

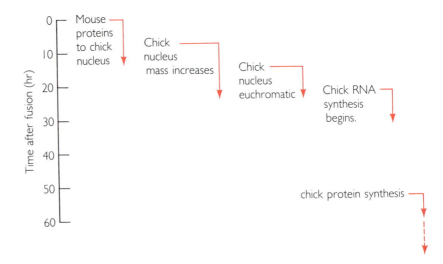

with their heterochromatic nuclei, are fused with mouse L cells in culture in which typical euchromatic nuclei are present. Several days after fusion, the erythrocyte nucleus is euchromatic and synthesizes RNA, and both chicken and mouse proteins are made by the hybrid cells.

The mechanism of euchromatization of the chicken nucleus appears to involve the uptake of proteins from the L cell cytoplasm. If the latter are labeled with radioactive amino acids prior to cell fusion, certain labeled nonhistones enter the chick nucleus immediately after fusion. Several hours later, euchromatization is microscopically detectable and RNA synthesis follows. From the timing of these events, it is tempting to propose that the cytoplasmic proteins entering the chick nucleus cause its decondensation and reactivation. Some of these have been identified, and they include RNA polymerases I and II and the proteins of the spliceosome.

The primary mechanism for chromatin condensation is not clearly known but may involve signals on the DNA itself. The nucleotides in DNA can be modified in situ. In eukaryotes, especially mammals, the most common modification is the **methylation of cytosine** (Figure 10–27), which occurs on about 2–8% of the cytosine residues on animal DNA and up to 30% on plant DNA. The enzyme DNA methylase is a nuclear protein and catalyzes the reaction when cytosines are in the sequence:

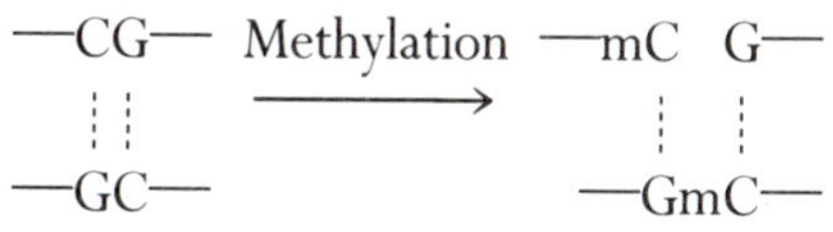

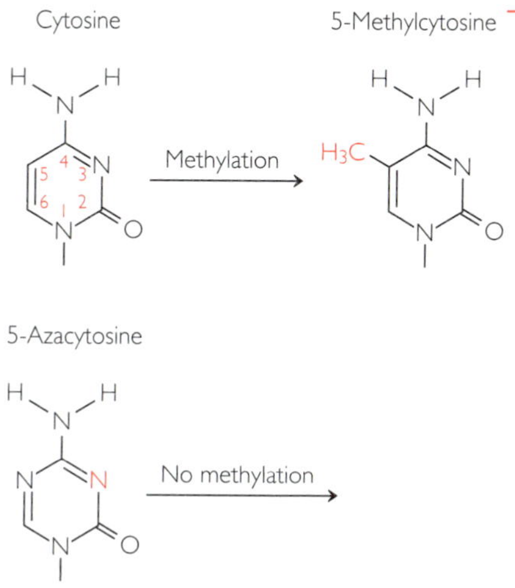

FIGURE 10–27

Cytosine on DNA can be methylated to form 5-methylcytosine. This modified base results in gene inactivation by heterochromatin formation. Azacytidine blocks DNA methylation and leads to an activation of previously inactive genes.

This provides a method of maintaining the methylation pattern of a DNA sequence after DNA replication and cell division:

$$
\begin{array}{ccc}
-\text{mC} \quad \text{G}- & \xrightarrow{\text{Methylation}} & -\text{mC} \quad \text{G}- \\
\vdots \quad \vdots & & \vdots \quad \vdots \\
-\text{G} \quad \text{C}- & & -\text{GmC}-
\end{array}
$$

$$
\begin{array}{cc}
-\text{mC} \quad \text{G}- & \xrightarrow{\text{Replication}} \\
\vdots \quad \vdots & \\
-\text{GmC}- &
\end{array}
$$

$$
\begin{array}{ccc}
-\text{C} \quad \text{G}- & \xrightarrow{\text{Methylation}} & -\text{mC} \quad \text{G}- \\
\vdots \quad \vdots & & \vdots \quad \vdots \\
-\text{GmC}- & & -\text{GmC}-
\end{array}
$$

(Parent DNA)   (Daughter DNAs)

This maintenance mechanism ensures that a gene that is methylated in a cell will also be methylated in the cell's progeny after division. On the other hand, methylation can be blocked if the cytosine analogue, 5-azacytosine (Figure 10–27), is incorporated into DNA during its replication. If this is done, a gene that is methylated in the parent cell will be methylated on only one strand of the DNA in the first generation progeny, and ultimately, after many cell divisions, the gene will be nonmethylated in most of the cells.

Methylation of cytosines has two major effects on chromatin, which may or may not be due to the same mechanism. First, if many residues are methylated, the region becomes heterochromatic. One example of this is the mammalian X chromosome. Like other parts of the genome, this chromosome has many "CG" sequences. Somehow, via a region called the X-inactivation center, these regions become heavily methylated on one X of a female during early embryology, and this is strongly correlated with the heterochromatinization of that X chromosome to form a Barr body. Another example of this general phenomenon is satellite DNA. These sequences are often heavily methylated and heterochromatic. It is remarkable that when these regions of DNA become methylated, they change the timing of their replication to later in the overall scheme of DNA synthesis.

The second effect of methylation on chromatin is on its function in transcription. Simply put, methylation is correlated with inactivation of transcription, while demethylation is correlated with activation. For example:

1. The β-globin gene is undermethylated and transcriptionally active in chick reticulocytes, while in the precursors to these cells, which do not make the globin mRNA, the gene is methylated.

2. The γ-globin gene is undermethylated in fetal cells, which transcribe it, while in a newborn infant the gene is methylated and "turned off."

3. If a gene is methylated in the test tube and then put into a host cell, that gene will be inactive transcriptionally, even if it normally would be active in that cell.

Methylation of the entire gene is not necessary to inhibit transcription. In fact, methylation of only a few cytosine residues in the promoter region is all that is needed in many cases. This appears to block the binding to DNA of proteins that are needed to activate transcription. This may be either a direct effect (the methyl group sterically interferes with binding) or an indirect one (the methyl group attracts certain proteins that in turn block binding of the transcription factor). In either case, the effect on transcription is strong.

Further Reading

Adolph, K. W. (Ed.). *Chromosomes and chromatin.* Boca Raton, FL: CRC Press, 1988.

Agutter, P. Nucleo-cytoplasmic transport of mRNA: Its relationship to RNA metabolism, subcellular structures and other nucleocytoplasmic exchanges. *Prog Mol Subcell Biol* 10 (1989):15–83.

Appels, R. Three dimensional arrangements of chromatin and chromosomes: Old concepts and new techniques. *J Cell Sci* 92 (1989):325–328.

Babu, A., and Verma, R. S. Chromosome structure: Euchromatin and heterochromatin. *Int Rev Cytol* 108 (1987):1–41.

Beato, M., Chalepakis, G., Schauer, M., and Slater, E. P. DNA regulatory elements for steroid hormones. *J Steroid Biochem* 32 (1989):737–747.

Berezney, R. The nuclear matrix: a heuristic model for investigating genomic organization. *J Cell Biochem* (1991):109–123.

Blackburn, E. Structure and function of telomeres. *Nature* 350 (1991):569–572.

Brachet, J. *Cell interactions: Molecular cytology,* Vol. 2. New York: Academic Press, 1985.

Burke, B. The nuclear envelope and nuclear transport. *Curr Op Cell Biol* 2 (1990): 514–520.

Busch, H. (Ed.). *The cell nucleus.* Multiple volumes. New York: Academic Press, 1983–.

Chandler, L. A., and Jones, P. A. Hypomethylation of DNA in the regulation of gene expression. *Dev Biol (NY)* 5 (1988):335–349.

Cook, P. The nucleoskeleton and the topology of replication. *Cell* 66 (1991):627–635.

Csordas, A. On the biological role of histone acetylation. *Biochem J* 265 (1990):23–28.

Dingwall, C., and Laskey, R. Nuclear targeting sequences: A consensus? *Trends Biochem Sci* 16 (1991):478–481.

Dworetzsky, S., and Feldherr, C. M. Translocation of RNA-coated gold particles through nuclear pores of oocytes. *J Cell Biol* 106 (1988):575–584.

Earnshaw, W. C. Mitotic chromosome structure. *BioEssays* 9 (1988):147–150.

Elgin, S. C. R. The formation and function of DNase I hypersensitive sites in the process of gene activation. *J Biol Chem* 263 (1988):19259–19262.

Elgin, S. C. R. Chromatin structure and gene activity. *Curr Op Cell Biol* 2 (1990): 437–445.

Gasser, S. M., and Laemmli, U. K. A glimpse at chromosomal order. *Trends Genet* 3 (1987):16–22.

Gerace, L., and Burke, B. Functional organization of the nuclear envelope. *Ann Rev Cell Biol* 4 (1988):335–374.

Georgiev, G., Vassetzky, Y., Luchnik, N., Chevnokhvostov, V., and Razin, V. Nuclear skeleton, DNA domains and control of transcription and replication. *Eur J Biochem* 200 (1991):613–624.

Getzenberg, R., Pienta, K., Ward, S., and Coffey, D. Nuclear structure and 3-dimensional organization of DNA. *J Cell Biochem* 47 (1991):289–299.

Goldfarb, D. S. Nuclear transport. *Curr Op. Cell Biol* 1 (1989):441–446.

Grunstein, M. Nucleosomes: regulators of transcription. *Trends Genet* 6 (1990): 395–400.

Gruss, C., and Sogo, J. Chromatin replication. *Bioessays* 14 (1992):1–8.

Haaf, T., and Schmid, M. Chromosome topology in mammalian interphase nuclei. *Exp Cell Res* 192 (1991):325–332.

Hamkalo, B., and Elgin, S. C. R. Functional Organization of the Cell Nucleus. San Diego: Academic Press, 1992.

Hanover, J. The nuclear pore at the crossroads. *FASEB J* 6 (1992): 2288–2295.

Heslop-Harrison, J., and Bennett, M. Nuclear architecture in plants. *Trends Genet* 6 (1990):401–405.

Hilliker, A., and Appels, R. The arrangement of interphase chromosomes: Structural and functional aspects. *Exp Cell Res* 185 (1989):297–318.

Holmquist, G. Evolution of chromosome bands: Molecular ecology of noncoding DNA. *J Mol Evol* 28 (1989):469–486.

Kessel, R. G. The annulate lamellae—from obscurity to spotlight. *Elec Microsc Rev* 2 (1989):257–348.

Kornberg, R., and Lorch, Y. Irresistible force meets immovable object: transcription and the nucleosome. *Cell* 67 (1991):833–836.

Lamond, A. Nuclear RNA processing. *Curr Op Cell Biol* 3 (1991):493–501.

Laskey, R. A., and Leno, G. Assembly of the cell nucleus. *Trends Genet* 6 (1990): 406–409.

Lewin, B. *Genes IV.* Oxford: Oxford University Press, 1990.

Lewis, J., and Bird, A. DNA methylation and chromatin structure. *FEBS Lett* 285 (1991):155–159.

Manuelidis, L. A view of interphase chromosomes. *Science* 250 (1990):1533–1540.

Marriott, S., and Brady, J. N. Enhancer function in viral and cellular gene regulation. *Biochim Biophys Acta* 909 (1989):97–110.

McKeon, F. Nuclear lamin proteins: Domains required for nuclear targeting, assembly and cell cycle regulated dynamics. *Curr Op Cell Biol* 3 (1991):82–86.

Miller, M., Park, M., and Hanover, J. Nuclear pore complex: Structure, function and regulation. *Physiol Rev* 71 (1991):909–940.

Nagl, W. Condensed interphase chromatin in plant and animal cell nuclei. *Plant Systems Evolution* Suppl. 2 (1979):247–260.

Nelson, W. G., Pienta, K., Barrack, E., and Coffey, D. S. Role of the nuclear matrix in the organization and function of DNA. *Ann Rev Biophys Biophys Chem* 15 (1986):457–475.

Newport, J., and Forbes, D. The nucleus: Structure, function and dynamics. *Ann Rev Biochem* 56 (1987):535–545.

Nigg, E. Mechanisms of signal transduction to the cell nucleus. *Adv Cancer Res* 55 (1990):271–306.

Paine, P., and Feldherr, C. Nucleocytoplasmic exchange of macromolecules. *Experimental Cell Res* 76 (1972):81–98.

Peters, R., and Trendelenberg, M. *Nucleocytoplasmic transport.* Berlin: Springer-Verlag, 1986.

Ringertz, N., and Savage, R. *Cell hybrids.* New York: Academic Press, 1976.

Scheer, U., Dabauvalle, M., Merkert, H., and Benevente, R. The nuclear envelope and the organization of the pore complexes. *Cell Biol Int Rep* 12 (1988):669–688.

Schroder, H., Bachmann, M., Siefert, B., and Muller, W. Transport of mRNA from nucleus to cytoplasm. *Prog Nucl Acid Res Molec Biol* 34 (1987):89–137.

Selker, E. U. DNA methylation and chromatin structure: A view from below. *Trends Biochem Sci* 15 (1990):103–107.

Silver, L., and Goodson, H. Nuclear protein transport. *CRC Crit Rev Biochem Mol Biol* 24 (1989):419–435.

Silver, P. How proteins enter the nucleus. *Cell* 64 (1991):289–297.

Smuckler, E., and Clawson, G. *Nuclear envelope structure and RNA maturation.* New York: Alan R. Liss, 1986.

Spiker, S. Plant chromatin structure. *Ann Rev Plant Physiol* 36 (1985):235–253.

Starr, C., and Hanover, J. Structure and function of the nuclear pore complex: New perspectives. *BioEssays* 12 (1990):323–329.

Svaren, J., and Chalkley, R. The structure and assembly of active chromatin. *Trends Genet* 6 (1990):52–56.

Van Holde, K. Chromatin. New York: Springer-Verlag, 1989.

Van Holde, K., Lohr, D., and Robert, C. What happens to nucleosomes during transcription? *J Biol Chem* 267 (1992):2837–2840.

Verheijen, R., Venrooij, W., and Ramaekers, F. The nuclear matrix: structure and composition. *J Cell Sci* 90 (1988):11–36.

Verma, R. S. (Ed.). *Heterochromatin: Molecular and structural aspects.* Cambridge: Cambridge University Press, 1989.

Wagner, P., Kunz, J., Koller, A., and Hall, M. Active transport of proteins into the nucleus. *FEBS Let* 275 (1990):1–5.

Zlatanova, J., and Yaneva, J. Histone H1-DNA interactions and their relation to chromatin structure and function. *DNA Cell Biol* 10 (1991):239–248.

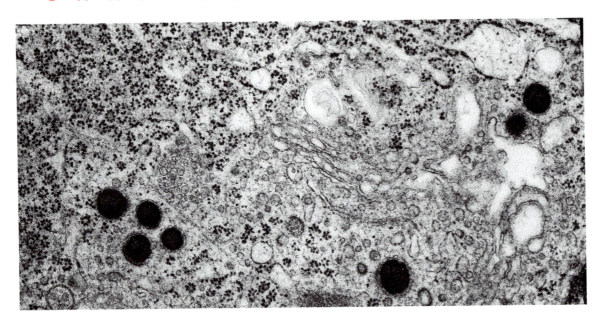

Nucleolus and Ribosomes

A Specialized Region Within the Nucleus

OVERVIEW

The existence of a nuclear matrix or scaffold (Chapter 10) implies that there is an order to what superficially is a disorganized jumble of DNA, RNA, and proteins. The nucleolus is the most striking example of this order. The nucleolus is associated with a specific chromosomal region, the nucleolar organizer, which has the DNA sequences coding for most of the cell's ribosomal RNAs (rRNAs). Within it, these sequences are transcribed and processed, and the initial steps are taken to combine them with proteins into functional ribosomes. As a nuclear domain, the nucleolus is structurally distinct and does not need to be in a separate membrane-bound compartment. In fact, ribosomes are made by plastids and mitochondria without the need for a separate nucleolus at all.

The immature ribosomes formed in the nucleolus are transported out of the nucleus via nuclear pores (see Chapter 10) and are then completed in the cytoplasm. Ribosomes are the "workbenches" where proteins are synthesized. In a number of respects, ribosomes are similar to microtubules: Both are not surrounded by a membrane; both are made up of insoluble proteins; both have remarkable self-assembling structures; and both can be thought of as frameworks onto which active proteins attach. In the case of microtubules, these are microtubule associated proteins (MAPs) such as dynein and kinesin, while the ribosomes have the apparatus for protein synthesis.

Nucleolus: Structure and Composition

Nucleoli are readily seen by **light microscopy** of interphase nuclei as dense, refractile bodies, about $0.5-3.0$ μm in diameter, but are not present in an organized form during mitosis. Diploid cells usually have one large or two or more smaller nucleoli, and often the combined surface area of the smaller nucleoli in one cell of a given tissue is the same as that of the single nucleolus in a neighboring cell. This indicates that the total nucleolar volume in a cell is regulated in some way. A reason for this regulation might be to keep the protein synthetic machinery in neighboring cells constant. Nucleolar volume is usually larger in cells that are active in protein synthesis than in inactive cells.

Electron microscopy reveals a distinct substructure of the nucleolus (Figure 11−1):

1. A granular component, composed of particles that are $15-20$ nm in diameter.

409

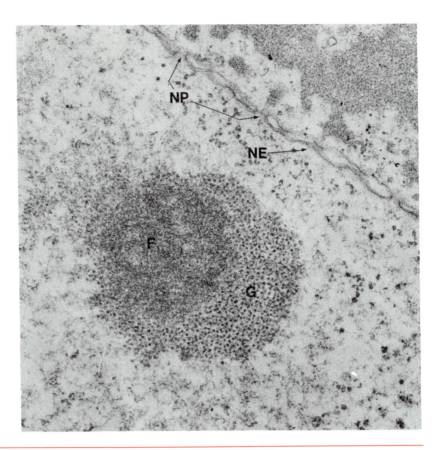

FIGURE 11-1

Nucleolus within a frog oocyte nucleus. Note the granular region (G) and fibrillar region (F). (NP = nuclear pore, NE = nuclear envelope.) (×80,000. Courtesy of Dr. O. L. Miller, Jr.)

2. A fibrillar component, composed of closely packed fibrils, 5–10 nm in diameter. This is often intimately associated with a layer of heterochromatin, suggesting that the fibrils are a euchromatic derivative of the denser chromatin.

3. A dense fibrillar component, which often surrounds the fibrils.

The nucleolus is not separated from the rest of the nucleus by a membrane but appears to be held in place within the nucleoplasm. This can be seen if living cells are examined under the microscope. Nuclei constantly rotate, but as this occurs, the nucleoli do not change their relative positions in the nucleus. Often, nucleoli are near the nuclear envelope, a situation advantageous to their role in producing ribosomes for export to the cytoplasm.

Another line of evidence favoring some kind of **nucleolar immobility** is the residual nucleolar skeleton that is seen in nuclear matrix preparations (see Chapter 10). The nature of the framework that enmeshes the nucleolus is not clear, but it will yield to strong forces. For instance, if a living cell is centrifuged, the nucleoli often are propelled to one end of the nucleus and may even penetrate the nuclear envelope and leave the nucleus.

A third indication that the nucleolus has some distinct framework comes from the discovery of **nucleolar-resident proteins** and a specific nucleolus-targeting signal. The resident proteins are nonribosomal and include B23, which participates in the packaging of the immature ribosomes, and nucleolin, which regulates RNA polymerase I transcription of the DNA coding for rRNA. A highly basic amino terminal sequence of 19 amino acids, of which half are either lysine or arginine, can target a protein to the nucleolus.

Much of the early information on nucleolar composition came from cytochemistry. At the light microscope level, the nucleolus stains strongly for protein and RNA. Microspectrophotometry (quantitation of dye binding under the microscope) indicates an approximate 3:1 mass ratio of these components, although in certain cases the protein content approaches 90%.

If cells are incubated with RNase and then examined with the electron microscope, the nucleoli lack both the granular and fibrillar zones, indicating that RNA is an important component of them. However, in what remains of the fibrillar zone are thin (3–5 nm), often circular, fibers that are digestible by deoxyribonuclease but not any other enzyme. This demonstrates that the fibers are made of DNA, and electron microscopy often shows them to have shorter fibers attached in a brushlike arrangement (Figure 11–2).

Experiments have been done in which cells are given a short pulse of radioactive uridine (an RNA precursor) and then chased with non-radioactive uridine, and they give consistent patterns at the light and electron microscope levels:

1. Light microscopy shows the bulk of nuclear label associated first with the nucleolus, and then apparently migrating to the cytoplasm. This indicates that the nucleolus can synthesize cytoplasmic RNA.

2. In electron microscopy, the period of nucleolar labeling is subdivided: The dense fibrillar component is labeled first, followed by the granular component, and then labeled granules are seen at the nuclear pores. Because the granules resemble ribosomes in size and shape, these observations are consistent with the idea that the fibrillar zone of the nucleolus represents DNA for ribosomal RNA in the process of transcription, and the granular zone is composed of ribosomes destined for export to the cytoplasm.

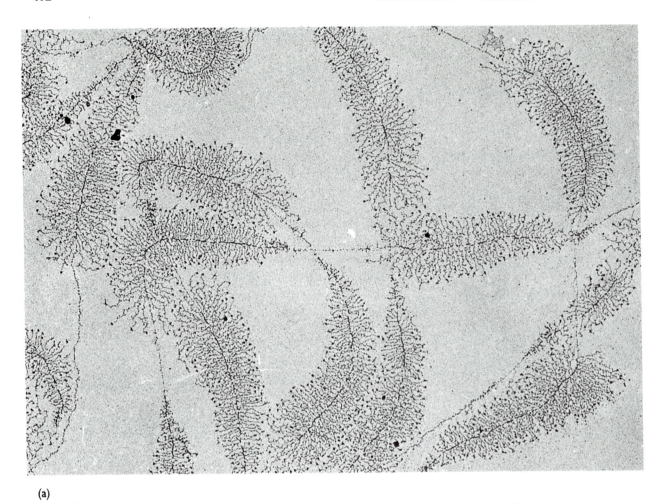

(a)

FIGURE II-2
Electron micrograph **(a)** and interpretation **(b)** of rRNA synthesis from DNA of the newt. (Photograph courtesy of Dr. O. L. Miller, Jr.)

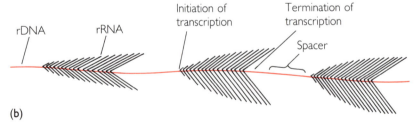

(b)

Ribosomes

Structure

In 1899, G. Garnier reported that exocrine gland cells contain a strongly basophilic component, that is, a substance stained by basic dyes. He termed this the ergastoplasm ("work fluid"), since it occurred in met-

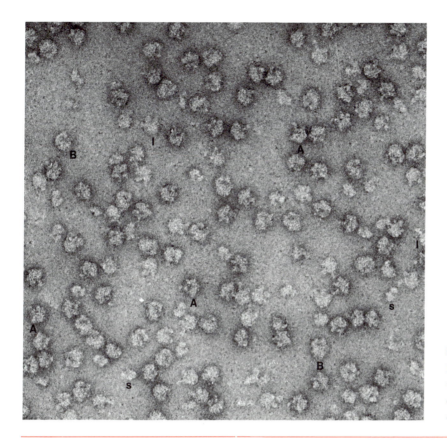

FIGURE 11–3

Ribosomes from *E. coli,* showing the small and large subunits. (×150,000. Photograph courtesy of Dr. J. Lake.)

abolically active cells. During the first half of the 20th century, specific cytochemical staining, ribonuclease sensitivity, ultraviolet spectroscopy, and, finally, isolation and chemical analysis showed that the basophilia was due to RNA.

Electron microscopy revealed that the ergastoplasm was actually composed mostly of numerous 20-nm–diameter granules, and these were often on the surface of the endoplasmic reticulum. Cell fractionation allowed their isolation, and when it was confirmed that the granules contained abundant RNA, R. Roberts called them ribosomes.

At 20 to 25 nm in diameter, ribosomes are not visible under the light microscope. Electron microscopy reveals that they are composed of two nonidentical subunits (Figure 11–3). In eukaryotes, the larger of the subunits has the dimensions of 11.5 × 14 × 23 nm, whereas the smaller is more round, with a diameter of approximately 20 nm. In prokaryotes, the subunits are somewhat smaller, with dimensions of 15 × 20 × 20 nm (large) and 6 × 20 × 22 nm (small).

Extensive X-ray diffraction and fine-structure microscopic analyses have been made of the prokaryotic ribosome, which is, except for size,

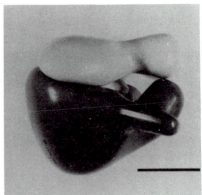

FIGURE II–4

Three-dimensional model of the *E. coli* ribosome. The small subunit is light, and the large subunit is dark. (Photograph courtesy of Dr. J. Lake.)

largely identical in overall function to its eukaryotic counterpart. A model consistent with these analyses is shown in Figure 11–4. The **large subunit** has three protruberances sticking out of its upper side, as well as a flattened area on one of its surfaces on which the small subunit sits. The **small subunit** is elongated, with a "head" and "body" separated by a constriction. The body occupies about two-thirds of the volume of this subunit.

The assembled ribosome, therefore, has distinct structural features. The relationships of some of these features to ribosomal function are described below.

Composition

Ribosomes can be isolated from cell homogenates by differential centrifugation. As the smallest organelles, they are sedimented only at very high forces and times (150,000 g for 90 min, as compared to 15,000 g for 20 min for mitochondria and 60,000 g for 30 min for endoplasmic reticulum). Treatment of homogenates with detergents removes ribosomes from the rough endoplasmic reticulum (see Chapter 6), and except for the presence of proteins for membrane binding, these ribosomes are indistinguishable from their counterparts in the cytoplasm.

Intact individual ribosomes from eukaryotic cytoplasm sediment at about 80S in analytical centrifugation. As noted in Table 5–9, the cells also contain 70S ribosomes in their mitochondria, and, in plants, there are 70S ribosomes in the plastids. These 70S particles are approximately the same size as prokaryotic ribosomes, but there are some differences between the two. For example, whereas prokaryotic ribosomes are relatively homogeneous in size, mitochondrial ribosomes are quite variable, ranging from 55S to 77S, and have a very high (75%) protein content (Table 11–1). Chloroplast ribosomes are more homogeneous at 67–70S and have a protein content typical of cytoplasmic ribosomes.

TABLE 11–1 Overall Composition of Ribosomes of Various Organisms

Organism	Mol. Wt. rRNA ($\times 10^6$) Subunit		No. Ribosomal Proteins Subunit	
Animals	**Small**	**Large**	**Small**	**Large**
Human (HeLa)	0.7	1.8	36	46
Rat liver	0.7	1.8	33	49
Rabbit reticulocyte	0.7	1.8	30	46
Plants				
Chlamydomonas	0.7	1.3	31	44
Euglena	0.7	1.3	32	43
Tobacco leaf	0.8	1.3	29	43
Bacteria				
E. coli	0.6	1.2	21	33
Organelles				
Plastids	0.6	1.2	22	34
Mitochondria	0.7	1.2	33	52

Isolation of intact ribosomes requires a Mg^{++} concentration of at least 1mM, and if it is less than this, cytoplasmic 80S ribosomes sediment at two peaks, 40S and 60S. In the case of the 70S type, the peaks are at 30S and 50S. Electron microscopy shows that the two peaks are the two ribosome subunits. (Sedimentation constants are not additive, since this property depends on shape and other parameters as well as molecular weight.)

Chemical analysis of the two subunits shows, not surprisingly from the cytochemical data, that they are composed entirely of RNA and protein (Tables 11–1 and 11–2). **Ribosomal RNA** accounts for almost all of the RNA in the particles, and since they are so numerous in the cell (an active liver cell has several hundred thousand), rRNA can be up to 80% of all of the RNA in the cell at a given time. The RNAs of the small subunit are similar in size across the plant and animal phyla, with a single RNA molecule per subunit; these RNAs sediment at 18S. In the large subunit, single copies of three different rRNAs are present: In eukaryotes, they sediment at 28S, 5.8S, and 5S.

Ribosomal proteins can be analyzed by solubilization in chaotropic agents such as urea and subsequent two-dimensional polyacrylamide gel electrophoresis. Proteins are designated from their subunit derivation (L or S) and appearance on the electropherogram (a number). Thus the ribosomal protein from the small subunit nearest the origin on electrophoresis is S1. Another nomenclature comes from their positions on purification in high-performance liquid chromatography.

TABLE II–2 Composition of Rat Liver Ribosomes

Parameter	Small Subunit	Large Subunit
Sedimentation coefficient	40S	60S
RNA		
Number of molecules	I	3
Sedimentation coefficient	18S	28S, 5.8S, 5S
No. of nucleotides	2000	5000, 158, 135
Mol. wt. (10^6)	0.7	1.75, 0.05, 0.04
Protein		
No. of molecules	33	49
Average mol. wt.	21,400	21,200
Range mol. wt.	11,200–31,100	11,500–41,800
Total mol. wt. ($\times 10^6$)	0.6	1.0
Total mol. wt. of subunit ($\times 10^6$)	1.3	2.8

Data of Dr. I. Wool.

Most ribosomal proteins are strongly basic at cytoplasmic pH, and this is due to their relatively high (10%) lysine content. The binding of these proteins to rRNA is similar to histones binding to DNA in that there is a strong ionic attraction, but in the case of the ribosome the overall negative charge due to the nucleic acid is not quite neutralized, so the particle has a net negative charge. Structurally, the proteins are mostly globular, with about 25% alpha-helix and 20% beta sheet regions, and average about 20,000 daltons in molecular weight.

There are more proteins in the large subunit than in the small one. However, the exact number of proteins varies between organisms (Table 11–1) and, sometimes, between laboratories investigating the same organism. Some proteins are apparently loosely associated with the ribosome during its biogenesis, and it becomes a semantic matter as to whether these are truly ribosomal. The same problem plagues studies of the stoichiometry of the proteins. The best estimates are that each ribosomal protein is present as a single copy in the particle.

Three-Dimensional Structure

Ribosomal RNAs range in size from 120 (5S) to 500 nucleotides (28S) long and have been sequenced from many organisms. Their most notable feature is a high degree of internal base pairing (the RNA folds back on itself) to give an elaborate three-dimensional structure. Thermodynamically, this base-paired clover-leafed structure is in a low free energy state and quite stable.

(a)

(b)

(c)

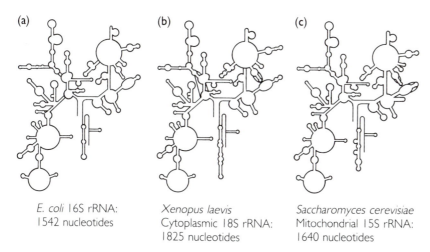

E. coli 16S rRNA:
1542 nucleotides

Xenopus laevis
Cytoplasmic 18S rRNA:
1825 nucleotides

Saccharomyces cerevisiae
Mitochondrial 15S rRNA:
1640 nucleotides

(d)

(e)

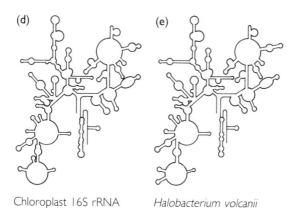

Chloroplast 16S rRNA
Maize: 1430 nucleotides

Halobacterium volcanii
16S rRNA: 1469 nucleotides

FIGURE 11–5

Comparison of the secondary structures of small subunit rRNAs from several sources. These rRNAs have different sequences, yet their three-dimensional structures are quite similar.

Model building using sequences of rRNA shows that at least half of the nucleotides are used in base-paired stretches. Moreover, there appear to be remarkable homologies in the general shape of the analogous rRNAs from prokaryotes and eukaryotes. For example, the small subunit rRNAs from the prokaryotes, eukaryotic cytoplasm, and eukaryotic organelles have very different nucleotide sequences, yet their base-paired structures are amazingly similar (Figure 11–5). This is an elegant example of convergent evolution, in which genes of different sequences come to code for gene products of similar structure and function. It implies that the rRNA must have a specific three-dimensional shape to interact with ribosomal proteins in a highly specific manner.

The topography of binding of the **ribosomal proteins** to each other and to the rRNAs can be studied by both dynamic experiments and examination of the intact particles. The experiments include in vitro reconstitution and in vivo assembly. Mapping techniques on intact ribosomes include cross-linking, neutron scattering, and immune electron microscopy.

Ribosomal proteins can be extracted in bulk from the isolated subunits by treatment in chaotropic agents (e.g., urea), which disrupt hydrophobic interactions and hydrogen bonding. Alternatively, strong acid can be used. In any case, the individual proteins are then separated by ion exchange chromatography. A remarkable feature of these proteins is that when added to the appropriate rRNA in the absence of the separating conditions noted above, the subunits spontaneously self-assemble. These reconstituted subunits can form fully functional ribosomes (Figure 11–6).

Reconstitution is both specific and sequential. A protein from the small subunit, for example, will not reconstitute with rRNA from the large subunit. In addition, prokaryotic ribosomal proteins will not bind to eukaryotic rRNA. In terms of sequence, certain proteins must bind to the rRNA first before others, which bind to the early-adding proteins. The results on rRNA binding can be confirmed by the use of affinity chromatography. If mixtures of ribosomal proteins are passed over a chromatographic column containing a specific rRNA as an immobilized ligand, only certain proteins will bind to the rRNA, and these are different from those binding to other rRNAs. For example, proteins L6 and L19 bind to 5S rRNA, and L8 and L19 bind to 5.8S rRNA from rat liver.

Extensive studies of reconstitution of prokaryotic ribosomes indicate three classes of proteins:

- Those that bind initially
- Those that bind later but are not essential for activity (i.e., structural)
- Those that are not essential for subunit formation but are essential for its activity

Ribosome reconstitution experiments provide some insight into the possible origins of cell organelles. Functional ribosomes, and even subunits, can be formed from mixing components of prokaryotic and chloroplast ribosomes, indicating a close homology between bacteria and plastids. But such functional associations do not form when mitochondrial and prokaryotic ribosome subunits are mixed, indicating little relationship between them. These data indicate that plastids are relatively homogeneous and close to their ancestor symbiont, whereas mitochondria may have had multiple origins and have diverged from their prokaryotic ancestors.

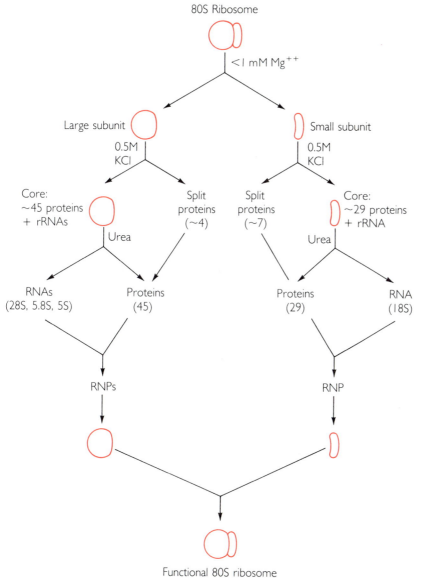

80S Ribosome

<1 mM Mg^{++}

Large subunit

0.5M KCl

Small subunit

0.5M KCl

Core: ~45 proteins + rRNAs

Split proteins (~4)

Split proteins (~7)

Core: ~29 proteins + rRNA

Urea

Urea

RNAs (28S, 5.8S, 5S)

Proteins (45)

Proteins (29)

RNA (18S)

RNPs

RNP

Functional 80S ribosome

FIGURE 11–6

Dissociation and reconstitution of a eukaryotic ribosome.

A complementary method to reconstitution is the limited digestion of intact ribosomes. To accomplish this, the protease trypsin is fixed onto a collagenous membrane and dipped into a solution of ribosomes or subunits. If the time and concentrations are just right, only a few surface proteins will be released from the particles. These can then be crudely mapped.

The second experimental approach to ribosome substructure involves following the assembly of the structure in vivo. If cells are labeled for a short time with a radioactive amino acid, newly synthesized labeled ribosomal proteins may be detected in immature-sized or biochemically incomplete ribosomes or subunits. These early-adding proteins can then be compared with those added later (present in the mature ribosome). In prokaryotes, the building up of the subunits by protein addition has this scheme:

$$\text{Small subunit: } 21S \rightarrow 30S$$

$$\text{Large subunit: } 32S \rightarrow 43S \rightarrow 50S$$

At each of these stages, specific new proteins are added. In most cases, these in vivo data corroborate in vitro reconstitution experiments.

In eukaryotes, these labeling experiments are complicated by the fact that immature ribosomes are assembled in the nucleolus, whereas ribosomal proteins are synthesized in the cytoplasm. This means that if a newly synthesized labeled protein associates with cytoplasmic ribosomes immediately, it is probably late-adding. If it associates with immature particles after a lag time, it must be early-adding.

Kinetic experiments indicate that, for HeLa cells (a human tumor-derived cell line), only 6 of 36 small subunit proteins and 7 of 46 large subunit proteins are added to precursor ribosomes in the cytoplasm. This corroborates analyses of which proteins are present in nucleolar ribosomes. They have 30 of the 36 small subunit ones and 39 of the 46 large subunit ones.

But methods like reconstitution and in vivo assembly sequences only give information as to when a protein adds to the ribosome. A dim outline of where a protein is located can emerge from such studies, but more specific information on ribosome topography has come from more sophisticated techniques.

1. In **immune electron microscopy,** antibodies to a ribosomal protein can be raised and then applied to isolated intact subunits. Because the antibodies are bivalent, subunits may be cross-linked into dimers if their antigenic target is exposed on the surface of the particle. Examination of these dimers by electron microscopy reveals the location of the antigen (Figure 11–7).

A map of some of the exposed proteins on the *E. coli* ribosome is shown in Figure 11–8. It should be noted that antigenicity does not necessarily mean that the bulk of the protein is exposed. Indeed, only a half-dozen amino acid residues are necessary for reaction with an antibody, and so the bulk of an antigenically localized protein could be buried internally in the ribosome.

2. Information on the internal topography of ribosomes has come from the use of **cross-linking** reagents, which have functional reactive

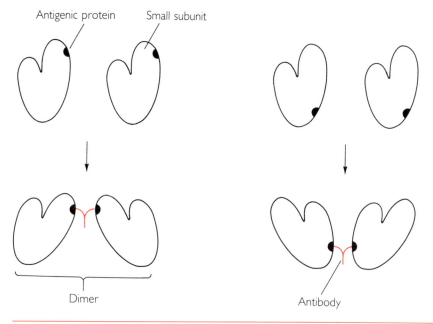

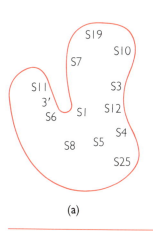

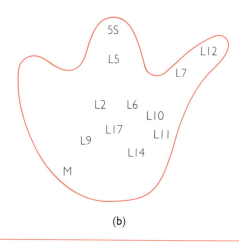

FIGURE 11–7

Immune electron microscopy. If ribosomes are incubated with an antibody to the protein whose antigenic site is exposed on the organelle surface, ribosome dimers will form. In these two cases, the proteins are on the small subunit.

FIGURE 11–8

Map of the *E. coli* small **(a)** and large **(b)** ribosomal subunits, showing proposed locations of proteins (S and L numbers) and other features. (M = membrane binding site; 5S = 5S rRNA; 3′ = end of small subunit rRNA.)

groups on two ends and can bind covalently to RNA or proteins. Incubation of ribosomes in such a reagent results in the cross-linking of neighboring macromolecules. The ribosomes are then split apart, and those molecules that are stuck together are identified. Both protein-protein and protein-RNA interactions have been described in this way.

3. Further data on the distances separating ribosomal proteins can be obtained from **neutron scattering.** In this method, ribosomal subunits are reconstituted with all but two of the proteins labeled with deuterium (the hydrogen isotope in "heavy water"). Placement of the subunits in a beam of neutrons results in scattering patterns that are distinctive for the unlabeled proteins and dependent on how far apart they are from each other.

These two latter techniques usually give complementary results. Protein pairs that are easily cross-linked show close distances on neutron scattering, and proteins not cross-linked are farther apart. Neutron scattering data also correlate well with the external mapping of the ribosome by the use of antibodies.

When all of the data on the internal topography of the ribosome are integrated, some interesting results emerge. The most information is for the large subunit of the *E. coli* ribosome. Proteins involved in peptide bond formation are located near each other, near the center of the subunit. Proteins involved in translocation of the mRNA are on the long, thin protuberance (Figure 11–8, proteins 7 and 12). Surprisingly, the bottom third of the subunit is relatively poor in proteins.

Function: Protein Synthesis

In terms of location and the proteins made on them, ribosomes can be divided into three general classes:

1. Free cytoplasmic ribosomes make some plastid and mitochondrial proteins (not coded for by the organelle DNA), all proteins bound for the nucleus and microbodies, proteins destined for the inner surface of the plasma membrane, and proteins that end up in the cytoplasm.

2. Ribosomes bound to endoplasmic reticulum (ER) make proteins that are secreted, membrane proteins that face the extracellular medium, and ER, lysosome, and Golgi-resident proteins.

3. Mitochondrial and plastid ribosomes make proteins coded for by the genomes of these organelles.

Regardless of their location, the function of ribosomes remains the same: They facilitate the translation of a copy of genetic information encoded in messenger RNA (mRNA) into a specific sequence of amino acids in a polypeptide chain. In an overall view, the three major ribosomal functions are initiation, translocation, and termination.

1. Initiation involves the formation on the small ribosomal subunit of a complex between an initiator transfer RNA (tRNA) and its cognate mRNA codon. The large subunit is then attached to the complex.

2. Translocation involves movement of the ribosome relative to mRNA, thereby exposing the second codon on the latter. Its anticodon-bearing tRNA, with its appropriate amino acid, is then brought to the ribosome, and a peptide bond is formed to link the two amino acids. This occurs in the large ribosomal subunit. Translocation occurs again as the next tRNA and amino acid are brought to the complex. This process is repeated until a termination codon, for which there is no anticodon-bearing tRNA, is reached.

3. This event signals **termination,** with the release of the two subunits, mRNA and newly synthesized protein.

This complex series of events is mediated by specific ribosomal and cytoplasmic macromolecules and is diagrammed in Figure 11–9.

The binding of mRNA to the ribosome is mediated by two factors. First, the methylguanine cap at the 5′ end of mRNA (Figure 10–17) facilitates binding of the mRNA to the 40S subunit. Second, there is often a nucleoside sequence in the part of mRNA 5′ to the initiator codon (the leader sequence) that is complementary to small subunit rRNA. Hydrogen bonding between the two RNAs facilitates the interaction between the ribosome and mRNA.

The ribosome then moves down the mRNA in an energy-requiring reaction, until the initiator codon, AUG, is in the small subunit. The initiating tRNA that binds to this codon carries the amino acid methionine. Although its amino acid and anticodon are the same as those for methionine residues internal in the protein, the initiator tRNA has a unique sequence and structure. Formation of the initiation complex is facilitated by about 10 initiation factors. The largest of these, eIF3, has a molecular weight of 720,000 and is microscopically visible as a bulge on the ribosomal surface. The three-dimensional location of mRNA appears to be in the "cleft" of the small ribosomal subunit (Figure 11–8).

At the time of arrival of the large subunit, the initiator tRNA occupies one of three sites for tRNA within the ribosome, the P, or peptidyl, site. Binding to the ribosome site apparently does not alter the three-dimensional shape of the tRNA molecule. The ribosomal proteins involved in this binding can be identified if a tRNA labeled with radioactive mercury is used. At the P site, these mercury atoms covalently attach to ribosomal proteins. Two proteins in the large subunit are heavily mercurated, implicating them in P-site tRNA binding.

The first event in the **translocation** cycle is the binding of the next tRNA, bearing its amino acid, to the second site on the ribosome, the A, or aminoacyl, site. This reaction requires several steps:

1. The amino acid binds to an elongation protein (T) and GTP. This protein was the first "G protein" (see Chapter 2) to be characterized, and it is visible as a bulge on the small subunit opposite the stalk emanating from the large subunit.

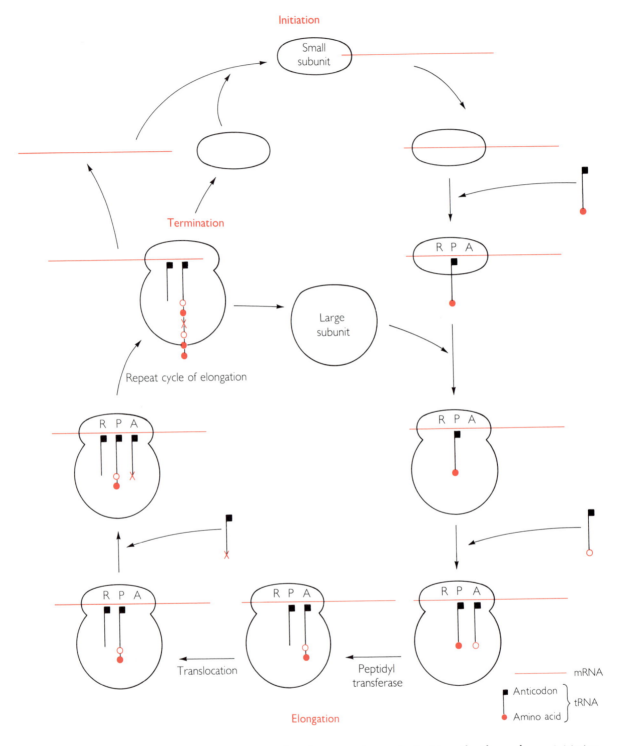

FIGURE II-9   The ribosomal cycle in protein synthesis. Note the three phases: initiation, elongation, and termination. R, P, A: tRNA binding sites on the ribosome.

2. The complex arrives at the ribosome, and GTP is hydrolyzed. This reduces the affinity of the elongation protein for certain large ribosomal subunit proteins.

3. The elongation-protein-GDP complex then exits from the ribosome, leaving the amino acid-bearing tRNA bound to its mRNA anticodon. The interaction of the codon and anticodon is facilitated by certain small subunit proteins.

The cleavage of GTP (step 2) happens only if the "correct" tRNA is present and binds tightly to mRNA by anticodon-codon hydrogen bonding. If the tRNA is "incorrect" and tight binding does not occur, the entire tRNA-protein-GTP complex leaves the ribosome. This **proof-reading** reduces the number of possible errors in the protein synthetic process. Because such errors would possibly make nonfunctional proteins, this step is an important one for cell survival. The overall error frequency in prokaryotes is 1 in 2500 amino acids.

Following aminoacyl tRNA binding to the A site, a **peptidyl transferase** activity in the large ribosomal subunit cleaves the amino acid ester bond to the tRNA at the P site. The carboxyl group of this amino acid reacts with the amino group of the A site-bound, tRNA-attached amino acid to form a peptide bond (Figure 11–10).

This key reaction may be catalyzed by a highly conserved region of the large subunit (28S) rRNA. Mutations in this region's DNA affect transferase activity, and in vitro, ribosomes still catalyze peptide bond formation after all proteins have been removed by extraction and proteolysis. An RNA-enzyme, or ribozyme, can remove an amino acid from tRNA.

Another elongation G protein catalyzes the actual translocation step, in which the mRNA is moved one base triplet (codon) length toward its terminal, 3′ end. This complex composes the stalk of the large subunit opposite the T factor on the small subunit. The hydrolysis of GTP is believed to provide energy. It results in the peptidyl-tRNA, which was in the A site, moving to the P site, and the cycle can begin anew with the arrival of another aminoacyl-tRNA at the A site.

The role of the third tRNA binding site on the ribosome, the E site, is not clear. It may be the site where deacylated tRNA moves after the A site and then is released from the ribosome (Figure 11–9). Or, it may be involved at the beginning of the process, ensuring the fidelity of base pairing between mRNA and tRNA. A fourth binding site for tRNA, the R site, has been proposed to be the place where aminoacyl tRNA binds initially, while complexed to T factor and GTP. After the latter is hydrolyzed and the former is released, the tRNA would move to the A site.

Amino acids are added to the nascent polypeptide by repeated cycles of peptidyltransferase and translocation. Meanwhile, the carboxy-

FIGURE 11–10

The peptidyltransferase reaction: The bond between the carboxyl group of an amino acid and tRNA is broken, and the carboxyl group then forms a peptide bond with the amino group of the amino acid attached to tRNA in the A site of the ribosome.

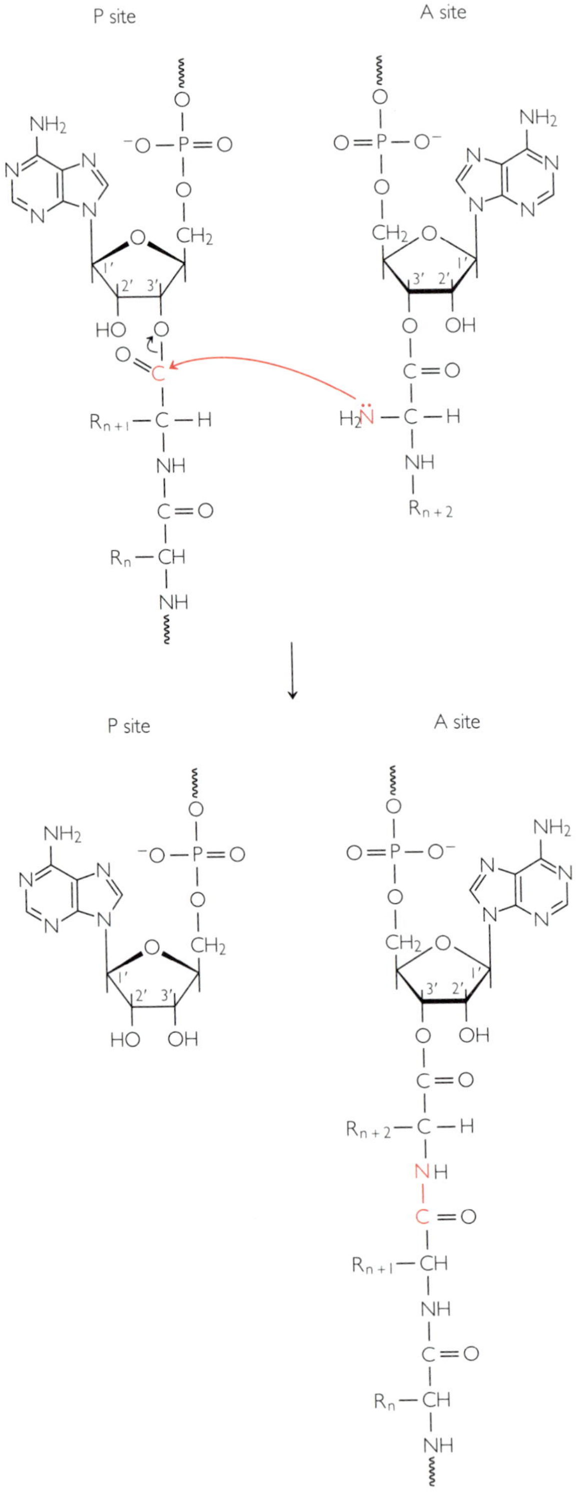

terminal end of the protein remains attached to the ribosome via the tRNA. When an mRNA **termination** codon is reached at the A site, no tRNA binds to it. This signals releasing proteins to hydrolyze the bond between the tRNA and the polypeptide, thereby releasing the latter to the cytoplasm or into the lumen of the ER. Additional releasing factors separate the remaining tRNAs, the two ribosomal subunits, and the mRNA so that these components can be reused.

How much mRNA reuse occurs is different in prokaryotes and eukaryotes. Prokaryotes live in a highly fluctuating environment, and their transcription must respond to stimuli very quickly. On the other hand, physiological homeostatic mechanisms ensure that the environment around a eukaryotic cell remains more stable. Their transcription apparatus need not respond as quickly and be retailored rapidly to new challenges.

Thus, prokaryotic mRNA can be degraded at the ribosome just after it is translated. The enzyme largely responsible for this is an exonuclease, which hydrolyzes bonds holding the polynucleotide together from the 5′ end. On the other hand, in eukaryotic mRNAs, the presence of the "cap" on the 5′ end confers protection against this nuclease, as it converts the 5′ end into a 3′ configuration. Thus, eukaryotic mRNAs typically have half-lives of hours or days.

Ribosome diameters are typically much smaller than the length of mRNA. For example, the mRNA for globin in reticulocytes is approximately 150 nm in length, sufficient for five ribosomes to be moving along it simultaneously, in a 5′ to 3′ direction, much like a cafeteria line. Such **polyribosomes** are the principal sites of protein synthesis in eukaryotes (Figure 11–11). They range in size from two ribosomes to over 10.

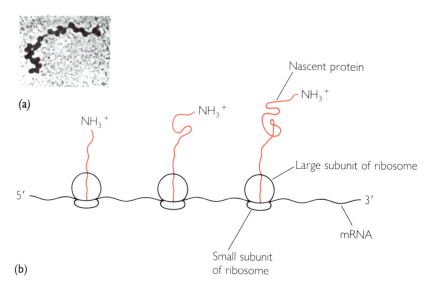

(a)

(b)

Nascent protein

NH$_3$$^+$

NH$_3$$^+$

NH$_3$$^+$

Large subunit of ribosome

5′

3′

mRNA

Small subunit of ribosome

FIGURE 11–11

(a) Polyribosomes from the silk gland of a silkworm. The protein being made is fibroin. (Courtesy of Dr. B. Hamkalo.) **(b)** An interpretative view of a polyribosome. More than one ribosome can be translating a single mRNA at one time.

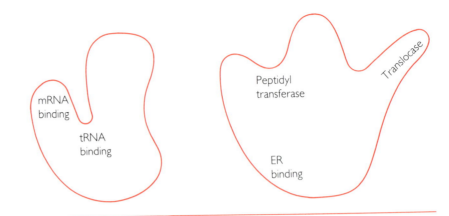

FIGURE 11-12

Functional view of the *E. coli* large and small ribosome subunits.

For a typically sized protein, the mRNA can accommodate four to six ribosomes. The total time taken for a ribosome to move from the 5′ end to the 3′ end of mRNA, which is the total time taken to synthesize a protein, is about one to two minutes. Thus, all of the events of elongating a polypeptide by adding an amino acid (Figure 11−9) occur in less than one second!

An outline of the architecture of the ribosome as a functional unit is given in Figure 11−12. (Compare this to the structural map in Figure 11−8.) The five regions shown identify broad ribosome functions, but the details of the relationship between structure and function are far from clear.

Antibiotics and Ribosomes

Virtually every step of the ribosomal cycle in protein synthesis can be inhibited by an antibiotic (Table 11−3). These molecules share the common characteristics of production by microorganisms and low (less than 1000) molecular weight, but otherwise, their structures generally give no clues to their functions (Figure 11−13). A distinctive feature of many antibiotics is their specificity for prokaryotic ribosomes, often due to their binding to ribosomal proteins. Because these proteins differ between 70S and 80S ribosomes, antibiotic specificity is possible.

An extensively studied example of ribosome inhibition is that by the antibiotic **streptomycin.** This molecule acts on the small subunit of 70S ribosomes by binding to two proteins (S3 and S5). Binding of the drug induces a conformational change in the subunit so that the mRNA-tRNA initiation complex at the P site and aminoacyl tRNA at the A site become much less tightly bound to ribosome. But some bacterial strains are resistant to this drug. In some of these cases, the mutation to resistance is in a different protein in the small subunit, S12.

TABLE 11–3 **Some Antibiotics That Bind to Ribosomes**

Drug	Specificity	Inhibited Step
Small Subunit		
Aurintricarboxylic acid	P,E	mRNA binding
Edeine A	P,E	mRNA binding
Neomycin	P	Miscoding
Pactamycin	P,E	Subunit association
Streptomycin	P	Initiation
Tetracycline	P,E	tRNA binding
Viomycin	P	Translocation
Large Subunit		
Chloramphenicol	P	Peptide bonding
Erythromycin	P	Translocation
Micrococcin	P	tRNA binding
Puromycin	P,E	Peptide bonding
Sparsomycin	P,E	Peptide bonding

P = prokaryotic ribosomes; E = eukaryotic ribosomes.

The facts that the drug binds to two proteins to cause misinitiation but that an entirely different protein causes resistance have led to the model that S3 and S5 are "fidelity proteins" and S12 is a "miscoding" protein. These antagonistic regulators usually balance each other's activities. However, if S3 and S5 are altered (streptomycin binding), the miscoding due to S12 predominates. But if S12 is altered (streptomycin resistance), the fidelity functions of S3 and S5 are not needed. These proteins are near each other in the small subunit, at the site where T factor with its tRNA binds.

Another well-characterized antibiotic is **puromycin,** and in this case the structure does indicate the function (see Figures 11–13 and 11–14). With its adenine-like base connected to ribose, this drug looks like the aminoacyl end of tRNA (tRNA molecules have an adenine at the end attached to the amino acid—compare Figures 11–14 and 11–10). Indeed, puromycin binds quite effectively to the ribosomal A site. Because the drug has a tyrosine-like residue, peptidyl transferase will add the growing peptide chain to this antibiotic. The puromycin-peptide is then transferred to the P site, but puromycin is much smaller than an aminoacyl tRNA, and it binds very weakly at the P site. At this point, the puromycin peptide falls off the ribosome, and protein synthesis is effectively stopped.

Tetracyclines are a widely used class of antibiotics in clinical medicine. Physicians prescribe them for use against both gram-positive

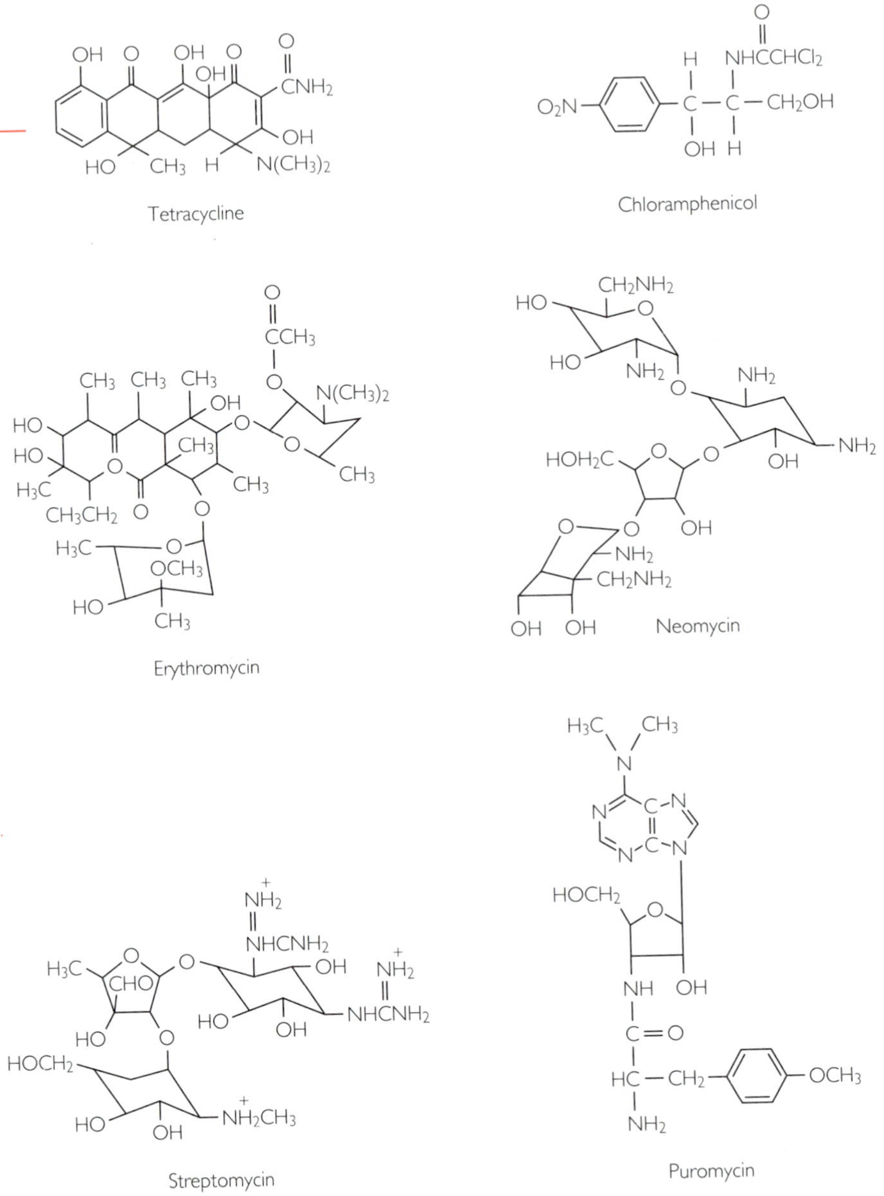

FIGURE 11–13

Structures of some antibiotics that bind to ribosomes.

Tetracycline

Chloramphenicol

Erythromycin

Neomycin

Streptomycin

Puromycin

and gram-negative bacteria that cause such diseases as pneumonia, listeria, gonorrhea, and cholera. (The "gram" designation refers to the ability of the bacteria to react with a certain dye.) In addition, these drugs are effective against chlamydia, another widespread venereal disease.

Because the tetracyclines bind to both prokaryotic and eukaryotic ribosomes (Table 11–3), the question arises as to how they can kill a bacterial infection without also inhibiting the host's protein synthesis.

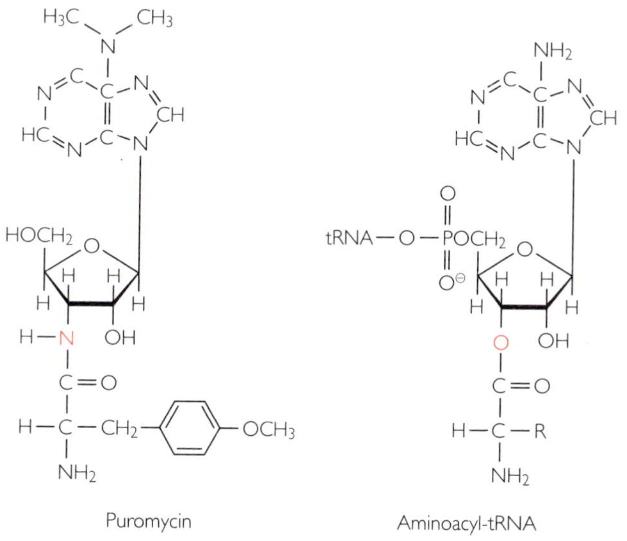

Puromycin

Aminoacyl-tRNA

FIGURE 11–14

Structures of puromycin and aminoacyl tRNA. The antibiotic looks very similar to the tRNA with an aromatic amino acid. However, the attachment of puromycin to its sugar (via nitrogen) cannot be hydrolyzed by peptidylsynthetase. (Compare to Figure 11–10.)

The answer lies at the plasma membrane: The microbes have a membrane protein that actively transports tetracycline into the cell, but mammalian cells lack this protein. Bacteria resistant to tetracycline often are so because they have a defect in the gene for the transport protein.

Nucleolus and Ribosomes

Nucleolar Organizer Genetics

During cell division, nucleoli disaggregate at prophase and re-form at telophase. In 1931, E. Heitz reported that the nucleolus re-forms at a specific region of a specific chromosome, which B. McClintock named the **nucleolar organizer.** In many plant tissues, this region occurs at a constricted area of a chromosome termed a satellite and is easily followed in mitosis.

The number of organizers determines the maximum number of nucleoli in a cell. These may fuse to form a single large nucleolus, or a nucleolar organizer may split in two by gene translocation. In this case, the two "half-organizers" both end up forming nucleoli. The genes associated with the nucleolar organizer can have mutations and can be mapped by genetic crosses. Biochemical analyses of organizer-altered strains of frogs and fruit flies have been instrumental in establishing its function.

Normal diploid cells of the African frog, *Xenopus laevis* contain two nucleoli. However, a recessive mutation termed anucleolate (nu) exists in which the cells do not contain a nucleolus. Heterozygotes (+/nu) have one nucleolus per cell. The nu mutation maps at the nucleolar organizer region on the *Xenopus* genome, and its effects in homozygotes are drastic — tadpoles die after one week.

The reason for early death in the anucleolate frogs is that they cannot synthesize 28S and 18S rRNA. If week-old embryonic frogs that are wild type (+/+ or +/nu) are exposed to a radioactive RNA precursor, they synthesize abundant amounts of 28S, 18S, and 5S rRNAs. But in the case of the anucleolate (nu/nu) mutant, only 5S rRNA is made. The mutant survives for a week on ribosomes presynthesized by its mother and stored in the egg. However, as the embryo grows, it requires more ribosomes, and in this respect it is found wanting.

A more direct demonstration of the relationship between the nucleolar organizer and ribosomes in *Xenopus* can be made by the use of nucleic acid hybridization. In this technique (see Figure 10–8), radioactive 28S and 18S rRNA are hybridized to denatured *Xenopus* DNA. However, instead of the reaction being done in solution, the DNA is in situ, that is, in chromosomes on a glass microscope slide. Hybrids between this DNA and the labeled rRNA can be detected by autoradiography (Figure 10–10), and the chromosomal location of the hybrids can be determined.

In +/+ frog embryos, the labeled rRNA binds to the nucleolar organizer region on chromosome 12. But in nu/nu frog embryos, there is no hybridization to this or any other region of the genome. This indicates that the nu mutation is a deletion of the genes for 28S and 18S rRNA and that these genes are at the nucleolar organizer. In humans, in situ hybridization reveals five pairs of chromosomes with nucleolar organizer regions, all located at the ends of chromosomes.

Mutations of the nucleolar organizer also exist in the fruit fly, *Drosophila melanogaster*. Both duplications (two organizers) and deletions (no organizer) occur, and these can be genetically manipulated to produce flies with 1, 2, 3, or 4 nucleolar organizers. If the DNA of these strains is extracted, the proportion hybridizing to 28S and 18S fruit fly rRNA can be determined. This fraction increases linearly with the number of nucleolar organizers present (Table 11–4). Once again, there is an exact correlation between the nucleolar organizer, which is probably nucleolar DNA, and the DNA sequences coding for the large rRNAs.

Ribosomal DNA

A remarkable result of the hybridization experiments on fruit flies, as well as on other organisms, is the rather high proportion of the genome coding for 28S and 18S rRNA. For instance, in fruit flies, each nucleolar

TABLE 11-4 Relationship Between the Nucleolar Organizer and Ribosomal DNA in *Drosophila*

Genetic Strain: No. of Nucleolar Organizers	Percent of Genome as Ribosomal DNA
1	0.140
2	0.270
3	0.410
4	0.535

DATA FROM: Ritossa, F., and Spiegelman, S. Location of DNA complementary to ribosomal RNA in the nucleolus organizer region of *Drosophila. Proceed Nat Acad Sci* 53 (1965):737–744.

TABLE 11-5 Frequencies of Ribosomal RNA Genes

	No. Genes/Haploid Genome	
Organism	18S–5.8S–28S rRNA	5S rRNA
Yeast *(Saccharomyces)*	140	150
Fruit fly *(Drosophila)*	120	170
Human *(Homo)*	200	2,000
Frog *(Xenopus)*	600	24,000
Sunflower *(Helianthus)*	3,350	
Newt *(Triturus)*	5,800	300,000
Larch tree *(Larix)*	13,400	

organizer (typical flies have one or two) is 0.14% of the total DNA (Table 11–4). From the total amount of nuclear DNA (Table 10–6) and the sizes of rRNAs, it can be calculated that the DNA coding for these molecules is moderately repetitive in the genome. Estimates for repeat frequencies for the three larger rRNAs range from several hundred to over a thousand (Table 11–5).

In addition to their repetition, the genes for the 28S, 18S, and 5.8S rRNA are clustered. There are three lines of evidence for this:

1. There is a single chromosomal location for the nucleolar organizer in many species.

2. Radioactive 28S, 18S, and 5.8S rRNAs hybridize to the same chromosomal region (the organizer) if they are used separately for in situ hybridization.

3. In electron micrographs of actively transcribing nucleolar DNA (Figure 11–2), the increasing lengths of the "feathers" are increasingly longer rRNA molecules in the process of transcription. The central element from which these transcripts arise is DNA. Often, these transcription units are seen in tandem arrays.

The structure of each of the tandem repeats of "ribosomal DNA" is shown at the top of Figure 11–15. The lengths of the transcribed and nontranscribed regions correspond to the lengths of the "feather" and "nonfeather" regions of nucleolar DNA. The order of the three genes— 5′–18S–5.8S–28S–3′—is the same in virtually all eukaryotes examined. The size of the transcription unit varies from 8000 base pairs in lower eukaryotes to 10,500 base pairs in birds to 13,000 base pairs in mammals, these variations being mostly due to variations in the lengths of the transcribed spacers between the individual genes. There is also heterogeneity in length in the nontranscribed spacers between the gene

FIGURE 11–15

Structure, transcription, and processing of a ribosomal gene cluster by RNA polymerase I in human (HeLa) cells. The large gene is transcribed as a unit, then cuts are made to remove the transcribed spacers and produce the mature rRNAs. The 5S rRNA is made by a different RNA polymerase at a different chromosomal location.

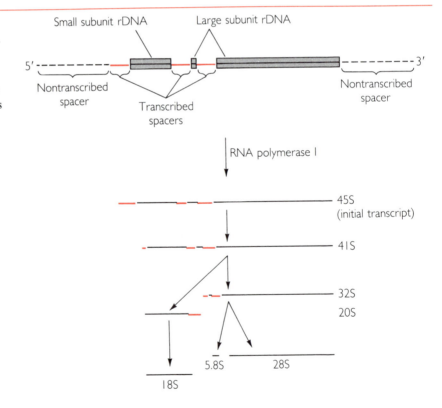

clusters. For example, in humans this region is 27,000 base pairs, over twice the length of the transcribed region.

The gene for 5S rRNA, which is a component of the large subunit, is not part of the gene cluster for the three other rRNAs. The eukaryotic exceptions to this rule are yeast and the slime mold *Dictyostelium*, where 5S DNA is closely linked to the cluster. In most organisms examined, 5S rRNA genes are located on a different chromosome than the 18S–5.8S–28S genes. The 5S genes are often highly repeated (Table 11–5) and occur in groups of two, separated by nontranscribed spacers of various lengths. One of the genes in the pair is a pseudogene, which resembles the 5S gene but is not transcribed.

Ribosome Biogenesis

The biogenesis of ribosomes involves the syntheses of rRNA and proteins, their assembly into particles in the nucleolus, and the export of these particles to the cytoplasm. Transcription of ribosomal genes involves all three cellular RNA polymerases:

RNA polymerase I: 18S, 28S, and 5.8S rRNA

RNA polymerase II: mRNA for ribosomal proteins

RNA polymerase III: 5S rRNA

Each gene cluster for the three larger **rRNAs** is transcribed into a single large precursor molecule. The enzyme catalyzing this nucleolar transcription is RNA polymerase I, which acts only on these genes. The large 45S RNA transcript then is processed to the rRNAs in several steps (Figure 11–15). Prior to processing, ribose moieties of the 28S and 18S sequences on the initial transcript are methylated, apparently to protect them from nuclease attack. Processing involves both the excision of the three rRNAs and the elimination of transcribed spacers. These events apparently occur at the fibrillar region of the organelle.

The rate and degree of transcription of rRNA genes are impressive. RNA polymerase I attaches to the 45S gene at a density of one polymerase per 100 base pairs and elongates the RNA at a rate of 30 nucleotides per second. Even so, the clusters of hundreds of rRNA genes are usually just enough to synthesize the RNAs for the millions of ribosomes that a growing cell requires. Between the transcribed units of the cluster is nontranscribed "spacer" DNA. This contains the molecular signals (promoters and enhancers) that stimulate transcription.

Although the gene clusters coding for rRNA are identical in most organisms, there are exceptions. In the *Plasmodium* protozoan that causes sleeping sickness in humans and other animals, two different unlinked rDNA sequences have been detected. While it is inside the mosquito vector, the parasite expresses one type of rRNA, and once it is in a person, the protozoan expresses the other rRNA. The significance of this change is not known, but it may relate to the specialization of ribosomes in differing programs of protein synthesis.

As noted above, the gene for 5S rRNA is unlinked from the larger rDNA cluster. It is transcribed by a different enzyme, RNA polymerase III, and is under different controls (see Chapter 10).

Ribosomal proteins are synthesized in the cytoplasm and then selectively taken up by the nucleus. For this to happen, they must have a nuclear targeting signal of basic amino acids (see Chapter 10). But then, within the nucleus there must also be a mechanism for targeting these proteins to the nucleolus. Such a signal has been characterized in virus-coded proteins, such as those that regulate the expression of the human immunodeficiency virus. These proteins appear to localize at the nucleolus of infected cells and contain in addition to the nuclear signal a sequence such as:

. . . gly-arg-lys-lys-arg-arg-gln-arg-arg-arg . . .

If it is attached to a nonnucleolar protein, this peptide causes the protein to be transported into the nucleus and to localize at the nucleolus. Conversely, if this sequence is deleted from the nucleolar protein, it fails to localize there. As is the case with the nuclear signal, it is not just the amino acids but also their three-dimensional context within the protein that appears to be important. The nature of the intranucleolar receptor for these targeted proteins is not known.

Each ribosomal protein is coded for by a unique gene, often a member of a large family consisting mostly of pseudogenes. The transcriptions of the mRNAs for all of these genes are done by RNA polymerase II and are strongly coordinated; there is no accumulation of ribosomal proteins in the nucleus, and indeed it is difficult to detect these proteins apart from the ribosome. This coordination is clearly seen when yeast cells are starved for an amino acid: All the ribosomal protein mRNAs are reduced in the cell. If glucose is added to yeast cells growing on ethanol, rRNAs and ribosomal protein mRNAs are both increased to the same extent.

The signals that control these transcription events have been called UAS sequences in yeast and are located upstream of many genes, including those for ribosomal proteins. A single nonhistone protein appears to bind to this sequence and may be involved in the regulatory events.

In addition, there is control of translation of the ribosomal genes. If mammalian cells are cultured without growth factors, most of the mRNAs for ribosomal proteins are untranslated, appearing in the cytoplasm bound up with proteins. The same phenomenon occurs when frog ooctyes mature. In both cases, environmental stimuli (addition of factors or embryogenesis) remove the translational inhibition, and the proteins get made.

Once the RNAs and proteins have been transcribed, a series of events essentially builds the ribosome:

1. A snRNA, U3, and an associated protein, fibrillarin, bind as a complex to the spacer at the $5'$ end of the large initial transcript.

2. Most of the ribosomal proteins bind to the pre-rRNA transcript, forming an 80S ribonucleoprotein particle.

3. As RNA processing occurs, this ribosomal precursor is split into the 40S subunit (containing 18S rRNA) and a 65S (containing 32S pre-rRNA) particle. The latter also contains 5S rRNA, added from its transcription elsewhere in the nucleus. The 5S RNA apparently binds to a specific ribosomal protein, L5, and is transported as a particle to the nucleolus, where it joins the complex.

4. The 65S complex is converted into the 60S subunit when the 32S rRNA is split into the 28S and 5.8S mature rRNAs.

5. The immature ribosomes are then bound to a protein, nucleolin, and rapidly transported to the cytoplasm, where the few remaining proteins are added to each ribosome.

6. The completed ribosomes are then incorporated into polyribosomes.

The cellular locations of these synthetic events are a matter of some controversy, but it appears that the initial transcription occurs at the fibrillar component of the nucleolus. Proteins are then added in the nucleoplasm, and the final processing occurs in the dense fibrillar component. The latter is not just a structure made up of complexed transcripts: If rRNA synthesis is blocked by the addition of an antibody to RNA polymerase I, the dense component remains more or less intact. It is probably part of the nuclear matrix.

Ribosomes are made constantly and in large members by cells, and their rate of synthesis generally reflects their need in protein anabolism. Rapidly growing cells generally produce more ribosomes per unit time than do quiescent cells (Table 11–6). Ribosome synthesis can be visualized in the light microscope by the stain ammoniacal silver nitrate, which binds to sulfhydryl groups on ribosomal proteins associated with active ribosomal DNA.

TABLE 11–6 Rates of Ribosome Synthesis

Cell or Tissue	Ribosome Synthesis/minute/cell
Erythroid	220
Resting fibroblasts	500
Rat liver	1100
Growing fibroblasts	1200
Hela cells	3100
L cells	4500

FROM: Hadjiolov, A. Biogenesis of ribosomes in eukaryotes. In Roodyn, D. *Subcellular biochemistry.* New York: Plenum, 1980, p. 48.

Silver nitrate stain has been used for over a century to visualize nucleolar organizer regions. Recently, it has come into use in pathology (Figure 11–16). Non-Hodgkin's lymphoma is a type of tumor that has different stages and grades, and the clinical management of the patient depends on an accurate diagnosis. A common method of staging and grading involves the determination of the percentage of tumor cells in the DNA synthetic S phase by flow cytometry (see Chapter 12). However, this is a costly procedure. Because the number of nucleolar organizer regions stained by the silver method correlates with the fraction of cells making DNA (Table 11–7), silver staining provides a rapid and inexpensive method of diagnosing these tumors.

In some instances the rRNA synthetic capacity of the cell is not enough to satisfy the demand for ribosomes. Amphibian and fish eggs, when mature, contain up to an incredible 1 trillion (10^{12}) ribosomes, which are used for protein synthesis after fertilization. These large cells come from the differentiation of a much smaller (and much less ribosome-laden) oogonium. The latter is a tetraploid cell, with 2400 copies of the main rRNA gene complex (Table 11–5). Tripling the peak efficiency of rRNA transcription to 22,500 molecules/min, it would take the cell over 50 years to synthesize the ribosomes it needs.

The oocyte solves this problem by the selective **amplification** of its rRNA genes. As it develops, this cell replicates these genes as extrachromosomal circular molecules. There is over a 1000-fold increase in ribosomal DNA and in the number of nucleoli (Figure 11–17 and Table 11–8). The great amount of this DNA after amplification composes a large fraction of the frog genome and is just sufficient for the synthesis of 10^{12} ribosomes.

TABLE 11-7 Use of Silver Staining of Nucleolar Organizers to Diagnose Lymphomas

Percent Cells in S	No. Silver-Stained NORs/nucleus	Diagnosis
1.4	1.2	Lymphocytic
2.8	2.3	Follicular
5.5	4.0	Centroblastic
14.7	7.5	Lymphoblastic
25.8	12.5	Immunoblastic

DATA FROM: Crocker, J., McCartney, J., and Smith, P. Correlation between DNA flow cytometric and nucleolar organizer region data in non-Hodgkin's lymphomas. *J Pathol* 154 (1988):151–156.

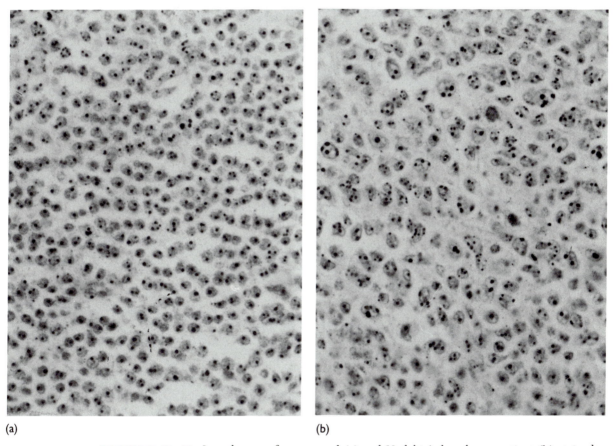

(a)   (b)

FIGURE 11-16 Lymphocytes from normal **(a)** and Hodgkin's lymphoma patient **(b)** stained with silver salts for the nucleolar organizer (NO). Note the increase in NO regions in the lymphoma cells. (Courtesy of Dr. J. Crocker.)

FIGURE 11-17 An isolated oocyte nucleus from the frog *Xenopus,* showing the many nucleoli that arose through rDNA amplification. (Courtesy of Dr. D. Brown.)

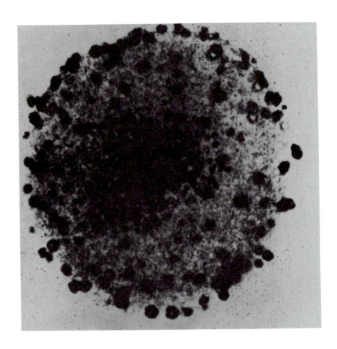

TABLE 11-8 Relationship Between the Number of Nucleoli and Ribosomal DNA in *Xenopus*

	Somatic Cell	Oocyte
Nuclear DNA ($\times 10^{-12}$g)	6	37
Chromosomal DNA ($\times 10^{-12}$g)	6	12
Ribosomal DNA	0.012	25
Percent genome as rDNA	0.2	68
Number of nucleoli	2	1000

DATA FROM: Dr. J. Gurdon.

Further Reading

Babu, K., and Verma, R. Structural and functional aspects of the nucleolar organizer regions in human chromosomes. *Int Rev Cytol* 94 (1985):151–176.

Boublik, M. Structural aspects of ribosomes. *Int Rev Cytol Suppl* 17 (1987):357–392.

Bourgeois, C., and Hubert, J. Spatial relationship between the nucleolus and the nuclear envelope: Structural aspects and functional significance. *Int Rev Cytol* 111 (1988):1–49.

Crocker, J. Nucleolar organiser regions. *Curr Top Pathol* 82 (1990):91–149.

Crocker, J., and Nar, P. Nucleolar organizer regions in lymphomas. *J Pathol* 151 (1987):111–118.

Deltour, R., and Motte, P. The nucleolemma of plant and animal cells: A comparison. *Biol Cell* 68 (1990):5–11.

Derenzini, M., Thiry, M., and Goessens, G. Ultrastructural cytochemistry of the mammalian cell nucleolus. *J Histochem Cytochem* 38 (1990):1237–1256.

Fischer, D., Weisenberger, D., and Scheer, U. Assigning functions to nucleolar structures. *Chromosoma* 101 (1991):133–140.

Ghosh, S. The nucleolus. *Int Rev Cytol Suppl* 17 (1987):573–597.

Goessens, G. Nucleolar structure. *Int Rev Cytol* 87 (1984):107–157.

Hadjidov, A. *The nucleolus and ribosome biogenesis.* New York: Springer-Verlag, 1985.

Hardesty, B., and Kramer, G. (Eds.). *Structure, function and genetics of ribosomes.* New York: Springer-Verlag, 1986.

Hatanaka, M. Discovery of the nucleolar targeting signal. *BioEssays* 12 (1990):143–148.

Hernandez-Verdun, D. The nucleolus today. *J Cell Sci* 99 (1991):465–471.

King, W. A., Chartrain, I., Kopecny, V., Betteridge, K., and Bergeron, H. Nucleolus organizer regions and nucleoli in mammalian embryos. *J Reprod Fertil Suppl* 38 (1989):63–71.

Larson, D. E., Zahradka, P., and Sells, B. Control points in eucaryotic ribosome biogenesis. *Biochem Cell Biol* 69 (1991):5–22.

Liljas, A. Comparative biochemistry and biophysics of ribosomal proteins. *Int Rev Cytol* 124 (1991):103–141.

Moore, P., and Capel, M. Structure-function correlations in the small ribosomal subunit from *E. coli. Ann Rev Biophys Biophys Chem* 17 (1988):349–367.

Nagano, K., and Harel, M. Approaches to a three-dimensional model of *E. coli* ribosome. *Prog Biophys Molec Biol* 48 (1986):67–101.

Nierhaus, K. H. The allosteric three-site model for the ribosomal elongation cycle: Features and future. *Biochemistry* 29 (1990):4997–5008.

Nomura, M., Gourse, R., and Baughman, G. Regulation of synthesis of ribosomes and ribosomal components. *Ann Rev Biochem* 53 (1984):75–118.

Scheer, U., and Benavente, R. Functional and dynamic aspects of the mammalian nucleolus. *BioEssays* 12 (1990):14–21.

Sollner-Webb, B., and Mougey, E. News from the nucleolus: rRNA gene expression. *Trends Biochem Sci* 16 (1991):58–62.

Spirin, A. S. *Ribosome structure and protein synthesis.* Menlo Park, CA: Benjamin-Cummings, 1986.

Tollervey, D., and Hurt, E. The role of small ribonucleoproteins in ribosome synthesis. *Mol Biol Rep* 14 (1990):103–106.

Walleczek, J., Schuler, D., Stoffler-Meillicke, M., Brimacombe, R., and Stoffler, G. A model for the spatial arrangement of the proteins in the large subunit of the *E. coli* ribosome. *EMBO J* 7 (1988):3571–3576.

Warner, J. Synthesis of ribosomes in *Saccharomyces cerevisiae. Microbiol Rev* 53 (1989):256–271.

Warner, J. The nucleolus and ribosome formation. *Curr Op Cell Biol* 2 (1990):521–527.

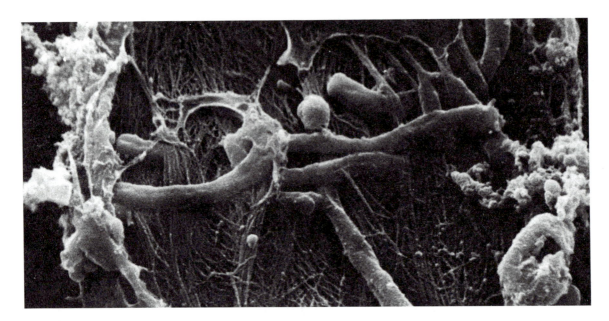

Cell Cycle

Cell Reproduction

There are two basic aspects to the cell theory, which was first elaborated over a century ago. The first—that the cell is the unit of biological structure and function—has been the subject of the previous chapters. It is the second—that the cell is the unit of biological continuity—that is discussed in this chapter.

Eukaryotic cells divide for one of three reasons:

1. Reproduction: *When a single-celled organism such as yeast,* Amoeba, *or* Chlamydomonas *divides, it is forming a new organism, and there are now two of them where there was once one. In multicellular organisms, reproduction may also be via single cells; in this case, a single sex cell acts as a "proxy" for the entire organism by acting as a sperm or an egg.*
2. Growth: *Most animals grow by the addition of new cells. A human is the product of many cell divisions of a fertilized egg. These divisions ultimately result in over 10^{13} cells.*
3. Replacement: *Tissues are constantly being degraded by autophagy (see Chapter 8) or a harsh environment. In both cases, cell division is essential to resupply the necessary pool of cells.*

Whatever the reason, the cell cycle is defined as the sum total of the division-related events that occur between the time a cell completes one cell division and the time it completes the next one. All cell cycles have two basic requirements: First, the genetic material in both the nucleus and organelles must replicate itself completely, and one copy of it must end up in each of the two cells that are formed; this is termed nuclear division or mitosis. Second, the cytoplasmic material and membranes must arrange themselves so that there will be two complete cells to receive this DNA; this is termed cytokinesis. Elaborate mechanisms ensure that both objectives are met.

A cursory glance at the growth rates of a human infant compared to an adult, or a plant seedling compared to a mature tree, indicates that the rate of cell division is not constant. There must be a set of controls that determine when a cell is to divide and, if it does, how quickly. These controls are turning out to be rather similar in organisms as distantly related as yeast, starfish, and humans, which is surprising given the complexity of the process.

Defining the Cycle

Cell Cycle Parameters

Most cells of an adult multicellular organism divide slowly, if at all; that is, their cell cycle times are quite long. For example, neurons in adult mammals and cortical cells in the plant stem are generally nondividing once they are differentiated. In contrast, certain cell types divide rapidly (Table 12–1). In higher plants, where most growth occurs by cell elongation, cell division takes place in localized regions called **meristems,** which are located at the tips of both roots and shoots. This allows for growth in length (primary growth). The vascular cambium, which lies between xylem and phloem, forms new cells of these types by division and is responsible for increases in thickness (secondary growth).

In human adults, there are also regions of constant cell division, mostly to replace cells that are lost. For instance, about 200 million erythrocytes are destroyed each day, and in the intestine, epithelial cells are constantly sloughed off to the extent of about 1 kg cells/day. These cells are replaced by the animal analogue of meristematic cells, **stem cells,** which occur in the bone marrow (for red blood cells) and crypts at the bases of the intestinal villi (for intestinal epithelial cells). In normal circumstances, the demands of these and other activities result in the occurrence of 25 million cell divisions at any time in the human body.

In an asynchronous population of cells that divide at the same rate, the total time for one cell cycle for a typical cell is the time taken for the overall population to double in cell number. Such estimates are reliable only for meristematic, or stem, cells or for cells in culture, since in these cases the populations are assumed to be homogeneous. Typical **cycle times** for these cells in both plants and animals range from 15 to 40 hours. For organs such as the liver or for tumors the situation is more complex, since the component cells are often cycling at widely differing rates. But one generalization can be made: Embryonic and regenerating tissues have shorter cycle times than adult differentiated tissues.

The concept of **cancer** as an uncontrolled growth leads to the assumption that cell cycle times of cancer cells are shorter than those of normal cells. However, data from human tissues contradict this idea (Table 12–2), for in many instances the tumor cells cycle at a slower rate than normal. The reason that tumors grow faster than the host is not that individual cells divide more quickly but that a greater fraction of the cells is cycling at a perceptible rate and fewer of the cancer cells die.

Some of the more widely used anticancer drugs are mitotic inhibitors, acting either on the mitotic spindle (Figure 9–8) or on DNA synthesis (Figure 12–11). Because some cancer cells are not dividing quickly, this means that these drugs are less specific for cancer than

TABLE 12–1 Major Sites of Mitosis in Adult Higher Plants and Mammals

Plants

Root apex (primary meristem)

Shoot apex (primary meristem)

Vascular cambium (secondary meristem)

Mammals

Blood cells (bone marrow)

Intestinal epithelium

Skin

TABLE 12–2 Cell Cycle Times of Some Normal Tissues and
Tumors in Humans

Tissue	Cell Cycle Time (hr)
Normal	
Colon, crypt epithelium	39
Rectum, crypt epithelium	48
Bone-marrow precursors	18
Bronchus, epithelium	220
Tumor	
Carcinoma of stomach	72
Acute myeloblastic leukemia	80–84
Chronic myeloid leukemia	120
Bronchus carcinoma	196–260

DATA FROM: R. Baserga.

originally proposed, killing normal cells that have faster proliferative rates than the tumor cells.

In general, a tumor arises from a single cell. This can be shown by biochemical markers, such as the enzyme glucose-6-phosphate dehydrogenase (G6PD). This enzyme is coded for by the X chromosome, and there are different detectable variants, termed A and B. Because of X chromosome heterochromatinization (see Chapter 10), a woman carrying both the A and B alleles will have the A expressed (and the B not expressed) in half of her white blood cells and the B expressed (and the A not expressed) in the other half of her white cells. However, a leukemia deriving from this tissue will show only one variant (A or B) in all of the cells, indicating that they all probably came from one cell.

Once a cell becomes cancerous it continues to divide beyond normal limiting stimuli. Unfortunately, when a physician can easily detect a tumor by touch (at 1 cm diameter) there have already been about 30 doublings, for a total of a billion cells. If an antitumor agent kills 99.9% of these cells (not an unusual "hit rate"), there will still be 1 million tumor cells surviving. Thus it is not surprising that combinations of antitumor agents are used.

Mitosis, defined as the appearance of cytologically detectable chromosomes, occupies only a part of the cell cycle. This can be shown if the idealized cell population (homogeneous, asynchronous) is examined by light microscopy. The mitotic index, M, is equivalent to the fraction of cells at any given moment that are in mitosis. If this fraction is, for example, 0.1, this means that a typical cell spends 10% of its cell cycle in mitosis. The remaining time is termed interphase.

Interphase in many cells can be further subdivided into three parts, which can be delineated in two ways. The first is the **percent labeled mitosis** technique (Figure 12–1).

1. The dividing cell population is incubated for a short time in radioactive thymidine, which is incorporated into DNA as it is made.

2. Following this pulse, the cells are "chased" in excess nonradioactive thymidine. The fate of these radioactively tagged cells can now be followed.

FIGURE 12–1 The labeled mitosis method to determine the lengths of cell cycle phases. Cells are incubated in radioactive thymidine (pulse) and then in nonradioactive thymidine (chase).

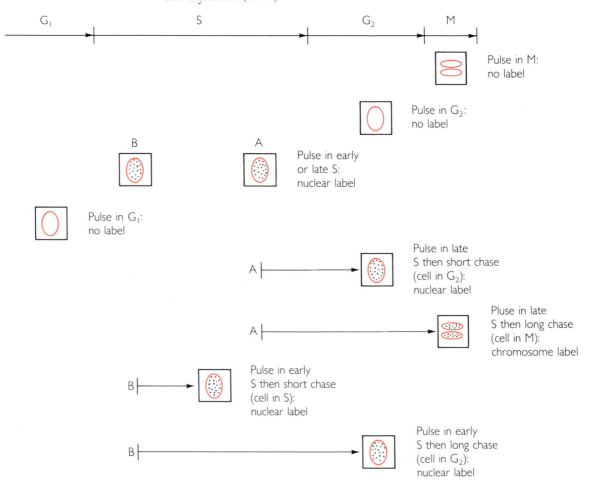

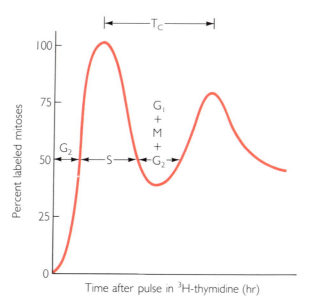

Time after pulse in ^{3}H-thymidine (hr)

FIGURE 12–2

Typical results of an experiment using the percent labeled mitoses method. (See Figure 12–1.) G_2 is defined as the time it takes for 50% of the cells labeled in S to reach M, as indicated by the appearance of labeled mitoses.

3. After various times of chase, cells are sampled for autoradiography, and the percentage of mitotic figures that are labeled is scored.

Initially (pulse) there are labeled interphase cells but no labeled mitoses. This shows that DNA must be made during interphase but not during mitosis. Also, an examination of the interphase nuclei shows only a fraction of them labeled, implying that DNA synthesis (S or synthesis phase) occupies only part of interphase.

As the time of chase prior to autoradiography lengthens, labeled mitoses are observed. These mitotic cells were in S phase during the labeling pulse, then passed through a non-DNA-synthesis G_2 ("gap") phase before mitosis. The chase time for 50% mitotic labeling is an estimate of G_2.

The S phase is the time between the 50% mitotic labeling points on the rise and fall of the wave of labeled mitoses (Figure 12–2). The total cell cycle time (Tc) is the interval between successive waves of labeled mitoses. Calculations of G_2, S, and T_c, coupled with the estimate of M, usually do not account for all of the time spent in the cell cycle. The remaining portion, another period of interphase when DNA is not synthesized, is calculated by subtraction and is termed G_1. Thus, the cell cycle is:

$$G_1 \rightarrow S \rightarrow G_2 \rightarrow M \rightarrow.$$

A second technique used to estimate the phases of the cell cycle is **flow microfluorimetry.** This is especially valuable for cells that are

FIGURE 12–3

Cell cycle phase estimates by flow microfluorometry. Because the DNA is stained with a dye that binds to it in proportion to DNA concentration, fluorescence is directly proportional to the DNA content of the cell (assume nucleus). The G_2 and M cells contain twice the DNA as the G_1 cells; the intermediate group cells are in S.

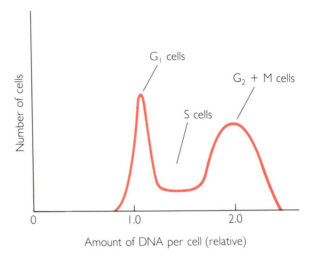

easily disaggregated. Cells are stained with a fluorescent dye that binds to DNA quantitatively and are then individually passed at a rapid rate (1000/sec) past a laser beam that is focused at the excitation wavelength of the dye. Under these conditions, the fluorescence emitted by each cell is directly proportional to its DNA content.

A frequency distribution of the population (Figure 12–3) shows G_1 cells with exactly half the DNA content of G_2 and M cells. Thus this method gives an accurate determination of G_1 and, with a mitotic index determined by microscopy, G_2. The length of S is then deduced by subtraction.

The durations of the four cell cycle phases have been determined for many tissues in a wide variety of physiological circumstances, and several generalizations (which have their exceptions) emerge:

1. Mitosis usually takes one to two hours; it is characteristically on the longer end of this scale in cancer cells.

2. S phase takes at least four hours; this is not surprising, given the prodigious task of the orderly replication of all of the nuclear DNA.

3. For a given cell type, variations in total cell cycle time (T_c) are reflected by variations in G_1. This phenomenon is illustrated by the developing rat neural tube (Table 12–3). As the embryo gets older, T_c lengthens, this increase being totally accounted for by the increase of G_1.

Approaches for Analysis of the Cell Cycle

Information on the events that occur during each of the four phases of the cell cycle is frequently obtained from asynchronous populations of

TABLE 12-3 Cell Cycle Parameters in the Developing Rat Neural Tube

Embryonic Age (days)	Cell Cycle (hr)	G_1	S	$G_2 + M$
12	13.1	3.7	6.9	2.5
14	12.9	3.5	6.9	2.5
16	17.0	6.8	7.7	2.5
18	20.8	11.2	7.1	2.5

DATA FROM: Schultze, B., and Korr, A. Cell kinetic studies of different cell types in the developing and adult brain of the rat and mouse: A review. *Cell Tissue Kinet* 14 (1981): 309–325.

cells. But it is easier to do biochemical analyses on a synchronous population, all of which are in, say, G_1. Fortunately, a few natural cell populations are synchronous. One well-studied example is the anther of the lily bud and flower, in which the pollen-forming cells undergo mitosis and meiosis in unison. A second example is the multinucleate plasmodial slime mold *Physarum*. A third system is the animal embryo during its cleavage stages. But in most other cases, synchronous eukaryotic cells must be selected or induced.

Selection synchrony can be achieved in two ways. The first uses the property of cultured mammalian cells of rounding up during mitosis. This decreases their adhesion to the substratum, as there are fewer fibronectin connections (see Chapter 13). Gentle washing of asynchronous cultures removes these mitotic cells, which then are put into a new medium and cycle synchronously.

The second method uses the fact that cells (usually) grow throughout interphase, so that newly divided G_1 cells are smaller than those at the end of G_2. These cells of different sizes can be separated by density gradient centrifugation, and usually the small cells are removed for synchronous culture. Both of these methods, although they generally result in only up to 80% synchrony, have the advantage that they do not overly perturb the cells. A disadvantage is that they cannot be used on solid tissues.

A variety of techniques are used to **induce cell synchrony.** The most common method involves treating the cells with an inhibitor of a cell cycle event, allowing all the cells to accumulate at the block and then releasing the cells by washing out the inhibitor. For example, cells treated with high (2 mM) concentrations of thymidine stop DNA synthesis because of feedback inhibition of the formation of deoxyribonucleotides.

If an asynchronous culture is incubated in high thymidine, G_2, M, and G_1 cells will progress through the cycle until they reach the G_1-S transition point; cells in S will be arrested in S. If the thymidine is removed and then reimposed after S is over, all cells will now arrest at the G_1-S boundary. Removal of the thymidine block will now lead to all cells synchronously entering S and the rest of the cycle. A problem with inhibitors is untoward effects on the cells. For example, aberrant DNA replication and chromosome structure can be detected in some thymidine-blocked and released cells.

A second general method for the induction of synchrony uses environmental perturbations, such as:

1. Light-dark cycles, which induce synchrony in unicellular green algae such as *Chlorella.*

2. Temperature cycles, which induce synchrony protozoa such as *Tetrahymena.*

3. Nutrient deprivation, in which cells are permitted to grow in a medium in which a particular essential nutrient is limiting. In time, the nutrient is depleted and the cells stop cycling, usually at that point in the cell cycle most sensitive to the missing nutrient. When it is added back, the cells are synchronous.

Nutrient-induced synchrony is shown in Figure 12–4, which describes data for the diatom *Nitschia angularis.* This unicellular alga incorporates silicic acid into its cell wall, and, indeed, its metabolism is regulated by this otherwise biologically rare substance. An asynchronous population has a T_c of about 12 hours. When silicon in the medium is used up, the cells accumulate at cytokinesis, as the cross-wall separating the two cells cannot be made. Addition of silicon to the medium leads to wall formation, cell separation, and synchrony.

A powerful approach to cell cycle analysis is the isolation and description of **cell cycle mutant strains.** These strains can provide two types of information:

1. They can aid in a biochemical description of a particular event in the cycle. For example, if a mutant blocked in DNA synthesis lacks a certain DNA polymerase, one can conclude that this enzyme is essential for S phase replication. If 16 different nonallelic mutants all affect DNA synthesis, there must be at least 16 separate functions involved.

2. Mutants can be used to determine the interdependence and timing of cell cycle events. There may be two mutants that fail to synthesize DNA: Chromosomes are formed in one but not in the other. Thus chromosome formation is dependent on the second gene but not on the first. In this way, cell cycle events can be arranged in pathways.

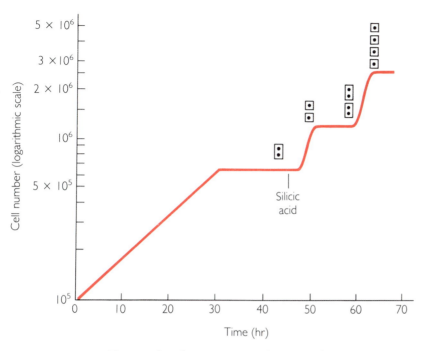

FIGURE 12-4 Silicon-induced starvation synchrony in the diatom *Nitschia angularis*. From 0 to 48 hours, these cells, which require silicic acid for cytokinesis, were denied this nutrient in the growth medium. After all of the residual silicic acid was used up, the cell population leveled off (at 30 hr). The cells then went through the cell cycle until they reached cytokinesis, where they remained as "double cells." Reintroduction of silicic acid at 48 hr led to a rapid completion of cytokinesis, and then the cells proceeded to divide synchronously (note the rapid doublings).

Mutants of the cell cycle are generally harmful to the cell. For a mutant strain to be grown and maintained, the mutation must be one that exhibits a normal phenotype under one set of conditions (the permissive condition) and expresses the mutant under different conditions (the restrictive condition). Such **conditional mutants** of the cell cycle usually grow normally at one temperature but show their mutant phenotype at an elevated temperature; that is, they are temperature-sensitive. This property comes from the fact that the mutant gene's protein product has an amino acid sequence that causes the protein to denature at a temperature where the wild type protein does not.

Budding yeast, *Saccharomyces cerevisiae,* goes through a cell cycle that is typical of eukaryotes in many aspects (there is a G_1-S-G_2-M progression) but has two important differences. First, as its name implies, it

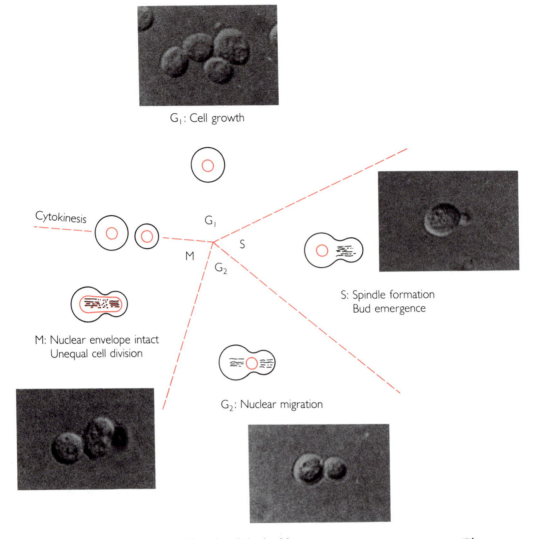

G₁: Cell growth

Cytokinesis

G₁

S

M

G₂

S: Spindle formation
Bud emergence

M: Nuclear envelope intact
Unequal cell division

G₂: Nuclear migration

FIGURE 12–5 The mitotic cell cycle of the budding yeast, *Saccharomyces cerevisiae*. (Photographs courtesy Dr. T. Schuster.)

forms a bud, and cell division is unequal; second, the nuclear envelope does not disaggregate during division (Figure 12–5). Fission yeast, *Saccharomyces pombe,* as its name indicates, has a more equal cell division. These two species have been intensively studied models of the eukaryotic cell cycle.

In budding yeast, geneticists have detected over 150 cell cycle mutants that map at 50 different genetic loci (Figure 12–6). Each mutant has an "execution point" during the cycle. If a synchronous yeast culture

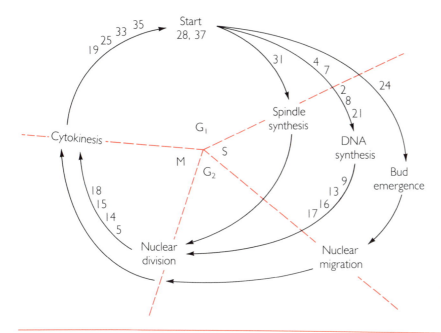

FIGURE 12–6

Temperature-sensitive mutations of the budding yeast cell cycle (see Figure 12–5). The pathways of spindle synthesis, DNA synthesis, and bud emergence are independent. The numbers indicate separate mutations (functions) along the pathways.

is maintained at the permissive temperature beyond this point, subsequent shift to the restrictive temperature has no effect and normal cell division occurs. But if the culture is at its restrictive temperature at the execution point, the abnormal protein is expressed and the cell cycle is arrested. Clearly, the execution point defines that period in the cell cycle when a gene product is used.

The ultimate consequence of a defective gene product is termed the terminal phenotype, which could be a binucleate cell, for example. A similar set of over 40 cell cycle mutants has been isolated from the fission yeast.

The use of mutant strains in ordering cell cycle events can be illustrated by budding yeast mutant strains cdc2 and cdc13 (cdc refers to cell division cycle). Strain 2 fails to complete DNA synthesis, but strain 13 does complete it; yet both strains do not undergo nuclear division, and their terminal phenotypes are identical. This suggests that DNA synthesis precedes nuclear division. Strains 7 and 24, on the other hand, have different phenotypes. In strain 24, bud emergence does not occur but DNA is synthesized, while in strain 7, the reverse is true. This means that these two events are on separate pathways.

In addition to yeast and other lower eukaryotes, temperature-sensitive cell cycle mutants can be isolated from cultured mammalian cells. There are mutant strains affecting the four cycle phases, as well as the stages of mitosis (Table 12–4). There are also mutant cell strains that are aberrant in the cell cycle phases, lacking virtually all of G_1 and/or G_2.

TABLE 12–4   Cell Cycle Mutants of Mammalian Cultured Cells

Organism	Cycle Stages Affected
Chinese hamster	G_1 of next cycle (2)
	Late G_1 (2)
	S
	Cytokinesis
Syrian hamster	G_1
	S (initiation)
	S
Hamster	Prophase
	Metaphase
	Anaphase
Mouse	G_1 (2)
	S (3)
	DNA joining
	G_2
	Mitosis (2)
	Cytokinesis
Rat	G_1
Green monkey	S

Specific Events in the Cell Cycle

G_1

The duration of the G_1 phase of the cell cycle is the most variable of the four phases (Table 12–3). In synchronous cells, it can be shown by the use of inhibitors and mutant strains that G_1 events are necessary for entry into S phase and the DNA synthesis. The nature of these events is not clear, but mutations of yeast offer promising leads (see below).

Nuclei and cells grow throughout the G_1 period, and in plant cells it is not uncommon for nuclear volume to double. Among the macromolecules that are synthesized in greater amounts during G_1 of cultured animal cells is fibronectin, which is involved in the reattachment of a rolled-up mitotic cell to the substratum. In addition, a number of enzymes involved in DNA synthesis are made late in this phase, presumably in preparation for S.

During G_1, a cell commits itself to one of three life-styles:

1. The cell can continue on the cycle and divide.

2. The cell can permanently stop division. Cells such as neurons, which are terminally differentiated and do not divide, contain the G_1 amount of DNA.

3. The cell can become reversibly quiescent. Such nondividing cells, which have a G_1 DNA content, are called G_0 cells. They differ from neurons in that they can be induced to proliferate by appropriate stimuli. An example of a G_0 population is the stem cells of the immune system, which enter S phase in response to antigenic stimulation.

Several lines of evidence point to G_0 as being a cell cycle phase quite different from an extended G_1. Biochemically, G_1 and G_0 cells differ in nonhistone proteins and RNA synthesis profiles. Hydroxyurea, a DNA synthesis inhibitor, is more potent in cells that were in G_1 than G_0 cells stimulated to divide. Fusion of G_0 cells, or cell extracts, with G_1 or early S cells blocks the latter from undergoing DNA synthesis, indicating that G_0 cells have some sort of inhibitor. This inhibitor appears to be a glycoprotein and may be a control point in releasing a cell from the quiescent G_0 state. It disappears rapidly from nuclei that have been reactivated from G_0 to enter G_1.

The transition from G_0 to G_1 and through G_1 has been studied in mouse fibroblasts in tissue culture, and this has allowed the characterization of a number of events (Figure 12−7):

1. There is an early increase in transport of various nutrients across the plasma membrane, as well as some changes in DNase sensitivity of chromatin. These events, and progression into G_1, do not occur unless the cell is given platelet-derived growth factor (see Table 12−6).

2. Once in G_1, epidermal growth factor is needed to stimulate the next series of events, which involve increases in polyribosomes and in some glycolytic enzymes.

3. A third growth factor, insulinlike growth factor I, is required for the events late in G_1, the syntheses of proteins needed for DNA replication.

4. A protein of molecular weight 68,000 must be made for cells to enter S phase. The function of this protein is not known, but it could be part of a protein kinase system (see below).

The "traditional" view of the cell cycle considers G_1 as an essential phase, during which events preparatory to DNA replication occur. However, these events could just as easily occur during G_2 or M. In the

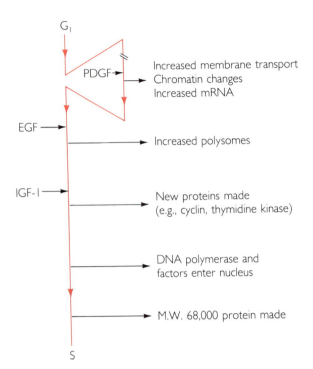

FIGURE 12-7

The subdivision of G_1 in cultured mammalian cells, and the various growth factors that control exit from G_0 and progress in G_1. (From A. B. Pardee.) (EGF = epidermal growth factor; PDGF = platelet derived growth factor; IGF-1 = insulinlike growth factor.)

bacterium *Escherichia coli,* DNA synthesis takes 40 minutes, but the cell cycle time is usually less than this. This means that these cells are continuously in S phase. Under certain adverse nutritional conditions, T_c is over 60 minutes, and in this case, G_1 occurs.

S. Cooper has proposed that in eukaryotes G_1 exists only because the cells grow slowly and must accumulate sufficient "division potential" for DNA synthesis. Clearly, a number of eukaryotic cell types have dispensed with G_1 entirely, including *Amoeba, Physarum,* early embryonic cleavage cells, and the precursors of bird erythrocytes. For example, in the first cleavages of the fertilized egg of the sea urchin, the cell cycle is less than an hour long, with S being 12 minutes and M the remaining 48 minutes.

A Chinese hamster cell culture line lacking virtually all of G_1 (termed a G_1^- line) provides insight into the nature of this phase. If these cells are placed in a nutrient-poor medium or treated with the protein synthesis inhibitor cycloheximide, they enter a G_1 state, just as *E. coli* does when its cell cycle is extended by slower growth. A subline of the G_1^- cells is also G_2^-. Therefore, with their S-M cell cycle, these cells resemble *E. coli.*

If budding yeast are given the DNA synthesis inhibitor hydroxyurea, S phase increases in length, but the total cell cycle time remains constant because G_1 shortens. If certain events had to occur during G_1, this shortening would not be tolerated by the cell. On the other hand, if total growth and "division potential" are the important parameters, the extended S phase permitted these to occur so that a long G_1 is not essential in a strict sense.

S: DNA Synthesis

The initiation of DNA replication requires that the cells have sufficient "division potential," which may or may not be acquired during G_1. In molecular terms, this potential is probably a diffusible cytoplasmic factor that acts at the G_1-S transition (see below). That such a factor exists is indicated from the fact that in both natural and artificially produced cell hybrids, all the nuclei synthesize DNA synchronously.

Also, if a cell is carefully treated with cytochalasin B (Figure 9–3) and centrifuged, a cytoplast without the nucleus can be obtained. Fusion of cytoplasts from early S phase cells with G_1 cells induces DNA to replicate in the nucleus of the latter. This indicates that the factor can travel through the cytoplasm to the nucleus. But G_2 nuclei cannot be induced by the factor to replicate again, ensuring that the DNA replicates only once in a single cell cycle.

The replication of chromosomal DNA can be studied by **fiber autoradiography.** In these experiments, cells in S phase are given a short pulse of radioactive thymidine, which is incorporated into DNA as it is synthesized. After the pulse, the cells are incubated for a time in nonradioactive thymidine, and then the DNA is carefully isolated to preserve as long strands as possible. The locations of radioactivity are then monitored by autoradiography.

The results of such an experiment (Figure 12–8) allow several conclusions to be drawn about replication. The first conclusion is that there are numerous, tandemly arranged replicating sections, each growing at the rate of about 3000 base pairs (1 μm) per minute. This shows that there is far more than one origin of replication per chromosomal DNA molecule (Figure 10–11). A simple calculation shows why this must be true: The average human chromosome has a single DNA molecule, 4 cm long but densely folded and packed. To replicate this molecule from a single origin at the stated rate would take about a month. Because S phase typically takes hours, there must be multiple origins.

The number of origins varies depending on how short a time the S phase lasts. For instance, in early embryonic cleavage, S is shortened to 30 minutes in the frog, *Xenopus.* This is accomplished by the addition of replication origins as well as their simultaneous replication. Because the origins must be on nucleosome-free DNA, this means that the nucleosomes must be disaggregated to a greater extent in the embryonic cells.

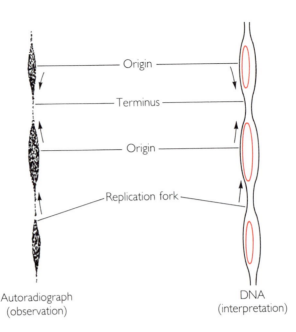

FIGURE 12–8

Fiber autoradiography *(left)* and interpretation *(right)* of replicating DNA from mammalian cells. The cells were given a brief pulse of radioactive thymidine during S phase, then a brief period of chase in nonradioactive thymidine. The results show several tandemly arranged replicating units, each making new DNA bidirectionally.

On the other hand, in mature tissues, replicons (stretches of DNA with an origin and terminus of replication at each end) may be fewer in number and replicate in a staggered fashion. For example, DNA associated with euchromatin replicates early, while DNA in heterochromatin including highly repetitive sequences replicates late in S. In general, DNA sequences coding for specialized cell functions replicate later in S than those for functions common to all cells.

The second conclusion that can be drawn about replication is that the replication origins grow in both directions on the DNA fiber. This shows that DNA replication is bidirectional. Indeed, before J. Huberman and A. Riggs demonstrated this for animal cells, it was thought that replication proceeded in one direction. Clearly, replicons can grow toward each other, but as indicated above, they need not be replicating at the same time.

The molecular biology of eukaryotic DNA replication is quite complicated, and its details are not completely known. As in prokaryotes, eukaryotic DNA is synthesized bidirectionally and semiconservatively. A number of proteins are involved, including DNA polymerase, RNA primase, nucleases, unwinding proteins, and ligases. Some of these can be defined through mutants affecting DNA synthesis. For example, mutants 9 and 21 of budding yeast both affect it (Figure 12–6) but involve different proteins. Mutant 9 affects DNA ligase, and mutant 21 affects thymidylate synthetase.

Chromosomal DNA may be in one large piece, but it is certainly not a linear molecule, neatly stretched out, as might be implied from the

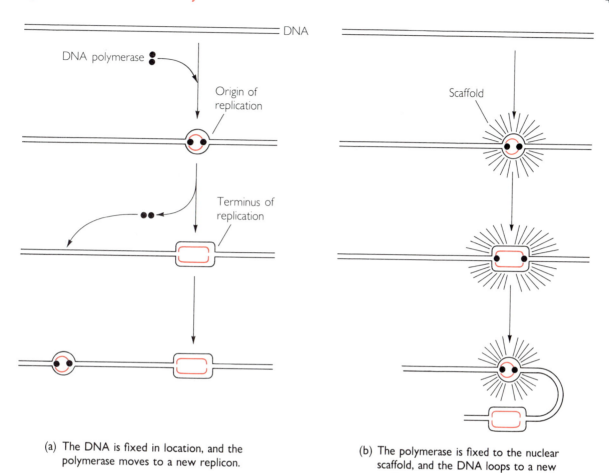

(a) The DNA is fixed in location, and the polymerase moves to a new replicon.

(b) The polymerase is fixed to the nuclear scaffold, and the DNA loops to a new replicon.

FIGURE 12–9   Two models for the topography of DNA polymerase in the nucleus.

fiber autoradiographs. Instead, the DNA is highly packed in the nucleus, an important packing agent being the chromosome scaffold or nuclear matrix (see Chapter 10). Because DNA replication is associated with this nuclear skeleton, the question arises as to whether this association comes after replication or is part of the replication machinery. Stated another way, there are two possibilities (Figure 12–9):

1. The replication enzymes are part of the nuclear skeleton and the DNA template passes through them; to get to a new origin, the DNA loops around to the scaffold, or

2. The replication enzymes are soluble and move along the DNA template; to get to a new template, the enzymes essentially detach from the DNA and move to the next origin.

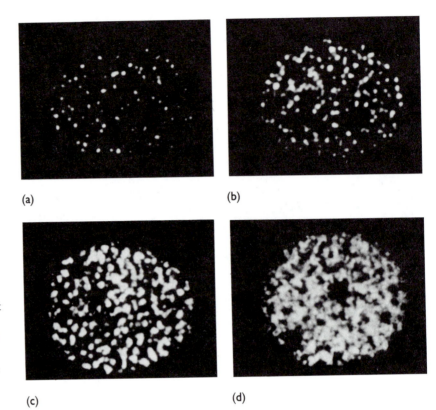

(a)   (b)

(c)   (d)

FIGURE 12–10

Sites of DNA replication in cultured mammalian cells. The nuclei were labeled with bromodeoxyuridine, and replicating sites were visualized with a fluorescent tag. During the first hour of labeling **(a–c)** there was a limited number of sites, each of which increased in intensity. Then, new sites were added. (Photographs courtesy of H. Nakamura.)

In prokaryotes, the second model seems to apply. But in eukaryotes, the first may be true, as shown when cells are labeled for a brief time with a DNA precursor (in this case, bromodeoxyuridine, which is followed with fluorescence). When the nuclei are examined, there are several hundred foci of labeling, which grow over time (Figure 12–10). This is much fewer than the actual number of origins of replication in use at the time, so each labeled spot must represent multiple (about 20) origins of replication. In fact, with careful technique, aggregates of 30 DNA polymerase molecules can be isolated from eukaryotic nuclei.

A number of **anticancer drugs** are specific for S phase cells (Figure 12–11). These agents generally inhibit DNA synthesis by interfering with nucleotide pools. Because cancer cells are not necessarily rapidly dividing (see Table 12–2), the drugs also affect those normal tissues that are likely to be in S phase. Thus, bone marrow, epithelia, and hair follicles are likely to be inhibited, as well as the targeted tumor cells.

G_2

The G_2 phase, a period for preparation for chromosome condensation and mitosis, is typically one to four hours long but can be altered in

Cytarabine (Cytosine Arabinoside)
[Cytosar-U]

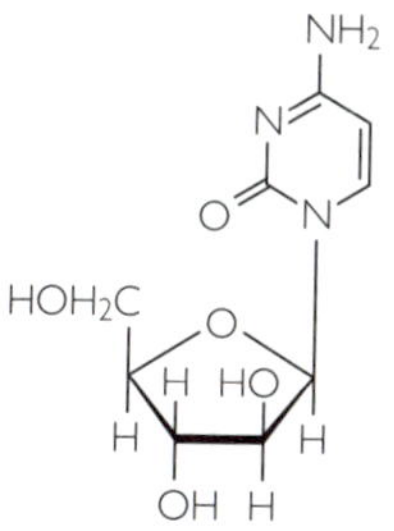

Inhibits DNA polymerase

Doxorubicin hydrochloride
[Adriamycin]

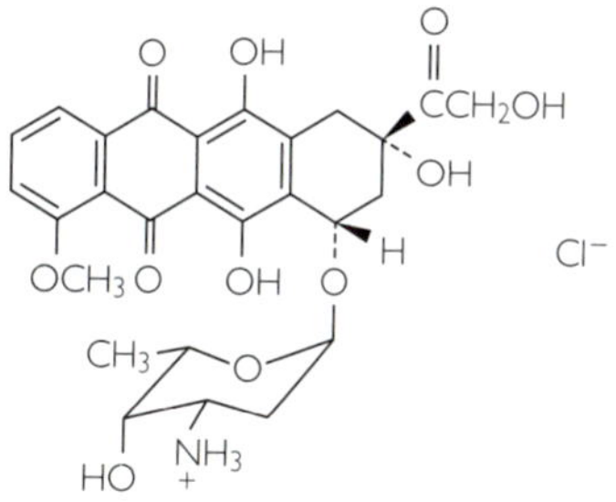

Inhibits DNA and RNA polymerases

Fluorouracil (5 FU)
[Adrucil, Fluorouracil]

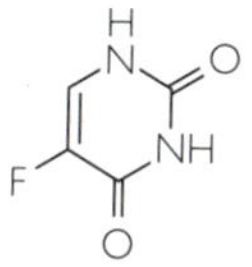

Inhibits thymidylate synthetase

Mercaptopurine
[Purinethol]

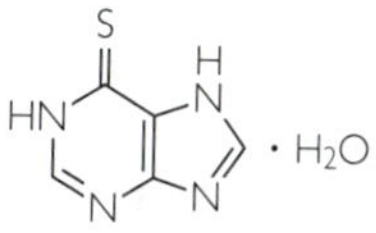

Inhibits purine synthesis

Methotrexate sodium
[Methotrexate, Mexate]

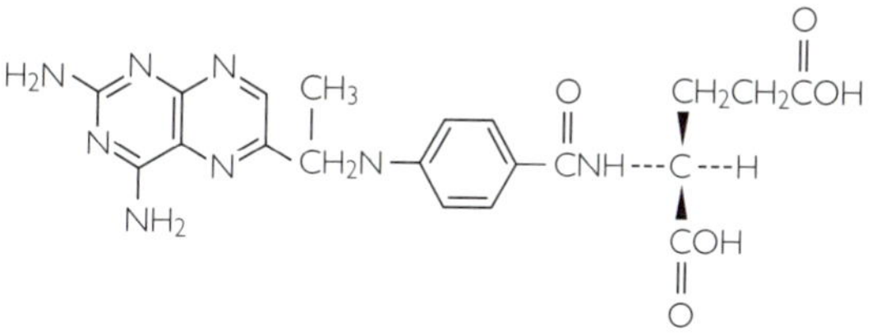

Inhibits dihydrofolate reductase and dTMP synthesis

Thioguanine

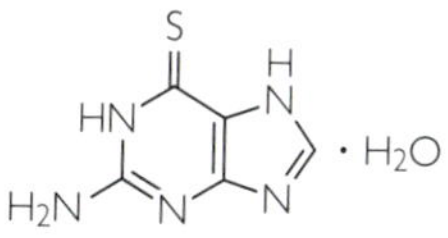

Inhibits purine synthesis

FIGURE 12–11 Anticancer drugs that affect the S phase of the cell cycle.

experimental conditions. For instance, exposure of cycling cells to sub-lethal doses of ionizing radiation or DNA-damaging chemicals causes a lengthening of the G_2 phase. This may be done to ensure the repair of the damaged DNA.

A number of anticancer drugs act to cause cells to accumulate in G_2. These include such widely used agents as cyclophosphamide, adriamycin, and mitomycin C. In all cases, the drug has the ability to cross-link adjacent DNA molecules, if given at the clinically effective dose.

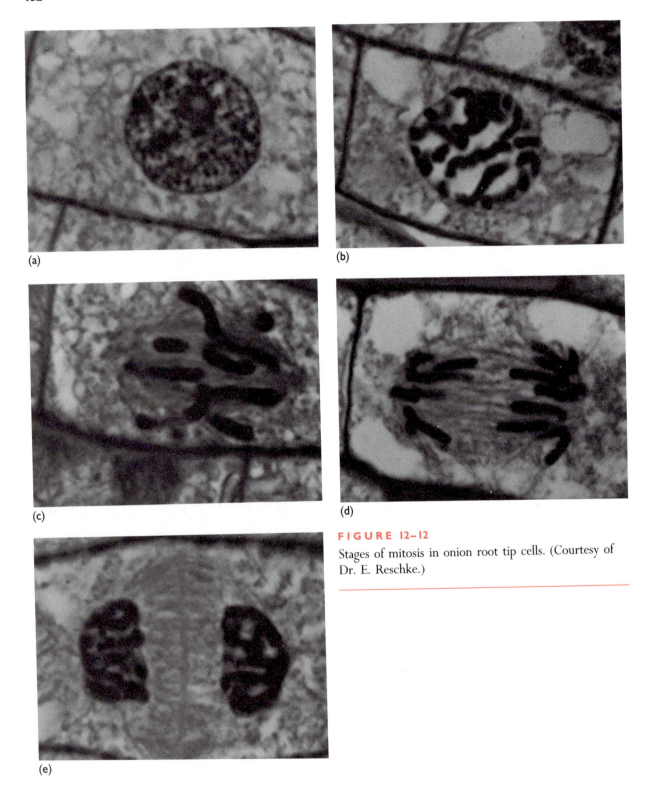

FIGURE 12–12

Stages of mitosis in onion root tip cells. (Courtesy of Dr. E. Reschke.)

When cross-linking occurs with newly replicated DNA, the two molecules produced cannot properly separate from each other preparatory to chromosome formation, and the cells enter G_2 and remain there.

A shortening of the G_2 period can be induced by cell fusion. If mitotic cells are fused with G_2 (or S or G_1) cells, there is a premature chromosome condensation in the latter. This occurs within 30 minutes, indicating that G_2 need only be a half hour in length to prepare for chromosome formation. In an analogous situation to the reactivation of quiescent cells in heterokaryons (Figure 10–23), there is a migration of proteins from the mitotic cells to induce chromosome condensation, with both histones and nonhistones involved. Presumably, these "condensation proteins" are made during G_2, since inhibition of protein synthesis during this period by drugs also blocks chromosome formation.

In synchronized cells, two events often observed to occur during G_2 are histone phosphorylation and tubulin synthesis. Phosphorylation of histones, especially H1, is apparently a prerequisite for the interaction of chromosomal fibers into higher-order structures (see Chapter 10). This event is catalyzed by a protein kinase that is activated at the G_2-S transition point (see below). Tubulin is a major component of the mitotic spindle.

The existence of cultured cells that divide yet are G_2-deficient indicates that, like G_1, the events that typically occur during G_2 need not happen at that time. They could occur earlier, during S phase, when RNA and protein synthesis continue.

Mitosis

Mitosis ("mitos": thread) begins with chromosome condensation and ends with decondensation. It is divided into five sequential periods: prophase, prometaphase, metaphase, anaphase, and telophase, although distinctions between the phases are somewhat arbitrary. They are illustrated in light and electron micrographs (Figures 12–12 and 12–13) and in diagrams (Figure 12–14). In general, mitosis takes one to two hours, with prophase being the longest and anaphase the shortest.

The three major events of **prophase** are chromosome coiling, nucleolar disintegration, and nuclear envelope breakdown. Chromosome condensation occurs in two stages, which are particularly observable in plants. Initially, the long threads of DNA and protein form a helical structure, which then folds on itself to form a secondary coil. Condensation may be controlled in part by Ca^{++}, since a chelating agent that is specific for this cation blocks condensation. Each shortened, thickened chromosome is divided longitudinally into two chromatids, which are held together at a region relatively poor in DNA but rich in protein, called the centromere.

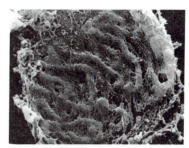
(a) Prophase

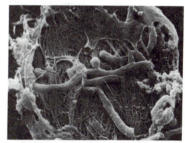
(b) Metaphase

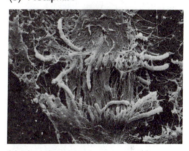
(c) Anaphase

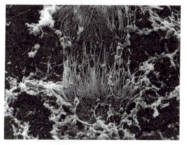
(d) Telophase

FIGURE 12–13

Scanning electron micrographs of mitosis in endosperm cells of the plant *Haemanthus katherinae*. (×10,000. Photographs courtesy of Dr. W. Heneen.)

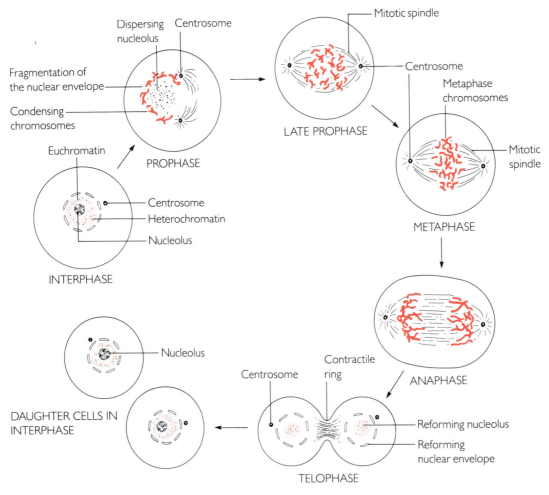

FIGURE 12–14   Diagrams of the stages of mitosis in an animal cell.

Centromere structure has been revealed in the mouse, where the normal process of centromere condensation in prophase is blocked by drugs such as 5-azacytosine (Figure 10–27). These elongated centromeres have three structural and functional domains:

1. A kinetochore domain on the outer surface to which spindle fibers attach (see Figure 12–19)

2. A central region, representing most of the centromere

3. A pairing domain along the inner surface, which is where the two chromatids are attached

That these regions occur in many mammals is indicated by the presence of certain common DNA sequences and proteins. It is of interest that

some of these centromeric proteins were identified as targets of autoantibodies made by patients with rheumatic diseases.

The arrangement of chromosomes during prophase appears to be
random but often is not. Staining of different regions of chromosomes
with different DNA probes shows that the telomeres (ends of the chromosomes) face the outside and are near the nuclear envelope, while the
centromeres face the center of the nucleus. This orientation may help to
begin the lining up of the chromosomes that is observed at metaphase.

Nucleolar dissolution, as described in Chapter 11, occurs at a specific chromosomal location, the nucleolar organizer. However, in many
protists, the nucleolus does not break down but instead divides during
anaphase. In most organisms, the nuclear envelope breaks down at the
end of prophase, and most of the envelope material remains near the
spindle throughout mitosis, with pore complexes and membrane fragments frequently observed in electron micrographs of anaphase cells.

Labeling with antibody shows that lamin B on the envelope (see
Chapter 10) appears on the endoplasmic reticulum (ER) surface during
mitosis and then returns to the nuclear envelope during telophase, an
indication that the ER is a "storage depot" for this lamin. On the other
hand, lamins A and C are rendered soluble in prophase by phosphorylation, possibly triggering nuclear envelope breakdown. However, nuclear envelope breakdown is not a requirement for mitosis, since the
envelope persists in many protists and fungi, such as yeast (Figure
12–5).

As the chromosomes condense inside the nuclear envelope, asters
with fibers appear on the outside of it. Once the envelope has disap-
peared, the spindle forms during **prometaphase.** In animal cells, this
occurs via centrosomes (Figure 12–15). These structures in animal cells
are composed of twin centrioles at right angles, surrounded by amorphous material. The centrosome is the major microtubule organizing
center (MTOC—see Chapter 9) during interphase. It has a special type
of tubulin, γ-tubulin, which may act as a nucleating center for the
growth of microtubules. Late in G_1 and into S, the centrosome replicates
once, and the pair is seen just outside the nuclear envelope. Replication
is apparently not just simple fission, but instead there seems to be some
sort of centrosomal determinant around which the new one forms. Centrosomal DNA has been detected, but its role is unclear.

Short microtubules radiate out in a starlike structure (aster) from
the pair. During prophase, the two centrosomes with centrioles may
migrate to opposite ends (poles) of the nucleus. Usually, this is achieved
by one of the organelles remaining stationary and the other migrating
180° around the nucleus. This event is the critical one in establishing the
plane of cell division.

The important chromosomal event of prometaphase is the attachment of the chromosomes to the spindle and their movement toward the

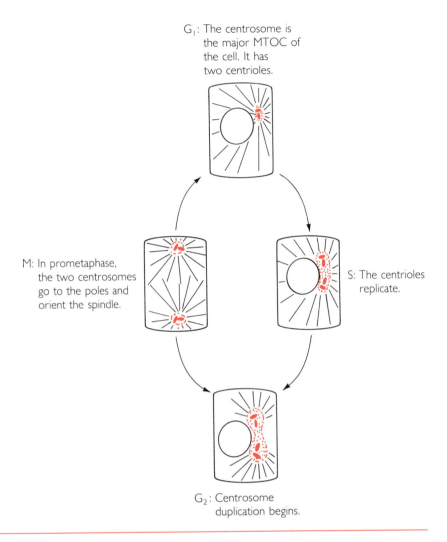

G$_1$: The centrosome is the major MTOC of the cell. It has two centrioles.

S: The centrioles replicate.

M: In prometaphase, the two centrosomes go to the poles and orient the spindle.

G$_2$: Centrosome duplication begins.

FIGURE 12–15

The duplication of the centrosome and spindle pole orientation.

center of the spindle. Attachment is via a region at the centromere called the kinetochore, a structure appearing as a coating on each chromatid. Because specific antibodies can be used to stain it, the kinetochore probably has specific proteins for chromatid attachment, and these proteins in turn are attached to the specific centromeric DNA sequences. Prometaphase chromosome movements can be observed in living cells, and rather than a smooth, gliding motion, the chromosomes move back and forth in short bursts along the spindle at about 0.05 μm/sec.

At **metaphase** the chromatids are attached to the spindle at the equatorial metaphase plate via their kinetochores. This arrangement is nonrandom: A spindle fiber from one pole attaches to the kinetochore of one chromatid of the pair that makes up a chromosome, and a fiber from the other pole attaches to the other chromatid. Occasionally, chromosomes are also arranged nonrandomly with respect to each other. Where there are distinctly larger and smaller chromosomes, the former tend to lie outside of the latter. In certain insects, such as fruit flies, the two homologous chromosomes of a diploid pair lie beside each other at metaphase.

Anaphase begins when the centromeres separate simultaneously. This is apparently a physical break, because the two chromatids at metaphase seem to have a strong force holding them together to counteract the attractive forces of the two centrosomes. In yeast, a mutation in the gene coding for topoisomerase II (see Chapter 10) results in a failure of chromosome separation. Because this enzyme unravels DNA coiled on itself, the mutant phenotype suggests that the replicated centromere DNAs of the two chromatids are interlocked. Their "unlocking" allows chromatid separation.

Separation allows one chromatid of each pair that makes up a metaphase chromosome to migrate to each pole of the spindle (anaphase A). In some cells, this is also a period when the spindle elongates, thereby contributing to the separation of the chromosomes (anaphase B). The rate of chromosome movement is about the same as that in prometaphase.

Because each chromatid contains a complete copy of the chromosomal DNA replicated during S phase, anaphase migration is a key event in accurately separating the two copies of the nuclear genetic material. If both chromatids of a pair migrate to the same pole, the error is termed nondisjunction. It has serious genetic consequences for the two new cells: One will have an extra copy of a chromosome and the other will lack it entirely. The latter cell does not survive in animals but can in plants.

Following anaphase, each pole of the spindle has a complete complement of the chromosomes of the cell. During **telophase,** two new nuclei are formed in what is essentially the reverse of prophase. The nuclear envelope appears to re-form from material that persisted since prophase as lamins A and C are dephosphorylated, causing them to polymerize. The persistence of lamin B in polymeric form associated with membrane fragments during mitosis may provide a site for the other lamins to polymerize in telophase. Reformation of the envelope does not require specific DNA sequences: If bacteriophage lambda DNA is injected into a cell without a nucleus, a structurally intact nuclear envelope forms around the viral DNA. Pore complexes are assembled into telophase envelopes, but their numbers are low. These structures increase well into G_1.

At telophase, the nucleolus re-forms at the nucleolar organizer, again largely from preexisting structures. Spindle dissolution proceeds during telophase so that the MTOCs become surrounded by progressively shorter spindle fibers. Following nuclear membrane formation, the centrioles in animal cells remain in their perinuclear position.

There is little information on how the chromosomes decondense to form the chromatin of the G_1 nucleus. During mitosis, both RNA and protein syntheses are severely limited, but some protein synthesis does occur during telophase. These proteins might be involved in the transformation from chromosomes into chromatin.

Cytokinesis

In all eukaryotes except land plants, the division of the cell contents into two separate, membrane-delimited cells occurs by **furrowing** (Figure 12–16b). This process usually begins in telophase or even late anaphase as an indentation of the plasma membrane at the location of what was the metaphase plate. This occurs at the former equator, which is not necessarily at the middle of the cell. Indeed, in lower organisms unequal cytokineses are common in the formation of certain specialized cells. As the furrow deepens, the cell assumes a dumbbell shape, and finally, the invaginating membranes fuse and two new cells are formed.

The plane of furrowing is apparently established by the spindle. If a sea urchin egg is flattened on a glass slide by pressing down on the coverslip, the mitotic spindle is reoriented to be parallel to the coverslip, and the cleavage plane is always perpendicular to the spindle. Astral centers at the centrosomes exert an influence; indeed, single asters can cause furrowing if injected into cells. Although it is tempting to propose that spindle microtubules originating from the astral centers somehow carry a "message" to the membrane (ions and polyamines have been suggested), convincing evidence for a direct role for the microtubules is lacking.

The mechanism of furrowing involves microfilaments. Just below the initial furrow, a contractile ring containing actin filaments with associated myosin is formed in a few seconds. But if cytochalasin B is applied to cells at cytokinesis, they stop furrowing and it regresses. Phalloidin, which inhibits actin polymerization, also stops furrowing. These observations indicate that active growth of the microfilaments is essential for the progression of the furrow. The myosin in the furrow is different from the myosin associated with interphase microfilaments and has been called myosin II. In addition, several as yet uncharacterized proteins also appear to be unique to the furrow and may play regulatory roles.

In higher **plant cells,** cytokinesis begins in early telophase with an aggregation of Golgi-derived vesicles at the region of the equatorial plate (Figure 12–16). These vesicles contain polysaccharides for the cell wall

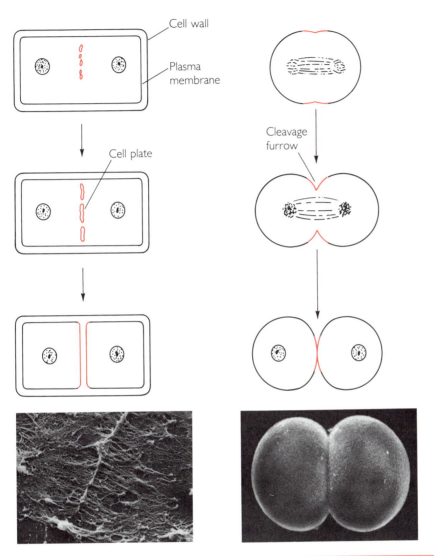

FIGURE 12–16

Cytokinesis in higher plant (left) and animal (right) cells. In plants, the vesicles from the Golgi secrete wall polysaccharides at the cell plate. (Electron micrograph of cell plate from endosperm cells courtesy of Dr. W. Heneen.) In animals, a cleavage furrow composed of actin filaments pinches in the membrane. (Electron micrograph of the furrow of amphioxus embryo courtesy of Dr. R. Hirakow.)

matrix. In addition, spindle fibers and ER cisternae congregate in this region as telophase proceeds. These organelles may be involved in the assembly of the wall cellulose and plasma membrane, respectively.

These wall materials coalesce, first at the equator of the cell and then growing out to the periphery in a structure termed the cell plate. Following telophase, the new cell walls and membranes are completed and the cells are separated. Plasmodesmata (membrane-lined channels through the cell walls connecting adjacent cells—see Chapter 14) apparently form during cytokinesis when ER cisternae, traversing the region of the cell plate, prevent the fusion of vesicles containing membrane and cell wall material.

Cytokinesis results in a distribution of nonnuclear cell organelles to the two new cytoplasmic compartments. In instances where a single large organelle is present, the structure grows and divides rather evenly. For instance, the single chloroplast in the green algae cell of *Spirogyra* elongates as the cell does and then divides transversely prior to cytokinesis.

But the situation for smaller organelles in cytokinesis is less clear. In general, members of each organelle population are distributed to each of the two new cells, but this process does not always result in equal numbers of, for example, mitochondria or Golgi bodies in the two cells following cytokinesis. Mechanisms for distribution, and for compensation in the event of unequal distribution, are not known.

Chromosome Movement

The elegance and importance of chromosome movement to the spindle poles during anaphase has made this an intensively investigated subject for over a century. Studies of spindle structure and function have been made at both the light and electron microscope levels, on intact as well as isolated spindles. Although a unified mechanism has not yet emerged from these studies, many interesting observations and proposals have been made.

Light microscopy reveals a characteristic pattern to the spindle. When it initially forms at the end of prophase, the spindle fibers as seen

FIGURE 12–17

High-voltage electron micrograph of a marsupial rat PtK cell at metaphase, stained with antitubulin antibody coupled with colloidal gold. (×13,000. Photograph courtesy of Dr. M. Morphew.)

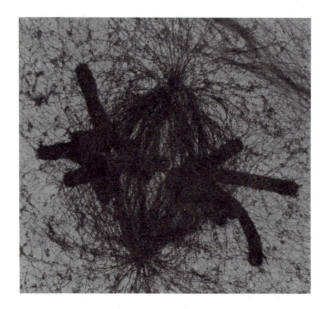

in light microscopy mostly run from pole to pole. Later, during metaphase, more fibers appear that are attached to the kinetochores of the chromatids and then converge at the poles. In anaphase, these kinetochore fibers shorten as the chromosomes move to their respective poles.

At the same time, the spindle itself may lengthen and separate the two poles to a greater extent. The attachment of spindle fibers to the kinetochore is quite strong (see below), and a fiber attached to one chromosome is independent of those attached to other chromosomes. This can be shown if a glass needle is inserted into a mitotic cell and snags a chromosome: That chromosome can be moved independent of the others.

Electron microscopy shows that the spindle fibers are actually composed of one or more microtubules (Figure 12–17). In fixed and sectioned tissues, a 25-nm–diameter component is the basic structural unit, having the morphological and biochemical characteristics of a microtubule (see Chapter 9). The long-standing observation that colchicine is an inhibitor of chromosome movement can be explained in terms of its effect on microtubules.

When it is formed, the spindle is actually two half-spindles (Figure 12–18). Each centrosome at the pole elaborates its own fibers, some of which attach to the kinetochores of the chromosomes and some of

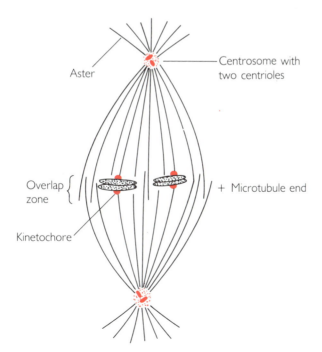

Aster

Centrosome with two centrioles

Overlap zone

Kinetochore

+ Microtubule end

FIGURE 12–18

Diagram of the components of the higher eukaryote mitotic spindle. At each pole is a centrosome containing two centrioles. These produce microtubules with their "+" ends distal to the poles, and some attach to the kinetochores on chromatids. There is an overlap zone of the two half-spindles where they meet.

which extend into the other half-spindle in an overlap region. Because the two halves are essentially running in the opposite directions (each has its "+" end extending to the equator), this may provide a way to distinguish the members of the chromatid pair, since one will migrate to each pole at anaphase.

Like other microtubules, those in the spindle have associated proteins or **spindle MAPs,** which can be observed in scanning electron micrographs as an irregular coating on the fibers (Figure 12−19). In spindles isolated from mitotic cells, MAPs of high (over 150,000) molecular weight can be detected. In addition, actin, myosin, kinesin, and dynein appear to be associated with some spindles. While the actin and myosin appear to be identical with the molecules associated with microfilaments, the dynein differs from that in cilia and flagella. Mutants of algae and humans that lack flagellar dynein have normal spindle dynein, and the two types are not immunologically cross-reactive. A final component of both isolated and intact spindles is membrane. Vesicles appear concentrated at the poles, while the spindle proper and individual spindle fibers are often surrounded by membrane fragments.

What these MAPs and other spindle components actually do to effect spindle function is not clear, but one way to attack the problem is to analyze the organisms with defined mutations in the spindle. Some of these are:

- The polo mutation in fruit flies, which causes defective centrosomes that form multiple poles
- The NDC21 mutation in budding yeast, which causes abnormal attachment of the spindle fibers to chromosomes
- The nuc2 mutation of fission yeast, which causes defective spindle elongation

Identification of the proteins defective in these strains will be valuable in unraveling the complexities of spindle function.

The composition and properties of the spindle have led to at least three proposed mechanisms for chromosome movement. The first mechanism involves **changing the microtubule length.**

Spindle fibers are elaborated from the polar centrosomes and radiate to the chromosomes with their "+" ends free. The dynamic instability of the microtubules, involving assembly and disassembly (see Chapter 9), results in continuous growth and shrinkage, and indeed the tubule itself has a half-life of less than a minute! The many "+" microtubule ends that are available are preferentially captured by the kinetochore during prometaphase. During anaphase, the kinetochore-end of the microtubules, which is not capped with GTP, can depolymerize,

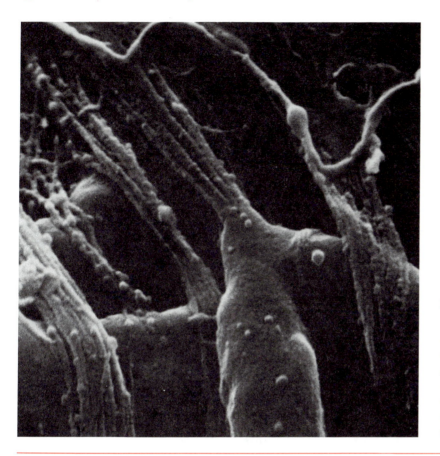

FIGURE 12–19
Scanning electron micrograph of a kinetochore region of a chromosome in a mitotic cell of *Haemanthus.* Kinetochore fibrils, presumably microtubules, extend from the chromosome. (×65,000. Photograph courtesy of Dr. W. Heneen.)

and it is this depolymerization that is supposed to provide the force for chromosome movement to the pole.

Suggestive evidence for this model has come from in vivo and in vitro studies:

1. If tubulin labeled with biotin is injected into a metaphase cell, the "+" ends of microtubules are labeled, and these may then be captured by the kinetochore. Anaphase movement results in these labeled subunits being lost from the microtubule, suggesting depolymerization at the kinetochore end.

2. If a spindle is made up of tubulin fluorescently labeled with rhodamine, it may be bleached at a region midway to the pole. During anaphase, the chromosomes move toward the bleached region, which itself does not move. If depolymerization occurred at the pole, the bleached area would move in that direction, and since it does not, the kinetochore must be the site for depolymerization.

3. In a remarkable in vitro experiment, isolated chromosomes were added to microtubules. The chromosomes attached themselves to the microtubules via kinetochores and proceeded to move along the microtubular network in a process reminiscent of anaphase. The microtubules shortened as the chromosomes moved, and this shortening occurred at the kinetochore end, as evidenced by labeling experiments. This movement did not require ATP, suggestive of a motor not involving MAPs such as myosin, dynein, or kinesin (see below). A similar experiment, in which chromosomes were attached to microtubules, which in turn were attached to a glass coverslip, also showed movement when the microtubules were depolymerized by reducing the concentration of tubulin.

The second proposal for chromosome movement envisions microtubules of opposite polarity that are thought to **slide past one another.**

Cross-bridges might provide the mechanochemical energy for sliding (see Chapter 9), and a chromosome attached to a microtubule sliding toward the pole would be pulled in that direction. The microtubules would still shorten because of disassembly, but this would not provide the force for chromosome movement.

Important evidence favoring this model is that cross-bridges are commonly observed between spindle microtubules. The presence of dynein in the spindle could provide mechanochemical energy for sliding of microtubules past each other or past the microtrabecular lattice. However, this model predicts that kinetochore-attached microtubules and their adjacent pole-to-pole neighbors should have opposite polarities. Such is not the case; they appear to be in parallel polarity.

Sliding is most likely to be important in the elongation of the spindle that occurs in anaphase B in some organisms. The microtubules in the central zone of overlap between the two half-spindles (Figure 12–18) run antiparallel to each other, and cross-bridges are common. During anaphase B, the overlap region becomes reduced, indicating some active sliding. The fact that ATP is needed for this, and not for anaphase A, supports the idea of a motor-driven sliding. But the identity of the motor is unknown.

The third idea for chromosome movement sees the microtubules as a framework on which the active components are **MAP motors.**

The kinetochore has a "collar" where the spindle microtubules are inserted (Figure 12–19). This region may also have a motor that powers chromosome movement. Evidence for this suggestion comes from a careful experiment on dividing lung cells of an amphibian, the newt, where it was shown that a single microtubule grew from the pole and was inserted into the kinetochore collar, after which the chromosome moved toward the pole. If there was a motor here, it had to be at the kinetochore. Evidence for the identity of the motor has come from a mutant of the fruit fly, *Drosophila,* in which a type of meiotic chromosome movement called distributive segregation does not occur. In this case, the defect is in the gene for kinesin. Cytoplasmic dynein has been shown to be associated with the kinetochore, indicating a possible role for it as a chromosome motor.

On the other hand, it has also been suggested that the active MAPs of the spindle are actin and myosin. Sliding of these microfilaments could generate the force for movement either directly on the chromosomes or indirectly via the microtubules. The latter would have to depolymerize to allow the chromosomes to move.

Evidence in favor of this model is that heavy meromyosin decoration shows that spindle actin is oriented properly for a poleward movement. In addition, if spindle fibers are irradiated with a microbeam at 270 nm and/or 290 nm, chromosome movement is blocked. It turns out that actin absorbs strongly at 270 nm and heavy meromyosin at 290 nm, but tubulin does not absorb appreciably at either wavelength.

The three mechanisms could involve **control by Ca^{++}**. This ion stimulates microtubule disassembly, dynein ATPase, and myosin ATPase. Its concentration in the cell can be regulated by sequestration within a membrane-bound compartment (such as the endoplasmic reticulum) and then controlled release by gated channels.

In the mitotic apparatus, membrane-bound vesicles could serve this role. If a high level of Ca^{++} is injected into a dividing sea urchin egg, there is a rapid, localized inhibition of spindle formation and depolymerization of microtubules. However, the Ca^{++} soon is sequestered, probably within vesicles, and the spindle reforms. This removal of Ca^{++} does not happen in the presence of uncouplers of oxidative phosphorylation, indicating an active pumping of Ca^{++} from the spindle into the vesicles.

All models must account for the **forces exerted by the spindle** on chromosomes. These can be measured by using a glass needle. The force to bend the needle is first calculated by applying known weights to the needle until it bends. Then, the needle is inserted into an anaphase cell, snags a chromosome, and stops it from going to its pole. When the needle is pulled out of the chromosome, it snaps back to its original position. The deflection of the needle is then related to the force required to cause this deflection (Figure 12–20). In this way, the maximum force, which causes chromosome movement to cease in the spindle, can be estimated and is about 7×10^{-5} dyn.

The force actually needed to move a chromosome can be calculated by Stoke's law:

$$F = nsv,$$

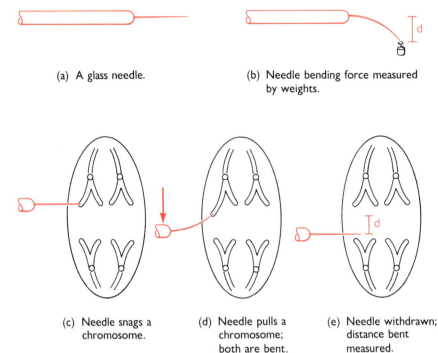

(a) A glass needle.

(b) Needle bending force measured
     by weights.

FIGURE 12–20

Measurement of the force due to
the spindle on an anaphase chromosome. (After R. B. Nicklas.)

(c) Needle snags a
     chromosome.

(d) Needle pulls a
     chromosome;
     both are bent.

(e) Needle withdrawn;
     distance bent
     measured.

where F is the force, n is the viscosity of the medium, s is a coefficient
based on the size and shape of the object, and v is the velocity. For a
large chromosome, this calculation results in a force of 10^{-8} dyn.

Two questions arise from these calculations. First, can the proposed models for force generation provide the measured force? The answer in all cases is affirmative, with MAPs such as dynein, myosin, and
kinesin providing the most power. Second, because the actual force on
the chromosome is several orders of magnitude greater than the theoretical force needed to move the chromosome, can the models provide
a regulator (governor) for the motor?

Control of the Cell Cycle

Control of the Sequence of Events

The cell cycle is an orderly series of events, as is the development of an
organism from a fertilized egg. In both cases, what is thought to underlie
the cellular events is the sequential expression of different genes. Determining these "genetic switches" during the cell cycle has been an
intensive area of research.

TABLE 12-5 Patterns of Protein Synthesis in Continuously Dividing Cells

Cell Type	Protein(s)	Pattern
Yeast	111	All increase through cycle
HeLa	90	Four peak M; two low G; rest: no change
Diatom (Cylindrotheca)	Over 600	Over 200 increase during G_1, early S
Several	Fibronectin	Peak G_1; low M
Several	Tubulin	Peak G_2
Sea urchin embryo	DNA topoisomerase	Peak S, low G_2 and M
Slime mold (Physarum)	Histone kinase	Peak G_2
Several	Histones	Peak S before DNA synthesis
Rat hepatocyte	α-Fetoprotein	Peak G_1
Several	Thymidine kinase	Peak S
Several	Cyclin	Peak late G_2-M; none G_1

DATA FROM: Dr. G. Stein and other sources.

Many studies have been made of the fluctuations of proteins during the cell cycle (Table 12–5), and they clearly show that some proteins become active and/or are synthesized at certain times in the cycle. For instance, enzymes involved in DNA synthesis are usually induced in G_1 and S, and tubulin for the spindle is made during G_2. However, on an overall basis, cell cycle proteins are a minority of the total proteins in the cell, and thus most proteins are not cycle regulated. It is estimated that in budding yeast, about 400 genes (of a total of over 4000) are directly involved in the cell cycle.

There are two possibilities for the determination of the sequence of cell cycle events:

1. Events are tightly dependent on one another; that is, the occurrence of event B depends on the prior occurrence of event A.

2. There is autonomy such that B can occur without A, and normally only time determines the fact that B follows A.

Two general methods can be used to find out which of these possibilities is correct. In the first, a specific inhibitor is used to block one event (A), and then the occurrence of the second (B) is monitored. In the second, a strain mutated in one event (A) is examined to determine if the second event (B) still occurs. Both of these methods have been widely used to study the cell cycle.

The microtubule **inhibitor** colchicine destroys the spindle and therefore blocks movement along the fibers. If dividing cells in a root meristem are exposed to this drug for several cell cycles, there is an increase in chromosome number. All events of the cycle except anaphase and cytokinesis occur, showing that cytokinesis depends on an intact spindle and/or chromosome migration but that the other events are not dependent on spindle integrity. This semi-independence has practical applications: Colchicine is used to increase chromosome numbers and cell size in crop plant breeding.

In contrast to the limited colchicine effect, DNA synthesis inhibitors characteristically arrest the cell cycle at the G_1-S boundary. This indicates that all subsequent events in the cycle are dependent on the completion of DNA synthesis.

Genetic analysis of the yeast cell cycle has clearly defined dependent and independent pathways (Figure 12–6). For example, the emergence of the bud (new cell) and DNA synthesis are independent events once the cycle is underway. On the other hand, duplication and separation of the spindle are on the same pathway and must occur in sequence. Mutants of mammalian cells (Table 12–4) for cytokinesis do not affect the subsequent cycle, while mutants of DNA synthesis, especially its initiation, generally lead to inhibition of mitosis. These observations are consistent with those made with inhibitors and show that there is a series of genetically programmed internal controls on cell cycle progression.

Internal Mitotic Inducers

The eggs of fish, amphibia, and molluscs are widely used to study the cell cycle because they are easily isolated in large quantities, are large in size, and can be induced to be synchronous. These oocytes are arrested in G_2 of meiosis and can be induced to resume the division cycle if treated with exogenous agents, such as the hormone progesterone, in the case of the frog. When the cycle does resume, the nuclear envelope breaks down, the chromosomes condense, and meiosis I is completed. The eggs are now mature and ready for fertilization.

If the cytoplasm of a mature egg is injected into an immature oocyte, all of the events of resumption of the cell cycle occur without hormonal stimulation. The molecule in the egg extract responsible for this transition is a protein called **maturation promoting factor** (MPF). Purification of active MPF reveals that it has two subunits, one 45,000 molecular weight (p45) and one 32,000 molecular weight (p32).

The larger subunit of MPF has enzymatic activity: It is a protein kinase, which adds phosphate groups to serine and threonine residues on cellular proteins. While all of the cellular targets of MPF are not yet known, three are most relevant to mitosis:

1. Microtubules are phosphorylated, and this is essential to their transition from interphase stability to metaphase instability.

2. Histone H1 is phosphorylated, and this leads to chromosome condensation.

3. Nuclear lamins are phosphorylated, and this leads to the disaggregation of the nuclear envelope.

The activity of MPF fluctuates with the cell cycle, being most active at the end of G_2 and falling rapidly when mitosis is over. But an antibody to MPF shows that the catalytic subunit is present in the cells throughout the cell cycle. This means that there must be some way to cyclically activate the kinase activity of MPF at the G_2-M transition point. The activator must be a newly synthesized protein, since cells will not progress beyond this point and into mitosis if protein synthesis is blocked with an antibiotic.

Such an activator has indeed been found, and it is appropriately called **cyclin.** As interphase proceeds from S and G_2, a cyclin (there are two types, A and B) is made and accumulates in the cell until there is enough of it to bind to MPF, which up to then has been inactive. Cyclin binding (it is the small subunit of active MPF) causes the kinase activity to become active, and mitotic induction follows. During mitosis, cyclin is degraded, and MPF is inactivated, and this allows the cell to exit mitosis and enter interphase. Indeed, if cyclin is altered so that its N-terminal amino acids are removed, it can still activate the kinase but cannot be degraded. A cell with this truncated cyclin will be stuck in metaphase.

The degradation of cyclin essentially completes the mitotic induction cycle (Figure 12–21). After mitosis, the cells enter G_1, and the process begins anew. This may be a common mechanism for progressing past the G_2-M transition, since proteins similar in sequence to MPF and cyclin have been detected in many organisms, ranging from humans to invertebrates to yeast. In the latter, the many cell division control mutations (see Figure 12–6) gave the first clues of the existence of the protein kinase system and are providing further information on its control.

In fission yeast, two genes have been identified by mutations that are the analogues of MPF (gene 2) and cyclin (gene 13). The proteins have been isolated, and their sequences and roles are indeed identical to those first isolated from eggs. But in addition, there are a number of mutant strains that modulate this system:

1. "Wee," named for its Scottish origin and small colony size when overproduced, codes for an inhibitor of the kinase. When wee is mutated to nonfunction, cells will progress into mitosis more readily. There

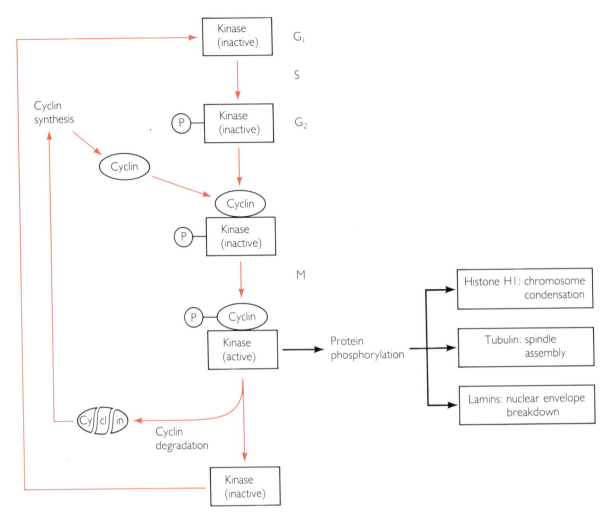

The cyclin-protein kinase cycle, which regulates entry of a cell into M phase from G_2. The kinase protein is always present. But it is only when it complexes with cyclin at the end of G_2 that it has kinase activity. Cyclin accumulates during G_1 and S to reach a peak at G_2. After M, it is broken down. Phosphorylated kinase is inactive, while dephosphorylated kinase is active if suitably bound to cyclin. The kinase is also called maturation promoting factor (MPF) or p34, the product of a yeast cell cycle gene.

is some evidence that this protein is itself a kinase, which phosphorylates the mitosis-inducing kinase to cause it to be inactive.

2. Gene 25 codes for an activator of the kinase. When this gene is nonfunctional, cells will stay in G_2. A homologous gene in fruit flies also leads to accumulation of G_2 cells when it is mutated. This class of gene apparently codes for a phosphatase, which dephosphorylates the mitotic inducer to activate it.

3. Suc1 codes for a protein required for the inactivation of the kinase when mitosis ends. The overproduction of this kinase inactivating protein leads to arrest of the cells in G_2, presumably because they cannot activate their kinase. A similar protein has been detected in mammalian cells.

While the kinase-cyclin system provides cells with a control point at the G_2-M transition, there is also evidence for control at the **G_1-S boundary.** The existence of the G_0 state and the various control points in mammalian cell G_1 attest to this (see above). But it has been best studied in yeast.

The multiple independent pathways of the cell cycle in budding yeast diverge from a single point during G_1 (Figure 12–6). A single gene, 28, has been identified through mutation as being one whose expression is required for this event and all subsequent ones. Cells in stationary phase enter a G_0 condition before expressing gene 28. Diploid cells that can undergo meiosis make the "decision" to do so before the gene 28 step, and haploid cells that mate also do so before the step. For these reasons, the function defined by gene 28 is termed "start."

Like cell division gene 2 of fission yeast, gene 28 codes for a protein kinase. Indeed, the two genes can substitute for each other in complementation experiments: If wild type gene 28 is put into cells mutant for gene 2, the cells will grow and divide even at the restrictive temperature for gene 2. A protein that activates the G_1-S kinase has been isolated through mutational analysis. Not surprisingly, this protein is very similar to cyclin; only in this case, the cyclin peaks in late G_1 rather than in late G_2. Several mutant strains of yeast have been identified that modulate this system, presumably by affecting the kinase-cyclin activities. Finally, proteins analogous to the G_1-S kinase and cyclin have been identified in a wide variety of higher organisms.

Two different cyclins, A and B, are involved in cell cycle transitions at the G_2-M boundary. However, only cyclin A is used for entry into S. G_1 cells prevented from making cyclin A (by antisense RNA) do not replicate their DNA. In addition, cyclin A appears to be involved in a mechanism that ensures the completion of DNA replication prior to mitosis.

The unraveling of the controls at the G_1-S and G_2-M boundaries has been a remarkable story in two aspects. First, it is the culmination and convergence of research done in many laboratories from the biochemical, genetic, and molecular viewpoints and attests to the value of all of these approaches to cell biology. Second, what appears to be a unitary mechanism, present in many diverse organisms and similar in scheme at both control points, is both unexpected and elegant.

External Mitotic Inducers

In addition to internal controls, the initiation of cell cycle events in higher eukaryotes is affected by external stimuli, such as hormones. One possible mechanism for controlling mitosis in a tissue is by **negative feedback.** This clearly must happen, since many tissues have limited cell division. For instance, when a wound heals, cells are stimulated not only to divide to accomplish the process but also to stop dividing when it is finished.

Endogenous factors called chalones (from the Greek word meaning "to relax or slacken") are hypothesized to be produced by differentiated cells in animal tissue to inhibit mitosis only in dividing cells of that tissue. A problem with this idea is that the cell fractions active in inhibition assays are usually low molecular weight substances, containing polyamines (spermine) and nucleotides, both of which are nonspecific mitotic inhibitors. But one protein, transforming growth factor (TGF), is a good candidate, since it is produced by epithelial tissues and inhibits cell division in some epithelial derived tumors. The potential inhibition of tumor growth is one of the reasons that the search for chalones goes on. Another is immunosuppression, since a specific inhibitor of immune cell division would be useful in preventing rejection of tissue transplants.

The search for cell cycle **inducers** generally uses tissue culture cells, since the medium surrounding them can be analyzed and changed. In higher plant cells, which can be grown in chemically defined media, hormones are involved in mitotic controls. For example, the synthetic auxin 2,4-dichlorophenoxyacetic acid (2,4-D) appears to be necessary for passage through G_1, and another hormone, cytokinin, is necessary for G_2.

In animal cell cultures the situation is more complex. Until the 1970s, most animal cell tissue culture media required the addition of blood serum to stimulate cell cycling. It has since been found that for growth and division, many mammalian cell lines require specific factors, or hormones. These molecules, which are active at concentrations as low as 10^{-10} M, include glucocorticoids, platelet-derived growth factor, fibroblast growth factor, nerve growth factor, and epidermal growth factor (Table 12–6). They can be purified from serum and appropriately substituted for serum to make defined tissue culture media, and they appear to stimulate the transition from G_0 to S phase.

Although growth factors act as mitogens (mitosis promoters) in cultured cells, in a number of cases their in vivo roles are less clear. However, functions for those that act on the blood system are well worked out. **Interleukin-2** is made by white blood cells (so it is termed a lymphokine) and acts to stimulate the division of T lymphocytes. When an antigen-presenting cell binds to a T cell, it stimulates the cell to synthesize IL-2 receptors, as well as more IL-2 to stimulate neigh-

TABLE 12-6 Some External Mitotic Inducers in Mammals

Name (Abbreviation) of Growth Factor	Target Cells Stimulated by Growth Factor
Epidermal (EGF)	Ectodermal and mesodermal derivatives
Erythropoietin (EPO)	Erythrocyte stem cells
Granulocyte colony stimulus (G-CSF)	Macrophages
Insulinlike (IGF) (Somatomedin C)	Skeletal muscle precursors; fibroblasts
Interleukin-2 (IL-2)	T lymphocytes
Nerve (NGF)	Embryonic neurons
Platelet-derived (PDGF)	Connective tissue and glia

boring cells. Because it stimulates the immune system, IL-2 is being used in clinical trials on cancer patients to prod their bodies to reject the tumor.

The in vitro action of IL-2 in stimulating T cells to divide has been invaluable in research on acquired immunodeficiency syndrome (AIDS). Before the discovery of IL-2, T cells, which are a target for the AIDS virus, could be removed from the body but would quickly die in culture in the lab. IL-2 stimulation allowed for the formation of sufficient T cells to permit an initial isolation of the virus.

Erythropoietin (EPO) is synthesized by the liver and kidney and stimulates the production of erythrocytes in bone marrow. If the number of circulating erythrocytes is low (a typical red cell has a finite lifetime of 120 days in humans), or if the blood oxygen level is low, EPO is made and released in increased amounts. This factor is being produced in quantity through biotechnology because of its potential as a drug to treat various conditions that lead to low erythrocyte counts in the blood.

Platelet-derived growth factor (PDGF) is made by blood platelets, by smooth muscle cells that are in blood vessel walls, and by the endothelial cells lining the blood vessel lumen. When a blood vessel is injured, a complex healing process occurs, an important aspect of which is the release of PDGF. The factor stimulates the division of smooth muscle cells and fibroblasts, which heal the wound. As noted above, PDGF is also needed for the G_0–G_1 transition in mammalian tissue culture cells.

The mechanism by which hormones and growth factors stimulate cell cycle progression is not clear. In higher plants, cytokinin resembles substituted purines in tRNA, and it is proposed that the hormone alters the machinery of protein synthesis. Although the role of auxins in plant cell elongation is beginning to be elucidated (see Chapter 13), their role in cell division is obscure.

Most of the growth factors for animal cells are polypeptides, which bind to specific receptors at the cell membrane. These receptors are often connected to tyrosine protein kinases (see Chapter 2). Following binding, the factor-receptor complexes aggregate and are internalized, usually by coated vesicles. The vesicles ultimately fuse with lysosomes, with the receptors possibly recycled back to the cell surface.

Two possibilities for the site of action of growth factors are at the cell membrane or by proteolytic activation in the lysosome. Most of the experimental evidence favors the former. For example, after internalization, the application of anti-epidermal growth factor receptor antibodies to fibroblasts can inhibit the transition to S phase. This indicates that continuous binding of the factor at the cell surface is a prerequisite for cell cycle stimulation.

The details of the relationship between cell surface binding and the nuclear events involved in the cell cycle are not simple. For example, within 10 minutes of PDGF binding to a cultured cell, the following events occur: tyrosine phosphorylations on a number of proteins, actin filament reorganization, stimulation of prostaglandin release, stimulation of polyribosome formation, and stimulation of production of the second messengers, inositol triphosphate and diacylglycerol (see Figure 2–21). Which of these are causes and which are effects is not clear.

The Meiotic Cell Cycle

Meiosis Versus Mitosis

Meiosis is a specialized cell cycle that occurs in organisms that reproduce sexually. Many of the aspects of this type of cell division are similar to mitosis, including:

1. There is a G_1 phase, followed by an S phase during interphase.

2. Chromosomes condense during prophase, are at an equatorial plate during metaphase, migrate to the poles during anaphase, and decondense during telophase.

3. Asters and a spindle form during prometaphase.

4. Migration of chromosomes occurs along spindle fibers attached to kinetochores.

5. Centromeres and centrosomes may be involved in spindle orientation.

But there are a number of differences between the two cell cycles. The most fundamental is that while mitosis results in the exact duplication of the genetic material, with an equal partitioning into the two resulting cells, meiosis is a **reduction division.** If an organism has two chromosomes in its cells, the result of a mitotic cell cycle is two cells with two chromosomes each. But after a meiotic cycle, the result is four cells with one chromosome each. In humans, a typical mitotic cell (e.g., in the bone marrow) has the diploid number 46 chromosomes (23 pairs), and its progeny also have 46 chromosomes. The meiotic products in a human each have 23 chromosomes.

A second difference is that whereas mitosis occurs in many cell types and tissues (although these are limited in an adult plant or animal—see Table 12–1), meiosis in higher animals occurs only in **germ cells** that form gametes. These cells become committed to their fate early in embryology, and indeed in a human the germ cell line can be detected as a group of 100 mitotic cells by the third week of gestation, long before these cells actually enter the meiotic cycle.

Plants have much more plasticity than animals and lack a defined germ line. For instance, in a bud at the shoot apex, meristematic cells can divide to form a flower and in this process undergo meiosis to make the sex cells. But under suitable environmental conditions, these same cells will continue in mitosis and form a leaf.

The "decision" to enter meiosis has been studied by both genetic and biochemical techniques. A number of yeast cell cycle mutants involve genes essential to meiotic commitment. For example, the wild type products of fission yeast genes 25 and 35 are required for a cell to enter the meiotic cell cycle in the presence of inadequate nutrients.

Mutants of another gene, cyr1, allow cells to enter meiosis even in the presence of adequate nutrients, indicating that the wild type protein in this case is a meiotic inhibitor. The protein coded for by this gene is adenylate cyclase, which forms cAMP. This leads to the conclusion that a low level of cAMP is necessary for a yeast cell to enter meiosis. This is also the case in frog oocytes. Entry into meiosis by these cells can be blocked by raising the intracellular cAMP by incubation in a phosphodiesterase inhibitor. Meiosis is resumed when the drug is removed.

Mammalian oocytes, which are held in prophase I of meiosis for many years, also respond to reduced cAMP levels to resume meiosis. In this case, the reduction is apparently an indirect response to the cyclic fluctuations in luteinizing hormone during the menstrual cycle, since the hormone leads to reduced intracellular cAMP in the oocyte. But exactly how cyclic AMP inhibits meiosis is not clear. One possibility is that the nucleotide acts as an activator of a protein kinase, which is needed for entry into, and completion of, the meiotic process (see above).

While the mitotic cell cycle has one period of DNA replication followed by a single cell division, the meiotic cycle has one period of DNA replication followed by **two cell divisions.** There is no intervening round of DNA synthesis between the two divisions. Thus, if a mitotic cell in G_1 has a 2C content of DNA, after S it has 4C, and after division each cell has 2C. But if a meiotic cell has 2C in G_1 and 4C after S phase, after the two divisions each of the four progeny cells has one C.

A fourth difference between the two cell cycles lies in the behavior of homologous chromosomes. In diploid organisms (most eukaryotes), there are two copies of the nuclear DNA in each somatic (nongamete) cell, packaged into two sets of homologous chromosomes. (They are called homologous because their genetic material, although very similar, is not identical: In a heterozygous situation, one chromosome carries a different allele for a genetic trait than the other.) During mitosis, each chromosome behaves independently of the others: If an organism has two homologues labeled A and A′, they are randomly arranged on the metaphase plate, and one goes to each pole of the spindle during anaphase. However, in meiosis, homologous chromosomes are **paired** during the first division and line up on the plate nonrandomly.

Meiosis results in the **recombination** of material between homologous chromosomes, resulting in part of the genetic variability seen in the progeny that come from the gametes formed by the meiotic process. In mitosis, such recombination is a much rarer event.

Finally, there is a sixth difference between the two cell cycles: the **time** taken to complete the cycle. Mitotic cell cycles range from hours to days (see Tables 12−2 and 12−3), but meiosis takes much longer, from weeks (as in the human testis), to months (as in many higher plants), to years (as in human females). The reasons for this are not clear.

Meiotic Division

The meiotic cell cycle has been extensively studied and begins, as does mitosis, with G_1 and S periods. During a relatively brief G_2 period, the commitment to meiosis occurs; up to this point, the commitment is reversible, and the cell can revert back to a mitotic state. Once meiosis is decided, G_2 is rather brief and the cell enters prophase. The molecular nature of the meiotic commitment during G_2 is not clear.

The chromosomal events of meiosis are shown in Figure 12−22 and outlined in Figure 12−23. Two divisions make up the process. In prophase I, the chromosomes form, and homologous pairing occurs. Following prometaphase, during which the pairs migrate to the equatorial plate, the homologous pairs are at the plate by metaphase. In anaphase I, homologous pairs separate from each other and migrate along the spindle, one of a pair to each pole. The centromeres of the two chro-

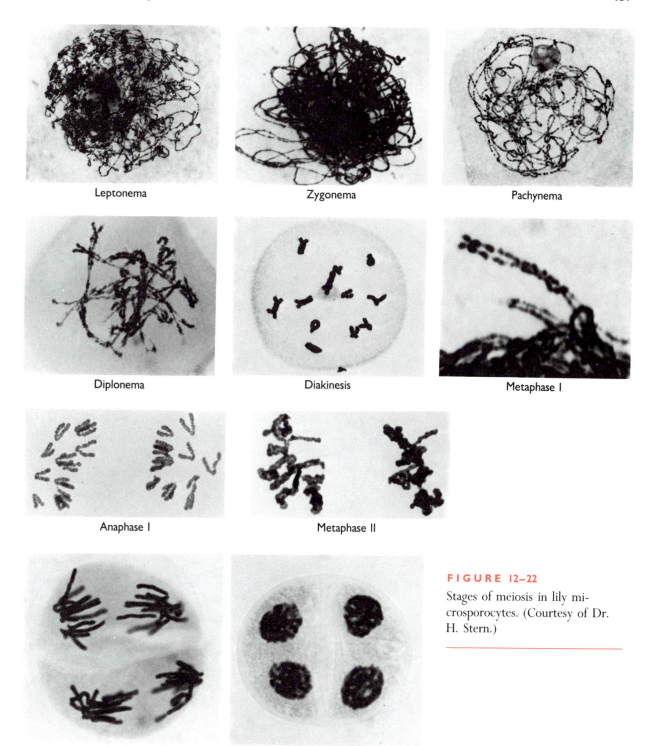

Leptonema

Zygonema

Pachynema

Diplonema

Diakinesis

Metaphase I

Anaphase I

Metaphase II

Anaphase II

Telophase II

FIGURE 12–22

Stages of meiosis in lily microsporocytes. (Courtesy of Dr. H. Stern.)

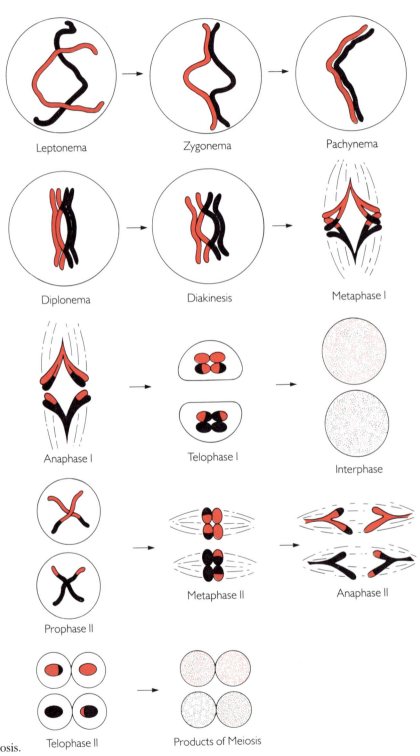

FIGURE 12–23

Diagram of the stages of meiosis.

matids of a homologous pair do not separate during this anaphase. Telophase I is followed without intervening DNA replication by meiosis II, which essentially is a mitotic division, with centromere separation during anaphase.

Prophase of the first division (prophase I) is quite complex when compared to mitosis. In addition to the events of nucleolar and nuclear envelope disaggregation, intricate chromosome coiling is observed. In the context of this coiling, two important events occur: First, homologous chromosomes pair, and second, there is recombination between chromosomes.

During the initial phase, **leptonema** (Greek "leptos": thread), the already duplicated chromosomes appear as single threads with beadlike thickenings termed chromomeres. After a few hours, the nucleus enters **zygonema** (Greek "zygon": adjoining). The homologous chromosomes appear paired longitudinally, and as this occurs a synaptonemal complex appears between the chromosomes.

The synaptonemal complex is a key structure in the lining up of homologous chromosomes. The necessity of the complex is seen in mutant strains of maize (corn) that lack it: In such strains, meiosis stops at pachynema. During leptonema, each chromosome of the pair produces lateral fibrous material emanating 20 to 30 nm from it. By zygonema, the central element has been produced, and the structure is clearly seen in electron microscopy (Figure 12–24).

The temporal association of this structure with chromosome pairing at the cytological level leads to the proposal that the complex both initiates and maintains homologous pairing of chromosomes. But only

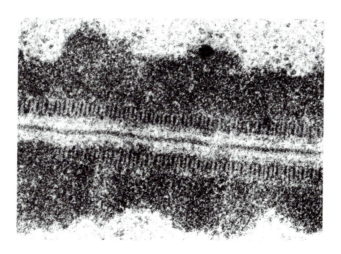

FIGURE 12–24

The synaptonemal complex of the Ascomycete *Neotiella*. (×61,000. Photograph by D. von Wettstein.)

the latter appears to be the case. Once formed, the complex is required for continued pairing and chiasma formation during pachynema.

This leads to the question of what initiates chromosome pairing. One possibility is that the chromosomes are already "prepaired" in interphase. There is some evidence that chromosomes are nonrandomly arranged in the nucleus, especially via attachments to the nuclear envelope or scaffold (see Chapter 10). Another possibility is that pairing begins at the molecular level by some recognition event involving similar DNA sequences before the complex is seen. A burst of DNA synthesis occurs at the beginning of zygonema, which may be involved in the pairing process.

At **pachynema** (Greek "pachus": thick), which may last for up to several weeks, the pairing of the chromosomes is completed, and each chromosome shows two chromatids. The synaptonemal complex at this stage is quite prominent in electron microscopy but gradually disappears late in this phase.

In the next stage, **diplonema** (Greek: double thread), the paired chromosomes appear to repel each other but are attached at a few sites along the pair. These sites, termed chiasmata (Greek, "chiasma": cross piece), are the cytological expression of recombination, in which a piece of one homologous chromatid is exchanged for the cognate piece of the adjacent chromatid. This breakage and reunion probably occurs in pachynema, since enzymes for the breakage and reunion of DNA have been shown to be induced during this stage. Diplonema can be very long: In women, 7 million oocytes begin meiotic prophase during the fifth month of fetal life, and many of these are held in diplonema until puberty, when one per month is released in ovulation and completes meiosis I. Because women can ovulate up to about age 50, diplonema in these cells can last over 50 years!

During diplonema, the chromosomes condense, so that by the next prophase stage, **diakinesis** (Greek: divided across), they appear short and thick, still connected by chiasmata. This is the stage when the nucleolus and nuclear envelope disappear, and the spindle begins to form.

In **prometaphase I,** the spindle is fully formed and becomes attached to the homologous chromosomes. They migrate erratically toward the equator of the spindle, so that by **metaphase I** they are arranged on the plate in a characteristic fashion. In contrast to mitosis, where the chromosomes are seemingly attached to the equator by their centromeres, meiotic homologous pairs appear attached by their chiasmata, while the centromeres are far apart, seemingly on their way to the poles. **Anaphase I** completes the separation of the homologous pairs, with one member of each pair migrating to each pole.

It is important to note the clear difference here between mitotic and meiotic division: In mitosis, the two centromeres of each homologue separate, so that each chromatid goes to a pole, but in meiosis I, the two centromeres of a homologue do not separate, and both chromatids of that homologue go to a pole together. The key to this difference is apparently the interaction of the kinetochores with the spindle microtubules, but the details have not been worked out.

In mitosis, the two kinetochores of a homologous chromosome lie back to back, with the kinetochore of one member of the pair of chromatids facing one pole and the kinetochore of the other member of the pair facing the other pole. This ensures that one chromatid will attach to a spindle fiber from one pole and the other will attach to a fiber from the other pole. In contrast, a meiotic chromosome at metaphase I has its two chromatids arranged so that the two kinetochores lie side by side, facing only one pole. This ensures that the two chromatids will attach to spindle fibers emanating from one pole only and migrate to that pole together (Figure 12–25).

During **telophase I,** the chromosomes unfold, and the nuclear membrane and nucleoli reappear. This is followed by **meiosis II,** where the chromosomal events resemble mitosis. In this case, the two kinetochores of a chromosome at metaphase face different poles, and the centromeres separate, one chromatid of each pair migrating to a pole. This results in four haploid cells.

The cytological result of meiosis is four cells, each with the haploid number of chromosomes. But in terms of the genes carried on the chromosomes, all of the products are not identical. First, which member of a homologous pair will end up in a given cell after meiosis I is random. For example, if the chromosome 1 pair has different alleles (A and A′) and the chromosome 2 pair also has different alleles (B and B′), and the only requirement is that a meiotic product have one copy of each chromosome, there are four possibilities: AB, A′B, AB′, and A′B′. Given many chromosomes (e.g., 23 pairs in humans) and many heterozygous genes (e.g., estimates are that at least 15% of all human genes are heterozygous in an individual) the possible combinations of chromosomes carrying different genes are great.

The second factor for variability in meiotic products is the exchange of genetic material during prophase I. This creates entirely new linkages of genes on chromosomes. Taken together with the variation due to chromosome migration, recombination results in a seemingly endless variety of meiotic products. Gametes, therefore, are not the same genetically as their parent, and this ensures the genetic variation needed for a species to change over the evolutionary time scale.

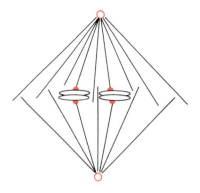

Mitosis: The kinetochore of one chromatid of a pair attaches to a fiber from one pole; the other attaches to a fiber from the opposite pole.

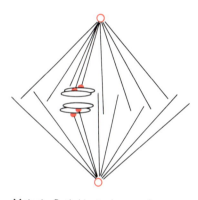

Meiosis: Both kinetochores of a chromosome attach to fibers from one pole, and both from the homologue attach to fibers from the other pole.

FIGURE 12–25

The differences in kinetochore attachment between mitotic metaphase and meiotic metaphase I.

Cell Death

Once a higher organism reaches its mature size, it stays at that size by carefully regulating the balance between cell proliferation and cell death.

This applies to tissues as well. For instance, in humans, red blood cells are made by constantly cycling stem cells in the bone marrow; these replace cells that, because of their lack of a nucleus or mitochondria, have a finite lifetime in the blood (about 120 days) and are destroyed by phagocytic cells in the liver and spleen. In the intestine, stem cells at the base of the mucosal lining constantly divide to replace cells that are sloughed off into the lumen, possibly because they are damaged by the contents of the gut.

But there is also a well-described phenomenon of programmed cell death in differentiated tissues, termed **apoptosis.** This process is an active one, requiring the expression of a number of specific genes. It occurs widely in both plants and animals. For example, when flowering occurs in soybeans, cells of the nitrogen-fixing nodules in the plant's root die; they are no longer needed by the plant because the need for fixed nitrogen goes down at this time. When certain fishes are transferred from salt water to fresh water, the so-called chloride cells of their gills die; in fresh water, specialized salt regulation is less of a problem for the fish.

Cell death is programmed into embryonic development. While an adult organism is certainly the product of many cell division cycles of a fertilized egg, cells die along the way. This has been shown in the nematode worm, *Caenorhabditis elegans,* where the fate of each cell division product can be followed. The adult worm, with 959 nuclei, could come from 10 cell divisions of a single cell to give 1024 cells, but this simple explanation is incorrect. Instead, there are over 200 mitoses with dozens of programmed deaths of cells. A similar pattern is true for most multicellular eukaryotes.

The molecular signals that stimulate apoptosis have been described in only a few cases. Generally, the cells that are stimulated to die are not actively cycling but are in G_0, although it remains to be seen if this is an absolute requirement. Glucocorticoids appear to stimulate the death of lymphocytes. On the other hand, when young mammals are weaned from the mother's breast, her serum prolactin levels go down, and this lack of a growth-promoting hormone promotes destruction of mammary gland tissue.

When lymphocytes are cultured in the absence of the cell division promoter interleukin-3 (see above), they quickly become apoptotic. The overexpression of a protein in the inner mitochondrial membrane prevents this planned cell death. The mitochondrial protein, termed bcl-2, is produced in large amounts by certain lymphomas (tumors of the white blood cells). Thus, instead of the usual case where a tumor appears to result from the stimulation of cell division, in this case it arises from an inhibition of cell death.

Whatever the signals, cell death by apoptosis always has several recognizable stages and takes about three hours (Figure 12–26).

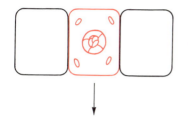

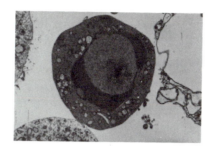

(a) A normal cell in contact with its neighbors.

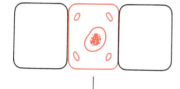

(b) Chromatin in the apoptotic cell condenses.

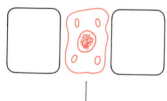

(c) The cell detaches from its neighbors.

(d) The contents of the cell break up into
     membrane-bound fragments.

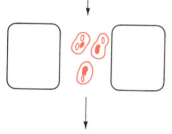

(e) The fragments are endocytosed by the
     neighboring cells.

FIGURE 12–26

Programmed cell death (apoptosis). (Photographs of normal fibroblast and fibroblast after apoptosis courtesy of Dr. A. Wyllie.
From: *J NIH Res* 3 (1991):68.
Reprinted with permission of the
J NIH Res.)

1. A specific endonuclease is induced to break down nuclear DNA. This is rapidly followed by chromatin condensation.

2. The cell loses membrane contacts with its neighbors and appears to shrink away from the rest of the tissue.

3. A ring of intermediate filaments appears around the nucleus.

4. The cell fragments into several membrane-bound bodies. These may contain nuclear pieces, as well as intact organelles and cytoplasm. Remarkably, there is no leakage of cytoplasmic or organelle contents to the extracellular medium.

5. The fragments are endocytosed by neighboring cells (or invading macrophages) and broken down in lysosomes. This essentially recycles the dead cell's constituents.

As in studies of cell proliferation, there are many things to learn about programmed cell death. These include a fuller biochemical description of the sequence of events and the external and internal controls that govern them. This knowledge will doubtless lead to an increased understanding of developmental processes, as well as the possibility of designing molecules that specifically promote the death of tumor cells.

Further Reading

Andrews, B., and Herskowitz, I. Regulation of cell-cycle dependent gene expression in yeast. *J Biol Chem* 265 (1990):14057–14060.

Baserga, R. *Biology of cell reproduction.* Cambridge: Harvard University Press, 1985.

Baskin, T. I., and Cande, Z. The structure and function of the mitotic spindle in flowering plants. *Ann Rev Plant Physiol Plant Mol Biol* 41 (1990):277–315.

Birky, C. The partitioning of cytoplasmic organelles at cell division. *Internat Rev Cytol* [Suppl.] 15 (1983):49–93.

Black, D. J. Antineoplastic drugs in 1990: A review. *Drugs* 39 (1990):489–501.

Bornens, M., Bailly, E., Gosti, F., and Keryer, G. The centrosome: Recent advances in structure and function. *Prog Mol Subcell Biol* 11 (1990):86–120.

Brachet, J. *The cell cycle: Molecular cytology,* Vol. I. New York: Academic Press, 1985.

Brinkley, B., Ouspenski, I., and Zinkowski, P. Structure and molecular organization of the centromere-kinetochore complex. *Trends Cell Biol* 2 (1992): 15–21.

Brooks, R., Fantes, P., Hunt, T., and Wheatley, D. (Eds.). The cell cycle. *J Cell Sci* [Suppl.] 12 (1989).

Bryant, J., and Francis, D. (Eds.). *The cell division cycle in plants.* Cambridge, U.K.: Cambridge University Press, 1988.

Bursch, W., Kleine, L., and Tenniswood, M. The biochemistry of cell death by apoptosis. *Biochem Cell Biol* 68 (1990):1071–1074.

Cande, Z., and Hogan, C. J. The mechanism of anaphase spindle elongation. *BioEssays* 11 (1989).

Carpenter, A. Distributive segregation: Motors in the polar wind? *Cell* 64 (1991):885–890.

Conrad, G., and Schroeder, T. Cytokinesis: Mechanisms of furrow formation during cell division. *Ann NY Acad Sci* 582 (1990).

Cross, F., Roberts, J., and Weintraub, H. Simple and complex cell cycles. *Ann Rev Cell Biol* 5 (1989):341–395.

Davidson, D. Cell division. In Steward, F. C. (Ed.), *Plant physiology: A treatise,* Vol. X, 1991, pp. 342–429.

Doonan, J. Cycling plant cells. *Plant J* 1 (1991):129–132.

Doree, M. Control of M-phase by maturation-promoting factor. *Curr Op Cell Biol* 2 (1990):269–273.

Draetta, G. Cell cycle control in eukaryotes: Molecular mechanisms of cdc2 activation. *Trends Biochem Sci* 15 (1990):378–383.

Ellis, R., Yuan, J., and Horvitz, H. Mechanisms and functions of cell death. *Ann Rev Cell Biol* 7 (1991):663–698.

Forer, A. Do anaphase chromosomes chew their way to the pole or are they pulled by actin? *J Cell Sci* 91 (1988):449–453.

Forsburg, S., and Nurse, P. Cell cycle regulation in the yeasts Saccharomyces cerevisiae and Schizosaccharomyces pombe. *Ann Rev Cell Biol* 7 (1991):227–256.

Goebl, M. G., and Winey, M. The yeast cell cycle. *Curr Op Cell Biol* 3 (1991):242–246.

Golubovskaya, I. Meiosis in maize: Mei genes and conception of genetic control of meiosis. *Adv Genet* 26 (1990):149–190.

Gorbsky, G. Chromosome motion in mitosis. *Bioessays.* 14 (1992):73–80.

Gustafson, J. P., Appels, R., and Kaufman, R. (Eds.). *Chromosome structure and function: The impact of new concepts.* New York: Plenum, 1988.

Hartwell, L., and Weinert, T. Checkpoints: Controls that ensure the order of cell cycle events. *Science* 246 (1989):629–634.

Hayles, J., and Nurse, P. A review of mitosis in fission yeast. *Exp Cell Res* 184 (1989):273–286.

Hochhauser, S., Stein, J., and Stein, G. Gene expressions and cell cycle regulation. *Internat Rev Cytol* 7 (1981):96–225.

Hockenberry, D., Nunez, G., Millman, C., Schreiber, R., and Korsmeyer, S. Bcl-2 is an inner mitochondrial protein that blocks programmed cell death. *Nature* 348 (1990):334–336.

Inze, D., Ferreira, P., Hemerly, A, and Van Montagu, M. Control of cell division in plants. *Biochem Sci Trans* 20 (1992):80–85.

Jackson, D. The organization of replication centers in higher eukaryotes. *BioEssays* 12 (1990):87–90.

John, B. *Meiosis.* Cambridge, U.K.: Cambridge University Press, 1990.

Kelly, T., and Stillman, B. Eukaryotic DNA replication. In *Cancer cells,* Vol. 6. New York: Cold Spring Harbor Laboratory, 1988.

Kirschner, M. Biochemical nature of the cell cycle. In *Important Advances in Oncology.* Ed. V. DeVita, S. Hellmand and S. Rosenberg. Philadelphia: J.B. Lippincott, 1992, pp. 3–16.

Konopa, J. G_2 block induced by DNA crosslinking agents and its possible consequences. *Biochem Pharmacol* 37 (1988):2303–2309.

Koshland, D., Mitchison, T. J., and Kirschner, M. W. Polewards chromosome movement driven by microtubule depolymerization in vitro. *Nature* 331 (1988): 499–505.

Lewin, B. Driving the cell cycle: M phase kinase, its partners, and substrates. *Cell* 61 (1990):743–752.

Lloyd, C. Actin in plants. *J Cell Sci* 90 (1988):185–188.

Lloyd, D. Biochemistry of the cell cycle. *Biochem J* 242 (1987):313–321.

Lohka, M. Mitotic control by metaphase-promoting factor and cdc proteins. *J Cell Sci* 92 (1989):131–135.

Mabuchi, I. Biochemical aspects of cytokinesis. *Int Rev Cytol* 101 (1986):175–213.

Maller, J. L. Mitotic control. *Curr Op Cell Biol* 3 (1991):269–275.

Mazia, D. The chromosome cycle and the centrosome cycle in the mitotic cycle. *Int Rev Cytol* 100 (1987):49–94.

McIntosh, J.R., and Hering, G. Spindle fiber action and chromosome movement. *Ann Rev Cell Biol* 7 (1991):403–426.

McIntosh, J. R., and Koonce, M. Mitosis. *Science* 246 (1989):622–629.

Mitchison, T. J. Microtubule dynamics and kinetochore function in mitosis. *Ann Rev Cell Biol* 4 (1988):527–549.

Moens, P. B. (Ed.). *Meiosis.* Orlando, FL: Academic Press, 1987.

Motlik, J., and Kubelka, M. Cell cycle aspects of growth and maturation of mammalian oocytes. *Mol Reprod Dev* 27 (1990):366–375.

Murray, A., and Kirschner, M. Dominoes and clocks: The union of two views of the cell cycle. *Science* 246 (1989):614–621.

Naeve, G., Sharma, A., and Lee, A. S. Temporal events regulating the early phases of the mammalian cell cycle. *Curr Op Cell Biol* 3 (1991):261–268.

Nicklas, R. B. The forces that move chromosomes in mitosis. *Ann Rev Biophys Biophys Chem* 17 (1988):431–449.

Pardee, A. B. G_1 events and regulation of cell proliferation. *Science* 246 (1989):603–608.

Pardue, M. L. Dynamic instability of chromosomes and genomes. *Cell* 66 (1991):427–431.

Parkinson, K., and Balmain, A. Chalones revisited: A possible role for transforming growth factor in tumor production. *Carcinogenesis* 11 (1990):195–198.

Pickett-Heaps, J., Tippit, D., and Porter, K. Rethinking mitosis. *Cell* 29 (1982):729–744.

Pluta, A., Cooke, C., and Earnshaw, W. Structure of the human centromere at metaphase. *Trends Biochem Sci* 15 (1990):181–185.

Prescott, D. M. Cell reproduction. *Int Rev Cytol* 100 (1987):93–129.

Quinn, C. M., and Wright, N. A. The clinical assessment of proliferation and growth in human tumors. *J Pathol* 160 (1990):93–102.

Rabinovitch, P. S., Kubbies, M., Chen, Y., Schinlder, D., and Joehn, H. BrdU-Hoechst flow cytometry: A unique tool for quantitative cell cycle analysis. *Exp Cell Res* 174 (1988):309–318.

Rattner, J. The structure of the mammalian centromere. *BioEssays* 13 (1991):51–56.

Sakai, H., and Ohta, K. Centrosome signalling in mitosis. *Cell Sign* 3 (1991):267–272.

Salmon, E. D. Cytokinesis in animal cells. *Curr Op Cell Biol* 1 (1989):541–547.

Satterwhite, L., and Pollard, T. Cytokinesis. *Curr Op Cell Biol* 4 (1992):43–52.

Schlegel, R., Halleck, M., and Rao, P. *Molecular regulation of nuclear events in mitosis and meiosis.* New York: Academic Press, 1987.

Schwartz, L. The role of cell death genes during development. *BioEssays* 13 (1991):389–394.

Schulman, I., and Bloom, K. Centromeres: An Integrated Protein/DNA complex required for chromosome movement. *Ann Rev Cell Biol* 7 (1991):311–336.

Smith, L. D. The induction of oocyte maturation: Transmembrane signalling events and regulation of the cell cycle. *Development* 107 (1989):695–699.

Ucker, D. S. Death by suicide: One way to go in mammalian cellular development? *New Biol* 3 (1991):103–109.

Whitaker, M., and Patel, R. Calcium and cell cycle control. *Development* 108 (1990):525–542.

Wick, S. M. Spatial aspects of cytokinesis in plant cells. *Curr Op Cell Biol* 3 (1991):253–260.

CHAPTER 13

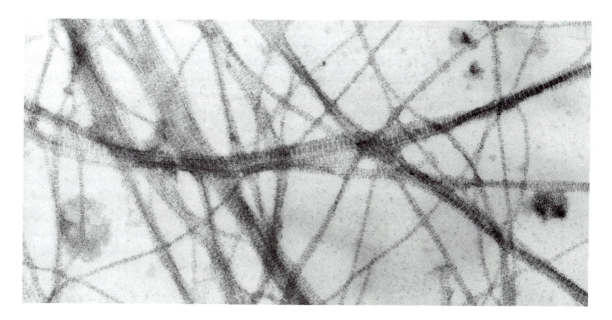

The Extracellular Matrix

An Organelle Outside of the Cell

Cells do not exist in isolation but are in contact either with other cells, with the environment in which an organism lives, or both. For example, a cell on the outside of human skin has one surface exposed to the outside air and its other surfaces exposed to cells adjacent and below it. A mesophyll cell within a leaf is exposed to the atmosphere inside that organ, as well as to other mesophyll cells.

Whereas some cells have only a plasma membrane surrounding their cytoplasm (e.g., the erythrocyte, Figure 2−30), most have an extracellular matrix (Figure 13−1). Typically, this matrix has a fibrous component (collagen in animals and cellulose in plants) within a more amorphous polymeric medium (e.g., proteoglycans in animals and hemicelluloses in plants). Although these materials appear to be only structural, it has become increasingly apparent that the matrix has dynamic functions and is truly a cell organelle.

The extracellular matrix gives shape and form to cells and tissues. Trees can stand tall because of their cellulose walls, and teeth are hard and sharp because of the dentin in the matrix of the odontoblasts. A

FIGURE 13−1

Cartilage cells of a chick embryo limb bud surrounded by an extensive fibrous extracellular matrix. (×4000. Courtesy of Dr. J.-P. Revel.)

related role for the matrix is to act as a rigid container for the cell, which can swell osmotically but will not burst because of the resistance of the matrix (e.g., turgor; see Figure 2–1). The presence of a matrix also confers some protection to the cell from pathogens, since the matrix components are more complex and stable molecules than the lipids and proteins that make up the plasma membrane. Even physical damage can be prevented by the matrix.

The cells that make up a multicellular organism are not arranged randomly but sort themselves into tissues. This process occurs during development and happens within and via the extracellular matrix. The interactions of cells with their matrix are specific and are now becoming well characterized in animal cells. Adhesion of cells to each other and specific intercellular recognition are important roles of the matrix that will be described in Chapter 14.

Animal Connective Tissue Matrix

Collagen

Collagen ("produce glue") is the most abundant animal protein, accounting for about 30% of all mammalian proteins by weight. In all of nature, its abundance as a protein is second only to RubisCO in plant chloroplasts (see Chapter 4). Collagen is responsible for the integrity of the fibrous tissues of bone, teeth, cartilage, skin, and tendons. It also occurs in invertebrates, where collagens from a nematode worm and fruit fly have been characterized.

The chemistry of collagen is quite distinctive in several respects:

1. It contains significant amounts of two otherwise rare amino acids, hydroxylysine and hydroxyproline (Figure 13–2). Depending on the collagen type, these account for up to 25% of the total amino acids and are essential for the structure and modification of the protein.

2. Glycine accounts for about one-third of the amino acids, usually occupying every third position of the sequence (-X-Y-gly-), with X often proline and Y hydroxyproline.

3. Short (two-sugar) oligosaccharide chains composed only of glucose and galactose are attached to the protein.

4. Perhaps most important from the viewpoint of the matrix, collagens can fold and cross-link into fibrils or networks.

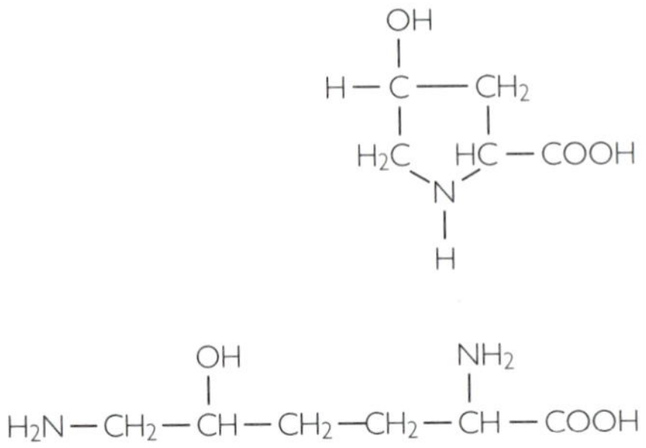

FIGURE 13–2

Two otherwise rare amino acids found in the extracellular matrix proteins.

4-hydroxyproline

5-hydroxylysine

Not all collagen molecules are identical. The collagen gene family (see Chapter 10) has over 20 members, scattered over seven human chromosomes. More than 10 different collagen isotypes have been identified in humans, each showing a preferential association with specific tissues (Table 13–1). These collagen isotypes differ in the identities of the three polypeptide chains that make up the protein, and with 20 different genes it is clear that there are many possible combinations. Five isotypes (I–V) have been well characterized in terms of structure and location; less is known about the other five types.

Type I, the first discovered and most abundant (about 90% of all human collagen), predominates in rigid, nonelastic tissues and contains two identical α1-1 chains and one α1-2 chain in each protein. Relative to the other types, it is poor in hydroxylysine and carbohydrates.

Type II is the only collagen in hyaline cartilage and is found with type I in some other tissues. It has three identical α1-2 chains. Its distinctive compositional features are high levels of hydroxylysine and carbohydrates.

Type III is always associated with type I. Its three identical α1-3 chains have disulfide links between them, a feature not found in most of the other types. Antibodies can be made to the different collagens, and immunocytochemistry indicates that an individual fibroblast in tissue culture synthesizes both types I and III. This shows that collagen isotype production is not cell specific.

The basal lamina (see below) contains **type IV** collagen. Its composition differs from those of the first three isotypes in the presence of 3-hydroxyproline; the other collagens contain the 4-hydroxy isomer. It also contains, in addition to the glucose and galactose present in other collagens, fucose, hexosamine, mannose, and sialic acid. In addition, this collagen, unlike the first three, does not form long fibers but instead is flexible and can form networks.

The other collagen isotypes are relatively minor components of certain extracellular matrices (Table 13–1), and each has its own distinctive characteristics. For example, type IX contains more than one

---

TABLE 13–1 Major Collagen Isotypes

Type	Composition	Characteristics	Tissue Locations
I	$\alpha 1\text{-}1_2$, $\alpha 1\text{-}2$	Low carbohydrate	Skin, bone, tendon
II	$\alpha 1\text{-}2_3$	>10 hydroxylysines per chain	Cartilage, cornea, vitreous body
III	$\alpha 1\text{-}3_3$	Has cysteine, low hydroxylysine	Fetal skin, blood vessels, organs
IV	$\alpha 1\text{-}4_3$	Extensive lysine hydroxylation and glycosylation	Basal lamina
V	$\alpha 1\text{-}5$, $\alpha 2\text{-}5$, $\alpha 3\text{-}5$	Elevated hydroxylysine, low alanine	Blood vessels, smooth muscle
VI	$\alpha 1\text{-}6$, $\alpha 2\text{-}6$, $\alpha 3\text{-}6$		Interstitial tissues
VII	$\alpha 1\text{-}7_3$		Anchoring fibrils for basal lamina
VIII	$\alpha 1\text{-}8_3$		Endothelial cells
IX	$\alpha 1\text{-}9$, $\alpha 2\text{-}9$, $\alpha 3\text{-}9$		Cartilage (minor)
X	$\alpha 1\text{-}10_3$		Hyaline cartilage

triple-helical region separated by nonhelical segments, is secreted as the mature form, and does not form fibrils by itself but rather with other fibrillar collagens. Despite these differences, most collagen isotypes are quite similar in their amino acid sequences, mode of assembly, submicroscopic structure, and properties.

The **assembly** of the collagen molecule is a beautiful example of the use of different cell compartments to perform a series of sequential steps (Table 13–2). Some of these steps have already been discussed in general in Chapters 6 and 7, while others are unique to the collagen system. It has been intensively studied in fibroblasts and embryonic cartilage, and the description that follows is for the fibrous type I collagen of these tissues (Figure 13–3).

Collagen genes are very complex, with over 50 introns, so the initial transcript of the gene is over 18,000 bases long. Splicing out of the introns within the nucleus results in a much smaller mRNA of about 6000 bases. Because type I collagen has both $\alpha 1$-1 and $\alpha 1$-2 chains, the syntheses of both must be coordinated, and this apparently occurs in some way at the transcriptional level, as mRNAs for both are usually made in equivalent amounts.

The spliced mRNAs leave the nucleus via the nuclear pores, and translation starts on free ribosomes in the cytoplasm. The emergence of a signal peptide causes translation to stop when the signal recognition

TABLE 13–2   The Events of Collagen Biosynthesis

Event	Location
Transcription, splicing	Nucleus
Transport to ribosome	Nuclear pores
Translation of preprocollagen	Ribosome
Cleavage of signal	RER membrane
Hydroxylation of proline	RER lumen
Hydroxylation of lysine	RER lumen
Glycosylation of hydroxylysine	RER lumen
Addition of oligosaccharides	RER lumen
Intrachain disulfide bonding	RER lumen
Triple helix formation	RER lumen
Transport to Golgi	Spike-coated vesicles
Modification of oligosaccharides	Golgi
Sulfation	Golgi
Exocytosis	Plasma membrane
Amino-terminal cleavage	Extracellular matrix
Carboxy-terminal cleavage	Extracellular matrix
Cross-linking	Extracellular matrix
Fibril formation	Extracellular matrix

ADAPTED FROM: Byers, P. Inherited disorders of collagen gene structure and expression. *Am J Med Genet* 34 (1989):74.

particle binds to the complex. Translation then resumes on the rough endoplasmic reticulum (RER), where the signal is cleaved by a peptidase as it enters the lumen. The signal peptides of collagens differ from those of other secretory proteins (see Table 6–6) in its length: Whereas the typical signals are 20 amino acids long (just enough to traverse the lipid bilayer of the endoplasmic reticulum [ER]), the collagen signals are about 100 amino acids. The significance of this is not known.

As it enters the ER lumen, the collagen polypeptide is **hydroxylated.** There are three hydroxylases, all of which act co-translationally on certain residues to form:

3-hydroxyproline, from proline

4-hydroxyproline, from proline

hydroxylysine, from lysine.

All three hydroxylases require ascorbic acid (vitamin C) for activity, although the exact role of this cofactor is not clear. Most organisms can make this compound, with the notable exceptions of guinea pigs and

(a) Collagen mRNA translation on RER. Certain lysine and proline residues are hydroxylated co-translationally and then glycolsylated.

(b) As they are completed, individual chains associate in threes to form helical preprocollagen.

(c) At the Golgi, procollagen is packaged and then secreted.

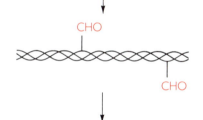

(d) In the extracellular matrix, procollagen is cleaved to form tropocollagen.

(e) Certain lysine residues on tropocollagen are deaminated.

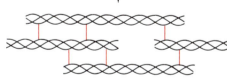

(f) Cross-links form between tropocollagen units.

FIGURE 13–3
Cellular events in collagen fiber formation.

primates (including humans). If people do not have vitamin C in their diet, collagen is not hydroxylated, and this results in scurvy (skin lesions, blood vessel fragility, poor wound healing, and defective gums). All of these symptoms are due to the inability of collagen to form a helical structure.

The newly synthesized **preprocollagen** is larger than the polypeptide in mature, secreted collagen in that the precursor has extensions at both ends of the molecule. The amino end has about 200 extra amino acids, and the carboxyl end has approximately 300 additional residues. These extensions differ in composition from the main collagen chains in that they contain cysteine and can form interchain disulfide bonds. In addition, the extensions, especially at the carboxy-terminus, are glycosylated with mannose and N-acetylglucosamine, two sugars generally absent from mature collagen (see above).

After the collagen polypeptides are released from the polyribosomes into the ER lumen, they self-assemble in threes into triple helices. At this stage, the molecule is termed **procollagen** (Figure 13–4). The aggregation of α-chains to form the triple helix occurs only very slowly if mature collagen chains are incubated together in the test tube. This indicates that the extension peptides, which are not present in the mature collagen, must have a role in lining up the chains or increasing the

FIGURE 13–4

Procollagen aggregates in the extracellular medium. These molecules still have extension peptides and have not been cross-linked. (Courtesy of Dr. J. Gross.)

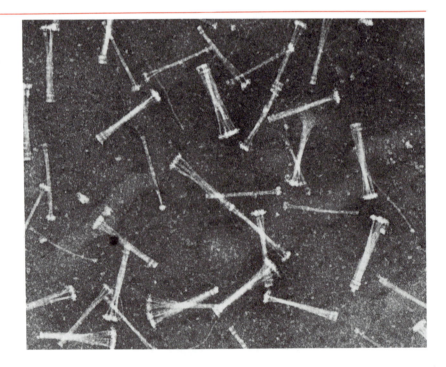

rate of helix formation. The extension regions are not helical themselves but are largely globular and linked to extensions on opposite chains first by hydrophobic interactions and then by disulfide bonds. These events line up the main collagen chains so that they can interact to form a helix.

Because type I collagen does not contain cysteine, the helix itself is not held together by disulfide bonds. Instead, hydrogen bonds are the primary attractive forces between the chains. The importance of the many glycine residues is that their "R" group is small (one hydrogen atom), and they can therefore fit inside the crowded helix. Indeed, glycine is the only amino acid that is small enough to occupy its position in the interior of the molecule. In addition, the two imino acids proline and hydroxyproline confer essential stability to the helix.

Like DNA, the collagen helix can be disrupted by heating, and this induces a random coil configuration. The temperature at which this occurs ("melting temperature" [T_m]) is directly proportional to the imino acid (proline and hydroxyproline) content. For instance, in poikilothermic animals that live at high temperatures, the imino acid content of collagen is high, and this allows a T_m above body temperature so that the helical structure is maintained. The triple helix is also stabilized by hydroxylation, and underhydroxylated collagen has a low T_m, a fact that explains the poorly formed collagen in scurvy.

Following hydroxylation, procollagen is **glycosylated.** This also appears to begin during translation and is completed prior to helix formation. A galactosyltransferase in the ER adds galactose to certain hydroxylysine residues, and this is followed by the addition of glucose to the galactose. The oligosaccharide chains on the extension peptides are not built up one sugar at a time but are added as a chain via a dolichol intermediate (Figure 1–19).

Hydroxylated, glycosylated procollagen is **secreted** to the extracellular medium. Most of it appears to pass through the Golgi region, where oligosaccharide modifications occur, but direct passage from the ER via vesicles is also detected. Secretion is dependent on the post-translational modifications, since underhydroxylated macromolecules are secreted slowly and tend to accumulate in the RER. Such slow secretion is seen in people with scurvy or cells treated with inhibitors of the hydroxylases. This means that the secretion process must somehow require a stable helix. In addition, colchicine incubation blocks both intracellular transport and secretion of procollagen, indicating some role for microtubules in these processes.

Once in the extracellular medium, procollagen is cleaved by specific N- and C-terminal peptidases, which remove the two extension peptides. The same peptidases act on all the collagen isotypes. The resulting molecule is termed **tropocollagen,** which is much less soluble than procollagen and self-assembles to form highly polymerized fibrils.

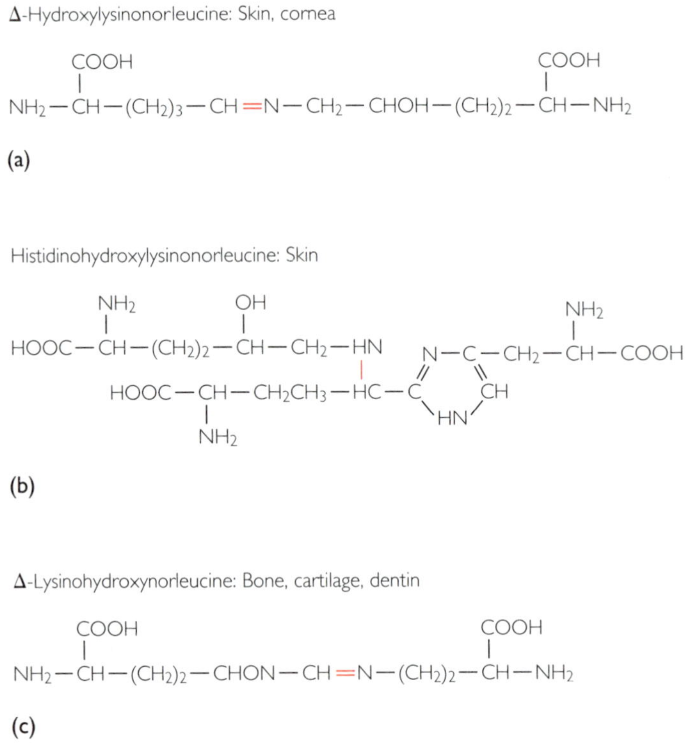

Δ-Hydroxylysinonorleucine: Skin, cornea

$$\underset{NH_2}{\overset{COOH}{|}}{-}CH{-}(CH_2)_3{-}CH{=}N{-}CH_2{-}CHOH{-}(CH_2)_2{-}\underset{NH_2}{\overset{COOH}{|}}{CH}{-}$$

(a)

Histidinohydroxylysinonorleucine: Skin

(b)

FIGURE 13–5

Three cross-links that form between collagen chains. Lysyl oxidase deaminates the amino groups of lysine and hydroxylysine and form reactive aldehydes, which are involved in the cross-linking.

Δ-Lysinohydroxynorleucine: Bone, cartilage, dentin

$$\underset{NH_2}{\overset{COOH}{|}}{-}CH{-}(CH_2)_2{-}CHON{-}CH{=}N{-}(CH_2)_2{-}\underset{NH_2}{\overset{COOH}{|}}{CH}{-}$$

(c)

These are then stabilized by two types of cross-linking reactions:

1. Disulfide bonds are formed spontaneously between and within the collagen chains that make up collagens III–VII.

2. An enzyme, lysyl oxidase, catalyzes the formation of unique cross-links in collagens I–III and V by catalyzing the oxidative deamination of the "R" amino groups on certain lysine and hydroxylysine residues on procollagen:

$$R\text{-}CH_2\text{-}NH_2 + O_2 + H_2O \rightarrow R\text{-}CHO + NH_3 + H_2O_2$$

The four sites for this reaction are similar on the four collagens, and the aldehydes produced are highly reactive. They will combine with the nitrogen-containing "R" groups of lysine, hydroxylysine, or histidine on an adjacent procollagen to form covalent cross-links (Figure 13–5).

Like the globin gene family (Figure 10–11), the human collagen genes can become mutated, and these mutations may result in human genetic diseases. Instead of the thalassemias that can result from defects in the globins, the disorders in case of α1 and α2 collagens are called **osteogenesis imperfecta** (OI—brittle bone disease). As a class, these diseases are either autosomal dominant or autosomal recessive in inheritance pattern and occur in about 1 birth in 10,000. If the disease is minor, abnormalities in the skin, tendons, and ligaments are most common. But if it is major, there can be severely malformed bones and even an early death.

At least four different OI mutations have been detected in each of the α1 and α2 collagen genes, and they all produce chain abnormalities. Some are amino acid substitutions, often resulting in another amino acid being put in the chain where glycine should be. Because glycine is essential for the formation of the collagen helix, such chains do not associate properly as procollagen I, and the disease is severe. In another type of OI, the patients have one of the exons deleted, and because these exons are repeated many times, the resulting procollagen is shorter but still structurally sound. The problem here is that although the α1 chain might be the right length, the α2 chain is shorter, and this causes the assembly of abnormal collagen.

Several pathological states are deficient in the extracellular steps of collagen modification (Table 13–3). In cattle and sheep, a genetic disease, dermatosparaxis, is due to a low level of procollagen amino-peptidase. The resulting mature collagen retains its amino-terminal extension peptide, and the skin is easily shred. An analogous genetic disease in humans, **Ehlers-Danlos syndrome** type VII, results in joint dislocations. These two diseases indicate that the conversion of procollagen to collagen is important for the function of the molecule.

Patients with Ehlers-Danlos syndrome type V are deficient in lysyl oxidase, and therefore their collagen is poorly cross-linked. Their skin is extremely extensible and easily bruised, and their heart valves are friable, leading to congenital heart disease. An indirect effect on cross-linking occurs in the deficiency of lysyl hydroxylase in patients with Ehlers-Danlos syndrome type VI. Their collagen is secreted but not adequately cross-linked, resulting in eye abnormalities, scoliosis, and hyperextensible skin and joints.

TABLE 13–3 **Some Human Diseases of Collagen Processing**

Disease	Effect
Ascorbic acid dietary deficiency	Low hydroxylation
Ehlers-Danlos syndrome, type V	Lysyl oxidase
Ehlers-Danlos syndrome, type VI	Lysyl hydroxylase
Ehlers-Danlos syndrome, type VII	Procollagen aminopeptidase

FIGURE 13–6

Collagen fibrils, showing the repeat structure. (Photograph courtesy of Dr. J. Gross.)

Chemicals can also prevent collagen cross-linking. When animals eat the seeds of the sweet pea, *Lathyris odoratus,* they develop lathyrism, with symptoms very much like Ehlers Danlos syndrome type V. The cause of this is that the pea contains a toxin, aminopropionitrile, that inhibits the lysyl oxidase. A drug, penicillamine, can do the same thing by chelating copper ions that are required for lysyl oxidase activity. This is clearly of possible use in the treatment of patients who overproduce collagen in fibroses and has had successful trials in animals.

Tropocollagen is a rod-shaped molecule, 300 nm in length and 1.5 nm thick. On the other hand, collagen fibers in the extracellular matrix are millimeters long and micrometers thick (Figure 13–6). Clearly, there must be extensive cross-linking of tropocollagen to form collagen. But both electron microscopy and X-ray diffraction data show that the cross-linked tropocollagen molecules are not in exact register but rather in a staggered array (Figure 13–2). Instead of the molecules being end to end, there is a 40-nm gap between the linearly arrayed protocollagen units. This gap may be a nucleation site for hydroxylapatite (calcium phosphate gel) during bone formation. In this way, the organic phase of bone may control the deposition of the mineral phase.

Collagen fibrils are the major component of scar tissues. Thus, an understanding of the **control of collagen synthesis** is a prerequisite to an understanding of normal wound healing and pathological fibrosis. Patients with Ehlers-Danlos syndrome type IV often die because of the rupture of a major blood vessel or the intestine, and their skin is 25% of normal thickness. Fibroblasts of these patients store procollagen type III in the ER and the ratio of type III:type I is very low in their secreted collagen. This indicates that the error in these people is in the processing of collagen and its exit from the cell. However, in their smooth muscle cells, the type III collagen is normally made and secreted, showing that different tissues have different regulatory mechanisms for collagen isotype processing.

In general, type III collagen is the first deposited in the extracellular matrix during wound healing and fibrosis, followed by increasing amounts of type I collagen. The "early" collagen, with a high type III:I ratio, is not as extensively cross-linked as the "later" fibers, which provide tensile strength. An elevation of prolyl hydroxylase generally accompanies fibrosis, providing the additional capacity for hydroxyproline synthesis.

The events of fibrosis are especially relevant in atherosclerosis, the narrowing of the blood vessel lumen that occurs in arteries of people who later have heart attacks or strokes. Hypertension (high blood pressure) is a major risk factor for atherosclerosis. When hypertension is induced in laboratory animals, an early event is an increase of prolyl hydroxylase, and this is followed by increased collagen synthesis.

Apparently, hypertension also injures the endothelium of the artery, which exposes collagen fibers of the extracellular matrix to the blood plasma, and platelets aggregate onto them. Platelets release platelet-derived growth factor (PDGF—see Chapter 12), which induces the proliferation of smooth muscle cells, and more collagen is produced. This thickens the inner wall of the blood vessel and forms a solid fibrotic cap, which overlies a lipid-filled core (Figure 13–7). The reduced diameter of the artery can lead to a blood clot that blocks it off entirely, and the tissue it is supposed to nourish becomes oxygen-starved.

Despite the solidity of its appearance, the collagen pool in a tissue is in a **dynamic state,** with both synthesis and degradation constantly occurring. Both intracellular and extracellular collagens can be broken down. If an animal is injected with radioactive proline, there is a rapid appearance in the urine of radioactive hydroxyproline, and because this imino acid is made posttranslationally, its appearance must signify degradation of newly synthesized collagen chains inside the cell. Depending on the organism and tissue, this can occur in 10–90% of procollagen. The collagenase that catalyzes this reaction occurs in both the Golgi and lysosome, but it is not known which of these organelles contributes to the intracellular hydrolysis.

Degradation of extracellular collagen occurs before the molecule is incorporated into a fibril. This may be due to secreted collagenase, which

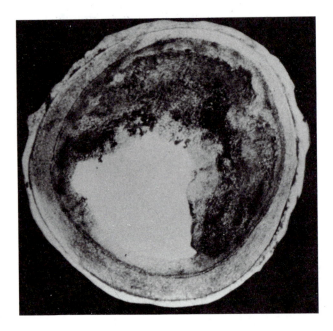

FIGURE 13-7

An atherosclerotic plaque blocking an artery. This cross section shows a much narrowed lumen through which the blood flows. A collagen-rich fibrous cap overlies a lipid core. (Photograph courtesy Drs. M. Brown and J. Goldstein.)

can be detected in the medium around cultured fibroblasts. Or, it may be a consequence of endocytosis, since it is estimated that a typical cell endocytoses its membrane every few hours.

The balance between collagen synthesis and degradation can be affected in different physiological states. For instance, after birth, the uterus loses a great deal of collagen primarily because of increased collagenase activity. In emphysema, tissue inflammation leads to increased collagenase (as well as elastase—see below), which breaks down the extracellular matrix in lung alveoli, destroying the integrity of the organ. A similar situation may occur in inflammation due to arthritis. Finally, metastatic tumor cells invade tissues different from the origin of the tumor. Because the extracellular matrix acts as a barrier to invasion, the tumor cells often secrete collagenase to digest away the fibers, as well as other enzymes to digest the other molecules of the matrix.

Elastin and Fibrillin

In addition to small amounts of collagen, the extracellular matrices of blood vessel walls, the trachea and bronchi, and extensible ligaments contain elastin. Like collagen, the amino acid composition of elastin has abundant glycine (33%) and proline (11%). But unlike collagen, there is little hydroxyproline, no hydroxylysine, and a large number of nonpolar, hydrophobic amino acids. Elastin contains about 5% carbohydrate and, in contrast to collagen, lacks cysteine, meaning that there are no disulfide bridges in the molecule or its precursor.

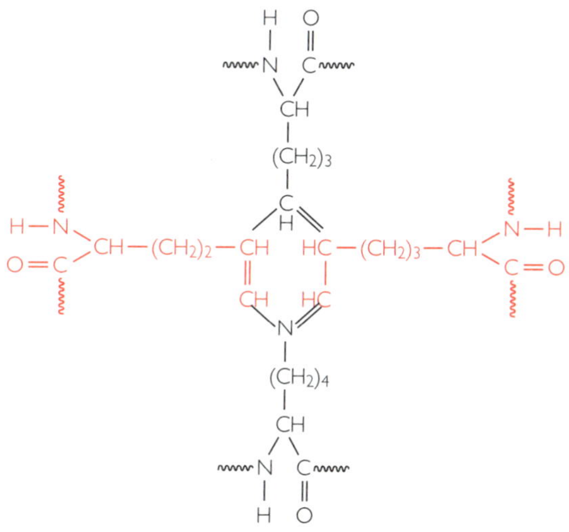

FIGURE 13–8
The cross-linking of four elastin chains via desmosine.

The intracellular and extracellular pathway of elastin synthesis is quite similar to that of collagen. Like collagen, the elastin gene is in many exons (at least 32 in cows). Collagen prolyl hydroxylase, or an enzyme similar to it, is involved, and proelastin is produced initially and then cross-linked in the extracellular matrix.

The polypeptide chain of elastin, termed **tropoelastin,** contains about 800 amino acids, about 75% of which are nonpolar. It does not form a regular helix but rather an interrupted one. The glycine-rich regions are helical and are involved in the stretching property of the molecule. Between the helices are lysine-rich, nonhelical regions that are involved in cross-linking. Elastin cross-links largely occur via desmosine, a molecule deriving from the "R" groups of four lysine residues on different elastin chains, three of which have been deaminated by lysyl oxidase (Figure 13–8).

The morphology of elastin fibers differs in various tissues. In the artery wall, there are thick, concentric lamellae, while in lung alveoli, thin fibers follow the outline of the air sacs, and networks, intermeshed with collagen fibers, occur in mesentery and skin. In all cases, the structure is less regular in appearance than that of collagen.

The distinctive property of elastin fibers is that they return to their original length after being stretched. Lathyrytic animals and Ehlers-Danlos type V patients, who have deficient lysyl oxidase activity, form few cross-links in elastin, and their blood vessels are not elastic. This shows that desmosine cross-links are important in the elastic properties of the matrix.

Elastic fibers are often associated with **fibrillin.** This large protein (DNA cloning suggests 1973 amino acids) is rich in the sulfur-containing amino acid cysteine, which indicates that there are many possibilities for intrachain and interchain disulfide bonds. In addition, there are repeated domains similar to epidermal growth factor (EGF). Fibrillin, as its name implies, forms fibrils 10 nm in diameter and has been detected by antibody staining in the extracellular matrices of many tissues, including skin, lung, kidney, cartilage, and tendon.

However, it is the presence of fibrillin in the lining of blood vessels that has led to its characterization. Marfan syndrome, a genetic disease, affects one person in about 20,000. Patients often have long, thin limbs, spindly fingers, and a "hollow" chest, suggestive of excessive growth of bones. In addition, the connective tissue lining the aorta is structurally weak and can rupture; indeed, this is a common cause of death in Marfan patients.

Taken together, the clinical picture indicates some defect in the extracellular matrix. First by a lack of staining by antifibrillin antibodies and then by molecular genetic methods, this defect has been shown to reside in the gene coding for fibrillin. In the process, an additional fibrillin gene has been characterized, and this may mean that, like the collagens, there is a family of fibrillins. Based on photographs and descriptions, it has been suggested that Abraham Lincoln, president of the United States during the Civil War, had Marfan syndrome.

Ground Substance: Proteoglycans

Connective tissue collagen and elastin and fibrillin fibers lie in the ground substance of the extracellular matrix. This substance, which accounts for about 25% of the weight of the matrix, is composed of **proteoglycans,** complex macromolecules containing up to 95% carbohydrate attached to a protein backbone.

The polysaccharide side chains of the ground substance are called glucosaminoglycans because most of them are derived from glucosamine. The chains vary in length from 30 to 30,000 repeating disaccharide units, such as chondroitin sulfate, dermatan sulfate, keratan sulfate and heparan sulfate, the sulfate group being added in the Golgi (see Chapter 7). In addition, the nonsulfated hyaluronic acid is part of proteoglycans, although present in small amounts. The tissue distributions of these polysaccharides are noted in Table 7–6.

Proteoglycans can be extracted from cartilage and separated by density gradient centrifugation into three fractions:

1. Two proteins of low molecular weight, termed link proteins ($<1\%$)

2. Proteoglycan subunits containing a core protein and polysaccharide chains (99%)

3. Hyaluronic acid (1%)

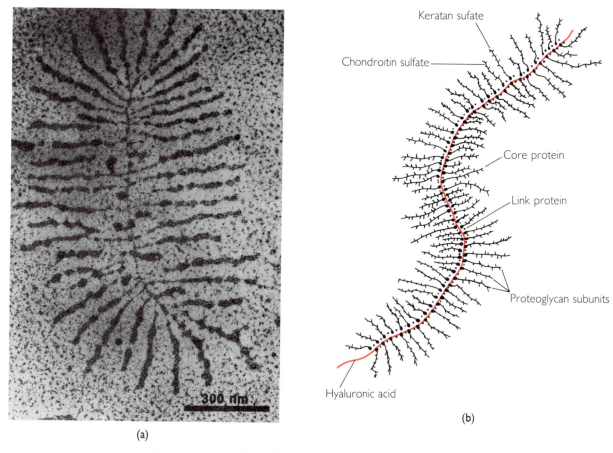

(a)

(b)

FIGURE 13–9 A proteoglycan aggregate from the extracellular matrix **(a)** and an interpretation of the molecular structure **(b)**. (Micrograph courtesy of Dr. J. Buckwalter.)

These fractions can be mixed together in the test tube to form huge aggregates of molecular weight above 30 million. Analyses of reconstitution experiments and electron micrographs of aggregates have led to a model for the proteoglycan shown in Figure 13–9. In the model, hyaluronic acid forms the 0.5–4.0 μm "backbone" of the featherlike aggregate. Bound to the filament at intervals are the link proteins and polypeptide chains from which radiate glucosaminoglycans of varying lengths.

Molecular cloning has led to the identification of 16 different core proteins, and each appears to be part of a unique proteoglycan in a particular location. For example, aggrecan has a core protein of 215,000 daltons molecular weight, has both chondroitin sulfate and keratosulfate attached, and occurs in cartilage; versican has a core of 36,000, chondroitin sulfate only, and occurs outside fibroblasts; and thrombomodulin has a core of 60,000, chondroitin sulfate only, and occurs outside endothelial cells lining blood vessels.

Being complex molecules, proteoglycans are assembled in a series of sequential events:

1. The core protein is assembled on the rough ER and then cotranslationally inserted into the ER lumen.

2. Dolichol phosphate is then used to attach N-linked, mannose-rich oligosaccharides.

3. In the Golgi, glycosyltransferases attach O-linked sugars, and the donor, phosphoadenosine phosphosulfate, is used to attach the sulfate groups.

4. Following some trimming of the sugars, the proteoglycan is secreted via a vesicle.

Several disorders of this pathway have been characterized. Lack of the core protein in chicks (nanomelia) and mice (cartilage-matrix deficiency) results in severe lowering of the volume of the extracellular matrix, skeletal deformities, and early death. Lack of the sulfate donor in mice (brachymorphism) and undersulfation of chondroitin sulfate in humans (spondyloepiphyseal dysplasia) result in skeletal and kidney abnormalities.

The composition of the ground substance changes with age of the tissue, with keratosulfate increasing and chondroitin sulfate decreasing. Growth hormone administration at any age induces the synthesis of proteoglycans of the "young" pattern (higher ratio of chondroitin: keratosulfate).

Proteoglycans are degraded intracellularly, primarily after endocytosis and insertion into lysosomes. Various hydrolytic enzymes first cleave the sugar chains, and proteases hydrolyze the protein core. The importance of these reactions is indicated by numerous lysosomal hydrolase deficiency diseases in humans (Table 8–3). The often severe pathologies of these disorders where an enzyme for ground substance breakdown is absent show that breakdown of proteoglycans is an essential activity. About 250 g of proteoglycans are degraded per day in a human adult.

In the extracellular matrix, proteoglycans serve a number of roles:

1. Because they are polyanions, proteoglycans tend to aggregate in the presence of divalent cations. This aggregation leads to a **high viscosity,** a property useful in the extracellular matrix surrounding the joints. In rheumatoid arthritis, hyaluronic acid

is generally less polymerized than normal, and this leads to a reduced viscosity, and joints are poorly lubricated. The administration of corticosteroids induces the hyaluronic acid to re-polymerize, and lubrication is restored.

2. The ground substance acts as a **sieve.** Large molecules such as serum proteins or infectious agents such as bacteria cannot penetrate the cross-linked, viscous medium.

3. Heparan and dermatan sulfates are potent **anticoagulants,** which can aid in wound healing. The sulfates also can be used as drugs.

4. Hyaluronic acid appears to control **cell aggregations** during embryogenesis, since mesenchymal cells fail to aggregate if treated with hyaluronidase.

5. Binding to heparan sulfate is required before fibroblast growth factor binds to its receptor on the surface of cultured cells. This indicates that the proteoglycan may orient the growth factor to its receptor and suggests a **channeling** role for heparan sulfate.

Collagen interacts only indirectly with the ground substance (via fibronectin, see below). Nevertheless, in the matrix, the proteoglycan aggregates surround the collagen fibers, providing, in the case of cartilage, a compressible gel. The collagen defines the strength and overall shape of the tissue. The higher the concentration of collagen, the more tensile the extracellular matrix.

Binding to the Cell Surface

The extracellular matrix does not exist as an isolated entity but is usually bound in some fashion to the surfaces of the cells it surrounds. In connective tissues, the two molecules that mediate this interaction are fibronectin, a matrix glycoprotein, and integrin, the receptor for this glycoprotein on the cell surface.

Fibronectin (Latin "nectere" to bend) was discovered as a protein in blood plasma that could cause cultured cells to attach to a collagenous substratum. It is a large molecule, with two identical chains of about 2500 amino acids grouped into several domains, which are repeated. As occurs with many other proteins, the different domains have different roles (Figure 13–10) in the overall function of the protein as a link between the cell and its surrounding matrix:

- A collagen-binding domain
- Two domains that bind to proteoglycans
- A sequence that binds to a membrane receptor

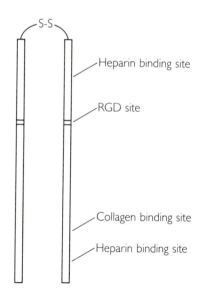

FIGURE 13–10

Molecular structure of fibronectin. The molecule consists of two chains connected by two disulfide bridges. There are binding sites for extracellular matrix components (collagen and heparin) and a cell membrane protein (integrin—to the RGD site).

The identification of the cell membrane receptor for fibronectin came from studies in which the molecule was cut into smaller and smaller fragments with proteases, and each was tested for binding to cells. Once the active fragments were found, it was possible to construct peptides that would mimic the cell in binding to fibronectin.

The critical peptide in binding was arginine-glycine-aspartate (RGD, after the single-letter amino acid abbreviations). This peptide blocked fibronectin binding to cells if presented in soluble form and was highly specific: A single amino acid change destroyed its activity. The RGD peptide could then be injected into an animal to make antibodies to the fibronectin receptor, and these were used in turn to purify the receptor from the plasma membrane.

The receptor, a member of a family of proteins called **integrins,** is an integral membrane protein that occurs in a wide variety of cells, including muscle, neurons, and fibroblasts. The proof that integrin is the receptor comes from an experiment in which anti-integrin antibody is applied to the cells and they lose the ability to attach to their matrix. In addition, when cultured cells are grown on a collagenous matrix, they attach to the matrix via regions called plaques, and an integrin is concentrated at these plaques.

Taken together, these data show that the integrin is the fibronectin receptor and that the interaction of receptor with its RGD region and ligand is the key to cell-matrix adhesion. Integrin spans the plasma membrane. On the extracellular side is the receptor activity for the matrix glycoprotein, while on its intracellular side, it interacts either directly or indirectly with actin filaments that come to a focal point at the plaques (Figure 13–11).

The use of anti-RGD antibodies has resulted in the identification of many proteins containing this sequence. Some of them are clearly involved in recognition, for example the integrins, epidermal growth factor receptor, thrombin (a blood-clotting protein), and even the receptor protein on the surface of *E. coli* for bacteriophage lambda.

However, the presence of RGD in other proteins—such as the enzyme hydroxyacyl CoA-dehydrogenase, which acts in fatty acid metabolism—is probably coincidental. Nevertheless, the widespread occurrence of this sequence indicates that it is evolutionarily ancient and must serve an important role. Viper venoms have RGD-containing proteins aptly named **disintegrins** because they bind to fibronectin and thereby cause cells to detach from their matrix. This is especially important in preventing platelets from aggregating at a wound. There may also be natural antiadhesive molecules that act during development. Tenascin, a large (molecular weight over 200,000) extracellular matrix glycoprotein, inhibits cell adhesion to fibronectin. When tenascin is injected into a frog embryo at gastrulation, the normal process of cell

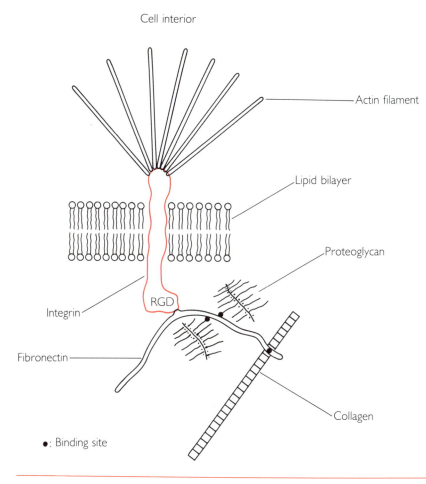

Cell interior

Actin filament

Lipid bilayer

Proteoglycan

RGD

Integrin

Fibronectin

Collagen

●: Binding site

FIGURE 13–11

Molecular model for the role of fibronectin in the attachment of the collagen extracellular matrix to the plasma membrane. The membrane receptor, integrin, binds not only to the matrix (via fibronectin) but also to the cytoskeleton (via actin filaments).

migration of mesodermal cells is blocked, an effect that is reversed by antitenascin antibodies. These experiments indicate that tenascin, or other antiadhesive molecules, may be important in the detachment-attachment cycles seen in embryology.

The fibronectin-integrin system is not the only way to attach a cell to its extracellular matrix. A proteoglycan, **syndecan** (from the Greek "syndein"—to bind together) has been shown to do this in epithelial tissues (Figure 13–12). Syndecan typically is located at the basal and lateral sides of these cells, which is appropriate for an adhesive molecule because the apical region faces the extracellular medium. Its protein core spans the lipid bilayer once and has both chondroitin and heparan sulfate attached to its large, extracellular domain. The latter can bind to collagen, and to complete the story, the cytoplasmic domain has a binding site for actin microfilaments. This means that the syndecan system has

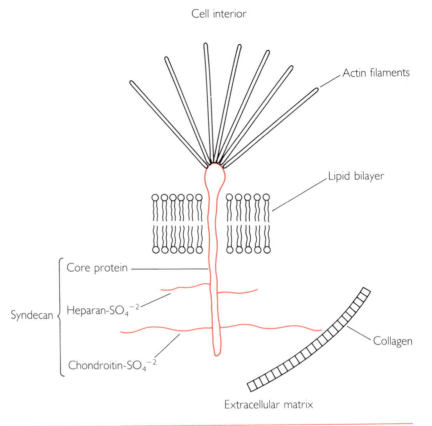

FIGURE 13–12

Molecular model for the role of syndecan, a proteoglycan that attaches the collagen extracellular matrix to the plasma membrane. Syndecan has a transmembrane core protein that binds both the matrix (via collagen) and the cytoskeleton (via actin filaments). Compare to Figure 13–11.

the same role as the fibronectin-integrin system, and indeed, the two may act together in tissues where both are present.

Not all types of cells are attached to an extracellular matrix. For instance, although the precursors to red and white blood cells in the bone marrow are attached to a matrix, the mature, circulating cells are not. Also, when a normal cell is transformed into a cancer cell, it loses its ability to be attached to the matrix, and such cells may travel to a new location in the body and their integrin receptors to attach the matrix at the new site.

To model this phenomenon, normal cells can be treated with anti-RGD antibody. This results in detachment from their substratum, a clear indication that the fibronectin-integrin system is holding them onto their matrix. But if already detached tumor cells are treated with the antibody, they fail to establish themselves in a new tissue. This means that attachment to the matrix is a key event in metastatic cell invasion and suggests that inhibiting this attachment could be valuable in preventing metastasis.

When normal cells become transformed into tumor cells in culture, one result is the loss of normal, flattened cell shape and adhesion to the substratum (extracellular matrix). One way to transform cells is to infect them with retroviruses. An essential protein product for transformation in virus-infected cells is that of the viral oncogene, and a number of these are tyrosine protein kinases, similar to the membrane protein receptors (see Chapter 2). Integrin has been shown to be phosphorylated by the tyrosine kinase, and this phosphorylation is directly correlated with the loss of adhesion to the matrix in transformation. The loss of function of integrin may also explain the observed loss of fibronectin from the matrix surrounding transformed cells.

Basal Lamina

Structure and Composition

In 1857, R. Todd and W. Bowman described the extracellular material on which animal digestive and respiratory epithelia rest as a fibrous "basement membrane." This structure was later also found around muscle, fat, and nerve cells and even in invertebrates, except for the sponges. Electron microscopy revealed that it is not a membrane but rather is a layer of dense material (Figure 13–13), and now the more correct term

FIGURE 13–13 The glomerular capillary wall of the kidney, showing the glomerular basement membrane between the capillary lumen and epithelial cell. ($\times$10,000. Courtesy of Dr. B. Hudson.)

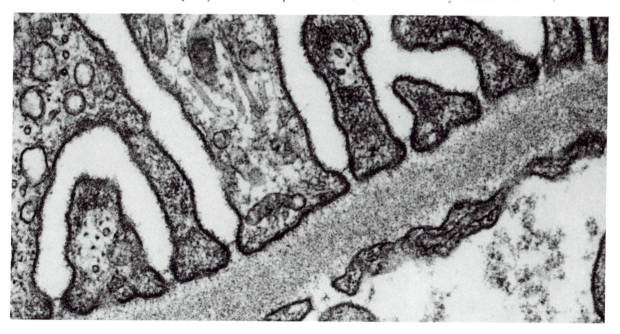

basal lamina is used. Ultrastructurally, the basal lamina is usually composed of 4-nm–diameter fibrils in a meshwork arrangement and a granular matrix.

As its name suggests, the basal lamina is situated at the base of the epithelia that line the circulatory, digestive, reproductive, respiratory, and urinary tracts, where it separates the epithelium from the connective tissue. In some cases, such as kidney glomeruli and lung alveoli, it surrounds the tissue. In many instances, it is interposed between a collagenous extracellular matrix and the tissue. Typically, there is a microscopically clear lamina rara adjacent to the tissue, and the lamina, 20 nm to 5μm thick, faces the extracellular matrix (Figure 13–14).

Basal laminae can be isolated from cell homogenates by careful extraction of cell contents with detergents and nucleases or by differential centrifugation. This has allowed the identification of the lamina's major constituents (Table 13–4):

1. Type IV collagen (Table 13–1) occurs only in the lamina and ac counts for as much as 60% of the protein content of the lamina. Type IV collagen is compositionally the highest isotope in hydroxylysine and

FIGURE 13–14

Structure of the basal lamina between a cell layer and a collagen extracellular matrix (the reticular layer).

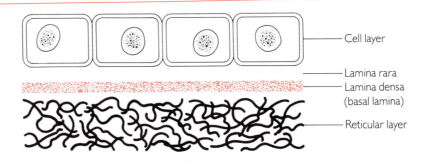

Cell layer

Lamina rara
Lamina densa
(basal lamina)

Reticular layer

TABLE 13–4   **Composition of the Basal Lamina**

Protein	Role
Collagenous proteins	
Collagen type IV	Structural component
Glycoproteins	
Entactin	Interacts with laminin
Laminin	Binds to proteoglycan and membrane
Proteoglycans	
Chondroitin sulfate	Proteoglycan
Heparan sulfate	Interacts with collagen and laminin

one of the highest in the hydroxyprolines. Instead of mature tropocollagen, the unit of lamina collagen is a procollagen. Electron microscopy of the lamina fibrils reveals that they are about 400 nm long, with a knob at one end and a kink at the other end. There are linkages between the ends of different fibrils, and these dimers can associate into extensive networks (Figure 13–15). Thus basal lamina collagen is quite different in its structure from interstitial collagen in connective tissues.

2. Heparan sulfate proteoglycan in the basal lamina has two forms. One has a high density, with short chains of both protein and the sugars, and acts as the major filtration component. The other has a much lower density, with especially long sugar chains, and acts as the connection with the cell, since it binds to laminin.

3. Entactin (nidogen) is a sulfated glycoprotein of molecular weight about 150,000 that occurs near the junction of the lamina and plasma membrane and binds to laminin. It can bind to both type IV collagen and to laminin and apparently promotes their interaction.

4. Laminin (not the same as the lamins associated with the nuclear envelope) is a very large (about 900,000 molecular weight) molecule, principally located in the lamina rara between the plasma membrane and the granular basal lamina (Figure 13–14). It is composed of three polypeptide chains that are held together by disulfide bridges to form a structure reminiscent of a bunch of flowers, with a cross with one longer arm.

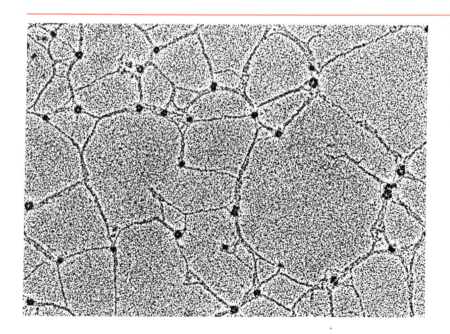

FIGURE 13–15
Type IV collagen network in the basal lamina. Note the knobs at the end of the collagen molecules that are involved in the extensive cross-linking. (Courtesy of Dr. H. Furthmayr.)

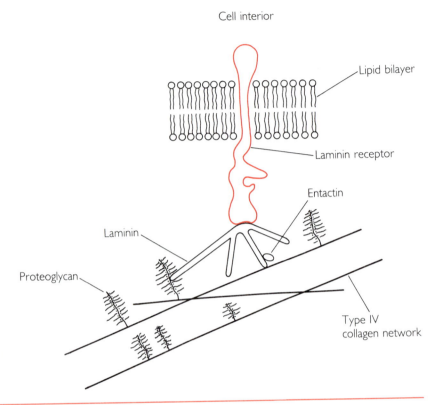

FIGURE 13–16

Molecular model of the basal
lamina (not to scale). The laminin
receptor is an integrin. Compare
to Figures 13–11 and 13–12.

The structural interactions between these components are reminiscent of the fibronectin-mediated connections between other cells and
their matrix (Figure 13–16; compare to Figure 13–11). In the case of
the lamina, the role of mediator binding a membrane protein to the
extracellular matrix is played by laminin. It has binding sites for type IV
collagen and the proteoglycans of the matrix, as well as an affinity for an
integrin-type receptor with the RGD peptide sequence. The integrin in
this case has a different polypeptide chain structure than the one that
binds the typical collagenous matrix.

Function

A major role for the basal lamina is in **structural support.** For example,
in Bowman's capsule of the kidney and in the lens capsule of the eye, the
lamina surrounds the internal structure, giving it shape and support. The
cross-linked procollagen is largely responsible for this property. In addition, disulfide bonds between collagen and the sulfated proteoglycans
add to the tensile strength of the lamina.

A second function of the basal lamina is **filtration.** This is especially evident in capillaries of the kidney glomeruli. Electron-dense high

molecular weight tracers such as ferritin can be used to show that the basal lamina is impermeable to large proteins. This is probably the mechanism by which serum proteins are excluded from the renal tubules and remain in the capillaries. Both collagen and the proteoglycans appear to be involved in this filtration property, and treatment of isolated filter pads with collagenase makes them permeable to proteins.

In the human disease Goodpasture's syndrome, patients make antibodies against their own proteins (autoantibodies), and one target is the basal lamina, especially the globular region of type IV collagen. The binding of these antibodies to the lamina leads to malfunctions in both the kidneys and the lung alveoli.

In diabetes mellitus, a defect in the amount or action of the hormone insulin leads to excessive blood glucose levels. The sugar can nonenzymatically glycosylate proteins, such as hemoglobin (glycosylated hemoglobin levels are monitored closely in diabetics). If collagen IV of the basal lamina surrounding endothelial cells of a blood vessel is heavily glycosylated, it becomes resistant to the normal processes of degradation.

The accumulated collagens tend to cross-link with each other, and the basal lamina thickens. Both leakiness to small molecules and trapping of large molecules (e.g., serum albumin) in the matrix follow, and this leads to damage of the endothelium. This microangiopathy is most pronounced in the retina and kidney, and malfunctioning of these organs is a common problem in diabetics. The drug aminoguanidine blocks nonenzymatic glycosylation of collagen, and so cross-linking does not occur. This drug may be useful in preventing blood vessel damage in diabetes.

The third role for the lamina and laminin, as for fibronectin, is in **morphogenesis.** Epithelial cells have a distinct polarity (see Figure 1–10), and this polarity is an important differentiating function of these cells. At the base of this tissue is a basal lamina, with laminin as a link between the cell and the lamina. In differentiation of embryonic mesenchymal cells into kidney epithelium, the expression of one of the chains of the laminin protein occurs at precisely the time where polarity develops. If antibody to the laminin chain is applied, differentiation does not occur. The mechanism for laminin triggering cell polarity is not clear.

A promising system for the investigation of basal lamina function is the mutant, emb-9 in the worm *Caenorhabditis elegans.* This mutant strain dies during late embryogenesis, and the wild type gene for emb-9 codes for type IV collagen, the fibrous component of the lamina. In the mutant, glycine in the gly-X-Y repeat is replaced with another amino acid, glutamic acid.

A related role of the lamina is its acting as a **barrier** to cell invasion. When a metastatic tumor cell derived from an epithelium lands on a tissue surrounded by a basal lamina, the cell cannot reach the tissue because it cannot penetrate the lamina. The reason for this is that the invading cell's integrin is not the one that binds laminin (for instance, it probably has the receptor for fibronectin). When the invasion is successful, however, the metastatic cell can be shown to have replaced its normal receptor with the integrin that does bind laminin. This allows the tumor cell to penetrate the host tissue extracellular matrix.

Plant Cell Wall

Structural Chemistry

Plant cells are surrounded by a distinct extracellular matrix, the cell wall. The wall has a morphogenetic function, as it forms a rigid skeleton that determines the shape of the cell, and it has a physiological function in that it acts as a restraint to cell expansion, the primary mode of growth of the plant.

Two types of cell wall are recognized, and both may occur in the same cell:

1. The primary wall is present in growing cells. It is capable of expansion along with the cell and is generally thin (1–5 μm).

2. The secondary wall is formed after cell growth ceases. It is often quite thick (over 10 μm) and is inextensible. Usually, cell walls of adjacent cells are separated by a middle lamella (Figure 13–17).

Isolation of the plant extracellular matrix can be done following homogenization of the tissue. Usually, the toughness of the wall makes

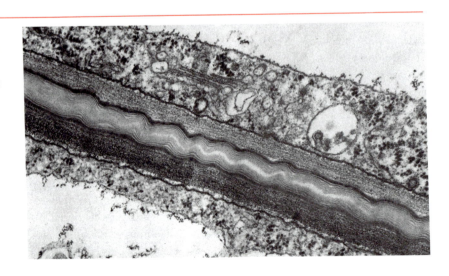

FIGURE 13–17

The cell walls separating adjacent leaf cells of the shrub *Atriplex*. The wavy structure is the middle lamella. (×63,000. Photograph courtesy of Drs. K. Platt-Aloia and W. Thomson.)

homogenization by standard methods for animal cells (e.g., sonication, ground-glass tubes) impossible, and the more traditional mortar-and-pestle technique must be used. The large size and insolubility of wall components makes it easy to pellet them in a centrifuge. But there is a problem with nonspecific adsorption of cytoplasmic components, so that repeated washing is needed to remove these contaminants.

Once a pure cell wall fraction is analyzed, the great contrast between it and the animal cell extracellular matrix is apparent: Instead of the many proteins in the animal tissue, the plant matrix is mostly carbohydrate polymers (Table 13–5 and Figure 13–18). These are more

TABLE 13–5 Carbohydrate Polymers of Plant Cell Walls

Polymer Category	Structure
Cellulose	β1-4, D-glucan
Hemicelluloses	β-glucans, glucomannans, xylans, xyloglucans, arabinoxylans
Pectins	Arabinans, galactans, galacturonans
Other polysaccharides	Arabinogalactans, β1-3-glucans, glucoronomannans
Glycoproteins	Hydroxyproline-rich

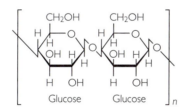

Cellulose

D-Galactose

D-Mannose

D-Xylose

D-Galacturonic acid

L-Arabinose

FIGURE 13–18

Some important carbohydrates of the plant cell wall.

difficult to work with because of their complexity, insolubility, and lack of approaches (for instance, the many methods of molecular biology are not appropriate for most of the plant cell wall). Nevertheless, considerable progress has been made.

The primary wall is composed by dry weight of 25–40% cellulose, over 50% other polysaccharides, and about 5% glycoproteins. An exception to the "rules" for building of plant cell walls occurs in green algae of the volvocine line: In this case, the entire wall is made up of glycoproteins, and there are neither cellulose nor hemicelluloses.

Cellulose is the most abundant organic molecule in the biosphere. Approximately 10^{15} kg of it are synthesized in the biosphere each year; this is over 100-fold more than its closest rival, chitin. It accounts for about one-half of the dry weight of green plants and also occurs in fungi, bacteria, and even animals (tunicates). It is the major stable reservoir of biologically fixed carbon and has been important to human civilization as a fuel and fiber (such as the paper of this book).

Cellulose is a β1-4 linked polymer of glucose, with, commonly, over 10,000 residues in each chain. The chains are extensively crosslinked, primarily by hydrogen bonds between the glucose hydroxyl groups. This bundles the chains into microfibrils, 10–30 nm in diameter and up to several micrometers in length.

X-ray diffraction shows the microfibrils to be highly ordered. In cellulose of the alga *Valonia,* the glucose chains are arranged in parallel array and in the same polarity, and this may be true of other species as well. The ordered nature of cellulose was predicted as early as 1850, when polarizing microscopy showed the cell wall to be strongly birefringent. A century later, the electron microscope revealed highly ordered arrays of microfibrils (Figure 13–19).

Like the collagenous matrix of animal cells, the plant cell wall is composed of fibrous and amorphous components. In this case, the cellulose microfibrils are in a matrix of other polysaccharides and glycoproteins:

1. The **hemicelluloses** are soluble in alkali and contain complex combinations of arabinose, xylose, galactose, glucose, glucuronic acid, and mannose. The two major hemicelluloses are xyloglucans, in which the backbone is the same as cellulose (glucose in β1-4 linkage), but there are xylose side chains attached to 60–75% of the glucose residues, and xylans, with a backbone of xylose and various side chains.

2. **Pectins** are typically polygalacturonic acid (up to 2000 residues in a chain) with other sugars on side chains. Pectins are especially abundant in the middle lamella between cells.

3. There are two types of cell wall **glycoproteins.** The first comprises wall-bound enzymes, such as invertase, cellulase, and peroxidase. In the

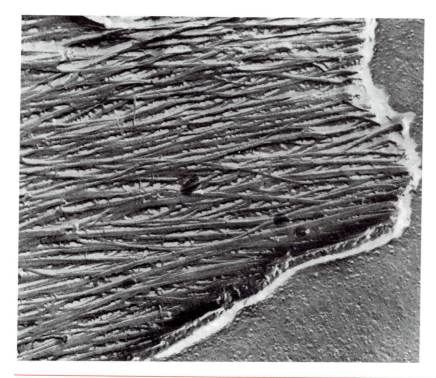

FIGURE 13–19

Cellulose cell wall microfibrils of the alga *Closterium,* as revealed by freeze-fracturing. (×125,000. Photograph courtesy of Drs. T. Giddings and L. A. Staehelin.)

case of the latter enzyme, the glycosylation is via an N-glycosidic bond (N-acetylglucosamine-asparagine). The second type of glycoprotein is structural and is termed extensin. It contains hydroxyproline, an amino acid also present in the fibrous proteins of the animal cell extracellular matrix (Figure 13–2). Carbohydrate linkages are by O-glycosidic bonds (galactose-serine and arabinose-hydroxyproline).

The **molecular architecture** of the primary cell wall can be approached in three ways. One is to chemically break down the wall into smaller and smaller pieces and analyze each one. The other is to break the wall into pieces with specific hydrolases (e.g., a $\beta1,3$-glucanase) and then analyze the fragments. A third way is to isolate cell wall polymers just after they are secreted into the wall but before they are cross-linked; these pieces of wall can then be compared to the intact wall to determine where they get attached.

These methods have led to the resolution of the structures of some of the wall matrix components (Figure 13–20), but how they are built into the wall is less clear. What is clear is that the wall is extensively cross-linked, with so many interactions between components that in many ways the entire wall can be considered as one huge macromolecule. There is a multitude of demonstrated linkages between the various wall polymers, with xyloglucans and arabinoxylans being the primary

FIGURE 13–20

Structures of two parts of hemicellulose fraction of the primary
cell wall of a sycamore cultured
cell.

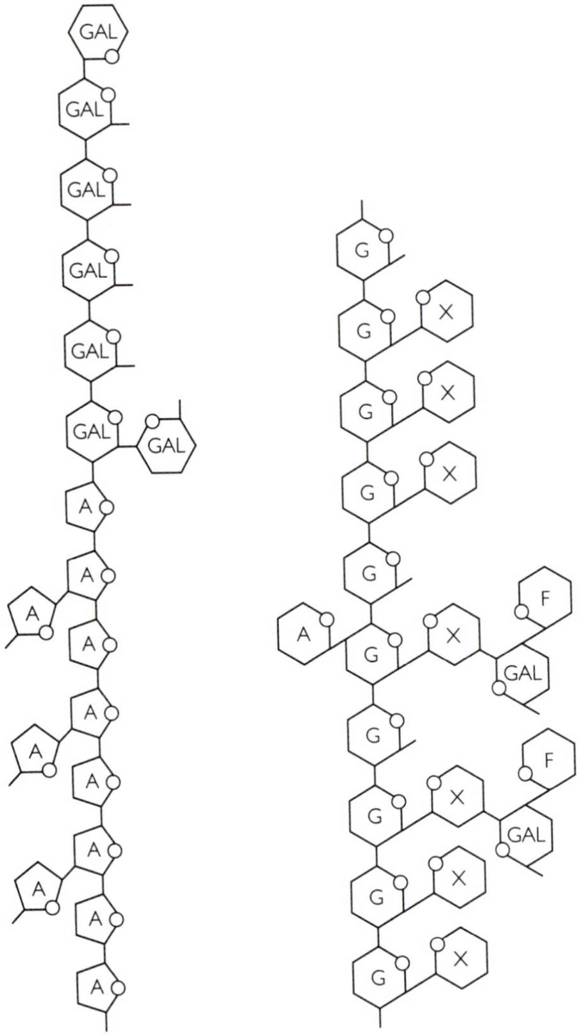

(a) Arabinogalactan
A = arabinose
Gal = galactose

(b) Xyloglucan
X = xylose
G = glucose
F = fucose
Gal = galactose

linkages to the cellulose microfibrils (Figure 13–21). Xyloglucan is attached to cellulose via hydrogen bonds, probably in an orientation
such that a single xyloglucan chain can cross-link several cellulose
microfibrils.

Several of the polymers can cross-link to themselves. For instance,
pectin can be cross-esterified or Ca^{++} bridges can form, both of which
increase the viscosity of the pectin. Indeed, chelation of this cation is
necessary to extract most of the pectin from the cell wall. The glyco-

protein extensin contains tyrosine residues that can form cross-links with adjacent extensins, and these bridged extensins could form looped structures that envelop cellulose microfibrils or pectin. Immune electron microscopy, using anti-extensin antibodies to stain the wall, shows that extensin does appear at the same places in the wall as cellulose.

But even this picture is overly simplified. Immune electron microscopy shows that the various wall polymers are not uniformly distributed in the plane of the wall. For instance, pectic substances are unusually concentrated near the air spaces in the corners of cell walls. This may be a defense mechanism, since the air spaces provide an entry point for fungal pathogens, and pectic substances form elicitors to block the fungi (see below).

In addition to the complexity of the information on wall architecture, it is not known how the wall (an extracellular matrix) is attached to the cell it surrounds. There is no clear candidate, such as the animal cells' RGD-integrin family of proteins.

Secondary cell wall architecture is not well worked out. Generally, the ratio of cellulose to matrix is higher than in the primary wall. For example, wood is about 60% cellulose, and its content rises to 98% in cotton hairs. The cellulose microfibrils are usually laid down in distinctly oriented layers (Figure 13–22). The middle (S_2) layer is usually the thickest and, because of its random ordering of microfibrils, the least birefringent in polarizing microscopy.

The reduced matrix of the secondary wall contains up to half of its dry weight as lignin. This is a complex phenolic macromolecule with a

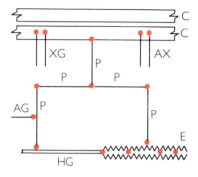

FIGURE 13–21

Speculative view of the primary cell wall of plants based on known cross-links (•) between molecules. (C = cellulose; XG = xyloglucan; AX = arabinoxylan; P = pectin; AG = arabinogalactan; HG = homogalacturonan; E = extensin [protein].)

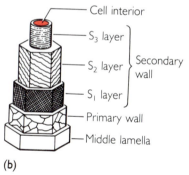

(b)

FIGURE 13–22

(a) Electron micrograph showing the cellulose microfibril orientation in the plant secondary cell wall. (×23,000. Photograph courtesy Dr. T. Itoh.) **(b)** Diagrammatic representation of the orientation of cellulose microfibrils in the cell wall layers of a xylem tracheid.

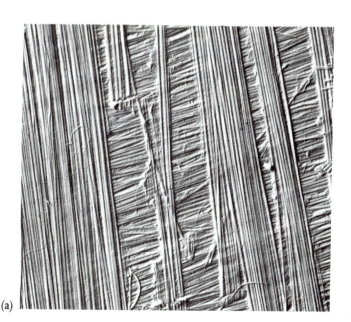

(a)

molecular weight greater than 10,000 and a monomeric unit of oxyphe-
nylpropane. Lignification of the matrix confers considerable tensile
strength to secondary cell walls with a relatively low cellulose content,
such as wood.

Biogenesis

Cellulose is synthesized via a nucleotide-sugar intermediate. Specifically, uridine diphosphate glucose (UDP-glucose), an "activated" form
of the hexose, is used to link two hexoses to form cellobiose, and this
disaccharide is then added to the growing polysaccharide. A lipid intermediate is involved in accepting the glucose and cellobiose units and
then donating them to the oligosaccharide.

These reactions appear to occur at the plasma membrane, after
which the β-glucan polymers assemble into crystalline microfibrils. This
assembly can be inhibited by agents such as Congo Red, which hydrogen bond to the polymers and prevent interchain hydrogen bonding
from forming. Cells treated with such inhibitors do not form organized
microfibrils.

Freeze-fracturing of plasma membranes of plant cells active in cellulose synthesis reveals ordered arrays of particles on the surface of the
membrane (Figure 13–23). The diameter of the rosettelike particle aggregate appears to correlate with the diameter of the cellulose microfibril. This is indicated from studies on the alga *Micrasterias,* where
there are broad microfibrils at the center of the cell and narrow ones at
the periphery, and the rosettes at the plasma membrane at the center of
the cell are wider than those associated with the edges. This strongly
suggests that the membrane assembly is the cellulose-synthesizing enzyme complex.

Microtubules are often present in the cytoplasm just inside cells
synthesizing cell wall materials, and there is often a parallelism between
the orientation of microtubules and cellulose microfibrils deposited in
the wall. This is especially noticeable during cell division, where a ring
of microtubules at prophase marks the region where the cell plate will
grow. In plant development, whenever the orientation of the plane of
cytokinesis changes, so does the orientation of the band of microtubules.
Experiments have been reported in which incubation of cells in colchicine disrupted the precise orientation of the newly deposited microfibrils
without affecting their rate of synthesis (see Figure 9–10). This indicates
that cytoplasmic microtubules are involved in cellulose wall microfibril
orientation.

The mechanism of this involvement is not known. One model,
which combines the synthesizing membrane particles and orienting cortical microtubules, is shown in Figure 13–23. The microtubules are en-

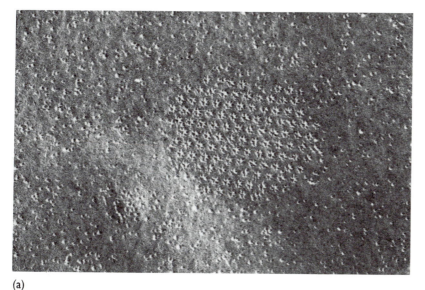

(a)

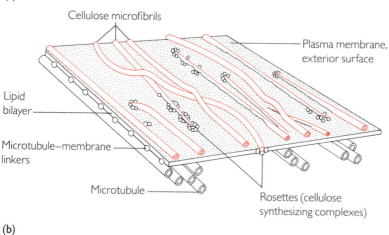

(b)

Cellulose microfibrils

Plasma membrane,
exterior surface

Lipid
bilayer

Microtubule–membrane
linkers

Microtubule

Rosettes (cellulose
synthesizing complexes)

FIGURE 13–23
(a) Surface of the plasma membrane of the plant *Micrasterias,* showing putative cellulose-synthesizing complexes ($\times$ 33,000). (b) Cross-sectional model of the cellulose-synthesizing complex. The rosettes are cellulose-synthesizing complexes at the plasma membrane. Microtubules act as a guide for the orientation of the rosettes as they form the microfibrils. (Courtesy of Drs. L. A. Staehelin and T. Giddings.)

visaged as "lining up" the cellulose-synthesizing complexes in the plane of the membrane, with the rosettes traveling between microtubular tracks, possibly via molecular motors.

The cell wall **matrix polysaccharides** are synthesized in the Golgi complex and secreted to the wall by vesicles (see Chapter 7). Glycosyltransferases for the biosynthesis of hemicelluloses and pectins are localized in the Golgi fraction of cell homogenates. This fraction also contains newly synthesized matrix polysaccharide polymers with compositions identical to those in the cell wall. Presumably, the linkages between polysaccharides are formed in the wall. The Golgi fractions involved here appear to be enriched in cis rather than trans Golgi, an indication that the polysaccharides may actually be transported to the wall without passing through the trans compartment.

The assembly of the hydroxyproline-rich cell wall glycoprotein follows a pathway similar to that taken by collagen in animal cells. Certain proline residues are hydroxylated after translation, and then the protein appears to be sequestered, possibly in the ER. It is then transported to the Golgi complex, where O-glycosylation occurs, and then is secreted to the wall, presumably via vesicles. Unlike the case with collagen, this plant glycoprotein is secreted at a normal rate if proline hydroxylation is inhibited. The sugars linked to other cell wall glycoproteins by N-glycosidic bonds appear to form via lipid (dolichol phosphate) intermediates.

Function

The cell wall is a **physical barrier,** which protects plant cells from being invaded by pathogenic viruses, fungal spores, or bacteria. In general, these organisms invade host plants by secreting enzymes that degrade the cell wall. The wall is generally permeable to large and small molecules (up to 60,000 molecular weight). In these instances, selectivity occurs at the plasma membrane.

The tensile strength of the cell wall in providing a **structural skeleton** for the plant is quite impressive. In growing tissues, both cell wall structure and turgor pressure (a hydrostatic skeleton—see Chapter 2) provide support for the organs. The importance of turgor is evident from the loss of structural stability when it is absent (wilting). However, in nongrowing tissues, especially wood, the cells providing support (xylem) are dead, and the only support for a tree is its cell walls.

Higher plants grow largely by cell expansion, and cell division occurs only in localized areas termed meristems, which largely occur at the growing root and shoot tips (Table 12–1). Following cell division, a plant cell typically elongates to up to 50 times its previous length, but because of its strength, the cell wall presents a **barrier to cell expansion.** Both osmotic pressure and growth of the cell contents exert a force analogous to that of an expanding balloon in a cardboard box, the primary cell wall. Clearly, the "box" (wall) must either add new material and grow along with the cell or expand, breaking bonds in its internal structure. Both events occur as plant cells grow.

Both kinetic and morphological observations show that there must be breakage of covalent bonds that link up wall polymers to allow for growth. For example, if responsive tissues (e.g., subapical stem) are incubated in the expansion-inducing hormone indoleacetic acid (auxin), growth occurs immediately with virtually no lag time. This would make addition of cell wall material for the early response unlikely.

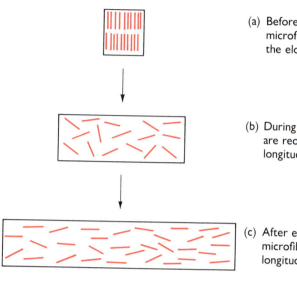

(a) Before elongation: Cellulose microfibrils are transverse to the elongation axis of the cell.

(b) During elongation: Microfibrils are reoriented toward a more longitudinal orientation.

(c) After elongation: New microfibrils are in a longitudinal orientation.

FIGURE 13–24

Diagram of the multinet theory of cell wall microfibril changes during cell elongation.

Microscopically, the orientation of cell wall cellulose microfibrils changes as the cell expands. Initially, microfibrils are deposited in a transverse orientation, perpendicular to the plane of cell elongation. During cell expansion, they are passively shifted toward the outside of the wall as new material is laid down. Moreover, their orientation gradually changes to a more longitudinal direction (Figure 13–24). P. Roelofsen has proposed in the "multinet" theory that this change in microfibrillar orientation is a passive one, driven by the force of cell expansion and due to changes in the forces within the wall that hold the microfibrils in a transverse array.

The identity of the bonds that break to allow cell expansion to begin is not known. From the architecture of the primary wall (Figure 13–21), it is clear that the whole wall is extensively cross-linked. An event that appears to correlate with the initiation of expansion is the acidification of the wall. This occurs immediately in response to the addition of auxin and requires the presence of an intact cell. The lowered wall pH could either activate a wall-bound glucosidase or act directly on an acid-labile component in the matrix.

A good candidate for a molecule that is altered during elongation is xyloglucan, which is attached to cellulose microfibrils. Xyloglucan is hydrolyzed by a cellulase that has been demonstrated in the wall and that is activated during auxin treatment. (Note that the backbone of xyloglucan is a β1-4 linked glucan, as in cellulose.) Apparently, the fibrillar structure of cellulose in the wall makes it relatively resistant to the cellulase, whereas xyloglucan is fully accessible . The activation of this en-

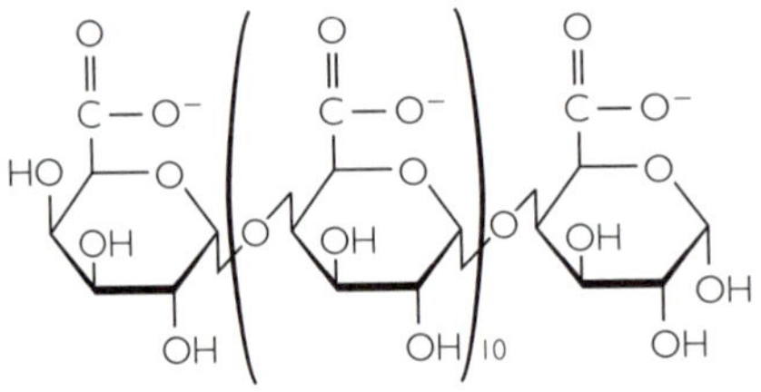

FIGURE 13–25

An elicitor from pectin, part of the cell wall amorphous component. This molecule can be released by fungal proteases or hydrolysis at a wound. It can diffuse throughout the plant to cause the synthesis of antimicrobials.

zyme could be direct (e.g., via decreased pH) or indirect (e.g., by release of soluble pectin products, resulting in increased activating polyanions).

Cell wall synthesis occurs during elongation, with the result that the wall retains a uniform thickness. Both matrix and cellulose materials are added throughout the length of the wall. As elongation ceases, the qualitative pattern of synthesized wall components changes. First, there is an increase in the synthesis and secretion of the hydroxyproline-rich glycoprotein component. The precise role for this is not clear, but a hypothesis of it being used for wall stiffening is not inconsistent with the data. Second, secondary wall is formed, and its much higher cellulose and, usually, lignin contents render the wall irreversibly inextensible.

The cell wall has a biochemical (as well as physical) role in regulating cell expansion. Xyloglucan fragments released by cellulase digestion can inhibit auxin-induced cell elongation and may act as a governor on the process, to prevent an excessive response to the hormone.

Cell wall components can serve as part of the **defense mechanism** of a plant against fungal pathogens. When a plant tissue is physically wounded, it is at risk for infection by fungal pathogens. One way that the plant prevents this is to synthesize antimicrobial agents called phytoalexins, such as enzymes that degrade the cell walls of fungi or proteinase inhibitors that kill insects. These molecules are made by tissues both at the wound and far from it, so there must be some way to "communicate" with the distant tissue. Partial acid hydrolysis of plant cell walls releases carbohydrate fragments (Figure 13–25) that have the ability to substitute for the wounded tissue in stimulating the synthesis of phytoalexins. These fragments are termed elicitors and are under intensive investigation as a natural way to make plants resistant to pests.

Finally, some cell wall polysaccharides can act as **reserve carbohydrate** for the plant. Seeds of such plants as nasturtium and tamarind have cell walls rich in xyloglucan, and it can account for 25% of the dry weight of the seed. When the seed germinates, enzymes are activated that digest the xyloglucan to monosaccharides, which can then be used by the embryonic plant.

Further Reading

Adair, W. S., and Mecham, R. P. (Eds.) *Organization and assembly of plant and animal extracellular matrix.* New York: Academic Press, 1990.

Albeda, S. M., and Buck, C. A. Integrins and other cell adhesion molecules. *FASEB J* 4 (1990):2868–2880.

Beck, K., Hunter, I., and Engel, J. Structure and function of laminin: Anatomy of a multidomain glycoprotein. *FASEB J* 4 (1990):148–160.

Brett, E. T., and Hillman, J. R. (Eds.). *Biochemistry of plant cell walls.* New York: Cambridge University Press, 1985.

Burridge, K., Fath, K., Kelly, T., Nuckolls, G., and Turner, C. Focal adhesions: Transmembrane junctions between the extracellular matrix and the cytoskeleton. *Ann Rev Cell Biol* 4 (1988):487–525.

Byers, P. Inherited disorders of collagen gene structure and expression. *Am J Med Genet* 34 (1990):72–80.

Cassab, G., and Varner, J. Cell wall proteins. *Ann Rev Plant Physiol* 39 (1988):321–353.

Cassiman, J.-J. The involvement of the cell matrix receptors or VLA integrins in the morphogenetic behavior of normal and malignant cells. *Cancer Genet Cytogenet* 41 (1989):19–32.

Cosgrove, D. Wall relaxation and the driving forces for cell expansive growth. *Plant Physiol* 84 (1987):561–564.

Ekblom, P., Vestweber, D., and Kemler, R. Cell-matrix interactions and cell adhesion during development. *Ann Rev Cell Biol* 2 (1986):27–47.

Emons, A., Derksen, J., and Sassen, M. Do microtubules orient cellulose microfibrils?. *Physiol Plant* 84 (1992):486–493.

Esko, J. Genetic analysis of proteoglycan structure, function and metabolism. *Curr Op Cell Biol* 3 (1991):805–816.

Foster, J., and Curtiss, S. The regulation of lung elastin synthesis. *Am J Physiol* 259 (1990):L13–L23.

Fry, S. *The growing plant cell wall.* Harlow, Essex: Longman, 1988.

Fry, S. C. Cellulases, hemicelluloses and auxin-stimulated growth: A possible relationship. *Physiol Plant* 75 (1989):532–536.

Furthmayr, H. Basement membranes. In Clark, R., and Henson, P. (Eds.), *The cellular and molecular basis of wound healing.* New York: Plenum Press, 1988, pp. 525–558.

Gallagher, J. The extended family of proteoglycans: Social residents of the pericellular zone. *Curr Op Cell Biol* 1 (1989):1201–1218.

Garg, H., and Lyon, N. Structure of collagen-fibril associated small proteoglycans of mammalian origin. *Adv Carboh Chem Biochem* 49 (1991):239–261.

Giddings, T., and Staehelin, L. A. Spatial relationship between microtubules and plasma membrane rosettes during the deposition of primary wall microfibrils in *Closterium. Planta* 173 (1988):22–30.

Gleeson, P. Complex carbohydrates of plants and animals: A comparison. *Curr Top Microbiol Immunol* 139 (1988):1–30.

Goetinck, P. Proteoglycans in development. *Curr Top Dev Biol* 25 (1991):111–130.

Gould, R., Polokoff, M., Freidmann, P., Huang, T., Holt, J., Cook, J., and Niewiarkowski, S. Disintegrins: A family of integrin inhibitory proteins from viper venoms. *Proc Soc Exp Biol Med* 195 (1990):168–171.

Guengerich, F. P. Reactions and significance of cytochrome P450 enzymes. *J Biol Chem* 266 (1991):10019–10022.

Hay, E. D. Cell Biology of the Extracellular Matrix. ed. 2. New York: Plenum, 1991.

Hay, E. Extracellular matrix, cell skeletons, and embryonic development. *Am J Med Genet* 34 (1989):14–29.

Heickendorff, L. The basement membrane of arterial smooth muscle cells. *Acta Pathol Micro Immun Scand Suppl* 9 (1989):5–25.

Hoson, T. Structure and function of plant cell walls: immunological approaches. *Int Rev Cytol* 131 (1991):233–266.

Jarvis, M. C. Self assembly of plant cell walls. *Pl Cell Environ* 15 (1992):1–6.

Jennings, O. E., and Barnett, A. H. New approaches to the pathogenesis and treatment of diabetic microangiopathy. *Diab Med* 5 (1988):111–117.

Knox, P. The cell surface in health and disease. *Molec Aspects Med* 7 (1984):177–311.

Kornblihtt, A., and Gutman, A. Molecular biology of the extracellular matrix proteins. *Biol Rev* 63 (1988):465–507.

Kuivaniemi, H., Tromp, G., and Prockop, D. J. Mutations in collagen genes: Causes of rare and some common diseases in humans. *FASEB J* 5 (1991):2052–2060.

Laurent, G. Dynamic states of collagen: Pathways of collagen degradation in vivo and their possible role in regulation of collagen mass. *Biochim Biophys Acta* 252 (1987):C1–9.

Lee, B., Godfrey, M., Vitale, E., Hori, H., Mattei, M.-G., Sarfarazi, M., Tsipouras, P., Ramirez, F., and Hollister, D. Linkage of Marfan syndrome and a phenotypically related disorder to two different fibrillin genes. *Nature* 352 (1991):330–334.

Lewis, N. G. (Ed.). *Biosynthesis and biodegradation of plant cell wall polymers.* New York: American Chemical Society, 1989.

Liesi, P. Extracellular matrix and neuronal movement. *Experientia* 46 (1990):900–907.

Linskens, H., and Jackson, J. F. *Modern methods of plant analysis: Plant fibers,* Vol. 10. Berlin: Springer-Verlag, 1989.

Lloyd, C. Toward a dynamic model for the influence of microtubules on wall patterns in plants. *Int Rev Cytol* 86 (1984):1–52.

Marcus, A., Greenberg, J., and Averyhart-Fullard, V. Repetitive proline-rich proteins in the extracellular matrix of the plant cell. *Physiol Plant* 81 (1991):273–279.

Mayne, R., and Burgeson, R. (Eds.). *Structure and function of collagen types.* New York: Academic Press, 1987.

McCann, M., Wells, B., and Roberts, K. Direct visualization of cross-links in the primary plant cell wall. *J Cell Sci* 96 (1990):323–334.

McDonald, J., and Melcham, R. Receptors for the Extracellular Matrix. San Diego: Academic Press, 1992.

McNeil, M., Darvill, A. G., Fry, S., and Albersheim, P. Structure and function of the primary cell walls of plants. *Ann Rev Biochem* 53 (1984):625–664.

Mecham, R. Laminin receptors. *Ann Rev Cell Biol* 7 (1991):71–92.

Mecham, R. Receptors for laminin on mammalian cells. *FASEB J* 5 (1991):2538–2546.

Mercurio, A. M. Laminin: Multiple forms, multiple receptors. *Curr Op Cell Biol* 2 (1990):845–849.

Mosher, D. (Ed.). Fibronectin. New York: Academic Press, 1988.

Poole, A. R. Proteoglycans in health and disease: Structures and functions. *Biochem J* 236 (1986):1–14.

Preston, R. *The physical biology of plant cell walls.* London: Chapman and Hall, 1974.

Price, R., and Hudson, B. *Renal basement membranes in health and disease.* New York: Academic Press, 1987.

Prockop, D., Constantinou, C., Dombrowski, K., Hojima, Y., Kadler, K., Kuivaniemi, H., Tromp, G., and Vogel, B. Type I procollagen: The gene-protein system that harbors most of the mutations causing osteogenesis imperfecta. *Am J Med Genet* 34 (1989):60–67.

Ricard-Blum, S., and Ville, G. Collagen cross-linking. *Int J Biochem* 21 (1989):1185–1189.

Roberts, K. The plant extracellular matrix. *Curr Op Cell Biol* 1 (1989):1020–1027.

Roberts, K., Johnson, A., Lloyd, C., and Woolhouse, H. The cell surface in plant growth and development. *J Cell Sci* [Suppl] 2 (1985).

Ruoslahti, E. Fibronectin and its receptors. *Ann Rev Biochem* 57 (1988):375–413.

Ruoslahti, E. Integrins and tumor cell dissemination. *Cancer Cells* 1 (1989):119–126.

Ruoslahti, E. Proteoglycans in cell regulation. *J Biol Chem* 264 (1989):13369–13372.

Savige, J. Hereditary abnormalities of renal basement membranes. *Pathol* 23 (1991):350–355.

Scott, J. Proteoglycan-fibrillar collagen interactions. *Biochem J* 252 (1988):313–323.

Staehelin, L. A., Giddings, T. H., and Moore, P. J. Structural organization and dynamics of the secretory pathway in plant cells. *Curr Top Plant Biochem Physiol* 7 (1988):45–61.

Sykes, B. Inherited collagen disorders. *Mol Biol Med* 6 (1989):19–26.

Tryggvason, K., Hoyhtya, M., and Salo, T. Proteolytic degradation of extracellular matrix in tumor invasion. *Biochim Biophys Acta* 907 (1987):191–217.

Turner, C., and Burridge, K. Transmembrane molecular assemblies in cell-extracellular matrix interactions. *Curr Op Cell Biol* 3 (1991):849–853.

Van der Rest, M., Foucher, E., Dublet, B., Eichenberger, D., Font, B., and Goldschmidt, D. Structure and function of fibril-associated collagens. *Biochem Soc Trans* 19 (1992):820–825.

Van der Rest, M., and Garrone, R. Collagen family of proteins. *FASEB J* 5 (1991):2814–2823.

Varner, J. E., and Lin, L.-S. Plant cell wall architecture. *Cell* 56 (1989):231–238.

Wilson, L. G., and Fry, J. C. Extensin—A major cell wall glycoprotein. *Plant Cell Environ* 9 (1986):239–260.

Yurchenco, P., and Schittny, J. Molecular architecture of basement membranes. *FASEB J* 4 (1990):1577–1590.

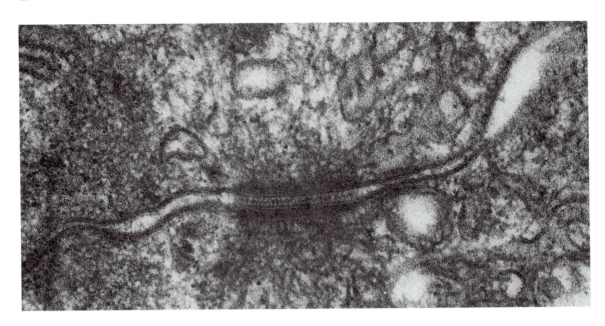

Cell Interactions

Cells in Association

When cells form a tissue, they are seldom separated by an extensive extracellular matrix. Instead, the membranes of adjacent cells come quite close to one another, and components of the membranes allow the cells to interact in specific ways.

A striking example of cell interactions happens during development, when a group of cells (e.g., part of the ectoderm) loses the ability to stick to the rest of the tissue and detaches, forming a new, separate tissue (e.g., neural tube). This implies that there must be some adhesive mechanism that keeps the cells in a tissue together. The search for the mechanism has revealed not one but several, ranging from membrane proteins that allow cells to recognize others to complex membrane structures that act as "glue" to keep cells together.

Somewhat analogous to development is the spread of cancer cells from their original site in the body to new organs. Once again, these cells start out as part of a tissue but lose their tissue-specific adhesion and travel by the bloodstream to different sites, a process called metastasis. And once again, the recognition mechanisms are on the plasma membrane.

Surprisingly, the search for cell-to-cell recognition mechanisms has uncovered a new property of adjacent cells: They can communicate with each other, sending ions and small molecules back and forth through protein-lined channels in the apposed plasma membranes. Even plants can do this, through more elaborate membrane-lined pores that go through their cell walls. This means that the analogy of a tissue to bricks held together by mortar is not quite the entire story. The individual cells in some respects are not alone but are parts of an interconnected system.

Junctions Between Cells

Animal Cells: Desmosomes and Tight Junctions

Desmosomes ("desmos": bond) are anchoring junctions that occur in many types of epithelia. Most commonly, sheets of cells are held together by dozens of desmosomes on their surfaces, much like rivets holding adjacent sheets of metal. The **ultrastructure** of the desmosome is quite complex (Figure 14–1):

FIGURE 14–1

Desmosomes between animal cells. **(a)** Electron micrograph of desmosome between two intestinal epithelial cells. (×50,000. Courtesy of Dr. J.-P. Revel.) **(b)** Diagram of a cross section of a desmosome.

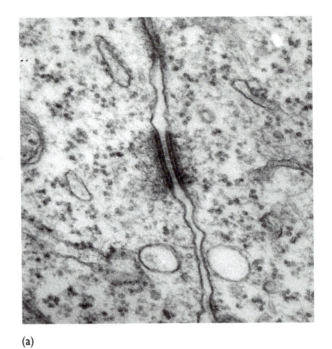

(a)

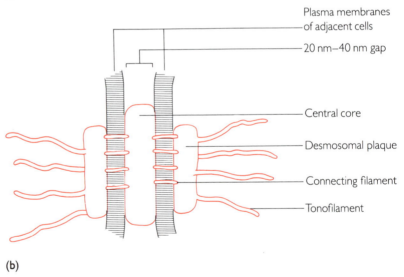

- Plasma membranes of adjacent cells
- 20 nm–40 nm gap
- Central core
- Desmosomal plaque
- Connecting filament
- Tonofilament

(b)

On the cytoplasmic side of each of the two plasma membranes is a desmosomal plaque, 15–20 nm in diameter and about 300 nm in length. Attached to the plaque are intermediate filaments that radiate into the cytoplasm, sometimes attached to organelles near the cell periphery. This means that the junction is intimately connected to the cytoskeleton. There are also occasional filaments that appear to traverse the plasma membranes and connect the cytoplasmic plaques with the intercellular core, the 20- to 40-nm space between the two cells.

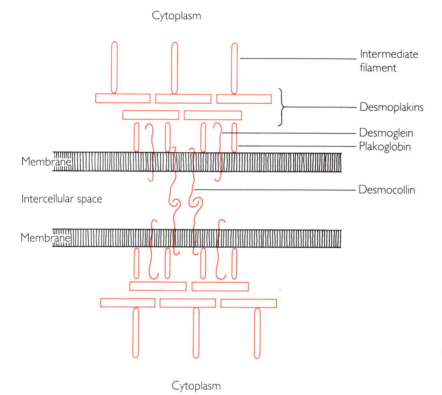

Cytoplasm

Intermediate
filament

Desmoplakins

Desmoglein
Plakoglobin

Membrane

Desmocollin

Intercellular space

Membrane

Cytoplasm

FIGURE 14–2

The molecular structure of a
desmosome. (After D. Garrod.)

The composition of the desmosome has been analyzed both cytochemically and biochemically. The filaments stain with antibodies to keratin and therefore are probably composed of intermediate filaments of this type (Table 9–4). Other intermediate filament types, including desmin and vimentin, are sometimes present. The central core stains with the dye, ruthenium red, which binds to carbohydrates, and it is digestible by the protease trypsin. Taken together, these data indicate that the core is glycoprotein in nature.

Desmosomes are insoluble in nonionic detergents that solubilize membranes, and this fact has made isolating them relatively straightforward. Analyses of the isolated junction reveal several proteins, and antibodies to them have been used combined with electron microscopy to map them at the junction (Figure 14–2):

1. Desmocollins (molecular weight 130,000) are the transmembrane glycoproteins that stick out from each membrane into the intercellular space and interact noncovalently to form the actual junction.

2. Desmoglein (molecular weight 165,000) and plakoglobin (molecular weight 83,000) are membrane proteins that extend into the cytoplasm. Desmoglein is a member of the cadherin family of cell recognition proteins.

3. Desmoplakins (molecular weight 215,000–250,000) are on the cytoplasmic side of the structure and interact with cytokeratins (intermediate filaments).

The presence of half-desmosomes in epithelial and some connective tissues indicates that two cells each contribute identical halves to form a desmosome. Hemidesmosomes are present at regions where the cell interacts with the basal lamina (see Chapter 13) or, in tissue culture, with the culture dish. Their role in anchoring cells is shown by the fact that these cells can be detached from their substratum only if the hemidesmosomes are disrupted by brief treatment with trypsin.

In certain epithelial tissues, such as those lining the mammalian intestine, and in cardiac muscle, desmosomes are modified to form an **adhering band.** Positioned at the same level in each cell of the sheet, the band holds it together more tightly than would individual "spot"

FIGURE 14–3   Tight junctions between animal cells. **(a)** Electron micrograph of tight junction between liver hepatocytes. **(b)** Diagram of a cross section of a tight junction. (×40,000. Courtesy of Dr. J.-P. Revel.)

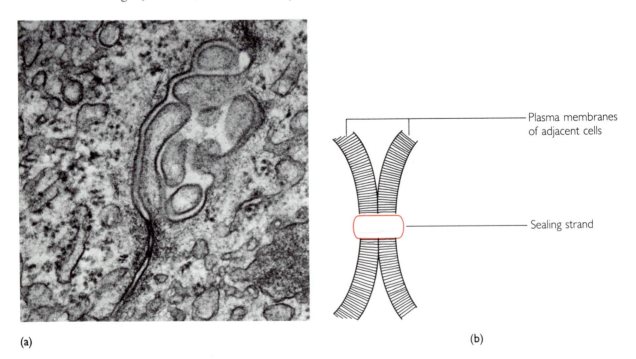

(a)   (b)

desmosomes. The gap between the cells is narrower (20 nm) in the band than in the individual desmosome. Also, the filaments are not keratin but stain with heavy meromyosin "arrowheads" and therefore are actin microfilaments. This contractile apparatus at the junctions of cells in a sheet may coordinate the movements characteristic of cardiac muscle and intestinal epithelium.

Tight junctions, as their name implies, are regions of very close intercellular contact (Figure 14–3). Indeed, the contact is usually so "tight" that no intercellular space is discernible by electron microscopy. Freeze-fracturing electron microscopy (see Chapter 4) reveals rows of particles interdigitating in the two apposed membranes, a zipperlike arrangement termed a sealing strand (Figure 14–4).

The chemical nature of the junction is not well worked out, but a major protein, ZO-1, is associated with it as well as a more abundant protein, cingulin. ZO-1 appears to be nearer to the actual junction of the two, but it is not certain if it is the actual junctional protein.

Tight junctions may encircle cells, or, like spot desmosomes, they may occur at intervals on the cell surfaces. Initially, it was thought that the membranes of the two adhering cells were actually fused at a tight junction. This would imply free diffusion of lipids between the two cell membranes, but such diffusion is not observed. Instead, the two membranes are kept separate, and the proteins of the two membranes seal them together.

FIGURE 14–4
Frozen replica of the sealing strand of a tight junction between two intestinal epithelial cells. Note the microvilli at the apices of the cells. (×20,000. Courtesy of Dr. J.-P. Revel.)

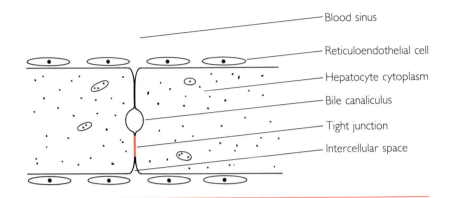

Blood sinus
Reticuloendothelial cell
Hepatocyte cytoplasm
Bile canaliculus
Tight junction
Intercellular space

FIGURE 14–5
Tight junctions between liver
hepatocytes separate the bile
canaliculus from the blood sinus.
Materials from either compartment cannot flow into the other
because of the block at the tight
junction.

The main function of tight junctions is to **prevent passage** of material between cells. This can be demonstrated if a sheet of epithelium
containing these junctions is placed with one surface on a solution of
ferritin, peroxidase, or lanthanum dyes. These molecules are too large to
enter cells directly but can diffuse into the intercellular spaces between
them. When the epithelium is examined with electron microscopy, the
markers are observed to have diffused into the intercellular spaces but
stopped at the tight junctions. This shows the barrier at the junctions to
the passage of large molecules.

The presence of tight junctions in epithelia lining the digestive and
urinary systems ensures that materials will cross these barriers by going
through, rather than around, the cells. For example, the intestinal epithelium is responsible for absorption of materials from the lumen of
intestine. Tight junctions between these cells apparently force amino
acids, carbohydrates, and so on to pass through the epithelial cells,
where control of absorption can occur. However, there is some evidence
that transport between the cells through the junction can occur on a
limited basis.

A related function of the junctions is the **separation of compartments** in a tissue. In the liver, hepatocytes are surrounded by blood
sinuses, while bile is secreted into canaliculi, which form between the
hepatocytes. Tight junctions between hepatocytes essentially prevent the
blood and bile from mixing (Figure 14–5). Other important examples of
these junctions serving as mixing barriers are the blood-brain barrier and
urinary bladder.

Tight junctions may contribute to **membrane asymmetry.** Epithelia characteristically show this asymmetry, with the apical and basal

ends differing in ionic permeability, membrane protein composition, and hormone receptor distribution. Fluorescent probes (Figure 1−6) can be used to label proteins on one surface of cultured epithelia, and photo-bleaching recovery can then be used to follow the diffusion of these molecules in the plane of the membrane (see Chapter 1). The results for many epithelial membrane proteins show that they diffuse in the plane of the membrane up to the tight junction and then stop, because the junction is an apparent barrier to their movement. Indeed, if adherent cells are physically disaggregated so that they no longer have tight junctions, their membrane proteins, which were formerly polarized, will freely intermix.

In the case of lipids, the apical region of the epithelial plasma membrane is relatively enriched in glycosphingolipids and depleted in phosphatidylcholine, a state that is maintained by a barrier to diffusion at the tight junction. Fluorescence studies of labeled lipids can show this in a way that is similar to the protein experiments.

The lipid diffusion barrier is apparently in the outer (exoplasmic) layer of the membrane bilayer, since fluorescently labeled lipids inserted into the outer layer do not diffuse into or out of the apical region, yet lipids in the inner layer can diffuse to some extent. The separation of the two leaflets is established when they are made in the Golgi complex.

Communication via Gap Junctions

Gap junctions occur between cells in virtually all multicellular animals, occupying as much as 25% of the area of the cell membrane. In cross-sectional view in electron micrographs (Figure 14−6), these junctions appear as narrow, 2-nm−wide spaces (thus the "gap") between plasma membranes of two adjacent cells.

Freeze-fracturing shows that the two lipid bilayers at the gap junction have hexagonal arrays of intramembranous particles (Figure 14−7), and X-ray diffraction analyses of these arrays in isolated junction preparations confirm the hexagonal arrangement. This has led to the model shown in Figure 14−6: "Connexons," which are hexameric, 8-nm−diameter, protein-lined aqueous channels, lie side by side in the two adjacent membranes, forming a pore 2 nm in diameter that connects the two cells.

Purification of gap junctions from liver and qualitative analysis of their proteins by gel electrophoresis reveals a major polypeptide of molecular weight 32,000 called **connexin.** Several experiments point to it as the junctional polypeptide:

1. An antibody against connexin localizes on immune electron microscopy at the gap junction.

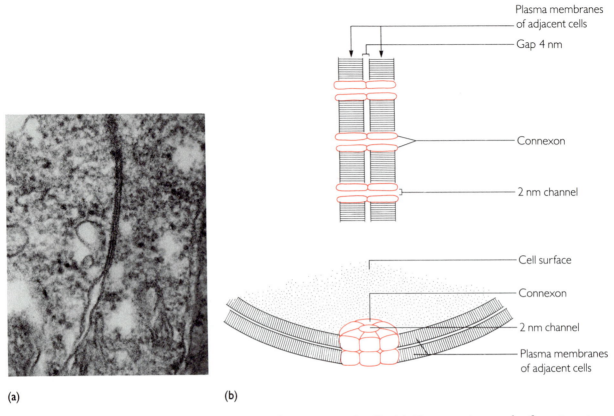

(a)   (b)

FIGURE 14–6   Gap junction between animal cells. **(a)** Electron micrograph of gap junction between liver hepatocytes. (×100,000. Courtesy of Dr. J.-P. Revel.) **(b and c)** Diagrams of gap junction structure.

2. If the junction protein is added to liposomes, they will form gap-junction–like structures.

3. If connexin's gene or mRNA is added to cells that normally do not have the junctions (e.g., oocytes), they will form the junctions.

4. Injection of an antibody to connexin into cells blocks the communicating functions of the junction.

Analyses based on the DNA sequence for this protein indicate that both amino- and carboxy-termini are on the cytoplasmic side of the membrane and that it has hydrophobic regions that allow it to cross the membrane four times (Figure 14–8). The connexins of different organisms and tissues differ from the liver protein largely in a carboxy-terminal extension: The heart protein is 43,000 daltons, and the lens protein is 70,000 daltons.

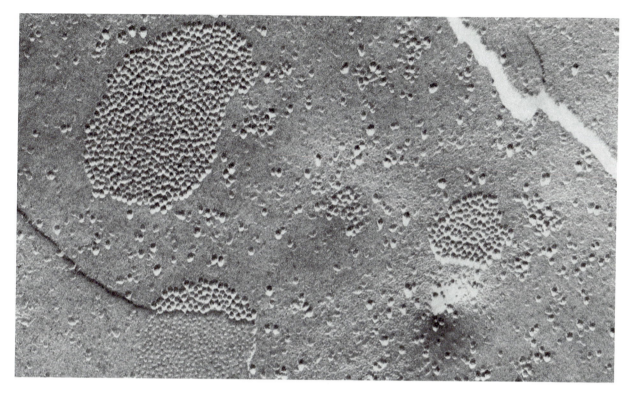

FIGURE 14–7 Freeze-fracture view of clustered particles at gap junctions. (×100,000. Courtesy of Dr. J.-P. Revel.)

The part of this protein that forms the actual channel along with its complement in the adjacent cell seems to be the transmembrane region. There is an unusual sequence here of polar amino acids surrounded by aromatic ones, and it has been conserved in all species examined. The polarity may indicate a water-lined pore.

The gap junction channel is used for the **intercellular transfer** of molecules and ions, and this transfer can be shown experimentally. If one cell of a pair connected by gap junctions is injected with a fluorescent probe, the transfer of fluorescence to the adjacent cell occurs at a much faster rate than predicted by diffusion. The size of the junctional channel can be estimated if probes of increasing molecular size are used. Size ranges of 1–2 nm have been estimated from such experiments on both invertebrates and vertebrates.

The existence of numerous 2-nm channels between cells implies the existence of considerable intercellular traffic of ions and small molecules. The former can be demonstrated by the phenomenon of **ionic coupling** (Figure 14–9).

FIGURE 14–8

Molecular model for one of the six subunits of the gap junction connexon (see Figure 14–7). The connexin protein spans the membrane four times, and the actual channel may be via the wide region, which has polar amino acids.

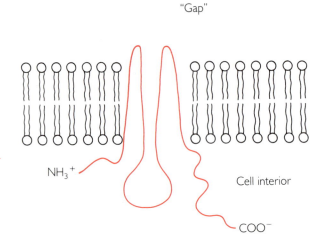

FIGURE 14–9

Demonstration of ionic coupling between two cells. Both cells have a constant negative potential. When cell one is stimulated, its potential changes (V_1). But cell 2, connected to cell 1 by gap junctions, also shows a change (V_2). The ratio of V_2/V_1 shows the degree of ionic coupling between the two cells.

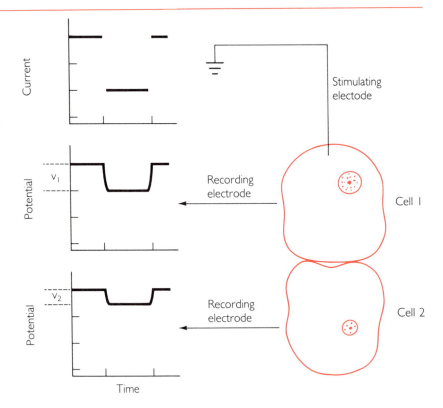

In this technique, recording microelectrodes placed in two cells connected by gap junctions typically show an internal negative resting potential. If electric current is applied to one of the cells, there is a depolarization (V_1) quite similar to that observed in stimulated neurons and muscle. But the gap-junction–connected cell also becomes transiently depolarized (V_2). The ratio of these two cells' depolarizations (V_2/V_1) is a measure of the ionic coupling between the two cells; values close to unity are indicative of a high degree of coupling. Typically, uncoupled cells have values less than 0.1.

While ionic coupling indicates that ions are passing through gap junctions from cell to cell, a different kind of evidence for this is obtained with the cardiac glycoside ouabain (Figure 2–12). Some lines of tissue culture cells are resistant to this drug, which inhibits the Na^+-K^+ pump. In normal cells, ouabain abolishes the high intracellular concentration of K^+, whereas resistant cells maintain their high K^+ levels.

If sensitive and resistant mammalian cells are cocultured, they form gap junctions. When these cells are incubated in ouabain, the level of intracellular K^+ is high in both the susceptible cell and the resistant cell. This means that the resistant cell "shares" its K^+ with the sensitive cell, probably via the gap junction.

The passage of small molecules though gap junctions can be demonstrated by the phenomenon of **metabolic cooperation.** Two lines of cultured Chinese hamster fibroblasts are "Don" and DA. The former contains the enzyme inosine pyrophosphorylase (IPP), and therefore it can incorporate the purine hypoxanthine into nucleic acids. But the DA line lacks this enzyme.

When cocultured, cells of these two lines form heterotypic gap junctions. If radioactive hypoxanthine is supplied to the cocultures, both cell types make radioactive nucleic acids, as detected autoradiographically. This "rescue" of the DA cells requires the continuous close presence, and indeed, touching, of the Don cells, and gap junctions must be formed between the two. The actual transfer involves labeled nucleotides formed from radioactive hypoxanthine in the Don cells to the DA cells via gap junctions. The small molecules involved in metabolic cooperation (Table 14–1) are well below the size exclusion limit for the gap junction of 1–2 nm.

The relationship between metabolic and ionic coupling and cell junctions can be investigated in the Chinese hamster cell lines noted above. In addition to Don and DA cells, a third line, A9, exists; like DA, it is IPP^-. However, when cocultured with Don cells, A9 cells are not metabolically corrected, nor do they form gap junctions. Heterotypic ionic coupling occurs between Don and DA cells but not between Don and A9 cells. These results, summarized in Table 14–2, show a direct

TABLE 14–1 **Some Molecules That Have Been Shown to Pass Through Gap Junctions**

Molecule	Molecular Weight
Cyclic AMP	329
Cytidine diphosphocholine	483
Glucose-6-phosphate	259
Inositol	180
Nucleotides	250–300
Phosphoribosyl pyrophosphate	386
Proline	115
Tetrahydrofolic acid	445

TABLE 14–2 **Correlation Between Metabolic Coupling, Ionic Coupling, and Gap Junctions in Fibroblasts**

Cells Touching	Cell Interactions			
	Metabolic	Ionic	Gap Jct.	Tight Jct.
Don: Don	−	+	+	+
DA: DA	−	+	+	+
A9: A9	−	−	−	+
Don: DA	+	+	+	+
Don: A9	−	−	−	+

ADAPTED FROM: Gilula, N., Reeves, O., and Steinbach, A. Metabolic coupling, ionic coupling and cell contacts. *Nature* 235 (1972):262–265.

correlation between metabolic cooperation, ionic coupling, and gap junctions.

The presence of hundreds of communicating gap junctions between its component cells indicates that a typical animal tissue is not composed of individual, semi-independent cells but rather is a "symplast" of interconnected cytoplasms. One probable role for these junctions is to equalize the distribution of metabolites in all of the cells of a tissue, especially if there is a limited blood supply.

For example, in the mammalian lens there are many gap junctions between cells (up to 25% of the surface area of the plasma membranes) but no blood vessels. Presumably, the gap junctions allow cells distant from the blood supply to maintain their metabolite levels. Indeed, experiments on metabolic cooperation show that it can occur over a long chain of cells. Gap junction transfer of molecular signals such as cyclic AMP could ensure that all cells of a tissue respond to a hormone, even if receptors for the hormone occur only on those cells most accessible to the blood.

The ability to transmit signals implies a role for gap junctions in **developmental signaling,** where concepts of cell interactions and morphogenetic fields have been proposed for over a century. The mammalian oocyte is surrounded by granulosa cells, which have long been thought to exert a nutritive and regulatory role on oocyte development. As the latter occurs, gap junctions between the two cell types allow transfer of nutrients and signals (e.g., cAMP). As is discussed in Chapter 12, high levels of cAMP are important in keeping the oocyte in meiotic arrest, and the cAMP supply to the oocyte is cut off only when gap junctions are lost on hormonal stimulation during the ovarian cycle. This results in the completion of meiosis.

While such correlations indicate that gap junctions are important in development, the use of an antibody against the gap junction protein has extended this to cause and effect. The rationale behind this approach is that if an antibody to the gap junction protein can selectively block gap junction function, and if it affects a developmental process, then that process must require intercellular communication via the junction.

For instance, the antibody can be injected into a specific cell in the eight-celled frog embryo and will block intercellular communication between that cell and its neighbors throughout development. The injected cell continues to divide normally, but the resulting tadpole shows specific deformations: If the injected cell is a determinant of the right-hand side, the tadpole has left-right asymmetry. In this way, not only can the role of gap junctions be delineated but the unraveling of the molecular signals in development can be started.

In the coelenterate *Hydra,* the head region inhibits the formation of another head in nearby tissues but not in distant tissues. Apparently, the head synthesizes a peptide of molecular weight about 500 daltons that inhibits head morphogenesis, and the gradient of this inhibitor determines whether a group of cells can differentiate to form another head. Thus, cells distant from an existing head would receive little inhibitor and would be "allowed" to form a new head.

Hydra has no circulatory system to deliver this inhibitor. Instead, it must rely on diffusion and entry into cells through the plasma membrane. With the small size of the inhibitor, it might be proposed that entry into cells is via gap junctions. S. Fraser tested this idea by using an antibody against gap junction protein (Figure 14–10):

1. If a piece of tissue just below the head of *Hydra* was removed and transplanted into the body of another *Hydra,* the piece

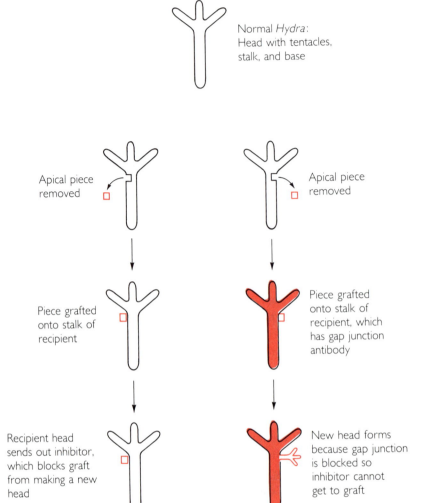

Normal *Hydra*:
Head with tentacles,
stalk, and base

Apical piece
removed

Apical piece
removed

Piece grafted
onto stalk of
recipient

Piece grafted
onto stalk of
recipient, which
has gap junction
antibody

FIGURE 14-10

Experiments with the coelenterate *Hydra*, to show the role of
gap junctions in the transmission
of a morphogenetic factor.

Recipient head
sends out inhibitor,
which blocks graft
from making a new
head

New head forms
because gap junction
is blocked so
inhibitor cannot
get to graft

would not develop into a new head, presumably because of the
inhibitor from the existing head of the recipient.

2. *Hydra* were incubated in dimethylsulfoxide to make them permeable to proteins and the anti-gap-junction protein antibody
   was added. This blocked cell communication as shown by the
   lack of transfer of an ionic dye between cells.

3. The antibody-loaded *Hydra* were given a transplanted piece of
   tissue as above. In this case, twice as many of the pieces formed
   heads. This shows that the inhibitor must have been entering
   cells through gap junctions.

A second function of gap junctions is **cellular synchronization.** Cardiac muscle cells are connected by junctions, and this allows them to beat synchronously. When individual cardiac muscle cells are placed in tissue culture, they beat asynchronously until joined by gap junctions, and this can be appropriately disrupted by antibody to the gap junction protein. A similar role is indicated for smooth muscle tissue: Uterine smooth muscle cells do not contract in nonpregnant or pregnant mammals, but they contract synchronously during the processes of labor and delivery of the newborn. This ability to contract is accompanied by the development of gap junctions between the muscle cells.

The role of gap junctions in the nervous system was first discovered in the 1950s, when they were observed as **low resistance connections** at certain synapses of the crayfish. Such direct electrical connections provide a faster bypass to the relatively (millisecond) slow synapse. For instance, the avoidance of predators by fish is mediated by specialized neurons called Mauthner cells. The sensory neurons connect to these cells via gap junctions, and these connections speed up the fish's response time by several milliseconds. This time might be brief, but it can be enough to save the life of the fish!

While normal cells can form gap junctions, many cancer cells cannot and do not show metabolic or ionic coupling. Treatment of certain normal tissue culture cells with carcinogens or tumor viruses causes them to lose intercellular communication. Occasionally, these tumor cells revert back to normal, and this reversion is accompanied by a restoration of gap junctions and communication. These observations suggest that the flow of metabolites and ions across gap junctions may be important in the maintenance of the noncancerous state.

Communication via Plasmodesmata

While the presence of communicating junctions in animal tissues is a relatively recent discovery, their existence in plants has been known for over a century. In 1879, E. Tangl described numerous thin cytoplasmic strands that connected various plant cell types. The term **plasmodesma** ("plasma": form; "desma": bond) was coined by E. Strasburger in 1901, and since then, plasmodesmata have been shown to be present in all groups of multicellular plants.

Although visible by light microscopy, plasmodesmatal structure is observed in more detail and in greater numbers under the electron microscope (Figure 14–11). There are several distinctive features (Figure 14–12):

1. The plasmodesma is actually lined with membranes of the two cells and contains a mixture of their cytoplasms. This is very different from

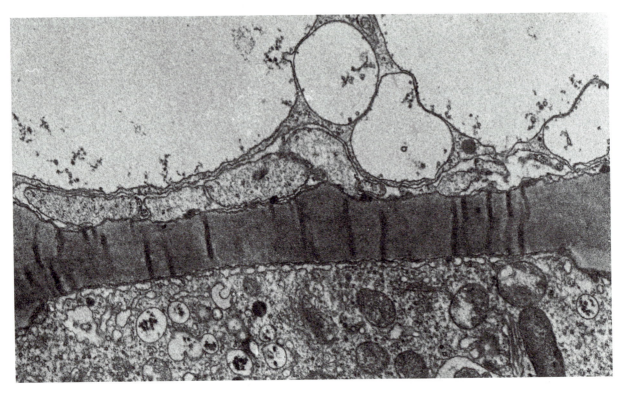

FIGURE 14–11
Plasmodesmata traversing the cell walls of adjacent cells in the Tamarix plant salt gland. (×26,500. Photograph courtesy of Drs. K. Platt-Aloia and W. Thompson.)

animal cell junctions, where the membranes come close together but do not fuse (compare Figures 14–6 and 14–12). Membrane fusion can be shown if a fluorescently tagged lipid is incorporated into the membrane of one cell. When it forms a plasmodesma with an adjacent cell, the tagged lipid diffuses rapidly in the plane of the membrane into the other cell's membrane. This means that localized lipid (and perhaps protein) asymmetry, which occurs in animal cells, is unlikely in plants.

2. In contrast to the gap junction, which has dimensions in nanometers, the plasmodesma channel can be quite long (several micrometers) and is quite wide (up to 50 nm).

3. A common, although not universal, feature of plasmodesmata is the presence of thickened endoplasmic reticulum (ER) termed desmotubules (Figure 14–12). The desmotubule-delimited structure occupies part

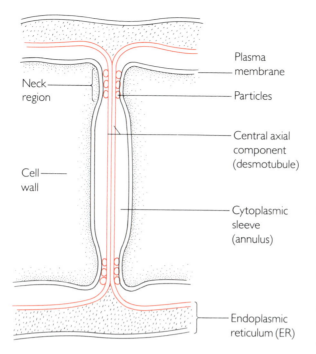

Neck region

Cell wall

Plasma membrane

Particles

Central axial component (desmotubule)

Cytoplasmic sleeve (annulus)

Endoplasmic reticulum (ER)

FIGURE 14–12

Diagrammatic section through a plasmodesma between two plant cells. Note the continuous membrane, as compared with the separation of the two cells' membranes in animal junctions (Figures 14–2, 14–3, and 14–6).

of the membrane-lined channel, the rest being cytoplasm. It is a closed channel, not providing ER continuity between the cells. In some cases, particles lie between the ER and plasma membranes, creating an annulus.

There are considerable variations in plasmodesmatal size and frequency between plants and tissues (Table 14–3). Assuming a pore diameter of about 40 nm and a plasma membrane thickness of 10 nm, the typical plasmodesma occupies a 60-nm–diameter area on the membrane. This corresponds to a maximum frequency of 280 per μm^2 membrane surface area. In the corn root meristem, therefore, the plasmodesmata occupy about 2% of the surface area of the typical cell, which makes over 5000 cytoplasmic connections with its neighbors!

Not surprisingly, these connections between plant cells have similar functions to the gap junctions that connect animal cells:

Ionic coupling between plant cells can be shown by techniques similar to those used on animal cells (Figure 14–9), and coupling ratios well above 0.5 have been obtained in both algae and higher plants. That this coupling may involve ion transport in the plasmodesmata is indicated by X-ray analytical electron microscopy, a technique that measures ion concentrations in situ. This method shows that concentrations of ions in plasmodesmata are generally higher than those in the surrounding cell wall, except in certain algae.

TABLE 14-3   Sizes and Frequencies of Plasmodesmata

Plant, Tissue	Pore Diameter (nm)	Pore Freq./µm²
Allium (onion) root meristem	40	7
Hordeum (barley) root cortex	46	1
Macrocystis (alga)	40	1
Nitella (alga) node	52	5
Tamarix salt gland	80	17
Triticum (wheat) leaf	50	8
Zea (corn) root cap	25	1
root meristem	25	5

DATA FROM: Gunning, B., and Robards, A. *Intercellular communication in plants: Studies on Plasmodesmata.* New York: Springer-Verlag, 1976, pp. 20–28.

Transfer of metabolites occurs via plasmodesmata and can be studied by microinjection of a fluorescent dye into one cell, followed by observations of its spread to other cells. In general, molecules up to 1000 molecular weight pass through the pores. There is some evidence that these molecules move through small pores formed by the particles present in some plasmodesmata (Figure 14–12). Often, these movements are at a greater rate than mere diffusion, implying some activating mechanism. Injection of a drug that raises intracellular Ca^{++} blocks plasmodesmatal transport in some cases, indicating that this cation may close the pores.

The rapid movement of molecules through plasmodesmata plays an important physiological role. Plants do not have a microvascular system of capillaries to transport hormones to distant cells, and simple diffusion is probably not adequate to do the job. Instead, plasmodesmata are used to ensure that all cells respond to the hormone at the same time.

In C_4 photosynthesis (Figure 4–12), the abundant plasmodesmata between mesophyll and bundle sheath cells are probably involved in the rapid movements of C4 acids (e.g., malate) to the bundle sheath and pyruvate to the mesophyll. A similar transport role may occur at the junctions of the nonvascular tissues and phloem, which conducts organic solutes throughout the plant. These junctions are characterized by numerous plasmodesmata.

The **spread of viruses** from one plant cell to another occurs via plasmodesmata. Although this is certainly important for the virus and from the viewpoint of plant pathology, it presents a conceptual problem, since viruses are much larger than the 1000 molecular weight size exclusion limit observed for solute passage. A class of virus-encoded "movement proteins" (MPs) has been described that may facilitate this process. Antibodies to MPs localize them at plasmodesmata, and when MPs are present, the size exclusion limit is greatly increased. They may act in a manner similar to the molecular chaperones that aid proteins in traversing membranes. That is, MPs may unfold the viral nucleic acid (usually RNA) to an extended configuration, allowing it to cross the small pores of the plasmodesmata.

Cell-Cell Adhesion

Invertebrates

The experimental analysis of selective cell adhesion dates from the work of H. Wilson. In 1907, he took a sponge and dissociated its cells from one another by passing the organism through a fine-meshed cloth. In sea water, the cells reaggregated into tiny spongelike clumps, with an organization similar to that of the "parent." When he mixed dissociated cells of two different sponge species, one colored red and the other green, at first mixed aggregates formed, but later they sorted out into largely red and largely green clumps of cells. These experiments demonstrated two phenomena commonly seen in animal tissues: preferential **cell adhesion** and topographical **sorting out** of cell aggregates.

Sponges can be dissociated into their individual cells by treatment in Ca^{++}- and Mg^{++}-free seawater, an indication that the cells are held together by ionic forces. Artificial seawaters can then be used to assay for reaggregation, and if the Ca^{++} is added back, reaggregation occurs after some time at 25°C. But this does not occur at 5° (Figure 14–13), showing that it is not only Ca^{++} but also some temperature-dependent (metabolic) process that induces reaggregation. A glycoprotein can be isolated from the original Ca^{++}- and Mg^{++}-free seawater after cell dissociation, which can stimulate aggregation even at 5°. Presumably, synthesis of this factor occurs in the 25° incubation.

The **aggregation factors** for the sponges *Microciona parthena* and *Goedia cydonium* have been purified. Both are very large (molecular weight about 22,000,000) glycoproteins that promote species-specific aggregation of sponge cells even in mixed cultures. After one to three days, these aggregates sort out their cells to make functional sponges. The aggregation factor structure is visible under the electron microscope as a "sunburst" composed of 4.5-nm–thick fibers arranged around an 80-nm–diameter circle with 15–25 radial arms about 0.1 μm long.

A small part of the arms (molecular weight about 47,000) of the *Goedia* factor has been isolated and shown to be the active component in cell binding. The target for binding on the cell surface is a glycoprotein (with over 80% sugar) termed the aggregation receptor. Binding of the

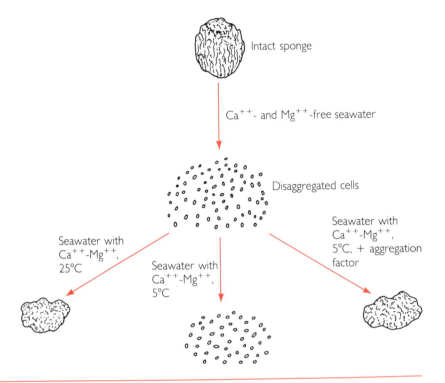

FIGURE 14–13

Dissociation and aggregation of sponge tissue.

47K fragment to the receptor is strong (affinity constant of 7×10^8 M^{-1}, and there are about 4 million binding sites per cell. The receptor has β-glucuronic acid at its terminus, and if this residue is cleaved off by a glucuronidase, binding of the factor is blocked. Because the terminal glucuronic acid can be reassembled onto the receptor via a transferase, this may provide a way for the cell to regulate adhesion to other cells.

Intercellular adhesion in the sponge is a complex process involving an aggregation factor and a receptor. An estimated 400 molecules of the factor are needed to bind for one cell to aggregate to another. Aggregation does not result in specific intercellular structures, such as desmosomes or gap junctions, but instead results in membranes separated by 10–50 nm. It would appear that the aggregation factor and receptor are enough not only to get the cells to recognize each other but also to keep them together.

Vertebrates

Vertebrate cells behave in a similar way to sponge cells in dissociation-reaggregation experiments. If vertebrate embryonic cells are dissociated mechanically or by treatment with trypsin to digest extracellular adhe-

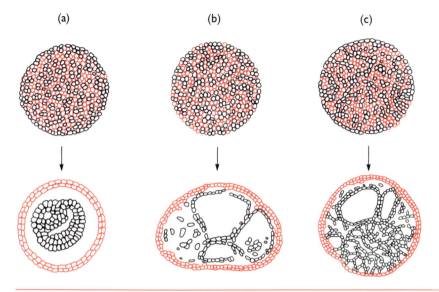

FIGURE 14-14

Cell arrangements seen with different combinations of vertebrate embryo cells following dissociation and reaggregation. **(a)** Epidermal and neural plate cells. **(b)** Epidermal and mesodermal cells. **(c)** Epidermal, mesodermal, and endodermal cells. (After Drs. W. Ham and M. Veomett.)

sive proteins, they reaggregate nonspecifically as to tissue. However, after some time, the cells sort out so that ectoderm, mesoderm, and endoderm are separated (Figure 14–14). These events even occur in mixed species experiments. For instance, if embryonic chick liver and cartilage cells are mixed with embryonic mouse liver and cartilage cells, interspecific liver and cartilage aggregates appear.

These mixed-tissue experiments can be done either on clumps of two embryonic tissues growing into each other or on disaggregated cells of the two tissues in a mixture. In both cases, the end result is the same (Figure 14–15): Tissues sort out in a specific hierarchy. From the outside of an aggregate to the inside, the order is liver, neural tube, heart, retinal epithelium, cartilage, and epidermal epithelium.

This almost invariant order indicates a hierarchy of adhesive strengths: The more distantly separated tissues are on the list, the weaker their adhesion to each other. This phenomenon is not exclusive to embryonic tissues: Many adult tissues can be disaggregated into cells that sort out in a specific arrangement. For example, liver cells will reassemble into an outer capsule, connective tissue strands, blood sinuses, and bile canaliculi—all reminiscent of the intact organ.

M. Steinberg has proposed a "differential adhesion hypothesis" to account for the sorting out of tissues in contact with each other. It proposes that each cell type has its own homotypic strength of adhesion, and the differences in adhesive strength of the cells to each other, as opposed to other cell types, lead to a maximum total adhesive strength only when tissues are arranged to maximize their individual adhesive

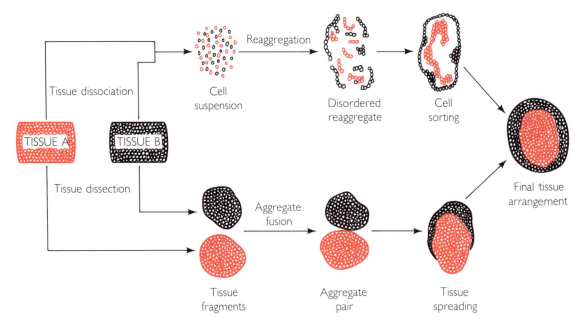

FIGURE 14–15   Animal cells sort into the same arrangement if they start out as dissociated cells *(top)* or two aggregates growing into each other *(bottom)*. This is consistent with the differential adhesion hypothesis, because the adhesive strengths of the two tissues are different, and this is what determines where they end up when mixed.

energies. Actual measurements of the adhesive energy between embryonic tissues confirm the differential hypothesis: "Inner" tissues such as cartilage form stronger intercellular bonds than "outer" tissues such as liver.

Cell Adhesion Molecules

Several methods have been used to identify the molecules involved in embryonic cell adhesion:

1. They can be purified directly from trypsin extracts following cell dissociation.

2. An alternative method is the use of monovalent antibody fragments (Fab′—antibodies have two antigen binding sites) that specifically inhibit cell aggregation. These fragments can be employed as reagents for identifying the adhesive macromolecules in cell membrane extracts.

3. Another method involves mixing the putative adhesive molecule with liposomes (artificial lipid vesicles) and determining if the liposomes will then adhere to each other.

4. Related to this is the insertion of the gene coding for a possible adhesive molecule into cells not normally expressing the molecule and again determining if new adhesive properties arise.

5. Correlative evidence for an adhesive function can been obtained if the physiologically or developmentally regulated expression of the molecule on the cell surface occurs at the same time as its proposed function.

Some of the better-characterized cell adhesion molecules (CAMs) by these methods are summarized in Table 14–4. Two trends are apparent. First, some of the molecules have a broad tissue distribution and are not specific for a particular tissue. Indeed, so-called primary CAMs such as NCAM and LCAM appear in cells of all embryonic germ layers. However, developmental studies show that only one or a unique pattern of CAMs is involved in a given tissue specificity at a given time in the life of a cell. On the other hand, some CAMs are quite specific in their properties and seem to be expressed only in a given cell at a particular time. There are probably many molecules in this class that have yet to be discovered. A second trend is that some CAMs require Ca^{++} for adhesion, whereas others do not. The Ca^{++} requirement is similar to that of the sponge aggregation factor (see above).

T A B L E 14–4 **Some Vertebrate Cell Adhesion Molecules**

Molecule	Ca^{++}-dependence	Major Tissue Distribution
Liver CAM*	Yes	Epithelia, others
P-Cadherin	Yes	Epithelia, others
N-Cadherin	Yes	Neurons, others
E-Cadherin	Yes	Epithelia
Uvomorulin	Yes	Blastomeres
Neural CAM	No	Neurons, muscles, others
Axonal CAM	No	Axons
MAG	No	Neuron-glia
ICAM	No	White blood cells, endothelium
LEC-CAM	No	Leukocytes, endothelium

*CAM: Cell adhesion molecule.

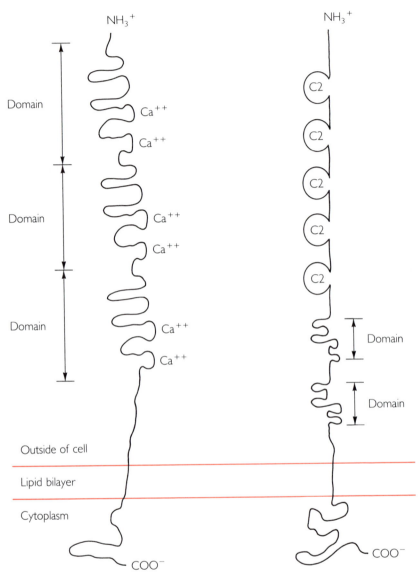

NH$_3^+$

Domain

Ca^{++}

Ca^{++}

Domain

Ca^{++}

Ca^{++}

Domain

Ca^{++}

Ca^{++}

NH$_3^+$

C2

C2

C2

C2

C2

Domain

Domain

Outside of cell

Lipid bilayer

Cytoplasm

COO$^-$

COO$^-$

FIGURE 14–16

Structures of the two major types of animal cell adhesion molecules (CAMs).

(a) Cadherins: There are three Ca^{++} binding domains, a membrane spanning region, and a cytoplasmic domain.

(b) Ca^{++} independent: There are domains (C2) similar to immunoglobulins and domains similar to fibronectin.

The **cadherins** are a general class of adhesive molecules that require Ca^{++} for activity. Molecular cloning of the DNAs for uvomorulin, LCAM, P-cadherin, and N-cadherin shows that these proteins are quite similar (Figure 14–16a). They are 723–748 amino acids long, with a signal peptide, a membrane-spanning hydrophobic region, and extracellular and cytoplasmic domains. The average amino acid similarity is about 50–60%, with the greatest conservation of sequence (80%) in the cytoplasmic region, which can bind to the cytoskeleton. The extracellular part has three similar domains that bind Ca^{++}.

If cadherins are introduced into cells previously without them, adhesion to other cells containing that cadherin is promoted. This indicates that the adhesion is **homotypic,** that is, that the "receptor" for the adhesive molecule on the second cell is another of the same molecule. However, if the receptor cell is coated with anticadherin antibody, adhesion is blocked. Although cadherins are glycoproteins, inhibition of glycosylation does not block adhesion, indicating that the sugars are not the adhesive determinants, as in the sponge factor receptor. Rather, it is the three-dimensional shape of the protruding proteins that allows them to precisely fit together.

In addition to intercellular adhesion, introduction of a cadherin into a cell causes profound alterations in cell shape. These changes are due to the interactions of the cytoplasmic domain of the cadherin with cytoplasmic proteins, and a class of proteins, termed catenins, has been identified as major caderin-binding molecules. Remarkably, plakoglobin, a component of the desmosome, is a catenin, and desmocolin and desmoglein, two other desmosomal proteins, are cadherins. In addition, other cytoskeletal structures, actin and intermediate filaments, interact with cadherins. Thus, these cell surface molecules have an important role in cell shape.

Cadherin expression changes during tissue development, and this apparently leads to the separation of different tissues that have the same origin:

1. In the embryonic neural retina, all cells equally express N-cadherin on their surface, but as the retina differentiates, the N-cadherin is lost from the cell surfaces. This lack of the marker differentiates these cells from their neighbors, to which they no longer adhere.

2. The embryonic neural tube forms from the ectoderm (Figure 14–17). Initially, all ectodermal cells express only E-cadherin, but at a certain point some of them lose this marker and begin to express N-cadherin. This difference allows these cells to detach from the ectoderm and bind homotypically, and they eventually form the neural tube. These observations confirm a theory put forth by J. Holtfreter in 1955, who proposed that the neural tube detaches because it loses a cell recognition factor.

The best studied of the Ca^{++}**-independent CAMs** is NCAM. This molecule exists in three major forms, of molecular weights 180,000 (180K), 140K, and 120K. Both molecular cloning and amino acid sequencing indicate that the smaller forms are truncated versions of the

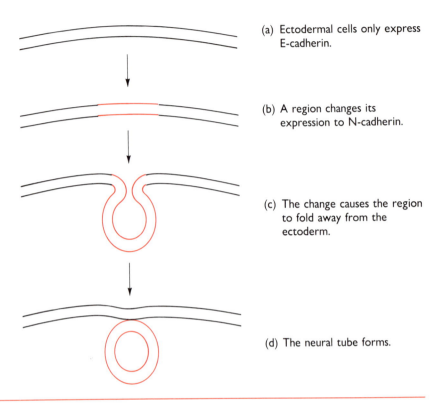

(a) Ectodermal cells only express E-cadherin.

(b) A region changes its expression to N-cadherin.

(c) The change causes the region to fold away from the ectoderm.

(d) The neural tube forms.

FIGURE 14–17

How the differential expression of cadherins leads to embryonic tissue differentiation.

larger form. Specifically, the 180K form, like the cadherins, is an integral membrane protein, which has an amino terminal glycosylated exoplasmic region, a membrane-spanning region, and a cytoplasmic domain. The 140K form has a shorter cytoplasmic domain, and the 120K form terminates before the hydrophobic region, attaching to the plasma membrane by a lipid anchor.

A second heterogeneity in NCAM is its degree of glycosylation. The sugars on the exoplasmic domain of NCAM are a polymer of sialic acid, and polymerized chains may lead to NCAM with 30% sialic acid, while a lesser degree of polymerization results in a 10% sialic acid content.

Like the cadherins, NCAM binding is homotypic. When the sialic acid content is lower, binding is still specific but stronger. Indeed, experimental removal of the sugars by glycosidases causes the cells to have visibly more close connections under electron microscopy. It is possible that the large amount of carbohydrate sterically inhibits the close membrane associations necessary for the formation of specialized junctions. Whatever the mechanism, the degree of sialic acid content varies during developmental situations. For example, when axons are bundled to form the optic nerve, they express the low sialic acid NCAM, presumably for a high degree of adhesion. However, when nerve branching occurs, the expression is changed to high sialic acid NCAM, with less self-affinity.

NCAM is involved in a number of specific functions during development:

1. In the formation of neuromuscular junctions, it appears first on the nerve and then on the muscle, after which the junction between the two is formed.

2. When axons leave the developing eye, they express NCAM, which is also on the glia that are at their target region in the brain. This homotypic interaction may be essential in directing the axon to its proper location.

The extracellular region of NCAM contains five adjacent domains that are homologous to each other (Figure 14—16(b)), as well as to part of the light chain of the antibody molecule, immunoglobulin G. This similarity is not so remarkable when it is considered that both NCAM and IgG have essential recognition functions, although one recognizes self and the other recognizes nonself. In fact, NCAM and the other Ca^{++}-independent CAMs are considered members of the immunoglobulin gene superfamily.

Adhesion and Intercellular Junctions

Adhesive macromolecules merely bring cells into contact with each other. The formation of intercellular junctions "cements" the relationship. Gap junctions, with intercellular communication, begin to form less than 30 minutes after adhesion of dissociated vertebrate cells, and, where appropriate, tight junctions begin at 30 minutes and desmosomes within 2 hours. Clearly, this must involve considerable membrane assembly and rearrangements.

The process of adhesion and cell junction formation has been studied in detail in canine kidney epithelial cells, which can serve as a summary of the many junctional relationships described in this and the previous chapters (Figure 14—18). These cells have three membrane domains:

1. The basal region sits on a basal lamina, to which it attaches in two ways: by the laminin receptor system (see Chapter 13) and by hemidesmosomes.

2. The apical region faces the lumen of the kidney and has many microvilli for transport.

3. The lateral region is adjacent to the neighboring cells, to which it is connected by no less than four types of interactions: tight

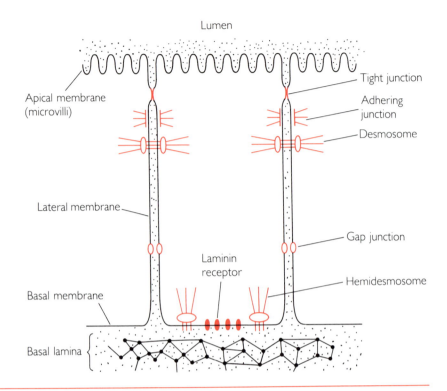

FIGURE 14-18

A simple epithelium, with a complex array of intercellular junctions.

junctions, desmosomes, adhering junctions (similar to desmosomes), and LCAM.

The key event in the formation of these cell-cell junctions is the homotypic recognition of LCAM by adjacent cells. Indeed, if an antibody to LCAM is present, the other three kinds of junctions do not form. Also, when this epithelium forms in the embryo, it does not do so until LCAM is expressed, and addition of LCAM to nonepithelial cells leads to the formation of gap and adhering junctions. The exact relationship between the CAMs and the junctions is not clear. The cell recognition event may bring the cells close together so that the junctions can be made, or the CAMs may even be part of the junctions themselves.

Plants

Studies of the adhesion of plant cells are complicated by the presence of the cell wall. In addition, although dissociated plant cells can grow in culture, they do not move as do animal cells, and therefore active sorting out is unlikely. Most of the experimental data on plant cell adhesion have come from single-celled algae, such as *Chlamydomonas*. Vegetative cells of this organism do not agglutinate, and when similar-appearing gametes

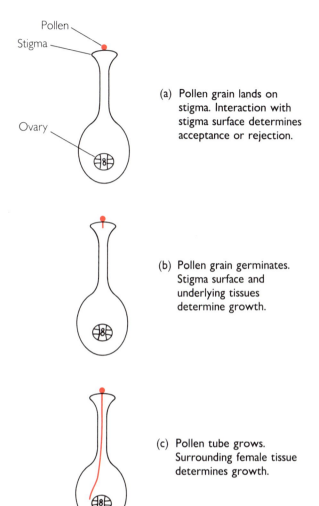

(a) Pollen grain lands on stigma. Interaction with stigma surface determines acceptance or rejection.

(b) Pollen grain germinates. Stigma surface and underlying tissues determine growth.

(c) Pollen tube grows. Surrounding female tissue determines growth.

FIGURE 14–19

Pollen-stigma interactions in a higher plant.

form, they may be of the (+) or (−) mating type. Only pairs of cells of opposite mating type adhere to one another, first by the flagellum and then by the body, and then cell fusion occurs to form a zygote.

Gametes shed into the medium adhesive molecules termed **isoagglutinins.** These complexes are visible under electron microscopy as pieces of flagellar membrane with fuzzy coats. The action of an isoagglutinin is species- and mating-type specific. For example, the complex released by *C. reinhardi* (+) cells agglutinates only *C. reinhardi* (−) cells and not those of any other species. The adhesion factors are held on the outer surface of the flagellar membrane and generally are glycoproteins that contain hydroxyproline.

The best-studied cell adhesion system in higher plants is the interaction of the male gamete-containing **pollen** grain with the female-containing ovary and its receptor, the **stigma** (Figure 14–19). When a wind-carried or insect-borne pollen grain lands on a stigma, a number of glycoproteins from the outer pollen wall, or exine, are released to the sticky hydrophilic surface of the "female" organ. In a compatible system, these pollen molecules interact with recognition factors on the stigma. Pollen then takes up water and germinates, and the pollen tube that will carry the sperm penetrates the stigma.

Reproductive incompatibility is common in most organisms and prevents interspecific matings. In animals, these mechanisms are either mechanical, behavioral, or genetic. But many plant species have both sex organs (anther and stigma) near each other in the same flower. This means that all plants in a population can mate with all others in that same population (as compared to single-sexed organisms, where mating only with the opposite sex is possible). But it also means, as Charles Darwin wrote in 1877, that there are "different forms of flowers on plants of the same species" and that self-fertilization is possible. This would tend to make the population more genetically homogeneous.

Plants prevent self-fertilization by using a self-incompatibility gene, or S gene, which is involved in the pollen rejection mechanism. S genes have been cloned from various plants and code for glycoproteins expressed on the pollen and stigma cell surfaces. Self-incompatibility occurs when the genes of pollen and stigma are the same. The mechanism by which adhesion of cells with apparently identical CAMs leads to rejection, rather than homotypic adhesion as in animals, is not known.

What is known is that in a reproductively incompatible system, such as with different species, the pollen-stigma protein interaction results in formation of the β-1,3-glucan polymer, callose, which prevents the grain from germinating. This rejection reaction can also occur while the pollen tube is growing within the female-bearing structure. The material in the female that reacts with pollen recognition factors is apparently a glycoprotein containing arabinose and galactose.

Grafting involves taking parts of plants of different genotypes and putting them together to form a new strain. In horticulture, the basic plant is termed the stock, and the new branch is called the scion. Acceptance or rejection of the scion differs from acceptance of tissue grafts in vertebrates, where cell surface markers (histocompatibility antigens) determine whether the host mounts an immune response and rejects the graft. In the plant, there is no immune system, but both stock and scion secrete glycoproteins into the area of the graft. Some of these are lectins, which may be involved in successful recognition events. Once this happens, cells at the graft differentiate to form thickened walls, and a vascular tissue connection provides a strong bond between graft and host.

Applications of Cell-Cell Adhesion

Invasion by Bacteria and Protozoa

The study of cell interactions is by no means simply an academic exercise. Many organisms have made use of cell recognition properties of plant and animal cells to permit binding and infection. What follows are a few representative examples of the types of interactions involved.

Many pathogenic and benign bacteria colonize the mucosal surfaces of animals (Figure 14–20). An essential event in successful colonization is the firm attachment of the microorganism to the mucosal membrane. This firmness is necessitated by the fact that the surfaces are usually washed by a flow of fluid due to ciliary action, such as mucus in the airway or urine in the ureter.

Neisseria gonorrhoeae, the bacterium that causes the venereal disease **gonorrhea,** infects the columnar epithelia of the male and female reproductive tracts. Attachment is via pili, protein filaments several μm long and 7 nm in diameter, and cells must have pili to infect. The pili are composed of subunits of a single protein, pilin, which contains about half of its amino acids as hydrophobic. Apparently, the pilin binds to a complex carbohydrate on the cell surfaces of susceptible cells but will bind much less to cells not commonly infected by the bacterium. This indicates that there is a receptor for the adhesive pilin on the human cell.

Because binding is a key event in a successful infection, preventing it by an antibody in the form of an antipilus vaccine has been attempted. However, this has not been successful because of the extensive and continuing variation of the pilin via somatic recombination between different coding DNA sequences. A second bacterial cell surface protein,

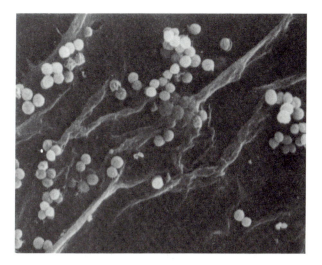

FIGURE 14–20

Staphylococcal bacteria adhering to human valvular endothelial cells. (×2850. Courtesy of Dr. L. Higgins.)

termed protein II, has also been implicated in adhesion to particular types of epithelia, especially buccal and Fallopian tubes. Unfortunately, it too undergoes extensive variation.

Projections microscopically similar to pili, termed **fimbriae,** are produced by many bacteria that infect humans. These projections are composed of hundreds of subunits of fimbrillin and bind to cell surfaces with certain sugars, depending on fimbrillin type. For example, type 1 fimbriae of such bacteria as *E. coli, Pseudomonas, Salmonella* and *Shigella* bind to mannose; fimbriae from the cholera bacterium, *Vibrio cholerae,* bind to fucose; and type 2 fimbriae from *Actinomyces* bind to the sequence galactose-glucose. Again, the identities of the cellular receptors are not known but are important to identify from the viewpoint of understanding and manipulating the infective process.

Seriously ill people are often prone to colonization of the upper respiratory tract by pathogenic microorganisms, such as *Pseudomonas* bacteria, which cause a form of **pneumonia.** In this case, the host cell surface may be changing to accommodate bacterial binding. Pretreatment of normal buccal epithelial cells with trypsin allows much more colonization by *Pseudomonas.* The major membrane protein released by the trypsin is fibronectin, a protein involved in cellular connections to the extracellular matrix (see Chapter 13). *Pseudomonas* binding to isolated buccal cells increases over time in vitro, and inhibitors of serine proteases, which are produced by damaged buccal cells, block this rise in bacterial adhesion. Thus, a hypothesis to explain the increased susceptibility to colonization by the patients is that their illness causes buccal cells to release proteases, which detach fibronectin from the membrane, and the loss of this molecule somehow allows bacterial attachment to the cell surface.

Bacteria of the genus *Rhizobium* infect the roots of legumes such as beans, clover, and peas. Successful infection results in a **symbiotic relationship** in which the plant supplies fixed carbon and nutrients to the microbe, and the latter fixes nitrogen into ammonia. Strains of *Rhizobium* are defined by the plant that they infect. The initial event is a species-specific attachment of the bacterium to a root hair cell exterior, mediated by specific receptors on the surface of the microbe and plant.

Plant lectins may mediate this specific cell interaction. If a lectin gene from a pea plant is introduced into clover, the clover will now be infected not only by the normal clover-specific *Rhizobium* but also by the pea-specific *Rhizobium.* On the opposite side, the *Rhizobium* bacteria also have species-specific nodulation genes (the "nod" system), and these can be transferred to alter or add to the plant targets of that bacterium. The role of the lectin may be to bind to both bacterial and plant receptors, and indeed, a lectin, trifoliim A, isolated from clover seeds and roots, binds to both. The specific sugar involved in lectin binding appears to be 2-deoxyglucose, since this molecule prevents both lectin binding and bacterial-root adhesion.

A model for the *Rhizobium*-clover attachment via a lectin bridge is shown in Figure 14–21. This has great implications for agriculture, because nitrogen-fixing bacteria reduce a host plant's dependence on fixed nitrogen in the soil (which must often be supplied as fertilizer). If plants that are normally weak hosts for the fixing bacteria, or not hosts at all (as are the major cereals, rice, wheat, and maize), could be genetically manipulated to act as hosts for efficient nitrogen-fixing bacteria, the need for nitrogen fertilizers would be reduced, and food production costs would be lower.

The malarial parasite, *Plasmodium falciparum*, infects human erythrocytes, and the initial event in this infection is binding to the erythrocyte membrane. The red cell membrane recognition marker for parasite binding is glycophorin (see Chapter 2). Antibodies to this protein block binding in vitro, and erythrocytes lacking this protein are not susceptible to infection. Following adhesion, a specific junction, very much like a gap junction in its concentration of intramembranous particles, forms between parasite and host, and the parasite is then internalized into the red blood cell. The infected cell then produces specific cell membrane proteins, called knob proteins because of the shape of the surface structure. The knobs act as the binding site to endothelial cells within the host person's organs, and many of the recurrent clinical problems of malaria patients are due to the ability of the parasite, inside of

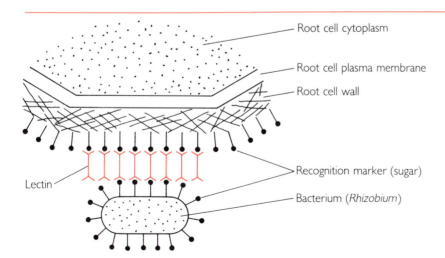

Root cell cytoplasm

Root cell plasma membrane

Root cell wall

Recognition marker (sugar)

Lectin

Bacterium (*Rhizobium*)

FIGURE 14–21

Model for cell recognition between a symbiotic nitrogen-fixing bacterium and a legume root hair cell. The lectin trifoliin acts as a bridge between the bacterium and the plant root. (After Dr. F. Dazzo.)

an infected cell, to "hide out" in this manner among the surface molecules on the endothelial cells that act as receptors for the knobs in ICAM. Obviously, studies of the details of these processes will be important in finding ways to prevent malaria.

When cells are infected with a virus, specialized white blood cells called **killer T cells** can bind to the infected cells and destroy them, along with the viruses. The proteins involved in this important intercellular interaction are on the cell surfaces on the two cell types: The killer cells have leukocyte function associated molecule (LFA), an integrin that binds heterotypically to ICAM on the target cell surface. The disease leukocyte adhesion deficiency is caused by a defect in the gene for LFA, and the result is often fatal childhood infections because the patients' killer T cells do not accumulate at the site of the infection. These children can survive if given repeated bone marrow transplants to supply T cells containing LFA.

Remarkably, ICAM is also the receptor for the **rhinovirus** that causes the common cold. This can be shown by a number of tests. For example, if the gene for ICAM is spliced into cells normally lacking it, the cells can now be infected by the virus. Also, rhinovirus infection is blocked by antibody to ICAM. That a cell adhesion protein also should be a target for viral infection is not surprising, given the prominence of this protein on susceptible cells (white blood cells, epithelial cells) and the rapid evolution of viruses in their interactions with hosts. Nevertheless, the identification of the binding site for rhinovirus may allow the development of drugs specific for this bothersome affliction.

A cell recognition protein quite similar to ICAM is involved in **multiple sclerosis.** This molecule, myelin associated glycoprotein (MAG), is specific for nerve tissue and has a structure similar to NCAM in its domains (Figure 14–22), with homologies to the immunoglobulins. There is strong evidence that MAG is the adhesion factor for the interaction between neuronal axons and the Schwann cells that surround them to form myelin. For example, an antibody to MAG will block the binding of neurons to oligodendrocytes (the class of glia to which Schwann cells belong).

In multiple sclerosis, there is a progressive degeneration of the nervous system over a period of years, and its hallmark is a loss of myelin. This results in aberrant transmission of impulses along the axons, since myelin normally acts as an insulator to modulate this transmission. Myelin loss is accompanied by a loss of MAG from the tissues, and this may result in the detachment of myelin from the axons. What causes the MAG loss is uncertain, but one possibility is that the patients make antibodies to it (multiple sclerosis is suspected as having an autoimmune cause).

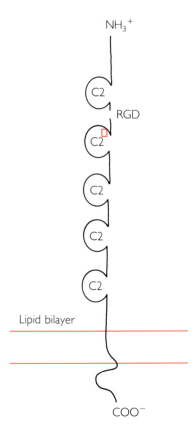

F I G U R E   1 4 – 2 2

The structure of myelin-associated glycoprotein (MAG), which is involved in the adhesion of axons to the glia that produce myelin. There is a typical Ca^{++}-independent CAM structure, along with the RGD sequence, which may bind to the adjoining cell's extracellular matrix.

Cancer Cells

The cell surface of cancer cells shows considerable changes from that of normal cells (Table 14–5), and a number of these alterations involve intercellular interactions. The loss of gap junctions and, concomitantly, intercellular communication was noted earlier in this chapter. A distinctive property of some cancer cells is their **rapid and uncontrolled rate of growth.**

The change in growth rate when a normal cell is transformed into a tumor cell can be seen in tissue culture. When normal fibroblasts divide and reach a limiting density (usually a monolayer), they stop growing, a phenomenon referred to as density-dependent growth inhibition. It is a consequence of cell-cell and cell-substratum contact, since the cells can divide if separated and subcultured. Also, addition of membrane fragments to sparse, growing cells causes them to stop dividing. In contrast, tumor cells lack this density-dependent inhibition and continue to divide to ever-increasing densities. Ultimately, their growth is limited only by the nutrients in the culture medium.

Cell membrane alterations are also involved in a second major property of some cancer cells, the ability to invade surrounding tissues and form secondary tumors, or **metastases** (Figure 14–23). This has severely adverse consequences for the patient. The process has several steps:

1. The tumor grows within host tissue and becomes vascularized with blood vessels so that it can be nourished.

2. Some cells become detached from the tumor and enter the blood vessels or lymphatic system.

3. These cells can interact with blood cells or are themselves carried to distant sites in the body.

4. At an organ, the tumor cells or derivatives adhere to the endothelial wall of the blood vessel (arrest).

TABLE 14–5 **Some Cell Surface Alterations of Cancer Cells**

Increased mobility of lectin receptors

Increased uptake of sugars

Loss of gap junctions and intercellular communication

Loss of inhibition of cell division

Loss of inhibition of cell movement

New antigens on cell surface

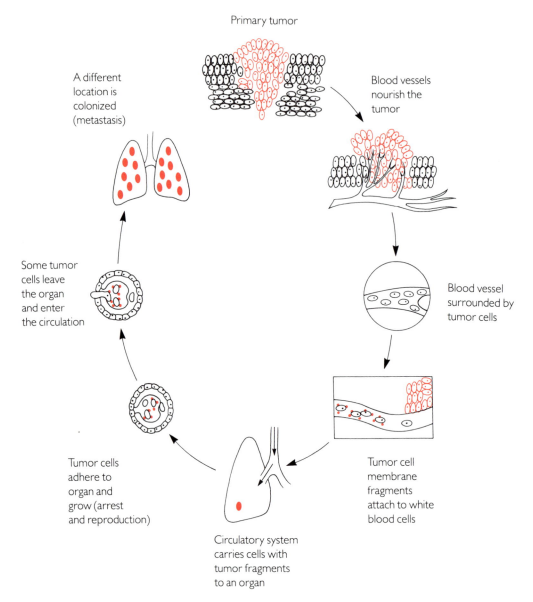

Primary tumor

A different location is colonized (metastasis)

Blood vessels nourish the tumor

Some tumor cells leave the organ and enter the circulation

Blood vessel surrounded by tumor cells

Tumor cells adhere to organ and grow (arrest and reproduction)

Tumor cell membrane fragments attach to white blood cells

Circulatory system carries cells with tumor fragments to an organ

FIGURE 14–23 Steps in the metastasis of a tumor from one site in the body to another. (After G. Nicolson.)

5. The tumor cells invade the extracellular matrix of the second host tissue.

6. The tumor cells then divide in the second tissue and form a new tumor (metastasis).

A number of these steps clearly must involve cell recognition. For instance, growth of a tumor within a host tissue requires that the tumor cell lose its host attachment and develop tumor-tumor attachments, much like the differentiation of tissues in the embryo (Figure 14–17). This is strongly indicated for colorectal cancer, where the deletion of a gene for a cell surface glycoprotein similar to NCAM is a prerequisite for tumor formation. In other cancers, there are changes in lectin specificity when the cells become tumors.

Most metastatic tumor cells are specific as to which organ or organs they successfully invade. For example, cells that prefer to grow in the lung can arrest and grow elsewhere in the body, but in time most will be in the lung. The specificity appears to reside in cell surface glycoproteins. In mice, metastatic melanoma cells selective for colonization in the brain have increased levels of ICAM (see above) exposed on their cell surface. Ovary-selective cells, derived from the same melanoma line, lack this marker but have instead two proteins of molecular weights 140,000 and 150,000. Dissociated metastatic cells that prefer lung tissue aggregate with dissociated lung cells but not with those of other tissues. This indicates that selective cell adhesion is involved in tissue localization.

Further evidence for the involvement of the tumor cell membrane in tissue recognition during arrest and metastasis comes from experiments with plasma membrane vesicles. These vesicles are shed spontaneously from metastatic cells in vitro and in vivo. Vesicles from a highly metastatic line of mouse tumor cells (F10) can be fused with cells of a poorly metastatic line (F1) and vice versa. The vesicle-fused cells are then injected into the bloodstream of a mouse.

The results of such an experiment (Table 14–6) show that the addition of the vesicle from the highly metastatic line causes the poorly metastatic cell to localize in the lung. This indicates that the tumor cell plasma membrane is responsible for its invasive properties. Invasion can be prevented by antibodies that are specific for the metastatic cell membrane and that do not cross-react with the normal cell membrane. In the case of metastatic rat pancreatic cancer, the identity of the membrane component responsible has been narrowed to a specific region of a membrane protein, termed CD44. This region occurs on the surfaces of metastatic cell lines but not on those tumor cells that do not metastasize. But if the latter cells are given an overexpressing form of the gene for this region, they become metastatic.

The molecular nature of the tumor cell membrane–normal cell interaction may involve known receptors for extracellular matrix glycoproteins. Most metastatic cells can bind to laminin and/or fibronectin and even to collagen and proteoglycans in some cases. The cell line de-

TABLE 14-6 Arrest and Retention in the Lung of Mouse Metastatic Cancer Cells Fused with Membrane Vesicles

Cells	Vesicles	% Injected Cells in Lung at:		
		2 min	I day	14 days
FI*	—	60.4	1.3	0.5
FI	FI	68.2	1.1	0.4
F10	—	98.4	14.3	1.5
F10	F10	99.2	11.4	2.2
FI	F10	80.3	6.2	1.1
F10	FI	99.4	12.6	1.9

*FI has low lung metastasis. F10 has high lung metastasis.

DATA FROM: Poste, G., and Nicolson, G. Arrest and metastasis of blood-borne tumor cells are modified by fusion of plasma membrane vesicles from highly metastatic cells. *Proceed Nat Acad Sci* 77 (1980):399–403.

scribed in the vesicle experiments has a laminin receptor, and the RGD peptide can prevent its metastasis. Indeed, laminin promotes the successful metastasis of a number of tumor cell types.

Once it has bound to the matrix, the tumor cell must somehow "burrow into" the host tissue, and this is done by digesting away the matrix. A variety of hydrolases is made, including collagenase, elastase, and stromelysin (for proteoglycans). These are made as part of a protease cascade, where an initial enzyme is induced and it activates others that have been stored in an inactive form.

There are naturally made protein inhibitors of the enzymes that digest the matrix. If these inhibitors are reduced, a cell is more likely to become metastatic than if they are in normal amounts. On the other hand, if the inhibitors are abundant, invasion is prevented. This control seems to apply to more than tumors: The decidua, a uterine tissue that surrounds the early embryo, has high levels of inhibitors, and this may regulate the rate of spread of the trophoblast. Clearly, these inhibitors are models for the design of drugs to prevent processes of tumor cell invasion into the extracellular matrix.

Further Reading

Albelda, S. M. Endothelial and epithelial cell adhesion molecules. *Am J Respir Cell Mol Biol* 4 (1991):195–203.
Beebe, D., and Turgeon, R. Current perspectives on plasmodesmata: Structure and function. *Physiol Plant* 83 (1991):194–199.

Bernatzky, R., Anderson, M., and Clarke, A. Molecular genetics of self-incompatibility in flowering plants. *Dev Genet* 9 (1988):1–12.

Beyer, E., Paul, D., and Goodenough, D. Connexin family of gap junction proteins. *J Membr Biol* 116 (1990):187–194.

Bongrand, P. *Physical basis of cell-cell adhesion.* Boca Raton, FL: CRC Press, 1988.

Bourrillon, R., and Aubrey, M. Cell surface glycoproteins in embryonic development. *Int Rev Cytol* 116 (1989):257–330.

Boyer, B., and Thiery, J. P. Epithelial cell adhesion mechanisms. *J Membr Biol* 112 (1989):97–108.

Brewin, N. Development of the legume root nodule. *Ann Rev Cell Biol* 7 (1991):191–226.

Britigan, B., Cohen, M., and Sparling, F. Gonococcal infection: A model of molecular pathogenesis. *New Engl J Med* 312 (1985):1683–1690.

Butcher, E. Cellular and molecular mechanisms that direct leukocyte traffic. *Am J Pathol* 136 (1990):3–9.

Cereijido, M., Ponce, A., and Gonzalez-Mariscal, L. Tight junctions and apical-basolateral polarity. *J Membr Biol* 110 (1990):1–9.

Chapman, G., Ainsworth, C., and Chatham, C. (Eds.). *Eukaryote cell recognition: Concepts and model systems.* Cambridge, U.K.: Cambridge University Press, 1988.

Dazzo, F. Bacterial attachment as related to cellular recognition in the Rhizobium-legume symbiosis. *J Supramolec Struct* 16 (1981):29–41.

Dedhar, S. Integrins and tumor invasion. *BioEssays* 12 (1990):583–590.

Diaz, C., Melchers, L., Hooykas, P., Lugtenberg, B., and Kijne, J. Root lectin as a determinant of host plant specificity in the Rhizobium-legume symbiosis. *Nature* 338 (1989):579–581.

Dickinson, H. Self-incompatibility in flowering plants. *BioEssays* 12 (1990):155–161.

Ebert, P., Anderson, M., Bernatzky, R., Altschuler, M., and Clarke, A. Genetic polymorphism of self-incompatibility in flowering plants. *Cell* 56 (1989):255–262.

Edelman, G. M. Morphoregulatory molecules. *Biochemistry* 27 (1988):3533–3542.

Edelman, G. M., and Thiery, J. P. (Eds.). *The cell in contact.* New York: John Wiley and Sons, 1987.

Finlay, B. Cell adhesion and invasion mechanisms in microbial pathogenesis. *Curr Op Cell Biol* 2 (1990):815–820.

Fleming, S. Cellular functions of adhesion molecules. *J Pathol* 161 (1990):189–199.

Garrod, D. Desmosomes, cell adhesion molecules, and the adhesive properties of tumor cells. *J Cell Sci* [Suppl] 4 (1986):221–237.

Gilula, N., Reeves, O., and Steinbach, A. Metabolic coupling, ionic coupling and cell contacts. *Nature* 235 (1972):262–265.

Gumbiner, B. Structure, biochemistry and assembly of epithelial tight junctions. *Am J Physiol* 253 (1987):C749–C758.

Gunning, B., and Overall, R. Plasmodesmata and cell to cell transport in plants. *Bioscience* 33 (1983):260–265.

Gunning, B., and Robards, A. *Intercellular communication in plants: Studies on Plasmodesmata.* New York: Springer-Verlag, 1976.

Guthrie, S., and Gilula, N. B. Gap junctional communication and development. *Trends in Neuroscience* 12 (1989):12–15.

Haring, V., Gray, J., McClure, B., Anderson, M., and Clarke, A. Self-incompatibility: A self-recognition system in plants. *Science* 250 (1990):937–941.

Heaysman, J., Middleton, C., and Watt, F. Cell behavior, shape, adhesion and motility. *J Cell Sci* [Suppl] 8 (1987).

Hunt, G. The role of laminin in cancer invasion and metastasis. *Exp Cell Biol* 57 (1989):165–176.

Hynes, R. O., and Lander, A. D. Contact and adhesive specificity in association, migration and targeting of cells and neurons. *Cell* 68 (1992):303–322.

Lander, A. D. Mechanisms by which molecules guide axons. *Curr Op Cell Biol* 2 (1990):907–913.

Liotta, L., Steeg, P., and Stetler-Stevenson, W. Cancer metastasis and angiogenesis: An imbalance of positive and negative regulation. *Cell* 64 (1991):327–336.

Marthy, H. J. (Ed.). *Cellular and molecular control of direct cell interactions.* New York: Plenum, 1986.

McClay, D., and Ettensohn, C. Cell adhesion in morphogenesis. *Ann Rev Cell Biol* 3 (1987):319–345.

Molitoris, B., and Nelson, J. Alterations in the establishment and maintenance of epithelial cell polarity as a basis for disease processes. *J Clin Invest* 85 (1990):3–9.

Moscatelli, D., and Rifkin, D. Membrane and matrix localization of proteinases: A common theme in tumor cell invasion and angiogenesis. *Biochim Biophys Acta* 948 (1988):67–85.

Muller, W., Diehl-Seifert, B., Gramzow, M., Friese, U., Renneisen, K., and Schroder, H. Interrelation between extracellular adhesion proteins and extracellular matrix in reaggregation of dissociated sponge cells. *Int Rev Cytol* 111 (1988):211–229.

Musil, L., and Godenough, D. Gap junctional intercellular communication and the regulation of connexin expression and function. *Curr Op Cell Biol* 2 (1990): 875–880.

Nicolson, G. N. Molecular mechanisms of cancer metastasis: Tumor and host properties and the role of oncogenes and tumor suppressor genes. *Curr Op Cell Biol* 3 (1991):75–92.

Nybroe, O., and Bock, E. Structure and function of the neural cell adhesion molecules NCAM and L1. *Adv Exp Biol Med* 265 (1990):185–196.

Ouassi, A., and Capron, A. Some aspects of protozoan parasite-host cell interactions with reference to RGD mediated recognition process. *Microbial Pathogen* 6 (1989):1–5.

Poste, G., and Nicolson, G. Arrest and metastasis of blood-borne tumor cells are modified by fusion of plasma membrane vesicles from highly metastatic cells. *Proceed Nat Acad Sci* 77 (1980):399–403.

Powell, D. W. Barrier function of epithelia. *Am J Physiol* 241 (1981):G275–288.

Prodi, G. (Ed.). *Cancer metastasis* New York: Plenum Press, 1988.

Quarles, R. H. Myelin-associated glycoprotein in demyelinating disorders. *Crit Rev Neurobiol* 5 (1989):1–25.

Quispel, A. Bacteria-plant interactions in symbiotic nitrogen fixation. *Physiol Plant* 74 (1988):783–790.

Ruoslahti, E., and Pierschbacher, M. New perspectives in cell adhesion: RGD and integrins. *Science* 238 (1987):491–497.

Schwarz, M., Owaribe, K., Kartenbeckj, J., and Franke, W. Desmosomes and hemidesmosomes: Constitutive molecular components. *Ann Rev Cell Biol* 6 (1990): 461–491.

Sharma, Y. Knobs, knob proteins and cytoadherence in falciparum malaria. *Int J Biochem* 23 (1991):775–789.

Sharon, N. Bacterial lectins, cell-cell recognition, and infectious disease. *FEBS Let* 217 (1987):145–157.

Smith, S., and Smith, F. Structure and function of the interfaces in biotrophic symbioses as they relate to nutrient transport. *New Phytol* 114 (1990):1–38.

Sobel, M. Metastasis suppressor genes. *J Nat Cancer Inst* 82 (1990):267–276.

Staunton, D., Merluzzi, V., Rothlein, R., Burton, R., Marlin, S., and Springer, T. A. A cell adhesion molecule, ICAM-1, is the major surface receptor for rhinoviruses. *Cell* 56 (1989):849–853.

Stevenson, B. R., Anderson, J., and Bullivant, S. The epithelial tight junction: Structure, function and preliminary biochemical characterization. *Mol Cell Biochem* 83 (1988):129–145.

Takeichi, M. Cadherins: A molecular family important in selective cell-cell adhesion. *Ann Rev Cell Biol* 59 (1990):237–252.

Van Meer, G., and Simons, K. Lipid polarity and sorting in epithelial cells. *J Cell Biochem* 36 (1988):51–58.

Yong, K., and Khwaja, A. Leukocyte cellular adhesion molecules. *Blood Rev* 4 (1990):211–225.

APPENDIX I

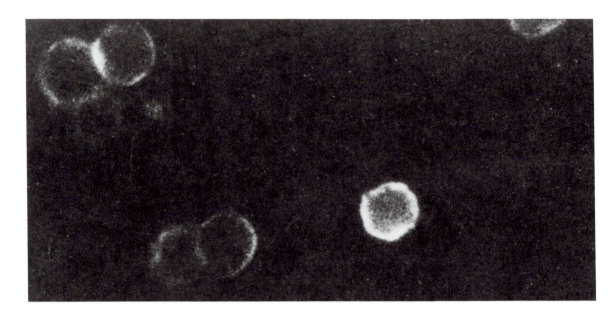

Methods

Microscopy

Theory

Although the use of lenses for magnification had been known for centuries before, modern microscopy began in Holland around 1600 when an eyeglass maker, Z. Jannsen, lined up two lenses to effectively multiply their individual magnifications. During the succeeding century, R. Hooke and A. van Leeuwenhoek performed and published the initial microscopic examinations of biological tissues. The **resolution,** that is, the ability to distinguish two points as being separate, of these early instruments was 1 μm. This was an improvement over the human eye, which can normally resolve to 200 μm.

The function of a lens is to bend light rays. This bending, or refraction, is dependent to an extent on the wavelength of the light. Thus, when white light, which is composed of many wavelengths, is bent by a lens, the different rays at different wavelengths will converge at different points. These varying focal points result in an image that is fuzzy, most typically with color fringes around the edges. This chromatic aberration severely limited the resolving power of the 17th-century microscopes. Over 200 years later, anachromatic lenses were developed. These are composed of several types of glass whose refraction-wavelength properties differ and cancel each other out; thus the light rays converge at a single sharp focal point.

The Swiss physicist E. Abbe developed the theory of microscopic resolution in the equation:

$$\text{R.L.} = 0.6\ \lambda/\text{N.A.}$$

where R.L. is resolution limit, λ is the wavelength of light, and N.A. is the numerical aperture. The latter term is defined as:

$$\text{N.A.} = n \sin \theta$$

where n is the refractive index of the medium around the object, and θ is the half-angle formed between the object and the lens (Figure A1–1).

Clearly, lowering R.L. can be achieved in three ways:

1. Lowering λ: blue light (400 nm) is better for resolution than red light (700 nm).

2. Raising n: immersion oil (n = 1.5) between the specimen and the lens is better than air (n = 1.0).

3. Raising sin θ: as lenses of shorter focal length are used, the distance between the object and the lens decreases (Table A1–1) and sin θ approaches unity.

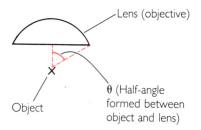

FIGURE A1–1

Determination of the angle θ for calculation of numerical aperture.

TABLE A1-1   Optical Properties of Objective Lenses

Magnification	Focal Length (mm)	Lens-Specimen Distance (mm)	N.A.
10	16	5.5	0.25
40	4	0.6	0.65
95	2	0.1	1.32

In combination ($\lambda = 400$ nm, $n = 1.5$, $\sin \theta = 0.99$), these parameters and an objective lens of maximal N.A. (Table A1–1) give a theoretical maximum resolution of the **objective lens** of about 0.2 μm. This is approximately 1000 times better than that of the unaided human eye.

However, there is another N.A. in the optical system of a light microscope, that of the **condenser,** which focuses light on the object. If this is taken into account:

N.A. microscope = (N.A. objective + N.A. condenser) / 2

Thus, the equation for resolution becomes:

R.L. = (1.2 λ) / (N.A. microscope)

In practical terms, this means that the N.A. of the objective used must be equaled or exceeded by the N.A. of the condenser system, or the resolving power of the microscope will be less than that possible. If the condenser system is partially closed (to allow less light into the system), this may improve the contrast between the specimen and the background, but because the N.A. of the condenser is reduced, the resolution of the microscope is lowered.

Light Microscopes

The **bright-field microscope** (Figure A1–2) detects objects by virtue of their absorbing more or less light than their surroundings. Contrast, which means the ability of an object to stand out relative to nearby objects as well as to the background, allows the objects to appear dark against a light medium. The object and its surroundings must be uniformly illuminated, and this is accomplished by the condenser, a series of lenses and diaphragms below the stage on which the object is located. The condenser lens, whose N.A. is typically equal to that of the objective lens (see above), can be focused by the operator.

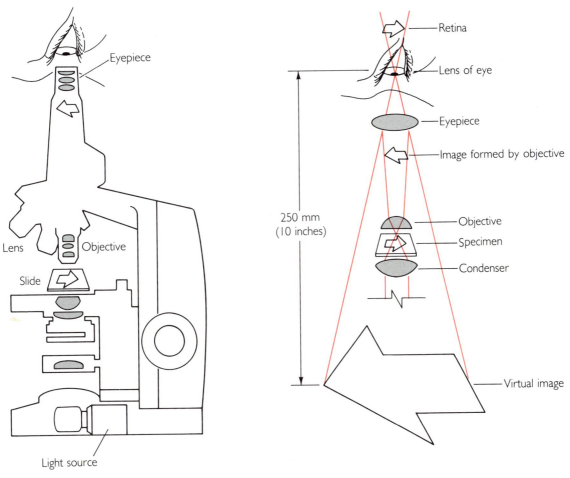

F I G U R E A I – 2 The path of light through the bright-field light microscope. (From: Abramowitz, M. *Microscope basics and beyond.* Olympus Corporation, 1985.)

Light focused on the object is either differentially transmitted (absorbed) or, for the background, fully transmitted to the objective lens. This lens resolves the object to produce a magnified image that is inverted and in focus in the microscope tube. This image is further magnified by a second lens, the ocular, which is stationary at the top of the tube beyond the image produced by the objective. The placement of the ocular is such that it produces a virtual image about 250 mm farther down the body of the microscope (Figure A1–2). Because the whole microscope is typically 260 mm high, the observer sees this image as if it were 10 mm from the bottom of the microscope (that is, at about the level of the object).

A major problem in the light microscopy of cells and tissues is their low level of contrast. The difference in light absorption between, for example, a nucleus and the surrounding cytoplasm is too low for them to be distinguished by conventional light microscopy. Special dyes are used to stain cells to differentiate cells and their components from the medium surrounding them. In addition, a number of modifications can be made to the standard light microscope to enhance contrast. These are summarized in Table A1−2.

The **dark-field microscope** is essentially a bright-field instrument in which the condenser illuminates the object with a hollow cone of light such that none directly enters the objective lens. If an object is not present, the field of view is dark. Instead, the object diffracts or scatters some light, and this enters the objective. Thus, the image formed is one of a bright object against a darkened environment. The contrast is quite vivid (as with stars against a dark sky), and organelles as small as mitochondria and lysosomes are easily detected.

In the **fluorescence microscope,** dyes are used that absorb light at one wavelength and emit light at a longer wavelength. For example, the dye fluorescein absorbs at 490 nm and emits at 520 nm (appears yellow-green), while rhodamine absorbs at 546 nm and emits at 580 nm (appears red). Filters are placed between the condenser and the object (to remove all but the excitation wavelengths) and between the objective and the ocular (to remove all but the emission wavelengths). Few biological molecules are naturally fluorescent in the visible light range. However, fluorescent dyes can be used to bind to cell components, and the sensitivity of this method makes it useful in detecting very small amounts of material. Thus, it can be used to visualize mitochondrial DNA and cell surface molecules (Figure 1−16).

TABLE A1–2   **Effects and Uses of Modified Light Microscopes**

Method	Effects	Uses
Dark-field	Cone illumination; object bright, background dark	Small aggregates; organelles
Fluorescence	Fluorescent molecules or dyes; very sensitive	Organelle DNA; antibody stains
Interference	Split light beam; enhances contrast	Living cells; chromosomes; cell organelle masses
Phase contrast	Retardation of light phase enhances differences	Living cells and organelles
Polarization	Visualization of birefringent structures	Microtubules; cellulose microfibrils

The **interference microscope** uses a series of semireflecting mirrors or birefringent prisms to "split" light from the source lamp into two equivalent beams, vibrating perpendicular to one another and traveling adjacent to one another through the object. The path of each is slightly altered by small differences in the part of the object through which it travels. When the beams are reunited by a prism at the objective, there may be interference due to the path difference, and one part of the object appears brighter than the adjacent part. The amount of interference can be quantified, and this is directly proportional to the dry mass of the object. Thus, this microscope can be used to evaluate changes in a dry mass of cells or nuclei.

When light passes through a cell, differences in materials in the cell cause some of the light to be diffracted, and their phase is altered. This phase change, which does not occur as greatly in light passing through the surrounding medium, can be converted into visible detail by the **phase-contrast microscope.** As in interference microscopy, a hollow cone of light is used for illumination, and this separates the diffracted from undiffracted rays. Above the objective lens, a phase plate further enhances the difference between the undiffracted and diffracted rays so that interference occurs and the object appears as bright and dark regions against a gray background. The phase-contrast microscope is widely used to observe living cells and organelles.

Elongated stacks of molecules, highly ordered in parallel fibers or crystals, are often birefringent because they refract polarized light in different ways depending on the direction from which the light strikes the specimen. The **polarizing microscope** is used to visualize these structures. It is essentially a bright-field microscope with two modifications: (1) a polarizer placed between the light source and the condenser to produce polarized light and (2) an analyzer placed above the objective lens. If the polarizer produces light vibrating in the plane of the crystalline object, the light will be refracted in that plane; if the polarized light is not in the same orientation as the object, it will be refracted in many directions. This microscope is useful in examining such highly ordered structures as microtubules and cellulose microfibrils.

Contrast in the light microscope can be improved using **video microscopy.** In this technique, a video camera is attached to the microscope and transfers the optical image electronically to an image processor. At this point, the image is a series of voltages, and the background voltage is subtracted from the object voltage (or vice versa, depending on microscope and object). The voltage of the object now is in great contrast to that of its surroundings, and this then is sent to a monitor, where it is converted into a visual image. Because the image processor enhances the contrast, other techniques to do this (staining, which kills the cell, or reducing the condenser's light, which reduces resolution) are not needed. The microscope can be used at its maximum resolution. This technique is used to observe small objects (e.g., microtubules—see Figure 9–6) in living cells.

Somewhat related to video microscopy in its use of computers is **confocal microscopy.** In this method, a pinhole device is put into the optical paths of the objective lens and condenser. This allows a single point in the specimen at a time to be used for both illumination and detection. A laser can be used as a concentrated light source, and as the beam moves over the field of view, a computer stores the point images. Background and out-of-focus images can be rejected, resulting in greater resolution of the specimen. This type of microscopy has been especially useful in studies of structures stained with fluorescent antibodies. In conventional fluorescence microscopy, there is often a flare around the specimen and a hazy background. Confocal microscopy largely eliminates these problems, with a resulting sharper image. It also is useful in viewing "optical sections" of thick slices of tissue, since the out-of-focus background is reduced. This has great potential for viewing living tissues and in clinical pathology, where frozen sections are usually cut before a thick biopsy specimen is analyzed.

Electron Microscopes

The practical resolution limit of the light microscope (200 μm) is imposed primarily by the lower limit of visible light wavelengths (400 nm). Visualization of smaller structures (e.g., membranes, ribosomes) can be achieved if the incident electromagnetic wavelength is further reduced. In 1924, F. deBroglie theorized that electrons could behave as waves, with the wavelength inversely proportional to the square root of the voltage used to generate the electrons:

$$\lambda = 12.3 / [(\sqrt{V})(0.1)] \text{ nm}$$

In the **transmission electron microscope,** the voltage is commonly 100,000, so the wavelength of the electrons is 0.004 nm. Aberrations inherent in the lenses of the instrument require its operation at very low numerical apertures (0.001). Thus, the resolution limit, according to Abbe's formula, is approximately 0.2 nm. Compared to the unaided human eye (200 μm or 200,000 nm), this represents a practical improvement of 1,000,000x.

Overall, the transmission electron microscope (Figure A1–3) is quite similar to the bright-field light microscope (Figure A1–2). However, because electrons rather than light rays are used, there are im-

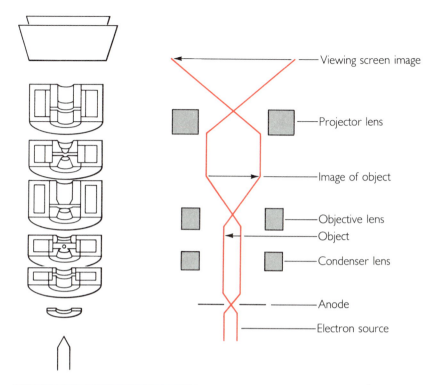

The path of electrons through a transmission electron microscope. The microscope is inverted so that a comparison can be made to the light microscope in Figure A1–2.

portant differences. First, the transmission system must be under **vacuum,** usually necessitating a dead specimen. Second, instead of a lamp as a light source, a hot tungsten filament (cathode) is used to emit **electrons,** which are accelerated and collimated into a beam by passing through a small orifice of high potential (anode). Third, instead of glass lenses, **electromagnets** are used as lenses to bend and focus the electron beam, and magnification can be varied by varying the current in the electromagnets. Fourth, because streams of electrons are not visible to the human eye, the final image is projected onto what is essentially a television screen; alternatively, the photographic film of a camera can be used to detect the image.

The transmission electron microscope, like the bright-field light microscope, operates on the principle of differential absorption. Thus, certain structures in the object absorb electrons, whereas others allow the beam to pass undisturbed to the detector. Staining, typically involving heavy metals (see below), is often required to enhance contrast.

The **scanning electron microscope** permits the visualization of specimens in three dimensions. In this case, a very thin electron beam quickly scans over the surface of the specimen. As is the case in all electron microscopes, the beam–specimen interaction causes the latter

TABLE AI-3   Sample Preparation for Microscopy

Technique	Light Microscopy	Electron Microscopy
Fixation	Aldehydes (formaldehyde, acrolein)	Aldehydes (glutaraldehyde), acids (osmic)
Embedding	Wax, plastic	Plastic
Sectioning	Steel knife	Glass or diamond knife
Staining	Dyes	Heavy metals

to emit weak secondary electrons. These are picked up by a scintillator-photomultiplier, which moves in concert with the scanning beam. The resulting image can be quite beautiful (Figure 12–19). Resolution with this instrument is inferior to that of the transmission microscope, largely due to electrons being scattered many times before reaching the detector. If the object is coated with an emitter of electrons (lanthanum or gold salts are typically used), resolution can be improved to 3 nm.

Sample Preparation

Most bright-field light microscopy and transmission electron microscopy require extensive sample preparation prior to viewing. Samples must be chemically killed and preserved (fixation), cut into thin sections following infiltration by a material that minimizes shearing (embedding), and then stained to bring out contrasts in tissues and cellular organelles (Table A1–3). Electron microscopy imposes even greater technical demands on sample preparation.

    Fixation has five main objectives: to stop biochemical reactions quickly; to minimize extraction of biochemicals from the tissue; to stabilize cell structures and protect them against physical damage; to minimize biochemical changes and allow specific staining reactions to occur; and, in the case of solid tissues, to keep them sufficiently soft to allow sectioning.

    The two major types of fixative are those that precipitate macromolecules (such as ethyl alcohol and various acids) and those that act as cross-linkers. The most widely used of the latter are aldehyde-based fixatives (Figure A1–4), which work by cross-linking macromolecules such as proteins, so that they remain in place in the cell. Buffered formaldehyde and acrolein are used for light microscopy of animal and plant tissues, respectively. Glutaraldehyde, with its two aldehyde groups, is used for electron microscopy. Often, the latter is used in combination with osmic acid, which fixes lipids and some carbohydrates.

    Embedding is necessary prior to cutting thin sections of a tissue, so that the cells can be infiltrated with a supporting medium that min-

HCHO   Formaldehyde

$H_2C = CHCHO$ Acrolein

$OHC — CH_2CH_2CH_2CHO$ Glutaraldehyde

(a)

$A'—NH_2 + HCHO \longrightarrow A'—NHCH_2OH$

$\downarrow B'—NH_2$

$A'—NHCH_2NH—B' + H_2O$

(b)

FIGURE AI-4

(a) Some aldehyde-based fixatives. (b) A possible mechanism of cross-linking of proteins (A' and B') by an aldehyde-based fixative.

imizes the shearing forces of the sectioning machinery. For light microscopy, paraffin wax is used, and the tissue is gradually taken through a series of intermediary solvents to replace its water with molten wax, which is then allowed to harden. This technique is adequate for conventional section thicknesses (about 10 μm). For the thinner sections required for more exacting light microscopy (2 μm) and electron microscopy (0.06 μm), a harder embedding medium is used. In this case, the tissue is dehydrated and infiltrated with an epoxy resin, which is then polymerized to form a hard block.

Instruments used for **sectioning** tissues are called microtomes. They are basically similar to the familiar delicatessen meat slicer: A sample is brought down onto a hard, stationary knife, and at each stroke, the sample advances a prescribed distance, giving the resultant slice a prescribed thickness. For light microscopy, the microtome employs a mechanical advance system and stainless steel knife. These are inadequate to cut the much thinner (less than 0.1 μm) sections needed for finer work. Thermal advance systems, employing metal alloys with known coefficients of thermal expansion, and glass or industrial diamond knives have been developed for the ultramicrotomes used for electron microscopy. Light microscope sections are mounted on glass slides, whereas for electron microscopy, sections are mounted on films on copper mesh grids.

A technique is available that circumvents conventional fixation and embedding. Very rapid freezing in organic solvents at $-110°C$ kills the tissue but prevents the formation of organelle-damaging ice crystals. The ice that does form can be removed by vacuum sublimation. Frozen tissues can be sectioned directly at low temperatures on a freezing microtome, or they can be viewed as whole mounts if sufficiently thin. The advantages of the freezing method are that it avoids chemical treatments and is fast. A disadvantage is its relatively poor structural preservation.

At the light microscope level, frozen sectioning is commonly used when rough, quick data are needed, as in the clinical pathology laboratory. In electron microscopy, it has come into use to examine structures in the cell that are often destroyed by chemical fixation.

The basic purpose of **staining** is to enhance the contrast between cells and the surrounding medium and between structures within the cells. For light microscopy, a large number of dyes are available (Figure A1–5). Hematoxylin, a natural product extracted from the tree *Hematoxylon,* does not bind well directly to tissues but instead is used with alum, an intermediary that promotes binding (mordant). Hematoxylin and safranin are both positively charged and apparently stain cells by binding to most proteins, nucleic acids, and phospholipids, which are negatively charged.

Colors are of little use in electron microscopy, which does not detect them. Instead, electron-dense substances that bind to cellular structures are used as stains to enhance contrast. Typically, these are the salts of heavy metals such as uranium, platinum, lead, iron, and tungsten. Staining can be negative: The heavy metal salt is allowed to surround the cell structures, which then appear light (relatively electron-lucent). If the material is coated with platinum at an angle, material deposited on one side of elevated particles appears as an electron-dense "shadow," and the height of the particle can be estimated.

If a stain is specific for a certain molecule or type of molecule, staining can give information on the chemical composition of the specimen. This study is termed **histochemistry.** The most direct method is to use a dye that binds only to a certain macromolecule. An example is

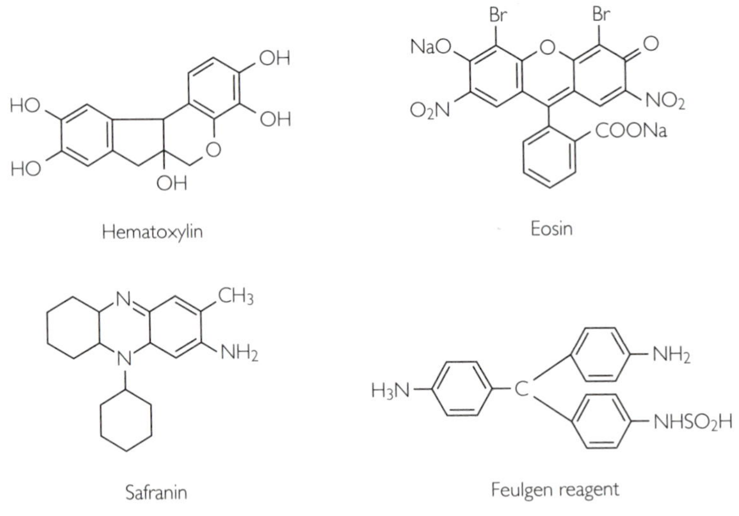

FIGURE A1–5

Structures of some common histological stains. Hematoxylin and eosin are used for animal tissues. Safranin is used for plant tissues. The Feulgen reagent specifically stains DNA.

Hematoxylin

Eosin

Safranin

Feulgen reagent

the Feulgen reagent (Figure A1–5), which reacts with DNA by binding to free aldehyde groups on deoxyribose sugar after purines are released from DNA by mild acid hydrolysis. Extensive control experiments indicate that the binding is specific and quantitative. Other specific stains include (1) the periodic acid-Schiff reagent for polysaccharides and glycoproteins and (2) acridine binding to nucleic acids. The latter is visualized through fluorescence microscopy.

Enzyme histochemistry is more indirect than specific staining. In this case, the sectioned tissue is supplied with exogenous substrate for a certain enzyme, and the product formed is then precipitated in the tissue to a dense or colored substance. In this way, the subcellular localization of an enzyme can be determined. For example, when supplied with esterified phosphate, a phosphatase catalyses the liberation of free phosphate. This can be combined with lead ions to form lead phosphate, which is then converted into lead sulfide, a black precipitate that is easily seen in light microscopy and, because of its heavy metal (lead) content, in electron microscopy as well (Figures 8–1 and 8–2).

Oxidation-reduction enzymes are often identified histochemically by the use of suitable electron acceptors or donors, which become dark or colored when their redox state changes. For example, peroxidases can be identified by supplying a tissue section with hydrogen peroxide and reduced diaminobenzidine. When the peroxide is reduced to water, the colorless benzidine becomes oxidized and turns dark brown (Figure 8–12).

The most specific technique for localization of a cellular molecule is **immunohistochemistry.** In this instance, an antibody is raised to a specific molecule, and the antibody is then used as a staining reagent. Because the antibody-antigen reaction is specific, the staining will also be specific. If the antibody is chemically coupled to fluorescein or rhodamine, its presence can be visualized by fluorescence microscopy (Figure 9–27). Coupling to electron-dense tracers such as ferritin or peroxidase permits fine structural localization of the antigen via the electron microscope. The production of a specific antibody requires that the target molecule be purified and that the antigenic site on the target molecule be exposed for the antibody to bind. This makes a negative result with this technique difficult to interpret.

Autoradiography

Autoradiography is a technique that detects the localization of a radioactive substance in situ. In histology, it is used to localize a macromolecule in the cell after its synthesis. A component monomer is supplied to cells that are synthesizing its polymer. The monomer contains radioactive atoms; thus the polymer made using it will be radioactive. The radioactive polymer is then localized by a process akin to photography.

FIGURE AI–6

Light microscope autoradiography for DNA synthesis.

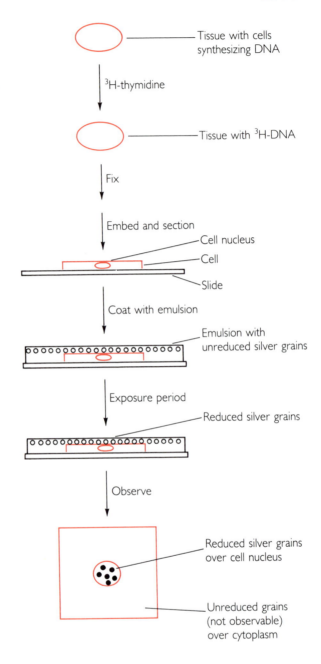

Perhaps the clearest way to explain this process is by using a specific example (Figure A1–6).

Suppose one wants to localize newly synthesized DNA in the cells of regenerating rat liver. Because the cells are in rapid division, DNA synthesis is occurring. The first choice to make is that of the precursor to supply the cells: In this case, thymidine, which is rapidly taken up by

TABLE AI–4 Isotopes Commonly Used to Study Cells

Isotope	Maximum Energy (MeV)	Half-Life (years)
^3H	0.018	12.3
14C	0.155	5570.0
32P	1.71	0.04

the cells and is specific for DNA, is used. For reasons to be explained, tritium (^3H) is the isotopic label on the thymidine. Following injection of the ^3H-thymidine, a period of time is allowed for it to be taken up into new DNA. The tissue is then fixed, embedded, and sectioned.

Slides (or grids) with tissue sections containing ^3H-DNA are then coated with a thin layer of very-fine-grain photographic emulsion. This step is done in the dark, as the emulsion is light-sensitive, and the coated slides are then left in the dark for a period of time ranging from days to weeks. Photographic emulsions are composed of silver halide crystals, which in photography are reduced to metallic silver by the energy from light. In autoradiography, the energy is from the radioactive disintegrations of the isotopic ^3H in the DNA. Because ^3H is a weak energy emitter (Table A1–4), its radioactive disintegrations travel only 1 μm and will reduce silver grains only directly above the site of ^3H-DNA (Figure 10–10).

Following the exposure period, the emulsion-coated slides are incubated (still in the dark) in photographic developer. This substance essentially supplies electrons to the tiny reduced silver grains, and they become precipitated with more silver and microscopically visible. Following fixation to stop these reactions, the slides or grids can be observed: Silver grains, 0.2–1.0 μm in diameter, appear over the region where the newly synthesized ^3H-DNA is located (the nucleus). In addition to DNA, RNA (^3H-uridine as precursor), and proteins (^3H-amino acids as precursor) can be followed in the cell autoradiographically. Pulse-chase experiments, in which the radioactive precursor is followed by a period in nonradioactive precursor, have provided valuable information on the temporal fate of newly made macromolecules (see Chapter 7).

Cell Isolation

Specific Cell Types

Studies of the biochemistry of cell function are often carried out on intact organs in situ. This has the advantage of minimal disturbance to the cell in its natural environment. However, the biochemical analysis of an entire liver or leaf is difficult to interpret at the cellular level because these organs are each composed of several distinct cell types, each of which may respond differently to a hormone or other perturbation. For this reason, studies of individual cell types, carried out in the test tube, are desirable complements to whole-organ analyses.

The easiest to isolate, and the most studied, cell types are those that exist in nature separated from other cells. The sea urchin sheds its eggs and sperm into the sea, and it is a relatively simple process to induce gamete shedding in the laboratory. Thus, it is not surprising that many studies of reproduction and development have been made on this system. Much information on membrane structure has been obtained from the vertebrate erythrocyte (see Chapter 2), since blood cells are easily obtained and can be separated from other cell types by centrifugation. For example, if unclotted blood in 2% dextran (a high-molecular-weight carbohydrate polymer) is left standing for one hour, the erythrocytes will settle to the bottom of the tube. The white blood cells in the supernatant are then purified by centrifugation.

The immediate problem in the isolation of cell types from intact animal organs is to separate the cells from each other. An extensive extracellular matrix composed of collagen and ground substance proteoglycans binds cells to each other (Chapter 12). Therefore, enzymes such as collagenase and hyaluronidase are commonly used to dissociate organs into single cells. Following this, the tissue is either filtered through an appropriate mesh or differentially centrifuged to isolate the desired cell type. Liver hepatocytes, intestinal mucosa cells, and adipocytes have been isolated and studied extensively.

The plant cell extracellular matrix is a wall composed largely of cellulose and other polysaccharides. In a technique analogous to the digestion of animal organs, plant organs can be treated with a mixture of hydrolytic enzymes, such as cellulases, hemicellulases, and pectinases in an osmotically stable environment. The resulting wall-less cells, or protoplasts, can be purified, and they provide a unique system for studies of the plant plasma membrane and for observations of cell wall formation. In addition, protoplasts can be fused to form hybrids and can be infected with foreign genetic material. There is currently great interest in using them for somatic cell genetic engineering of agriculturally important crops.

Cell Culture

When isolated cells or organs are placed in a nutrient medium, survive, and grow, it is termed a cell or organ culture. Newly dispersed cells are termed a primary culture. After these grow and proliferate, they are either physically or enzymatically (if attached to a substratum) removed and reestablished in a fresh medium as a serially propagated cell strain. Those strains that can be repeatedly and indefinitely repropagated are termed established cell lines. Many such lines have been derived from animal and plant sources. The human line derived from a cervical tumor (HeLa), and the cells derived from the stem of the sycamore tree are two often-used examples of cell culture lines.

It is important to note that the lines do not always resemble the parent tissue from which they were derived. For example, most HeLa cell lines have lost the ability to form tumors. In addition, cell lines often display unusual chromosome characteristics (aneuploidy). Nevertheless, cellular biologists tend to treat established lines as "organisms." They offer the advantages of easy care, low expenses once established, and homogeneity.

In the early 1900s, pioneers in the new field of animal cell culture used clotted lymph to keep cells alive. This was interpreted as a general need for physiological solutions. Chick embryo extracts and, more recently, blood serum have replaced the clotted lymph as a source of factors needed for cultured cell survival. Virtually all cell lines have some requirement for serum in the medium, but the exact chemical components of serum that the cells need have not been defined. Hormones, antitoxins, micronutrients, and binding carriers for other medium nutrients are all possibilities. In some cell lines, mixtures of hormones and growth factors can fully replace serum so that a chemically defined medium is possible.

The other environmental requirements for animal cell culture are complex (Table A1–5), and hundreds of media have been developed. Bulk ions and trace elements are needed for the metabolic functions of

TABLE A1–5 **Requirements for Animal Cell Culture**

Bulk ions	Na^+, K^+, Ca^{++}, Mg^{++}, Cl^-, PO_4^{-3}, HCO_3^-
Trace elements	Fe^{++}, $Zn,^{++}$, Se^{++}, Cu^{++}, Mn^{++}, Mo^+
Sugars	Glucose or galactose
Amino acids	Arginine, cysteine, glutamine, histidine, isoleucine, leucine, lysine, methionine, phenylalanine, threonine, tryptophan, tyrosine, valine
Vitamins	B series
Other nutrients	Choline, inositol, purines, serum
Temperature	Optimum organism temperature (e.g., 37°C)
pH	7.15–7.45
Carbon dioxide	2–5% in air
Humidity	Close to saturation

TABLE AI–6   Requirements for Plant Cell Culture

Salts	NH_4^+, K^+, Ca^{++}, Mg^{++}, Fe^{++}, Zn^{++}, Co^{++}, Mn^{++}, Cu^{++}, NO_3^-, Cl^-, SO_4^-, BO_3^-, I^-, MoO_4^{-2}
Sugars	Sucrose, inositol
Vitamins	Thiamine
Hormones	Auxin, cytokinin
Temperature	27°C
pH	5.7

all cells. In the case of the trace elements, these requirements are often unknown, since the elements are often unknowingly supplied as contaminants in other reagents. Hexose sugars, such as glucose or galactose, are used as an energy and carbon skeleton source. Most cell lines require 13 amino acids, and certain lines require additional ones as well (e.g., serine, proline). The water-soluble B vitamins and choline and inositol are general requirements of most cell lines.

Plant cell lines have chemically defined requirements (Table A1–6). These include salts of macro- and micronutrients, a carbon source, and two hormones. Amino acids and most vitamins are not needed, since the plant cell has the biochemical machinery to manufacture these molecules. Plant cultures are usually started as a callus, those undifferentiated cells that grow out as a solid mass from a wounded organ. Callus tissue can be subcultured to solid or liquid media. While the latter are more convenient for biochemical studies, the solid cultures are of interest in that they can be induced to differentiate to form organs and, indeed, entire plants. These effects can be achieved through simple manipulations of the hormones in the culture medium.

Cell Fractionation

Cell Disruption

Histochemical analysis of cells provides valuable but limited information. For example, quantitation is limited to those components that bind a dye stoichiometrically (e.g., the Feulgen reagent on DNA). Biochemical analyses are generally performed on disrupted cells, with a number of methods available for cell disruption.

The simplest method for animal cells is **osmotic lysis.** If cells are placed in a medium whose osmotic concentration of solutes is less than that of the cell interior, water will enter the cell by osmosis. Because cell membranes have limited elasticity, they will rupture because of the internal pressure (as in pumping air into a balloon). A drawback of this method is that the cell organelles also rupture, leading to an unnatural mixing of macromolecules (e.g., leakage of lysosomal hydrolases). In addition, osmotic lysis works well with dissociated cells but not with solid animal tissues or plant cells. A related method for membrane rupture is **sonication.** In this case, ultrasound causes molecular vibrations and membranes are easily broken. This method is commonly used to break up membrane-bound organelles (see Figure 3–11).

Intact tissues are usually disrupted by physical means. The simple **mortar and pestle** have been in use for centuries and are still widely employed for the homogenization of plant tissues. An important modification of this is the motor-driven rotating teflon or ground glass **pestle and close-fitting glass tube.** In this instance, the tissues are broken open by shearing. When tissues are too tough for the pestle homogenizers (e.g., plants, muscles), high-speed **blenders** with rotating steel blades are employed.

The choice of **medium** in which cells are disrupted is crucial. Typically, media are buffered at or near physiological pH, and they should be isosmotic to the cells' cytoplasm, so that organelles do not burst. The ionic strength of the medium is also an important consideration, as many molecules in the cell have electrostatic interactions with other components. An example of the problems encountered is the isolation of chromatin from cell nuclei. If the ionic strength of the homogenizing medium is too high, intrinsic chromosomal proteins will no longer bind to DNA or the chromosome; if the ionic strength is too low, cytoplasmic molecules can nonspecifically adsorb to the nuclear material when cells are disrupted.

Because many cellular molecules are labile at room temperature, homogenization and further cell fractionations are carried out at **ice-bath** temperatures. Even with this precaution, many small molecules break down easily, and care must be taken in measuring such substances as nucleotides and hormones. If a tissue is rich in lysosomal hydrolases, their action on other cellular constituents can be prevented by the use of specific inhibitors.

Centrifugation

Cellular organelles and macromolecules differ from each other in size and density. However, the differences between them are not very great, and the organelles are quite small. To separate cell components in a homogenate by unit gravity sedimentation would be a very time-consuming task. These separations can be hastened by putting the homogenate into a tube and applying a centrifugal force, causing the particles to migrate down the tube.

The velocity of a particle in a centrifugal field is defined by the equation:

$$v = \frac{\phi(d_p - d_m)w^2x}{6\pi nr}$$

where v is the radial velocity in cm/sec; ϕ is the volume of the particle in cm^3; d_p and d_m are the densities, in g/cm^3 of the particle and medium, respectively; w is the angular velocity in radians/sec; x is the radial distance of the particle from the center of rotation; n is the viscosity of the medium; and r is the radius of the particle.

This complex equation is very informative as to the parameters that can be manipulated in a cellular biology experiment using centrifugation. Clearly, a particle will move more quickly if its volume is larger; if it is much denser than the medium; if the centrifuge goes faster; if the particle is farther from the center of the spinning axis; if the medium is not viscous; and if the particle is compact and dense.

The sedimentation coefficient (S value) relates the velocity of the particle to its unit acceleration. Or,

$$S = \frac{v}{w^2x}$$

As an example, suppose a ribosomal fraction from liver is sedimented in a centrifuge tube, starting 7 cm from the center of rotation (x) and in a centrifuge at 70,000 revolutions per minute. Then

$$w^2x = (70{,}000 \text{ rpm} \times 2\pi \text{ radians/rev} \times 1/60 \text{ min/sec})^2 \times 7 \text{ cm}$$
$$= 3.76 \times 10^8 \text{ cm/sec}^2$$

If the particle sediments at 1.04×10^{-2} cm/sec (v), its sedimentation coefficient is:

$$S = \frac{1.04 \times 10^{-2} \text{ cm/sec}}{3.76 \times 10^8 \text{ cm/sec}^2}$$
$$= 276 \times 10^{-13} \text{ sec}$$

The S value is defined as 10^{-13} sec, so the coefficient of the ribosome cluster is 276 S under these conditions.

The theory of centrifugation suggests a practical approach to the separation of cell components. Because the organelles have different densities (Table A1–7), they should behave differently in a centrifugal field. To further improve separation, the medium in which centrifugation occurs can be set up as a density gradient, with the densest medium at the bottom of the tube and the lightest medium at the top. Thus, as the particle travels through this medium, the expression ($d_p - d_m$) is constantly changing.

TABLE AI–7 Densities of Various Cell Organelles

Organelle	Density (g/cc)
Lysosomes	1.22
Microsomes (smooth)	1.12
Microsomes (rough)	1.20
Mitochondria	1.17
Nuclei	1.32
Peroxisomes	1.27
Ribosomes	1.58

If the particle is denser than the densest part of the medium, the expression will be greater than zero and v will be positive: This difference in velocities is the basis of **rate-zonal** centrifugation. If the densest portion of the medium is denser than the particle, the latter will travel down the gradient until it reaches a position where its density is the same as that of the medium ($d_p = d_m$); at this point its velocity will be zero. This is the basis of **isopycnic** ("equal density") centrifugation.

A comparison between rate-zonal and isopycnic centrifugation is shown in Figure A1–7. Rate-zonal centrifugation is the method of choice for separating larger particles, such as organelles, and larger molecules with measurable differences in density, such as ribosomal RNAs. The medium used is usually a sucrose solution or, in cases where higher densities are desired for whole-cell separations, carbohydrate polymers such as dextrans. A flow sheet for the differential isolation of crude cell organelle fractions from a rat liver homogenate (Figure A1–8) shows that, in this instance, fractions are actually allowed to go to the bottom of the tube. This is not always the case: With molecular separations the centrifugation run is often stopped before pelleting, and the gradient is analyzed at that point.

Isopycnic centrifugation is used to separate macromolecules with small density differences (e.g., 1.110 g/cc vs. 1.115 g/cc). The medium is usually a high-density salt solution, such as cesium chloride. The centrifuge is run until equilibrium ($d_p = d_m$), and then the gradient is analyzed.

Analysis of a centrifugation experiment in which the fractions are not pelleted is straightforward. The completed gradient is fractionated (i.e., the contents of the tube are divided into many fractions), often by puncturing the bottom of the tube. Each fraction is then analyzed for the substance in the gradient (e.g., RNA, protein, or an enzyme). The results are typically plotted out as an "amount vs. fraction number" graph (Figure A1–9).

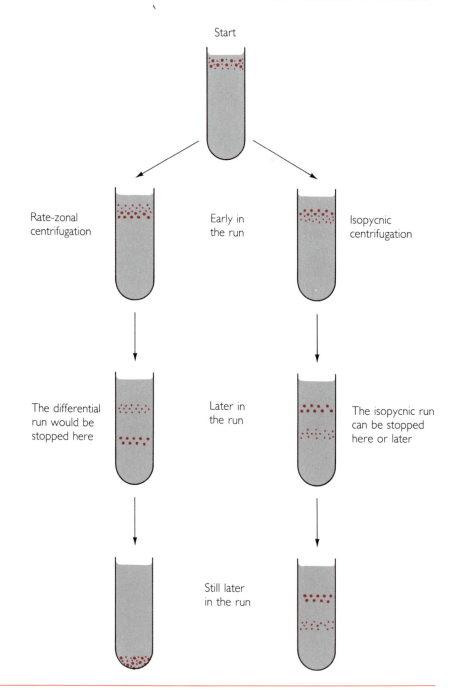

Rate-zonal
centrifugation

Early in
the run

Isopycnic
centrifugation

The differential
run would be
stopped here

Later in
the run

The isopycnic run
can be stopped
here or later

Still later
in the run

FIGURE AI–7

Rate-zonal versus isopycnic
centrifugation of ●●●●● large
particles of low density
and ∴∶∴∶∴ small particles
of high density.

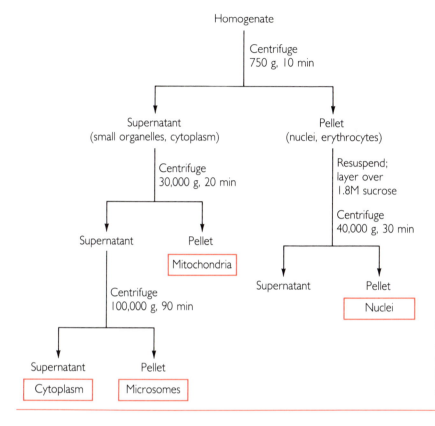

Homogenate

Centrifuge
750 g, 10 min

Supernatant
(small organelles, cytoplasm)

Pellet
(nuclei, erythrocytes)

Centrifuge
30,000 g, 20 min

Resuspend;
layer over
1.8M sucrose

Centrifuge
40,000 g, 30 min

Supernatant

Pellet

Mitochondria

Supernatant

Pellet

Nuclei

Centrifuge
100,000 g, 90 min

Supernatant

Pellet

Cytoplasm

Microsomes

FIGURE AI–8

Scheme for the fractionation of organelles from a rat liver homogenate by differential centrifugation.

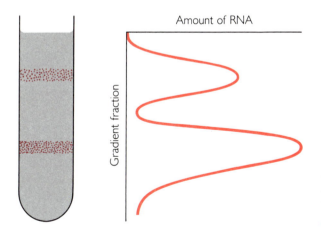

Amount of RNA

Gradient fraction

FIGURE AI–9

Graphical interpretation of a centrifugation experiment. In this case, dissociated ribosomes from cucumber seedlings were centrifuged in a gradient of 5–20% sucrose. The two peaks represent ribosome subunits.

Biochemical Methods

Quantitation: Spectroscopy

The measurement of the amounts of various molecules in cell fractions isolated by centrifugation or in samples purified biochemically (see below) is usually done by spectroscopic techniques. The energy of light is inversely proportional to its wavelength.

$$E = \frac{hc}{\lambda}$$

where E is the energy of light, h is Planck's constant, c is the speed of light, and λ is the wavelength. If a biological (or other) molecule absorbs light at a given wavelength, some of its electrons may go to an excited state. When these electrons return to their ground state, they emit the energy they had absorbed as heat or light:

$$E = \frac{hc}{E_2 - E_1}$$

where E_2 is the energy level of the molecule after absorption of light and E_1 is the energy level before absorption.

The probability of a mole of a molecule absorbing light at a given wavelength is termed the molar extinction coefficient. Different biologically important molecules have different absorption maxima (Table A1–8). This forms the basis of quantitative assays for these molecules when in solution in unknown amounts. According to Beer's law:

$$\text{absorbance} = \log (I_o/I_r) = \epsilon cl$$

where I_o is the intensity of light incident on the molecule, I_r is the intensity of light leaving the molecule after absorption, ϵ is the extinction

---

TABLE AI–8   **Absorption Maxima and Extinction Coefficients of Biological Molecules**

Molecule	Absorption Maximum (nm)	Extinction Coefficient ($\times 10^3$ mole^{-1}cm^{-1})
Adenosine	260	14.9
DNA	258	6.6
RNA	258	7.4
Tryptophan	280	5.6
Uridine	260	10.1

coefficient, c is the concentration of the molecule, and l is the path length of the light through the solution (typically 1 cm). Thus, for analysis of gradient fractions for RNA (Figure A1−9), the absorbance of each fraction at 258 nm is determined (Table A1−8). Because ϵ and l are known, c, the concentration of RNA, can be calculated.

Often, other substances are present in a solution that absorb at the maximum wavelength of absorption of the desired molecule. For example, if the quantitation of RNA in a nuclear fraction (Figure A1−8) was attempted by absorbance at 258 nm, it would be complicated by the presence of DNA, which also absorbs strongly at this wavelength. In the absence of further purification, the RNA must be analyzed by a dye-binding technique. In this case, a dye (orcinol) binds stoichiometrically to RNA and, when bound, absorbs strongly at a certain wavelength (600 nm). Specific colorimetric assays, sensitive to the μg level, are available for a variety of other biological molecules, such as DNA, protein, lipids, and carbohydrates.

Because a given enzyme is present in the cell in an extremely low concentration, direct quantitation is difficult. Instead, the enzyme is indirectly quantitated through measurement of its catalytic activity. The assumption, not always valid, is made that the activity of an enzyme is directly proportional to its concentration in solution. Assays are specific for each enzyme and commonly involve spectroscopy.

Isotopic Tracers

Chemical assays using spectroscopy are commonly sensitive to 10^{-7} M concentrations. However, the concentrations of cellular constituents can be in the range of 10^{-14} to 10^{-6} M. Clearly, for analysis of small amounts of tissues or molecules, standard spectroscopic assays are not sufficiently sensitive, but the use of radioactive isotopes can provide the desired sensitivity. Compounds containing such isotopes (e.g., ^{3}H-thymidine, see above) can be supplied to cells in extremely low amounts (10^{-15} M). The sensitivity of instruments designed to detect isotopic radiation is such that the fate of the tracer can be easily followed.

In a typical experiment, a known quantity of the radioactive molecule is supplied to cells, for example, ^{3}H-thymidine to regenerating rat liver as in the autoradiographic experiment described earlier. Following the experimental period, the tissue is homogenized, its DNA is purified, and its radioactivity is determined. This level of ^{3}H-DNA can then be compared to that of other tissues in other physiological situations.

The unit of radioactivity measurement is the Curie, that amount of radiation emitting from one gram of radium (3.7 × 10^{10} disintegrations per second). If the specific activity (amount of radioactivity per unit concentration) of the tracer molecule is known, the determination of the radioactivity in the final cellular product can be used to calculate the concentration of that product. For example, if the specific activity of the ^{3}H-thymidine is 1 Curie/millimole (3.7×10^{10} dps/mmole), and the radioactivity in the ^{3}H-DNA formed by the cell is 3.7×10^4 dps, the amount of thymidine tracer incorporated into the DNA is 10^{-6} mmoles or 1 nmole.

Detection of radioactivity commonly is done in a liquid scintillation spectrometer. Samples are put into glass or plastic vials containing a fluor, and when radioactive disintegrations occur, their energy excites electrons in the fluor. When these electrons return to their ground state, their absorbed energy is emitted as photons, which, via secondary fluors in the solution, can be detected and converted electronically by the spectrometer into a quantitative readout of disintegrations per unit time.

Immunochemistry

Antibodies are proteins produced by the immune systems of higher animals in response to the appearance of foreign substances, or antigens. Typically, the antibodies, or immunoglobulins, are present in the bloodstream. Each antigen elicits the formation of an antibody that is highly specific for that antigen. The antibody molecule is usually Y-shaped, with the two terminal arms each containing a site for binding to the antigen (Figure A1–10). For the organism, this impressive and complex system is a front-line defense against disease. For the cell biologist, antibodies are important analytical reagents.

The use of fluorescent antibodies in histochemistry has already been described. For biochemical analyses, immunochemistry can provide the sensitive quantitation lacking in other techniques. For example, if a pure human enzyme is available, an antibody can be made to it by injecting the enzyme as antigen into a different animal or by a newer method, monoclonal antibody production. If a known quantity of this antibody is exposed to a human cell extract, some of the antibody molecules will bind to their antigen (the human enzyme). Following binding, the number of free antibody molecules remaining can be quantitated. This gives an estimate of the number of bound antibody molecules and, hence, the amount of antigen present. Using this principle, assays have been developed that can reliably measure 10^{-15} M of such molecules as hormones and cyclic nucleotides.

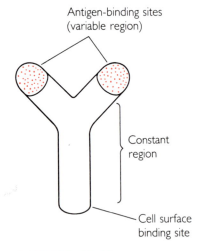

Antigen-binding sites (variable region)

Constant region

Cell surface binding site

FIGURE A1–10

General structure of an antibody (immunoglobulin G).

Separation Techniques

A large number of techniques have been developed for the biochemical fractionation of cells and cell organelles. One of the most important of these is chromatography, in which molecules in solution are placed in a system containing two phases—one mobile and the other stationary. The solutes then separate from each other by distributing themselves between the two phases on some chemical basis.

In **adsorption** chromatography, the mobile liquid with solutes is in contact with a solid stationary phase that has properties that allow it to bind the solutes with differing affinities. These solutes, separated from other molecules, can then be eluted from the solid matrix separately because of their different affinities of binding. Solid support matrices employed include silica gel (for the separation of steroids), activated carbon (for amino acids and carbohydrates), and calcium phosphate gel (for proteins and nucleic acids). Bulk separations can be done by stirring the solution mixture with a slurry of solid support and then filtering out the matrix with its bound solutes. Finer separations are done by packing the matrix into a hollow glass or plastic tube (column) and collecting fractions as the elution of the solutes proceeds.

Two specialized adsorption chromatography column techniques are widely used in the purification of biological macromolecules. **Ion-exchange** chromatography employs a solid support to which a charged group has been covalently attached. For example, if the group has a negative charge, positively charged solutes (at the pH used in the experiment) will ionically bind to the matrix, thereby effecting a separation of these solutes from the negatively charged solutes. The bound molecules are then separated from each other when the column is eluted with a solution of increasing ionic strength. Weakly bound (weakly charged) solutes are eluted first as they are replaced, or exchanged, for cations in the eluting solution. Later, strongly bound solutes are eluted (Figure A1–11). This type of chromatography is employed in protein and nucleic acid purifications.

The second specialized technique is **affinity** chromatography. In this case, an antibody or cofactor specific for a particular solute is covalently bound to the matrix. Even in a very complex mixture, the target solute will bind in a highly specific manner to its ligand. Clearly, this method can provide an essentially one-step purification, and enzyme purifications have been readily achieved by this technique.

Small molecules are commonly separated by **partition** chromatography. The solid support in this case does not exert a selective effect on the solutes, but rather their differential solubilities in a solvent passing over the matrix effects the separation. The matrix serves to hold in place the solutes as they come out of solution and precipitate. Paper was one of the first chromatographic matrices used. Later, thin-layer chromatography was developed, in which the matrix coats a glass or plastic plate as a very thin layer. This has the advantage of greater speed of

A mixture of positively and negatively charged solutes is applied to a column whose matrix has a negative charge.

The positively charged solutes bind to the matrix. The negatively charged solutes pass through the column.

Elution (ion exchange) begins with a gradient solution of high ionic strength.

As elution proceeds, loosely bound solutes are eluted first; tightly bound solutes are eluted later.

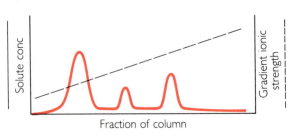

FIGURE AI–12
Principles of gel-filtration chromatography.

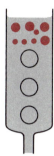

A mixture of large and small molecules is applied to a column containing beads of small pore size.

The small molecules can enter the beads and are retarded. The large molecules cannot enter the beads and pass through the column quickly.

The large molecules have left the column, but the small molecules remain.

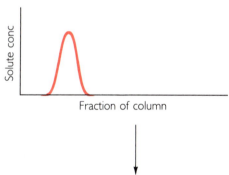

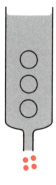

The small molecules are now eluted.

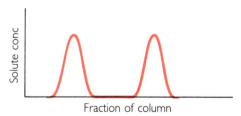

solute separation than paper and comparative ease of elution of the separated solutes.

Small molecules that are themselves volatile or can be made into volatile derivatives can be separated by **gas-liquid** chromatography. In this case, the mobile phase is a gas, which contains the solutes, and the stationary phase is a liquid, which coats a solid matrix. Separation is achieved by the partitioning of the solutes between the gas and liquid phases. This method is useful in the analysis of alcohols, amines, and fatty acids.

Gel filtration chromatography separates solutes on the basis of their sizes and shapes. The matrix is composed of a polymer such as dextran, agarose, or polyacrylamide, and careful regulation of polymerization can result in gels with pores of defined size that allow molecules of certain dimensions to enter. Clearly, the smallest molecule will enter the most pores and be quite slow to pass through the column. Larger molecules will enter fewer pores and travel through the column more quickly. Thus, a separation is achieved (Figure A1–12). This method of chromatography is used in protein and nucleic acid purifications.

Electrophoresis is the movement of electrically charged substances in an ionic solution with an electric field, performed with an inert support. Because many biological molecules are differently charged, this technique can be used to separate and analyze them. The electrophoretic mobility of such a molecule is:

$$\text{mobility} = QV \, / \, 6\pi nr$$

where Q is the electric charge of the molecule, n is the viscosity of the medium, r is the radius of the particle, and V is the applied voltage.

Several types of inert media are used in electrophoresis. Amino acids and small peptides can be resolved on paper. Cellulose acetate (paper with exposed hydroxyl groups acetylated) is used for rapid separation of proteins in the clinical laboratory. Finer separations can be achieved on gels made into slabs or tubes with media such as agarose, polyacrylamide, or starch. Small amounts (micrograms) of proteins or nucleic acids can be separated in a few hours and visualized by staining (Figure 2–28). Gels can also be autoradiographed by the use of X-ray film, making possible the detection of tiny amounts of material if it is labeled with a radioactive tracer.

The separated molecules in the gel can be analyzed after elution from the gel matrix. This can be achieved by cutting out the slice of gel of interest and incubating it in a buffer, often with the help of a current (electroelution). An alternative method is to overlay a slab gel with a piece of nitrocellulose paper and blot the gel contents onto the paper, which then essentially becomes a replica of the gel. The paper is then easily manipulated, much more so than possible with the friable gel matrix.

If the location of a specific protein in the gel is desired, an antibody to that protein is applied to the paper, with suitable staining (see above); this is termed the Western blot. If the location of a specific DNA sequence is desired, the DNA in the paper can be denatured and then hybridized to a suitable probe, and the hybrid can be detected autoradiographically; this is termed the Southern blot. Similarly, an RNA sequence can be located in the gel blot by hybridization; this is the Northern blot.

Gel electrophoresis matrices contain pores and in this respect are similar to the matrices used in gel filtration chromatography. Thus, there is a separation based on molecular size as well as on molecular charge. Conditions can be manipulated so that only size is the separation parameter: If the solutes are coated with the detergent sodium dodecyl sulfate (SDS), they become denatured and uniformly negatively charged because of the strongly charged detergent. Thus, SDS gel electrophoresis can be used to estimate molecular weights of proteins and nucleic acids.

Molecular Methods

Amplification of DNA

The development of methods based on research in molecular biology has provided cell biologists with powerful techniques for the analysis of the structure and function of DNA and proteins. These techniques make up the burgeoning field of biotechnology, and a detailed discussion of them is beyond the scope of this book. An important aspect of this work is the ability to amplify the amount of any DNA sequence. The purpose of this amplification might be to characterize the gene for a protein of interest, to use the gene sequence to characterize the protein product of the gene, to insert the gene into a new cell to test its function, or to modify the gene in the test tube, followed by insertion into a cell to test for the function of the modified protein product.

The first, and most widely used, method of amplifying a DNA sequence is **cloning** via the creation of recombinant DNA. In general, this technique involves using a naturally occurring (or derived) piece of DNA in a host cell as a vector, which is isolated and into which the target sequence is inserted. The vector-target recombinant DNA is then reinserted into the host and is propagated along with the host (Figure A1–13).

There are many **vectors** for the creation of recombinant DNA. These include bacterial plasmids and viruses, as well as plasmids and

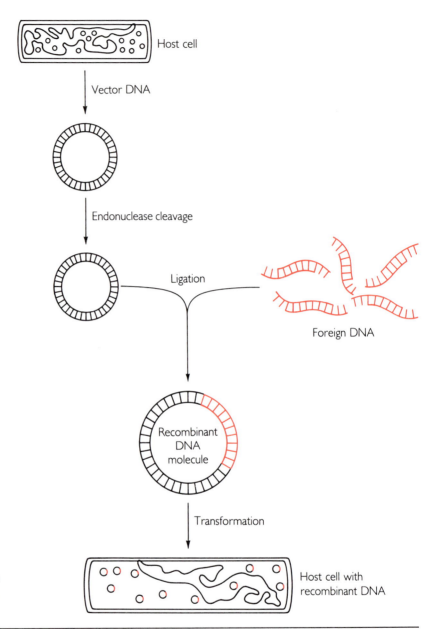

Host cell

Vector DNA

Endonuclease cleavage

Ligation

Foreign DNA

Recombinant
DNA
molecule

Transformation

FIGURE AI–13

Cloning of DNA by insertion into
a vector and reproduction within
a host cell.

Host cell with
recombinant DNA

viruses from eukaryotic plant and animal cells. Almost all of the vectors in current use are derived from naturally occurring DNA molecules
that have been modified for experimental use and include the following
properties:

1. They are usually smaller than the host chromosome(s), facili-
tating their isolation.

2. They contain an origin of DNA replication, so that they do not need to be integrated into the host genome to be propagated. Often, this origin acts independently of the host origin(s), so that multiple copies of the vector can be generated.

3. They contain unique sites for cleavage by restriction endonucleases. There are over 400 of these enzymes, which are synthesized by microorganisms to cut foreign DNA at specific DNA sequences. For optimal use, a vector has only one site for a given restriction enzyme.

4. They contain DNA sequences that code for proteins that provide markers for the presence of the vector in the host cell and for the presence of the inserted target sequence.

These properties are illustrated by the vector pBR322 (Figure A1–14):

1. This vector (4400 base pairs) is an extrachromosomal plasmid of *E. coli* and is much smaller than the bacterial host chromosome (about 1,000,000 base pairs).

2. This vector contains an origin of DNA replication that is independent of the host chromosome, and there can be several thousand vectors per cell.

3. This vector is cleaved once by a number of restriction endonucleases. For example, the enzyme Bam H1 cleaves DNA:

$$5'—GGATCC—\xrightarrow{\text{Bam H1}}—G \qquad GATCC—$$
$$3'—CCTAGG— \qquad —CCTAG \qquad G—$$

This occurs once in the plasmid, turning the circular molecule into a linear one. Target DNA with the complementary ends

$$GATC———$$
$$———CTAG$$

will bind to the cleaved vector by base pairing to circularize it.

4. This vector contains DNA sequences that code for host resistance to the antibiotics ampicillin and tetracycline. Any *E. coli* with this plasmid will normally be resistant to these antibiotics. However, the Bam H1 site is within the tetracycline resistance gene (Figure A1–14). If a target DNA sequence is inserted into this gene, the insertion event inactivates the gene, and this means that the resultant recombinant plasmid will confer only resistance to ampicillin on its host. Thus, host cells containing the normal versus the recombinant plasmid can be differ-

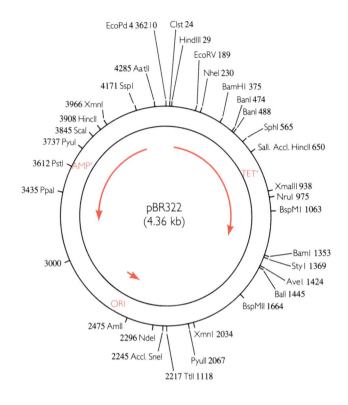

pBR 322: plasmid vector
          host: *E. coli.*
          AMP^r: ampicillin resist.
          TET^r: tetracycline resist.

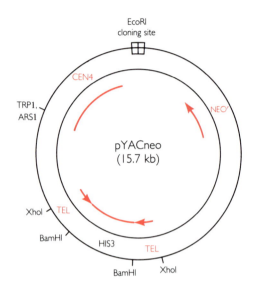

pYACneo: yeast artificial chromosome
          host: yeast.
          NEO^r: neomycin resistance.
          TEL: telomere sequence.
          CEN: centromere sequence.

FIGURE AI–I4 Two widely used vectors for DNA cloning. (Ori: origin of DNA replication; numbers: base pairs on the map; abbreviations at lines [e.g., Pvu II]: restriction endonuclease sites.)

entiated from each other by selection of media that contain antibiotics.

There are many different vectors and selection methods for both prokaryotic and eukaryotic host cells. However, the target DNA need not be from the host species. Because the creation of the recombinant vector is chemical and occurs in the test tube, it is perfectly feasible to clone a mammalian target in a bacterial host or vice versa.

Cloning allows amplification of a target DNA, but it is laborious and time consuming. A more convenient method when a specific target sequence is available is to amplify the DNA totally in the test tube by the **polymerase chain reaction** (Figure A1–15). By chemical synthesis, small primers are synthesized that are complementary to the ends of the target sequence. The target is heated to denature it; then it is cooled, and the primers are hybridized to the ends of the separated strands. The primers are then extended via DNA replication complementary to the

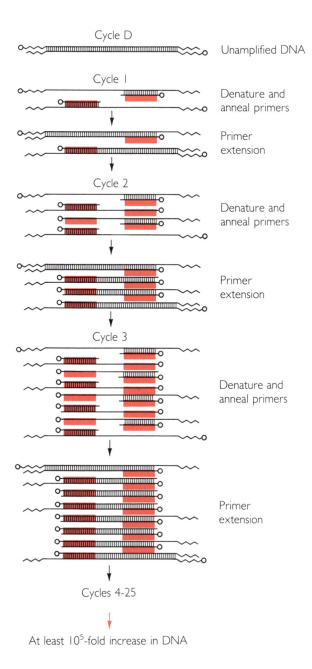

FIGURE A1–I5

Polymerase chain reaction (PCR) method of amplifying a DNA fragment in the test tube.

target sequence in a reaction catalyzed by DNA polymerase, thus making two strands. These in turn are doubled in a second round of denaturation, annealing of primers and DNA synthesis, and the process is repeated many times. The use of a heat-stable DNA polymerase isolated from bacteria that live in hot springs obviates the need to add new polymerase after each denaturation step (since this would denature most DNA polymerases).

Manipulation of DNA

Once a target DNA sequence has been amplified, it can be studied or used for experiments. The most obvious studies are intrinsic ones of the sequence itself. This can be done by DNA **sequencing,** for which several methods are available. These methods are easily automated, and the resulting sequences can be scanned by computer for stretches of molecular biological interest, such as controlling elements for transcription and binding sites for proteins.

An important result of DNA sequencing is the ability to infer the sequence of the protein coded for by the DNA. The traditional approach to analyze cell functions genetically has been to look for an altered phenotype (e.g., a lack of membrane transport of a ligand) and then try to isolate the affected protein by laborious comparisons of normal versus mutated cells. If the affected gene can be isolated (e.g., via characterization of a chromosomal deletion), amplified, and sequenced, the protein product can be deduced from the genetic code. This deduced sequence can then be used to make an antibody, which then binds to the true protein in tissue extracts. This technique has been termed **reverse genetics** and has been used to characterize the proteins that are deficient in such disorders as muscular dystrophy and cystic fibrosis.

Once a DNA sequence coding for a protein has been isolated, it can be used experimentally to test its function in new physiological situations. Because high concentrations of Ca^{++} tend to permeabilize membranes, a target DNA can be introduced **(transfected)** into a cell quite different from its host. With a low but measurable efficiency, the target is inserted into the host cell chromosome and is replicated along with the host. If the molecular conditions are right, the target DNA will be expressed to protein in the new cell. For example, if the DNA coding for a putative gap junction protein is inserted into cells that normally lack the protein, the cells will form gap junctions. Clearly, transfection is important in establishing the physiological role of a protein.

Chemical and enzymatic methods can be used to change specific nucleotides in the sequence of an isolated DNA. This in vitro **mutagenesis** provides a powerful method for testing hypotheses on the roles of specific regions on a protein. For example, it is hypothesized that the hydrophobic signal sequence is needed for transport of a protein across the membrane of the endoplasmic reticulum. If the DNA coding for such a protein is isolated and the signal region targeted, it can be modified to change the hydrophobic amino acids into hydrophilic ones. If transfection of this mutated DNA into cells abolishes trans-ER movement, the hypothesis is strengthened.

Further Reading

Abramowitz, M. *Microscope basics and beyond.* Olympus Corporation, 1985.

Abramowitz, M. *Contrast methods in microscopy.* Olympus Corporation, 1987.

Ausubel, F., Brent, R., Kingston, R., Moore, D., Seidman, J., Smith, J., and Struhl, K. *Current protocols in molecular biology.* New York: Wiley, 1987.

Bradbury, S. *An introduction to light microscopy.* Oxford: Oxford University Press, 1984.

Colowick, S., and Kaplan, N. (Eds). *Methods in enzymology.* New York: Academic Press (multi-volume series).

Freifelder, D. *Physical biochemistry.* San Francisco: W. H. Freeman, 1982.

Prescott, D. (Ed.). *Methods in cell biology.* New York: Academic Press (multi-volume series).

Sambrook, J., Fritsch, E.F., and Maniatis, T. *Molecular cloning: A laboratory manual.* Cold Spring Harbor Press, 1989.

APPENDIX 2

Outline of Biochemistry

Small Molecules
    Inorganic
    Organic
Large Molecules
    Carbohydrates
    Lipids
    Nucleic Acids
    Proteins
Energy and Enzymes
Metabolism
    Energy
    Protein Synthesis

Small Molecules

Inorganic

Cells are composed of the same chemical elements as the physical world, but the proportions of these elements in living tissues differ from those elsewhere (Table A2–1). Carbon, hydrogen, oxygen, nitrogen, sulfur, and phosphorus account for about 99% of the atoms in a typical cell. In addition to these major elements, cells contain trace amounts of virtually all of the other nonradioactive elements. These can be localized in tissue sections by very sensitive methods, such as electron microprobe analysis. Each trace element has a specific biochemical role(s), usually as a cofactor in an enzyme reaction (Table A2–2). Because of these specific roles, a dietary deficiency of a trace element can have serious consequences for cellular function, and their study is an important aspect of plant and animal nutrition.

Elements exist in cells as part of chemical compounds or as ions. The latter are charged species that are formed when salts, whose atoms are held together by ionic attraction, dissolve in water:

$$NaCl \rightarrow Na^+ + Cl^-$$

Because water is itself charged (see below), shells of water molecules form around the Na^+ and Cl^- ions. Such shells do not form around uncharged species, and they generally fail to dissolve in water. Because much of biochemistry occurs in water solution, many elements in cells

---

T A B L E A2–I **Elemental Compositions**

Element	Relative Abundance (atom%)		
	Universe	Earth's Crust	Human Body
H	90.79	0.22	60.3
C	9.08	0.19	10.5
O	0.057	47.10	25.5
N	0.042	Trace	2.4
Si	0.026	28.51	Trace
Fe	0.0047	4.52	0.0006
Mg	0.0023	2.21	0.01
S	0.0009	Trace	0.132
Al	0.00005	7.91	Trace
P	0.0003	Trace	0.134
Ca	0.0002	3.57	0.226
K	0.00002	2.51	0.036

TABLE A2–2   Roles of Trace Elements in Cells

Element	Role
Co	Part of vitamin B_{12}
Cu	Cofactor of several enzymes in respiration
I	Part of thyroid hormone
Mg	Component of chlorophyll
Mn	Cofactor for phosphokinases
Mo	Cofactor for nitrogenase
Zn	Cofactor for carbonic anhydrase

exist either as solitary ions (e.g., Cu^{++}, Fe^{++}, I^-, Cl^-) or as ionic compounds (e.g., BO_3^{-2}, MoO_4^{-2}, SO_4^{-2}, PO_4^{-3}.

Water makes up about 80% of the molecules in a cell and has several properties that make it a suitable cellular medium. First, it is a solvent that can dissolve substances such as ions and wastes but leave as insoluble molecules such as lipids, which form cellular structures such as membranes. Second, its high heat capacity and high heat of vaporization, relative to those of other solvents, allow water to absorb the heat released by a biochemically active cell without raising the cellular temperature. Third, water plays an active role in certain biochemical reactions, such as photosynthesis and hydrolysis.

Although its atoms are not primarily held together ionically, water can form ions to a measurable extent:

$$H_2O \leftrightarrow H^+ + OH^-$$
$$\text{hydrogen ion   hydroxide ion}$$

(The convention H^+ is used, although these ions are bonded to water to form hydronium ions, H_3O^+). One liter of water at standard temperature and pressure contains 10^{-7} moles of each of these ions. The **pH scale** depicts the negative logarithm of the hydrogen ion molar concentration:

$$10^{-7} \text{ M: } -\log(10^{-7}) = \text{pH } 7.0$$

Salts that donate hydrogen ions (protons) to a water solution are termed acids:

$$HX \rightarrow H^+ + X^-$$

Salts that reduce the hydrogen ion concentration in a solution are termed bases:

$$YOH \rightarrow Y^+ + OH^- \text{ (the } OH^- \text{ combines with } H^+)$$

Cells typically function optimally at pH 7, or neutral pH. Plant cells tend to be slightly below this value (pH 6.5), and animal cells are often slightly above neutrality.

Significant deviations from the optimum cellular pH lead to significant aberrations in cellular functioning. Thus it is not surprising that cells have a mechanism to maintain a constant pH. This is achieved by the use of **buffers,** which are salts of acids or bases that are weak (tend to dissolve into ions only minimally):

$$NaHCO_3 \quad \rightarrow \quad Na^+ + HCO_3^- + H^+$$

Sodium bicarbonate bicarbonate ion

$$HCO_3^- + H^+ \underset{\longleftarrow}{\longrightarrow} H_2CO_3$$

In this instance, bicarbonate ion tends to combine with H^+, thus removing it from solution. Clearly, the amount of buffering is proportional to the amount of the buffer salt in solution. Other important cellular buffers include amino acids and phosphates.

Organic

The ionic bond that holds a salt together when it is not in solution is fairly weak and is broken when the compound dissolves in water. In contrast, **covalent bonds,** which involve the sharing of electrons by two atoms, do not break easily and form the basis of the stable larger molecules that exist in cells. The carbon atom has four electrons that it can share with other atoms, and these four are generally shown as lines around the atomic symbol:

Carbon can form covalent bonds with other carbon atoms. This is an especially important property, as long chains of carbon atoms occur in the structures of cellular macromolecules such as proteins and nucleic acids. Other atoms that form covalent bonds include hydrogen (one electron to share), oxygen (two), and nitrogen (three).

In a covalent bond between two carbon atoms, the electron is shared equally. Such is not always the case when dissimilar atoms form covalent bonds. For example, in water the shared electrons tend to spend more time near the oxygen atom. The "oxygen end" of this bond thus tends to be more negative than the "hydrogen end":

$$\overset{+}{H}\!-\!\overset{--}{O}\!-\!\overset{+}{H}$$

This unequally shared bond is termed a **polar covalent bond.** In the case of water, polarity explains how it forms a shell around ions: The negative "ends" of the molecules bind to positive ions, and the positive "ends" bind to negative ions. In addition, water molecules can form weak attractions to each other:

$$\overset{+}{H}\!-\!\overset{--}{O}\!-\!\overset{+}{H}$$
$$\overset{+}{H}\!-\!\overset{--}{O}\!-\!\overset{+}{H}$$

A weak attraction between polar molecules where hydrogen is one of the atoms involved is termed a **hydrogen bond.**

Cellular molecules with polar covalent bonds tend to dissolve in water to a greater extent than nonpolar molecules. The two most important polar covalent bonds in the cell are

$$\overset{+}{>}\!C\!=\!\overset{-}{O} \quad \text{and} \quad \overset{+}{H}\!-\!\overset{-}{N}\!-$$

Large Molecules

Carbohydrates

As their name indicates, carbohydrates are compounds made up of covalent linkages between carbon, hydrogen, and oxygen in the ratio $C_1:H_2:O_1$. Thus a general formula for a carbohydrate is $(CH_2O)_n$. The simplest carbohydrates are **monosaccharides;** in cells, these sugars usually have $n = 3-7$. The presence of polar C—OH groups renders the molecules water-soluble.

There are many ways to arrange the atoms in C=O, C—H, and C—OH groups on the carbon skeleton. For a given "n," these differently structured compounds are termed isomers. There are a number of isomers with the general carbohydrate formula $C_6H_{12}O_6$, and two of them are shown in Figure A2−1. Each differently structured isomer has different chemical properties. For example, fructose ($C_6H_{12}O_6$) is sweeter and less stable and has a lower melting point than glucose ($C_6H_{12}O_6$). In solution, monosaccharides exist in a ring configuration, which is more stable than the open-chain form (Figure A2−1).

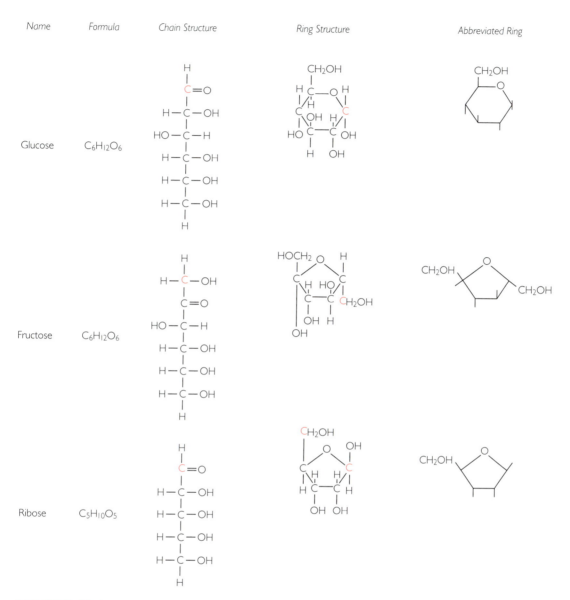

Name	Formula	Chain Structure	Ring Structure	Abbreviated Ring

FIGURE A2–1 Structures of monosaccharides.

Disaccharides are composed of two simple sugars, covalently linked through an oxygen atom when water is removed from adjacent C—OH groups on neighboring monosaccharides (Figure A2–2). Familiar examples include sucrose (glucose-fructose), a product of green plant photosynthesis; maltose (glucose-glucose), in cereal grains; and lactose (glucose-galactose), the milk sugar. Monosaccharides, specifically glucose, are an important source of extractable chemical energy in cells (see

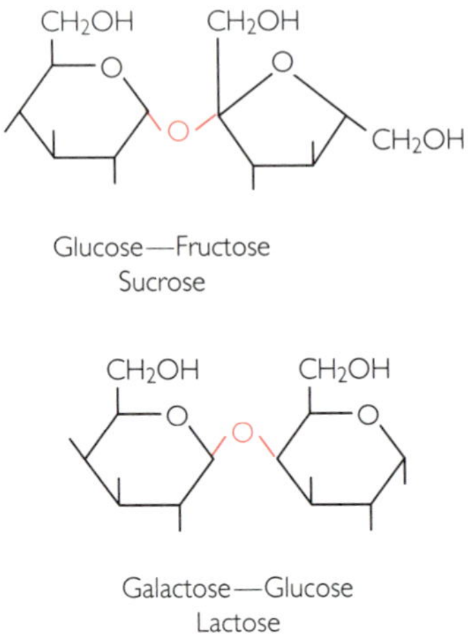

Glucose—Fructose
Sucrose

Galactose—Glucose
Lactose

FIGURE A2–2

Structures of disaccharides.

below). The initial event in the cellular use of disaccharides for energy is hydrolysis: the addition of water to break the covalent bond holding the two monosaccharides to each other.

Polysaccharides are composed of many simple sugars, covalently bonded to each other. Their large size usually renders them insoluble in water. Homopolysaccharides contain the same residue, repeated many times. For example, cellulose, the plant cell wall component, has thousands of glucose molecules, linked in a specific fashion (1,4 linkage), and starch, the plant storage polymer of glucose. The component molecules are linked differently (1,4 and 1,6 linkages), and the properties of the molecule are quite different (Figure A2–3). Glycogen, the storage form of glucose in higher animals, is similar to starch but more extensively branched. Chitin, the surface exoskeletal constituent of some invertebrates, is a homopolymer of the amino sugar N-acetyl-glucosamine. Heteropolysaccharides include the animal cell extracellular matrix components chondroitin sulfate, heparin, and hyaluronic acid (see Chapter 13).

Lipids

Lipids, or fats, are composed of the elements carbon, hydrogen, and oxygen, but unlike carbohydrates, there is no set ratio for these three elements. Most lipids are derived from the three-carbon sugar alcohol, glycerol (Figure A2–4). Fatty acids are bound to the glycerol via dehydration at its three hydroxyl (OH) groups. These long chains of carbon

FIGURE A2–3

Structures of polysaccharides.

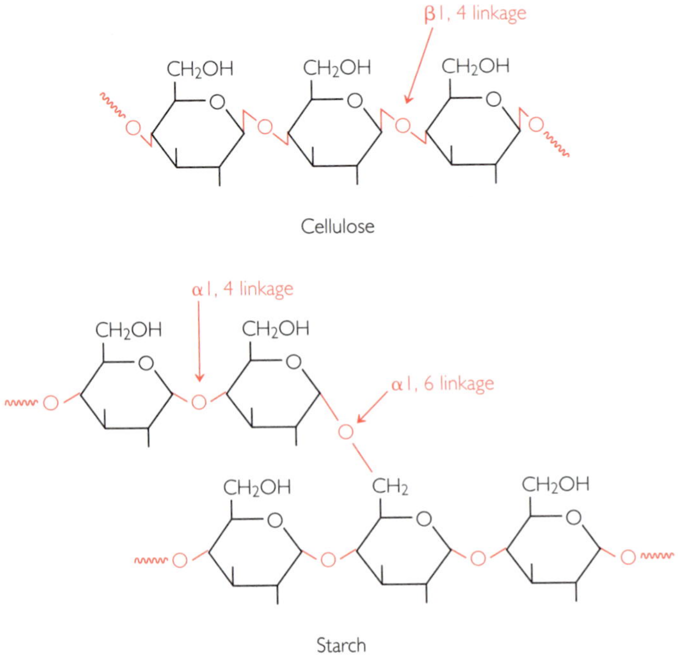

Cellulose

Starch

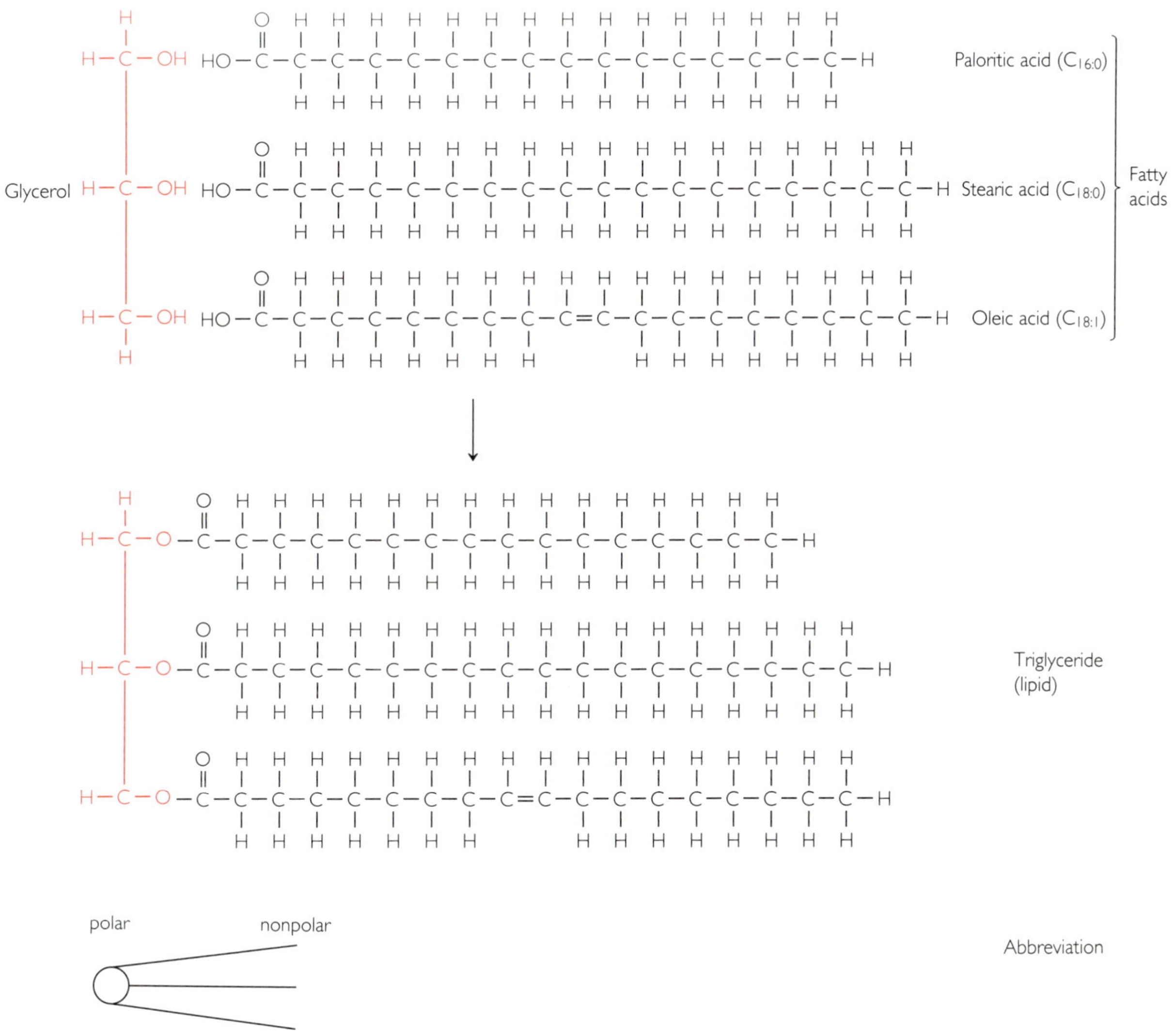

FIGURE A2–4 Structure of a lipid.

atoms (usually C_{16} to C_{20}) have, at one end, a charged COOH group that ionizes to COO^- and H^+. The other carbon atoms in the chain are substituted with hydrogen atoms, attached to the carbons by covalent bonds that are nonpolar. Thus these **fatty acids** have two distinct regions: a charged end and a long nonpolar chain. The charged end is hydrophilic (attracted to water), whereas the long chains are hydrophobic (repelled from water).

When up to three fatty acids are attached to glycerol (Figure A2–4), the charged species no longer exists; however, the C═O bond is polar, and so the lipid molecule retains its hydrophobic and hydrophilic regions. This property is important in cell membrane structure (see Chapter 1). The polarity of one end of a lipid can be increased if a highly polar species such as phosphatidylcholine is bound to the glycerol instead of to one of the fatty acids (Figure 1–4).

In addition to simple glycerides, there are other important forms of cellular lipids. Sphingolipids contain sphingosine, a long-chain amino alcohol, while glycolipids contain sugars attached either directly to glycerol or to other groups. Both of these lipids are important in some cell membranes (see Chapter 1). Steroids are complex hydrophobic molecules composed of multiple ring structures (Figure 1–5). Cholesterol is a component of membranes, and a number of animal hormones are steroids. Vitamins A, D, E, and K are also classified as lipids, largely on the basis of their insolubility in water.

Nucleic Acids

Like the polysaccharides, nucleic acids are polymers, with the nucleotide as the basic building block in this case. Because there is a limited number of different nucleotides, nucleic acids are limited heteropolymers.

Each nucleotide has three covalently linked components. Two components—a phosphate group and a five-carbon sugar (Figure A2–1)—are invariant in a given polymer. In ribonucleic acid (RNA), the sugar is ribose; in deoxyribonucleic acid (DNA), the sugar is 2-deoxyribose, in which an "OH" group at the second position in ribose is replaced by an "H."

Attached to the sugar is a nitrogenous base (Figure A2–5). The base may be a double-ringed purine (adenine [A] or guanine [G]) or a single-ringed pyrimidine (uracil [U], cytosine [C], or thymine [T]). DNA is composed of nucleotides with the bases A, T, G, and C, whereas RNA contains A, U, G, and C. The covalent linkage of nucleotides to each other occurs via dehydration between the sugar of one nucleotide and the phosphate of the next. Thus, a nucleic acid is composed of a sugar-phosphate backbone with attached bases. Because nucleic acids are informational macromolecules, and because the bases are their variable component, the sequence of bases clearly provides the information in a nucleic acid (see below).

When two nucleotides come near each other in a correct orientation, hydrogen bonds can form between −NH groups on one nitrogenous base, and −CO groups can form on the opposite base. Three hydrogen bonds can form between guanine and cytosine, and two can form between adenine and thymine or uracil (Figure A2–5). These weak interactions have important structural and functional consequences.

In typical DNA, the number of G residues equals that of C, and the

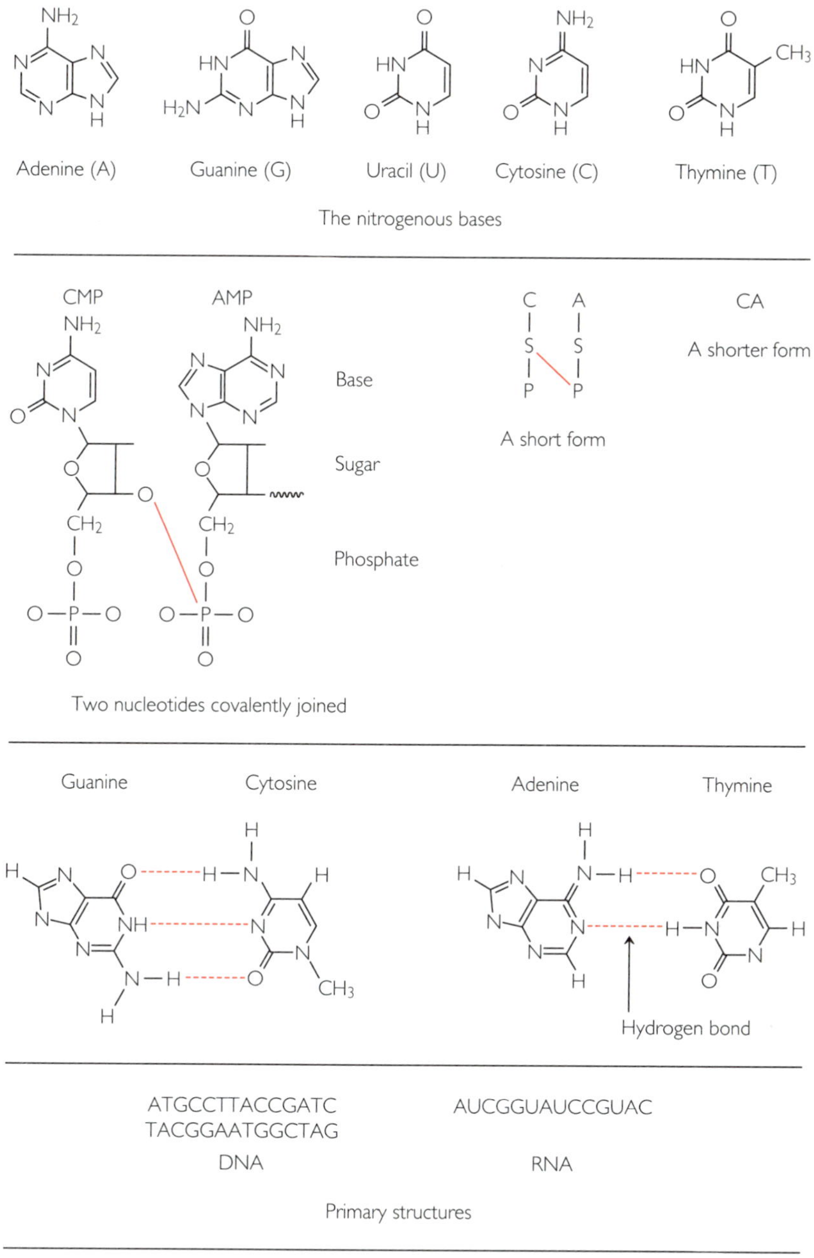

FIGURE A2–5

Chemistry of nucleic acids.

Adenine (A) Guanine (G) Uracil (U) Cytosine (C) Thymine (T)

The nitrogenous bases

CMP AMP Base C A CA
  Sugar S S A shorter form
  P P

  Phosphate A short form

Two nucleotides covalently joined

Guanine Cytosine Adenine Thymine

Hydrogen bond

ATGCCTTACCGATC AUCGGUAUCCGUAC
TACGGAATGGCTAG
DNA RNA

Primary structures

number of A residues equals that of T. Eukaryotic DNA is double-
stranded, with the strands arranged such that the bases are on the inside
of the molecule with an A on one strand always opposite a T, and a G
on one strand always opposite a C. This means that hydrogen bonds can
form between the base pairs and, therefore, between the two strands.

Taken together, these bonds form a strong force to maintain the two strands of DNA in their correct configuration. RNA is single stranded. However, the single strand can fold back on itself if there is hydrogen bonding due to base pairing. This confers a three-dimensional structure that is important in RNA function (e.g., binding to proteins).

Proteins

Proteins are polymers of **amino acids.** As their name indicates, the latter are molecules that contain an amino (NH_2) group and a carboxyl (COOH) group. Two adjacent amino acids are covalently joined by the elimination of water (Figure A2–6). The number of amino acids in a protein chain typically varies from 5 to 1000, with the median number being about 150.

Proteins are heteropolymers. There are 20 major "R" groups in the basic amino acid structure (Table A2–3). Some of the amino acids have nonpolar "R" groups, some have polar groups, and still others have groups that ionize in solution and are charged.

The precise order of amino acids in a protein chain is termed the **primary structure.** With 20 amino acids and variable length of chains, the number of possible different proteins is very large, and this variability in structural information can be used in cells for functional specificity. Thus a given protein type has a precise sequence of amino acids and a precise biochemical function.

The link between sequence and function is that the amino acid sequence determines a protein's three-dimensional structure. Hydrogen bonds between the NH group of one amino acid and the CO group of the residue three amino acids down the chain cause the chain to twist into a helical or pleated sheet shape (Figure A2–6). This is termed **secondary structure.** Disulfide bonds, covalent linkages between sulfur-containing amino acids, cause the chain to bend. The chemical affinity of hydrophilic amino acids for the outside of the protein near water, and the tendency for hydrophobic amino acids to be on the inside, away from water, cause further folding of the protein. The result can be a molecule of exquisitely specific three-dimensional shape **(tertiary structure).** This is important in the function of proteins as enzymes.

Protein structure is sensitive to those environmental conditions that disrupt the chemical forces shaping it. Heat (80°C) destroys hydrogen bonds; thus it is not surprising that heated proteins (for example, egg white) exhibit structural alterations. The degree of ionization of the charged groups on certain amino acids (Table A2–3) depends on the pH of the medium. For example, when the hydrogen ion concentration is low (pH high), a group such as NH_2 will not be ionized, but a COOH group loses its proton and becomes ionized. Ionization can covert an uncharged, hydrophobic residue into a charged, hydrophilic residue.

F I G U R E A2–6

Some aspects of protein structure.

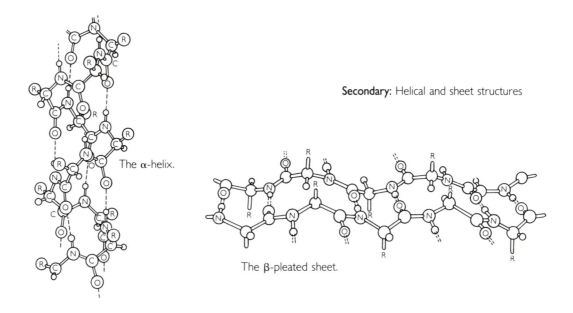

Primary: Sequence of amino acids

The peptide bond. The amino terminus is at the left.

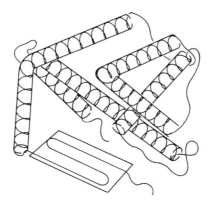

The α-helix.

Secondary: Helical and sheet structures

The β-pleated sheet.

Tertiary: Helical and sheet structures folded

TABLE A2–3　"R" Groups of Amino Acids

Class	Name (abbrev.)	"R" Group
Nonpolar	Glycine (G)	$H-$
	Alanine (A)	CH_3-
	Valine (V)	$(CH_3)_2-CH-$
	Leucine (L)	$(CH_3)_2-CH-CH_2-$
	Isoleucine (I)	$CH_3-CH_2-(CH_3)-CH-$
	Methionine (M)	$CH_3-S-CH_2-CH_2-$
	Phenylalanine (F)	⬡ $-CH_2-$
	Tryptophan (W)	⬡ CH_2-
	Proline (P)	$-CH_2-CH_2-CH_2-$
Polar	Serine (S)	$HO-CH_2-$
	Threonine (T)	$CH_3-HOCH-$
	Cysteine (C)	$HS-CH_2-$
	Asparagine (N)	$H_2N-CO-CH_2-$
	Glutamine (Q)	$H_2N-CO-CH_2-CH_2$
	Tyrosine (Y)	$HO-$⬡$-CH_2-$
Charged	Aspartic (D)	$HOOC-CH_2-$
	Glutamic (E)	$HOOC-CH_2-CH_2-$
	Lysine (K)	$H_2N-CH_2-CH_2-CH_2-CH_2-$
	Arginine (R)	$H_2N-CNH_2-NH-CH_2-CH_2-CH_2-$
	Histidine (H)	⬡ CH_2-

This can have considerable effect on where the residue is located (near or away from water), with concomitant effects on the three-dimensional structure of the protein.

Energy and Enzymes

According to the mechanistic view of life adopted by most biochemists, living systems obey physical-chemical principles, and among the more important of these principles are the laws of **thermodynamics.** According to the first law, energy can be neither created nor destroyed; it can only be changed during the various activities of cells. Thus, the chemical energy stored in a molecule of glucose can be converted, via many intermediates, into the mechanical energy of muscle movement. The second law states that during any change, the entropy (unusable energy) of a system increases. This means that the universe is being slowly converted into unusable energy. For cells, it means that a constant infusion of new energy is needed to maintain order and build complex systems.

Some biochemical reactions require energy, and others release energy. In exergonic reactions, the energy level in the initial reactant(s) is greater than that of the final products, and some usable energy is released:

$$Glucose + O_2 \rightarrow CO_2 + H_2O + energy.$$

In endergonic reactions, the reactants are lower in energy than the products, so that energy must be supplied:

$$CO_2 + H_2O + energy \rightarrow glucose + O_2.$$

Two general principles govern energy metabolism in cells. First, energy is released or used in small "packets." This is more efficient than trapping or using large amounts of energy. However, it means that there are many intermediate steps in any major conversion (e.g., between glucose and CO_2). Second, biochemical energy is released and used chemically without being converted into an intermediate form, such as heat. Again, this is efficient. Its consequence is that energy-requiring and energy-releasing reactions are coupled in the cell, so that they occur at the same place and time. An endergonic reaction is always coupled with an exergonic reaction. Schematically:

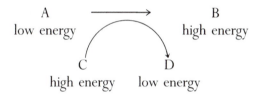

In order to begin, most chemical reactions, whether endergonic or exergonic, require an input of energy termed the activation energy. This is needed so that reactants can collide with one another in the correct orientation. In chemistry, heat, which causes molecules to move faster and collide more often, is a common source of activation energy (e.g., a flame on dynamite). This is hardly appropriate for cells, where protein structure is destroyed by heat. Instead, cells use specific molecules to align reactants correctly, thus lowering the activation energy and speeding up the reaction. These catalysts are termed enzymes.

Nearly all **enzymes** are proteins, the few exceptions being certain RNA molecules. Because each protein has a specific three-dimensional shape, each is ideally suited to align specific reactants. The region where the reaction occurs on an enzyme is termed the active site.

The degree to which an enzyme speeds up a biochemical reaction is remarkable. When compared to uncatalyzed reactants in solution, enzymatic reactions are often 10,000–1,000,000 times faster. Virtually every biochemical reaction in the cell has its own specific enzyme. Enzyme structure, and hence function, are sensitive to those environmental conditions which alter protein structure. Thus the rate of cellular metabolism is sensitive to variations in temperature and pH.

Enzyme-coupled reactions may release or require energy and, as noted above, this is done in cells via coupled reactions. One of the "energy carriers" which participate as energy donors or recipients is adenosine triphosphate (ATP). This molecule is composed of the ribonucleotide of adenine, with two additional phosphate groups (Figure A2−7). The bonds by which each of the two terminal phosphate groups are attached to the molecule release a considerable amount of energy when hydrolyzed and are termed "high-energy phosphate bonds." Thus, the release of the terminal phosphate from ATP to form adenosine diphosphate (ADP) releases energy, whereas energy is required to convert ADP to ATP. Schematically, an endergonic reaction could be:

$$
\begin{array}{ccc}
\text{A} & \xrightarrow{\text{enzyme}} & \text{B} \\
\text{low energy} & & \text{high energy} \\
\\
\text{ATP} & \text{ADP} + \text{P}_i \\
\text{high energy} & \text{low energy}
\end{array}
$$

A second method to energetically couple biochemical reactions is through **oxidation and reduction.** Oxidation describes the removal of an electron from an atom. The removal may be complete, as in ionization to form a positive ion, or, commonly in biochemistry, the removal is incomplete, as in the addition of oxygen via a polar covalent bond:

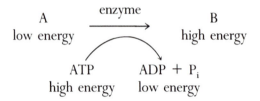

reduced central carbon   →   oxidized central carbon

In general, reduced compounds (electron added or closer to the atom) have a higher energy level than oxidized compounds. Thus, oxidation can be thought of as an exergonic process.

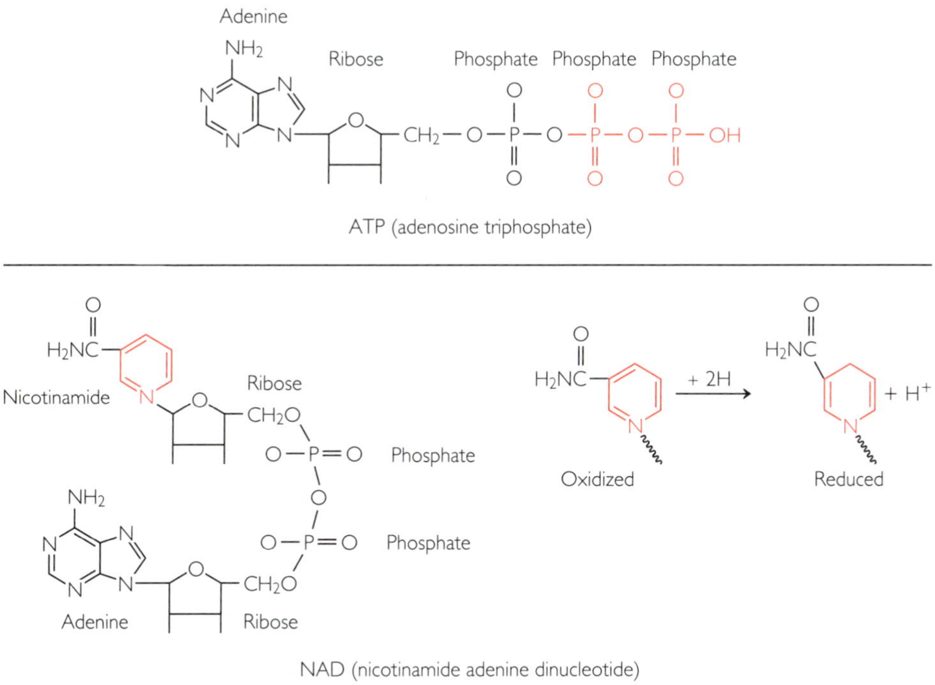

FIGURE A2–7

Structures of two biochemical energy carriers.

The energy, or reducing power, released when a molecule is oxidized can be used in a coupled system to reduce another molecule:

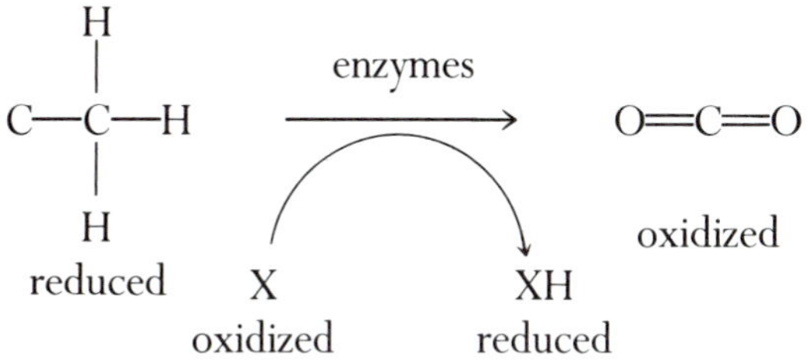

Molecules that act as donors or recipients of reducing power in biochemical reactions are redox coenzymes, which are relatively (to proteins) small molecules derived from vitamins. Two important examples are nicotinamide adenine dinucleotide (NAD—see Figure A2–7) and flavin adenine dinucleotide (FAD).

Metabolism

Energy

Metabolism is defined as the sum total of the chemical changes that occur in organisms. These changes are usually directed to either the breakdown of molecules to yield energy **(catabolism)** or the use of energy to synthesize complex cellular constituents **(anabolism).** The principal catabolic pathway in cells involves the oxidation of glucose to carbon dioxide. The anabolic pathway to be discussed is protein synthesis.

The oxidation of glucose to lower-energy carbon dioxide follows the two principles of cellular energy metabolism: There are many enzyme-catalyzed steps (18), and energy is released in coupled reactions. In the first stage of oxidation, the six-carbon molecule of glucose is converted into two molecules of three-carbon pyruvate. Two of the steps in **glycolysis** require energy and use the conversion of ATP to ADP to supply it. Three of the steps release energy: In two cases, ADP is converted to ATP, and in the third, NAD is reduced to NADH (Figure A2−8). Thus, from an energy standpoint, glycolysis results in:

$$\text{glucose} + 2\text{ATP} + 4\text{ADP} + 2\text{NAD} \rightarrow$$
$$(2)\ \text{pyruvate} + 2\text{ADP} + 4\text{ATP} + 2\text{NADH}$$

$$\text{Net: glucose} + 2\text{ADP} + 2\text{NAD} \rightarrow$$
$$(2)\ \text{pyruv ate} + 2\text{ATP} + 2\text{NADH}$$

Pyruvate is fully oxidized to carbon dioxide in the **citric acid cycle** (Figure A2−9). In a preparatory step, three-carbon pyruvate is converted to two-carbon acetate, with the concomitant release of carbon dioxide and reduction of NAD. This two-carbon molecule is then combined with a four-carbon carrier (oxaloacetate) to form the six-carbon citrate from which the name of the cycle is derived. Citrate is then sequentially converted to oxaloacetate, with the release of two molecules of carbon dioxide and the reduction of coenzymes, as well as, in one step, the formation of ATP from ADP.

The oxidation of one molecule of pyruvate can be summarized:

$$\text{pyruvate} + 4\text{NAD} + \text{FAD} + \text{ADP} \rightarrow$$
$$3\text{CO}_2 + 4\text{NADH} + 4\text{FADH} + \text{ATP.}$$

Because there are two molecules of pyruvate per molecule of glucose, the equation becomes, on a per glucose basis:

$$2\ \text{pyruvate} + 8\text{NAD} + 2\text{FAD} + 2\text{ADP} \rightarrow$$
$$6\text{CO}_2 + 8\text{NADH} + 2\text{FADH} + 2\text{ATP.}$$

Adding glycolysis and the citric acid cycle, the overall equation is:

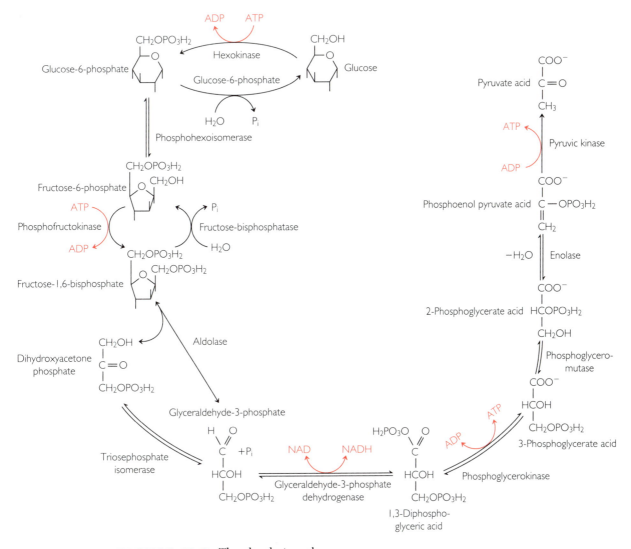

FIGURE A2–8 The glycolysis pathway.

$$\text{glucose} + 4\text{ADP} + 10\text{NAD} + 2\text{FAD} \rightarrow$$
$$6\text{CO}_2 + 4\text{ATP} + 10\text{NADH} + 2\text{FADH}.$$

The reducing power of NADH is relatively large, and there are few biochemical reactions that require this much energy. The energy released on hydrolysis of ATP to ADP is in a more appropriate range for anabolic reactions. The conversion of reducing power energy into high-energy phosphate energy occurs via electron transport and oxidative phosphorylation. Both of these processes occur, as does the citric acid

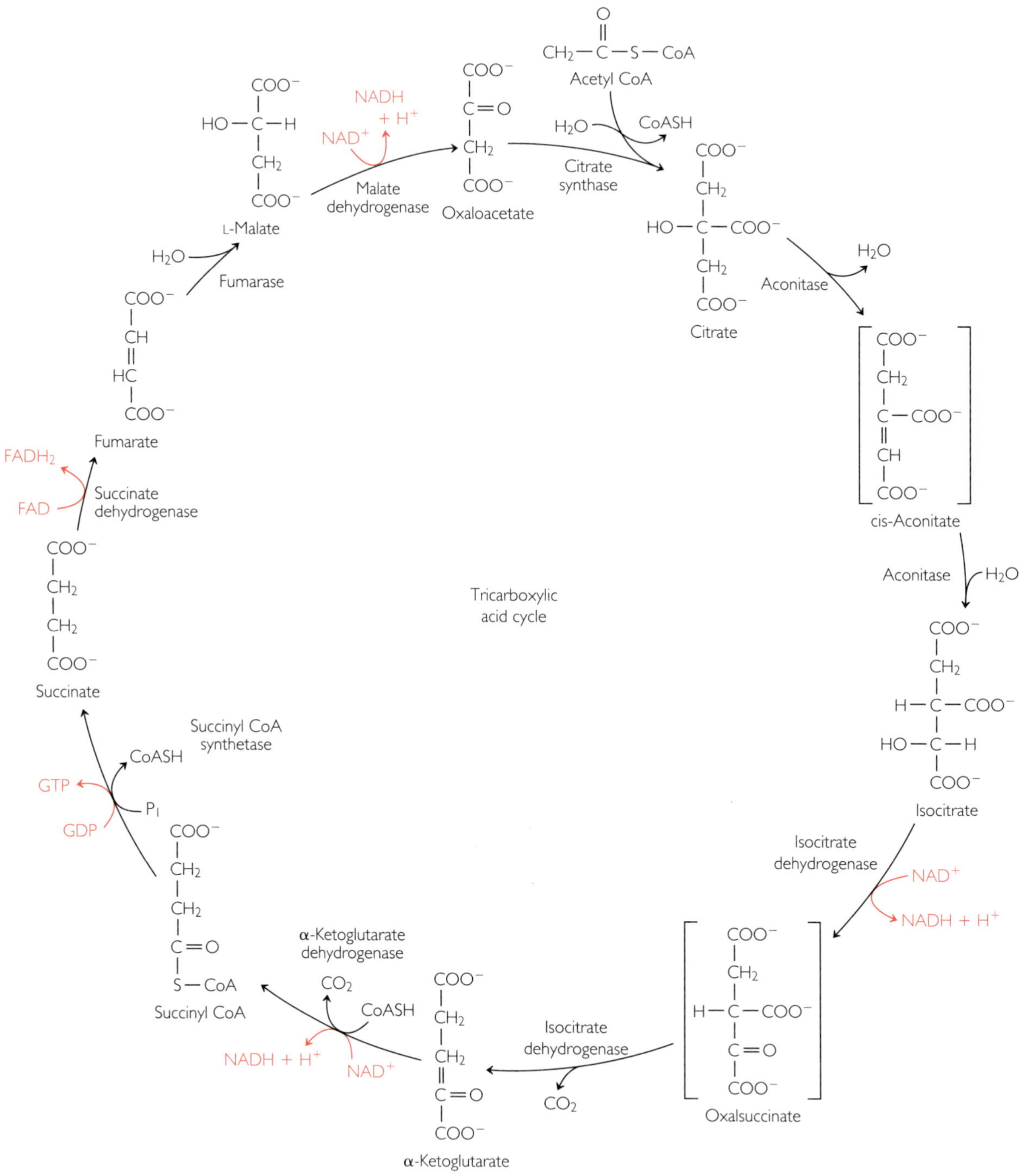

FIGURE A2–9 The citric acid cycle.

cycle, in the mitochondria of eukaryotic cells (see Chapter 3). Glycolysis occurs in the cell cytoplasm.

   Electron transport involves a chain of oxidation-reduction catalysts that carry electrons from reduced coenzymes (such as NADH and FADH) to oxygen:

$$\text{NADH} \xrightarrow[\underset{O_2 \qquad H_2O}{}]{\text{several steps}} \text{NAD}$$

Thus, while the coenzymes are oxidized, oxygen becomes reduced. However, the reducing energy level of NADH is higher than that of water, and thus some energy is released in the process. This energy can be used to phosphorylate ADP to produce ATP. During the sequential passage of electrons down the chain from NADH to oxygen, each intermediate carrier has a lower energy level when reduced than the previous one (Figure 3–7). In three steps, this energy difference is sufficient to drive the synthesis of ATP.

$$\text{NADH} \xrightarrow[\underset{3ADP \qquad 3ATP}{}]{\overset{O_2 \qquad H_2O}{}} \text{NAD}$$

   The entry of the FADH into the chain is at a lower energy level than NADH, and only two molecules of ATP are formed when its electrons are passed to oxygen. Thus, the equation for glucose oxidation becomes:

$$\text{glucose} + 4\text{ADP} + 30\text{ADP} + 4\text{ADP} \rightarrow$$
$$6\text{CO}_2 + 4\text{ATP} + 30\text{ATP} + 4\text{ATP}$$
$$\text{glucose} + 38\text{ADP} \rightarrow 6\text{CO}_2 + 38\text{ATP}.$$

Actually, because of some differences with the oxidation of the NADH formed in glycolysis, the ATP yield is closer to 36 molecules per glucose in some animal cells. In any case, the transfer of reducing energy from glucose to the energy in the phosphate bonds of ATP is remarkably efficient (about 40%).

   When oxygen is not present, the electron transport chain cannot function. From an energetic point of view, it makes little sense to completely oxidize pyruvate through the citric acid cycle, since ATP energy cannot be formed from the oxidation of reduced coenzymes. In fact, in the absence of oxygen, eukaryotic cells reduce the citric acid cycle and electron transport. But the NADH formed in glycolysis must be reoxidized, and in plants this is done by the formation of ethyl alcohol:

$$\text{Pyruvate} + \text{NADH} \rightarrow \text{ethanol} + \text{NAD} + CO_2.$$

In animals, lactate is formed:

$$\text{Pyruvate} + \text{NADH} \rightarrow \text{lactate} + \text{NAD}.$$

These conversions represent the termini of glycolysis in the absence of oxygen. The net ATP yield of glycolysis is two molecules per glucose. When this is compared with the 36 ATP molecules formed in the presence of oxygen, the advantages of the mitochondrial functions of the citric acid cycle and oxidative phosphorylation are evident.

The biochemical pathway of glucose oxidation lies at the center of the many pathways of anabolism and catabolism (Figure A2–10). Most of these pathways are reversible. For example, the carbohydrate glucose can be converted, via intermediates of glycolysis and the citric acid cycle, into amino acids. Likewise, proteins can be converted, by essentially the reverse pathways, to glucose. The rates of these conversions, and whether molecules will be fully oxidized to produce energy, are under a set of strict controls. Key enzyme activities in the pathways are regulated by the cellular levels of such metabolites as ADP, ATP, and glycolytic intermediates.

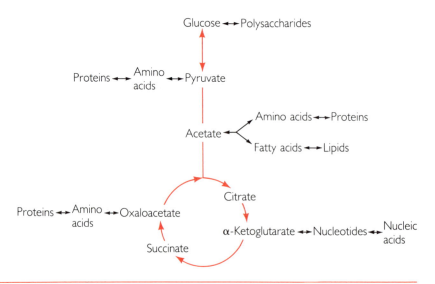

FIGURE A2–10

Metabolic interrelationships of carbohydrates, lipids, proteins, and nucleic acids.

Protein Synthesis

Protein synthesis is an anabolic process in which amino acids are linked in a specific order to form a specific macromolecule of defined structure and function. This presents two challenges: The first is thermodynamic, since the formation of the complex from simple components goes against the tendency to increasing randomness (entropy); energy is required. The second is informational, since there must be a mechanism to pre-determine the sequence of the amino acids.

The need for energy is solved by the coupling of protein synthesis to the hydrolysis of ATP. More specifically, each amino acid becomes attached to adenosine monophosphate (AMP) via a high-energy bond:

$$aa_1 + ATP \rightarrow aa_1{-}AMP + PP_i$$

$$aa_2 + ATP \rightarrow aa_2{-}AMP + PP_i$$

where aa_1 and aa_2 are two amino acids. The amino acid is then transferred to a molecule of RNA, transfer RNA (tRNA), with the high-energy bond being maintained:

$$aa_1{-}AMP + tRNA_1 \rightarrow aa_1{-}tRNA_1 + AMP$$

$$aa_2{-}AMP + tRNA_2 \rightarrow aa_2{-}tRNA_2 + AMP$$

The two tRNA molecules, with their attached amino acids, go to the ribosome where the energy in the aa-tRNA bonds is used in the synthesis of a peptide bond (Figure A2−6):

$$aa_1{-}tRNA_1 + aa_2{-}tRNA_2 \rightarrow aa_1{-}aa_2 + tRNA_1 + tRNA_2$$

The mechanism by which **amino acid ordering** is determined (that is, $aa_1{-}aa_2$ and not $aa_2{-}aa_1$) is best understood by following the diagrams in Figure A2− 11. Determination of amino acid sequence begins with DNA. This stable molecule is the genetic material, passed on from cell to cell during cell division and generation to generation during reproduction. The DNA blueprint for a specific protein is copied into a molecule that can be transported to the site of protein synthesis, the ribosome. DNA is double-stranded (Figure A2−5), and a region on one of the two strands is copied by complementary base pairing into messenger RNA (mRNA), which goes to the ribosome.

The role of mRNA at the ribosome is to ensure the correct ordering and length of the amino acid chain. This is done via base pairing with nucleotides on tRNA. Sequential triplets of nucleotides on mRNA each code for a certain amino acid (Table A2−4). Each tRNA molecule has an anticodon composed of a triplet of nucleotides. When an mRNA codon indicates a certain amino acid (e.g., proline), the tRNA that binds

FIGURE A2–11

Protein synthesis.

Gene ──────────────→ Phenotype
DNA   protein
base sequence   amino acid sequence
3 bases   1 amino acid

<u>Example</u>: Synthesis of the tripeptide, val-lys-pro

<u>Transcription</u>: DNA blueprint on one strand is copied by base pairing to messenger RNA(mRNA)

DNA   ATGGTTAAGCCTTGA
          TACCAATTCGGAACT ◄── Strand copied

mRNA   AUGGUUAAGCCUUGA

<u>Translation</u>: mRNA is translated into protein

1. mRNA binds to small ribosome subunit. The "start" signal is AUG.

AUGGUUAAGCCUUGA

2. The first tRNA, with an anticodon pairing with the triplet mRNA codon, is brought in.

3. The large ribosome subunit arrives. Then, the next tRNA is brought in. Each tRNA carries an amino acid.

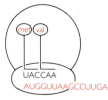

4. Peptide bond forms beween the two amino acids. The initiator tRNA is released.

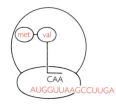

5. The mRNA "shifts" one triplet-base codeword.

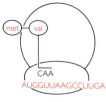

6. As in steps 3–4, the next tRNA, with its amino acid, is brought into the complex. Usually, the initial "met" is released.

7. The mRNA shifts after the second tRNA is released.

8. As in steps 3–4, the next tRNA and amino acid come in.

9. Step 7 is repeated.

10. "UGA" has no tRNA anticodon. This is recognized as a "stop" signal, and the complex is dissociated.

TABLE A2–4 Messenger RNA Codons

First Nucleotide	Second Nucleotide				Third Nucleotide
	U	C	A	G	
U	phe	ser	tyr	cys	U
	phe	ser	tyr	cys	C
	leu	ser	termin	termin	A
	leu	ser	termin	trp	G
C	leu	pro	his	arg	U
	leu	pro	his	arg	C
	leu	pro	gln	arg	A
	leu	pro	gln	arg	G
A	ile	thr	asn	ser	U
	ile	thr	asn	ser	C
	ile	thr	lys	arg	A
	met	thr	lys	arg	G
G	val	ala	asp	gly	U
	val	ala	asp	gly	C
	val	ala	glu	gly	A
	val	ala	glu	gly	G

to the mRNA codon carries that amino acid. This implies that tRNA, with its "correct" anticodon, must carry the "correct" amino acid. Such specificity is indeed the case. The enzyme, aminoacyl-tRNA synthetase, which recognizes prolyl-tRNA, attaches proline only to those tRNA molecules that have the anticodons for that amino acid.

Although the mechanics of the ribosome events are complex (see Chapter 11), the concept of protein synthesis is elegantly simple. Because the four different nucleotides in mRNA (A, U, G, and C) can be arranged into 64 triplet "codewords" (Table A2–4), there are some redundancies. Thus proline has four mRNA codewords (CCU, CCC, CCA, and CCG). This means that more than 20 different tRNA molecules (one for each amino acid) are needed so that all codewords can be bound with an appropriate tRNA. Actually, three codewords do not have a tRNA with the appropriate anticodon (UAG, UAA, UGA). These are "punctuation marks" for terminating the protein chain.

Because each is catalyzed by a specific enzyme, the many steps in protein synthesis occur rapidly. A typical protein containing 150 amino acids can be synthesized in the cell in approximately two minutes. Because of the highly specific nature of protein structure and function, there is little room for error in the sequence of amino acids, especially at the active site of an enzyme. The actual error rate in the synthesis mechanism is quite low. However, an alteration in the nucleotide sequence of DNA (mutation) can lead to an altered amino acid sequence in a protein. For example, a change in a single nucleotide (T to A) of the 867 that are the coding DNA for human globin leads to a single amino acid change in the 287 amino acids of the protein chains. This results in the human genetic disease sickle cell anemia.

All somatic cells of a given organism contain virtually the same DNA sequences. This means that all cells have the capability of forming all the proteins of that organism. Analysis of different tissues' protein contents indicates that, while all cells do synthesize the proteins necessary for such vital functions as energy metabolism, specific cell types make their own specific proteins. For example, liver cells synthesize albumin, and skin cells synthesize hair proteins, but not vice versa. The mechanism by which such selective expression of DNA information occurs is a major subject of molecular genetics.

Further Reading

Abeles, R. H., Frey, P. A., and Jencks, W. P. *Biochemistry*. Boston: Jones and Bartlett, 1992.

Conn, E., Stumpf, P., and Doi, R. *Outlines in biochemistry*. New York: Wiley, 1987.

Lehninger, A. *Principles of biochemistry*. New York: Worth, 1983.

Metzler, D. *Biochemistry*. New York: Academic Press, 1977.

Stryer, L. *Biochemistry*. New York: W. H. Freeman, 1988.

Voet, D., and Voet, J. *Biochemistry*. New York: Wiley, 1990.

Zubay, G. *Biochemistry*. Reading, MA: Addison-Wesley, 1987.

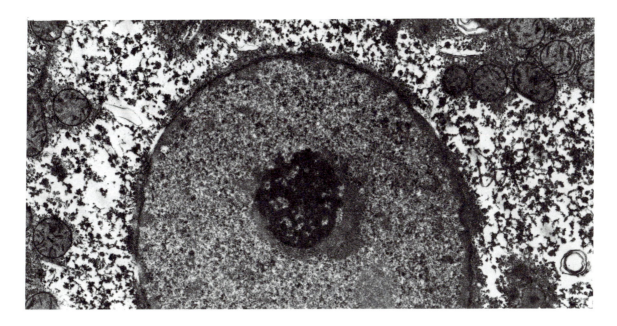

A Guide to the Literature on Cell Biology

The Problem

As this book has pointed out, cell biology is both interesting and important to study. Thousands of scientists around the world are doing this, and the information they publish is in many journals and books. To grasp the magnitude of this information explosion, here are the numbers of research papers published in the 7-year period of May 1985–1992, as indexed by the U.S. National Library of Medicine and other databases:

Membrane	127,726
Endoplasmic reticulum	6,715
Golgi	6,126
Mitochondria	18,568
Chloroplast	13,012
Ribosome	5,711
Cytoskeleton	5,754
Nucleus	35,921
Cell cycle	9,864
Basal lamina	1,840
Cell adhesion	9,446
Nucleolus	1,895
Cell	369,401

Faced with the daunting prospect of keeping up with such a huge field, most scientists rely on one or more of the following:

1. A selected list of "core journals"

2. Some books that provide good reviews

3. Computerized information monitoring

Journals

There are a number of journals that are most useful for their **reviews** of current literature:

American Scientist—occasional articles on cells, written for a general scientific audience

Biochimica et Biophysica Acta—some reviews on membranes and cancer cells

Bioessays—short, clearly written commentaries

Critical Reviews in Biochemistry and Molecular Biology—extensive and authoritative

Current Opinion in Cell Biology—written by authorities and accurately described by the title; very useful guide to recent papers

FASEB Journal—well-written and authoritative reviews, generally on subjects related to mammalian physiology

Journal of NIH Research—lively commentaries on research carried out at the NIH and under its sponsorship

Nature—commentaries on meetings and research; rapid publication of papers on "hot topics"

Naturwissenchaften (in German)—very well-written summaries of research

Physiological Reviews—authoritative, lengthy articles on topics related to physiology (usually membranes)

Recherche (in French)—excellent lively articles

Science—similar to Nature (see above)

Scientific American—written for general scientific audiences but still useful as an introduction to a topic

Trends in Biochemical Sciences—see BioEssays

Trends in Cell Biology—see BioEssays

Trends in Genetics—see BioEssays

Trends in Neurosciences—see BioEssays

Core journals are directly concerned with basic studies on cell biology

Cell—some reviews and commentaries; papers on "hot topics," generally with a molecular bent

Cell and Tissue Kinetics—cell cycle studies on animal cells

Cell and Tissue Research—wide range of papers on organelles

Cell Biology International Reports—rapid publication of some "hot topics" papers; some useful reviews

Cell Motility and Cytoskeleton—authoritative studies; excellent reviews

European Journal of Cell Biology—well-reviewed general research

Experimental Cell Research—wide range of studies, many on cell cycle

Journal of Cell Biology—authoritative source for well-designed, solid research

Journal of Cell Science—see Journal of Cell Biology; tends to more structural studies

Journal of Membrane Biology—excellent specialized journal; very good monthly reviews

Molecular Biology of the Cell—studies on cell regulation

Photosynthesis Research—informative brief reviews; specialized studies tending to chemical orientation

Plant Cell—molecular orientation; brief and lively reviews

Plant Journal—see Plant Cell

Plant Physiology—well reviewed and authoritative; some brief reviews

Proceedings of the National Academy of Sciences (U.S.)—studies on "hot topics"

Tissue and Cell—structural studies

Specialty journals publish some important papers on cell biology:

American Journal of Anatomy

American Journal of Clinical Pathology

American Journal of Pathology

American Journal of Physiology

Anatomical Record

Annals of Internal Medicine

Archives of Biochemistry and Biophysics

BioTechnology

Biochemical Journal

Biochemistry

Biochemistry and Cell Biology

Biochimica et Biophysica Acta

Biochimie

Biophysical Journal

British Medical Journal

Cancer Cells

Cancer Research

Carbohydrate Research

Comparative Biochemistry and Physiology

Cytobios

Development

Developmental Biology

EMBO Journal

European Journal of Biochemistry

FEBS Letters

Histochemical Journal

Journal of American Medical Association

Journal of Biological Chemistry

Journal of Cellular Biochemistry

Journal of Electron Microscopy

Journal of Experimental Botany

Journal of Lipid Research
Journal of Molecular Biology
Journal of Receptor Research
Journal of Ultrastructure and Molecular Structure Research
Lancet
Life Sciences
Lipids
Molecular and Cellular Biochemistry
Molecular and Cellular Biology
Molecular Biology and Medicine
New England Journal of Medicine
Nucleic Acids Research
Plant and Cell Physiology
Planta
Proceedings of the Royal Society
Protoplasma

Books

Many books published each year are scholarly reviews of cell biology topics. Some are summaries of conferences, while others are solicited reviews of a topic. In some cases, they either repeat material available elsewhere or are out of date. The better books are usually not collections of articles edited by several people but are single-authored works by leaders in the field.

On the other hand, there are some excellent books that appear regularly as compendia of **reviews** of current research. These series are usually up to date and written by authorities:

Advances in Carbohydrate Research and Biochemistry
Advances in Membrane Fluidity
Annals of the New York Academy of Sciences
Annual Review of Biochemistry
Annual Review of Biophysics and Biophysical Chemistry
Annual Review of Cell Biology
Annual Review of Medicine
Annual Review of Physiology
Annual Review of Plant Physiology
Biological Membranes
Biology of the Extracellular Matrix
Cellular Organelles

Ciba Foundation Symposia

Cold Spring Harbor Symposia on Quantitative Biology

Current Topics in Developmental Biology

International Review of Cytology: A Survey of Cell Biology

Methods in Cell Biology

Modern Cell Biology

UCLA Symposia on Molecular and Cellular Biology

Computerized Databases

Modern computer technology is essential to keep abreast of the huge volume of published work on cell biology. As has been the case for many years, the papers that are published in the journals noted above are indexed in such databases as Biological Abstracts and Index Medicus (for all but some plant journals). These databases are generally available in academic libraries and for each paper contain the key descriptive words, title, abstract, and bibliographical information. It is possible to look up the word *mitochondrion* in the index and then hunt down all papers published on this subject, with a few subheadings.

The personal computer has improved on this immensely. These databases are available in central computers, which can be reached through the telephone lines via a personal computer and modem. Access is available worldwide. For instance, the computer database of Biological Abstracts is called BIOSIS Previews, and the Index Medicus is called Medline. Once the database is accessed, almost any combination of topics can be searched for, as long as they have appeared in key words.

For example, the word *mitochondrion* can be combined with *human* and *disease* to get all of the papers published on human mitochondrial diseases. Once the database has been narrowed down, the titles and/or abstracts of the papers can be displayed on the computer screen and loaded into the personal computer for storage and later analysis. The entire process usually takes only a few minutes, compared to the hours or days needed to do the job by hand with the bound indexes in the library. Hard copies of each paper can even be ordered, but this is somewhat expensive at present. It is far more feasible and inexpensive to do the literature search by computer, get the relevant references, and then read them in the library.

The **major computerized database** vendors are:

1. Dialog
   3460 Hillview
   Palo Alto, CA 94304

This organization has the largest selection of databases. For those willing to use only a selected number of databases (including all those used by the journals and book series in cell biology), and to use the service only in the evenings and weekends, Dialog offers "Knowledge Index," which is relatively inexpensive ($35 to enroll and about $2–4 per search). This service also has an educational component for those interested in teaching database searching.

2. BRS Colleague
   Maxwell Online
   8000 Westpark Dr.
   McLean VA 22102

This organization supplies most of the widely used databases, including all of those used by the cell biology journals. There is a $95 initiation fee and a $20 monthly minimum.

3. Compuserve Information Service
   5000 Arlington Rd.
   Columbus OH 43220

This is the largest online service, with over 500,000 subscribers. It has many information, business, and entertainment sources. Of particular interest is the availability of "Paperchase," an excellent way to access Medline. There is a $40 initiation fee and no minimum billing requirement.

In a special class as an information source is the weekly "Current Contents—Life Sciences." This is not a book or journal but simply is a reproduction of the tables of contents of all of the biological journals that appear in a given week. It is useful in scanning papers that may be of interest. Authors' addresses are included for ease of requesting reprints. A subject index is also included in each issue. This database is now available on computer diskette, and this allows the information in each issue to be manipulated into larger databases.

Problems

Chapter 1

1. A poikilothermic organism living in the Arctic would have, compared to that organism living in the temperate climate zone, a plasma membrane richer in
   a. cholesterol
   b. long chain fatty acids
   c. protein
   d. unsaturated fatty acids

2. The thickness of a typical plasma membrane is approximately
   a. 1 nanometer
   b. 9 nanometers
   c. 100 micrometers
   d. 1 millimeter

3. In the fluid mosaic model for membrane structure
   a. carbohydrate is on the outer membrane surface
   b. lipids but not proteins can diffuse in the plane of the membrane
   c. proteins occur only in the inner leaflet of the membrane
   d. the polar ends of phospholipids face the interior of the membrane

4. In the typical plasma membrane the ratio by weight of lipid to protein is approximately
   a. 1:10
   b. 1:5
   c. 2:3
   d. 10:1

5. Covalent attachment occurs between each pair of membrane molecules except
   a. carbohydrate and lipid
   b. carbohydrate and protein
   c. glycerol and fatty acids
   d. lipid and protein

6. With regard to diffusion in the plane of the membrane
   a. it occurs least readily at high temperatures where lipids are fluid
   b. lipids diffuse independently of proteins
   c. membrane proteins diffuse only when attached to antibodies
   d. neither proteins nor lipids can diffuse

7. A fatty acid is labelled with a fluorescent probe at the C-2 position (where C-1 is the polar carboxyl group). A second fatty acid is labelled at the C-16 position. Both fatty acids are separately incorporated into membrane lipids, which are inserted into membranes. Fluorescence measurements will show
   a. both probes with equal mobility
   b. neither probe with any mobility
   c. the C-2 probe with greater mobility than the C-16 probe
   d. the C-16 probe with greater mobility than the C-2 probe

8. Intrinsic (integral) membrane proteins differ from extrinsic (peripheral) ones in that the intrinsic ones
   a. are larger
   b. are solubilized by detergents
   c. contain carbohydrate
   d. contain mostly hydrophilic amino acids

9. A protein which spans the lipid bilayer
   a. cannot diffuse in the plane of the membrane
   b. cannot have any attachment to cytoplasmic components
   c. cannot have bound carbohydrate
   d. usually has both hydrophobic and hydrophilic regions

10. Give the order of these events, which occur in the biogenesis of a membrane protein:

1. Protein synthesis on free ribosomes is stopped after the hydrophobic signal sequence is translated;
2. Signal sequence is cut from the protein by a peptidase;
3. Signal sequence embeds in ER membrane lipids;
4. Signal sequence leads the synthesizing complex to the ER.

   a. 1234
   b. 1342
   c. 1432
   d. 4132

Chapter 2

1. Diffusion across the plasma membrane is more rapid if a substance is
   a. a protein
   b. hydrophilic
   c. high in its oil:water partition coefficient
   d. large and globular in shape

2. You wish to design a new drug which will act as an ionophore to deliver Ca^{++} across the nerve cell membrane. This drug would most likely be
   a. hydrophobic on the outside and hydrophilic on the inside
   b. insoluble in lipids
   c. smaller than 0.001 nanometers in diameter
   d. soluble in proteins

3. Gated channels
   a. are always open
   b. can be opened by a chemical ligand or voltage change
   c. make up the Na^+-K^+ pump
   d. only occur in the postsynaptic element

4. A difference between diffusion and facilitated transport is that facilitated transport
   a. is concentration independent
   b. occurs across plasma membranes
   c. requires membrane proteins
   d. utilizes a substance moving with its concentration gradient

5. A difference between active transport and facilitated transport is that active transport
   a. needs ions to work
   b. occurs against a concentration gradient
   c. occurs only in neurons
   d. utilizes membrane proteins

6. Active transport occurs in all of the following except
   a. Ca^{++} across the muscle sarcoplasmic reticulum
   b. glucose across the intestinal epithelial membrane
   c. H^+ across organelle membranes
   d. Na^+ across the neuronal membrane

7. In receptor-mediated endocytosis, molecules such as LDL
   a. bind to receptors after they have become clustered
   b. enter the cell via coated pits
   c. internalize into the cell by diffusion
   d. utilize receptors common to many ligands

8. "G" proteins in membranes
   a. act as tyrosine protein kinases
   b. bind only to GTP but not to GDP
   c. can bind to both a receptor and an effector
   d. cannot act at a gated channel

9. The Na^+-K^+ pump
   a. can bind certain cardiac glycosides such as digoxin and ouabain
   b. has only a single conformation in the membrane
   c. occurs only in neuronal membranes
   d. pumps more Na^+ into the cell than K^+ out of the cell

10. If an erythrocyte membrane lacks the protein, spectrin, one would expect a red blood cell with
    a. a lower ratio of intrinsic:extrinsic membrane proteins
    b. a membrane with less fluidity
    c. an abnormal shape
    d. impaired transport of some ions across its membrane

Chapter 3

1. The outer mitochondrial membrane
   a. does not contain lipids
   b. is permeable to molecules up to about 5000 molecular weight
   c. sets up a proton gradient
   d. swells and contracts readily

2. The number of sites for ATP synthesis in the mitochondrial electron transport chain is
   a. 1
   b. 2
   c. 3
   d. 4

3. NADH transport into the liver mitochondrion
   a. is balanced by release of NAD
   b. is inhibited when O_2 is present
   c. occurs only indirectly, via shuttle carriers
   d. requires an ATP driven carrier

4. Reagent X attaches readily to proteins. In separate experiments, X is presented to intact mitochondria (M) and to inside-out submitochondrial particles (SMP). The ATP synthetase, F1, is then isolated in each case and the labelling pattern of X on F1 determined. This pattern will show
   a. label equally in both M and SMP
   b. label in M but not SMP
   c. label in SMP but not M
   d. no label in either M or SMP

5. Uncouplers of electron transport from ATP synthesis, such as dinitrophenol, act by inhibiting
   a. ATP transport
   b. electron transport
   c. proton gradient
   d. oxygen consumption

6. Succinate dehydrogenase catalyzes this reaction in the Krebs citric acid cycle:

   succinate + FAD $\rightarrow$ fumarate + FADH

   The P:O ratio of mitochondria given succinate as substrate is
   a. 1.0

b. 2.0
c. 2.5
d. 3.0

7. Submitochondrial particles are isolated by sonication of intact mitochondria and supplied with external NADH. According to the chemiosmotic hypothesis
a. ATP will be made on the inside of the particles
b. NADH will become reduced
c. the pH will increase on the inside
d. the pH will decrease on the inside

8. In the chemiosmotic hypothesis for energy generation in mitochondria
a. a pH gradient drives the hydrolysis of ATP to ADP
b. ATP must be continually translocated into the mitochondrion
c. electron transport carriers set up a proton gradient
d. one proton pump is involved

9. In a patient with muscle weakness and fever due to a mitochondrial myopathy, ATP synthesis is reduced but electron transport is normal. The inner mitochondrial membranes of the patient's muscle might show
a. absence of cristae
b. defective Ca^{++} pump
c. missing porin
d. open anion channel

10. In a typical cell active in energy metabolism
a. mitochondria are stationary in the cell
b. the P:O ratio is less than 1.0
c. there are about 1000 mitochondria
d. the inner mitochondrial membrane is far from the outer one

Chapter 4

1. The outer envelope of the chloroplast is
a. a double layer, containing pores

b. permeable to proteins of 100,000 molecular weight
c. the location of Rubisco
d. the location of photosystem II

2. Chlorophyll in the membrane of the chloroplast is
a. all directly active in photosystems I and II
b. embedded in the lipid via its hydrophobic side chain
c. not associated with membrane protein
d. present entirely as chlorophyll a

3. If thylakoids are treated with a detergent to solubilize lipids and then the proteins gently isolated, the following complexes will be observed, except
a. ATP synthetase (CF)
b. core complex II
c. cytochrome oxidase
d. light harvesting complex

4. Chlorophyll a has a photochemical absorbance maximum (A) at 675 nm and a fluorescence emission maximum (E) at 690 nm. A carotenoid pigment which is hypothesized to channel light to the active chlorophyll a would have the absorbance-emission characteristics as follows:
a. A at 500 and E at 480
b. A at 580 and E at 675
c. A at 675 and E at 690
d. E at 675 and A at 690

5. The effect of a herbicide which blocks electron transport from photosystem II to photosystem I will be
a. enhancement of the dark reactions
b. failure to oxidize NADPH
c. increased ATP synthesis in the chloroplast
d. lack of reduction of ferrodoxin

6. Which statement best describes the locations of the photosynthetic photosystems in the chloroplast?
   a. I in stacked lamellae and II in unstacked lamellae
   b. I in lamellae and II in stroma
   c. I in unstacked lamellae and II in stacked lamellae
   d. I in stroma and II in lamellae

7. In the chemiosmotic synthesis of ATP by chloroplasts,
   a. addition of ATP synthetase (CF) to membrane vesicles will allow the synthesis of ATP if only ADP is added
   b. CF must be very close to the two photosystems in the lamella
   c. proton ionophores have no effect on ATP synthesis
   d. the pH of the stroma rises due to proton pumping

8. Chloroplasts are disrupted and the stroma separated from the lamellae. The isolated stroma will fix CO_2 if it is supplied with
   a. ATP and NADPH
   b. carotenoids
   c. light
   d. oxygen

9. The enzyme which catalyzes the initial fixation of CO_2 in C3 plants has the following properties except
   a. can act as an oxygenase in photorespiration
   b. essential to the photochemical light reactions
   c. most abundant protein in nature
   d. present in the chloroplast stroma

10. Which of the following pairs is incorrect?
   a. amyloplasts–store starch in potatoes
   b. chloroplasts–contain lipid deposits
   c. chromoplasts–store pigments in flowers
   d. etioplasts–formed in intense light

Chapter 5

1. The mold, Neurospora, was incubated in radioactive choline for 5 days. Autoradiography of the mitochondria showed a grain count of 12 per mitochondrion. The cells were then incubated in nonradioactive choline for 2 cell divisions. Autoradiography would show a grain count per mitochondrion of
   a. 2
   b. 3
   c. 6
   d. 12

2. In corn, gene B codes for green chloroplasts; allele b codes for white chloroplasts. A cross between male BB and female bb gave green chloroplasts in the offspring; a cross betweeen male bb and female BB gave green chloroplasts in the offspring. These data show that
   a. both B and b are inherited through the chloroplast
   b. both B and b are inherited through the nucleus
   c. B is inherited through the chloroplast and b though the nucleus
   d. B is inherited through the nucleus and b through the chloroplast

3. The reason that a mutation carried on human mitochondrial DNA is inherited through the mother only is that
   a. egg cells do not contain mitochondrial DNA
   b. maternal mitochondrial DNA is degraded in the zygote
   c. paternal mitochondrial DNA does not enter the egg
   d. sperm cell mitochondria lack DNA

4. Plastid core complex II has 9 protein subunits. Plastids were isolated and incubated in radioactive amino acids. Antibodies to each subunit were used to precipitate the subunits separately and assay each for radioactivity. Subunits 1,2,4,5,7, and 9 were heavily labelled;

subunits 3,6, and 8 were not labelled. These data show that

a. all 9 subunits are made in the plastid
b. 3 of 9 subunits are made in the plastid
c. 6 of 9 subunits are made in the plastid
d. none of the 9 subunits is made in the plastid

5. The mitochondrion cannot make all of its proteins because it lacks

a. all of the tRNAs
b. ribosomes
c. RNA polymerase
d. sufficient DNA

6. Proteins made in the cytoplasm and then transported to the chloroplast

a. are smaller upon intitial translation than the final product
b. are made on cytoplasmic free ribosomes not on the ER
c. contain a hydrophobic signal cleaved at the chloroplast surface
d. contain a general organelle signal, rather than one for plastid

7. Ethidium bromide (EB) inhibits transcription of circular DNA; rifampin (RF) inhibits transcription of linear DNA. Chloramphenicol (CP) inhibits translation on 70S ribosomes; cycloheximide (CX) inhibits translation on 80S ribosomes. A mitochondrial protein coded for by the nuclear genome, but, uniquely, made in the mitochondrion, would have its synthesis inhibited by

a. CP or RF
b. CP or CX
c. CX or EB
d. EB or RF

8. The molecular biology of protein synthesis in the nucleus-cytoplasm and mitochondrial systems use the same

a. genetic code
b. linearity of DNA
c. size of ribosomes
d. structures of amino acids

9. The role of a molecular chaperone in the assembly of Rubisco in the chloroplast is to

a. act as a signal for targeting the large subunit to the chloroplast
b. bind to subunits to stop them from folding prior to assembly
c. inhibit translation on the ER of chloroplast mRNAs
d. prevent the Rubisco from denaturing when heated to 100°C

10. The endosymbiotic origin of plastids from bacteria is suggested by all of the following except

a. antibiotic sensitivity of plastid ribosomes
b. circular DNA in plastids
c. introns in plastid DNA genes
d. ribosome size in plastids

Chapter 6

1. The signal sequence of a protein which is processed in the ER

a. contains hydrophobic amino acids
b. is coded for by a gene separate from the nuclear genome
c. is added post-translationally to the protein
d. makes the precursor run further than the mature protein on SDS gel electrophoresis

2. A cell which is very active in the synthesis and secretion of proteins would be expected to have

a. equal amounts of smooth ER and rough ER
b. more rough than smooth ER
c. more smooth than rough ER
d. neither smooth nor rough ER

3. Smooth ER is isolated from liver cells as intact microsomes (M). A portion of this cell fraction is then treated with ultrasound (which makes the ER membrane permeable to proteins), forming leaky microsomes (L). Both of these fractions are incubated with an antibody for the active site of cytochrome P450 reductase. The antibody will bind to
   a. both L and M
   b. neither L nor M
   c. L only
   d. M only

4. Rough and smooth ER differ in that smooth ER contains
   a. less fatty acid desaturase activity
   b. lower density
   c. phospholipids
   d. ribophorins

5. What is the order of these events in secretory protein synthesis?
   1. Signal cleaved by peptidase
   2. Signal recognition particle binds to signal
   3. Signal sequence translated
   4. Synthesizing complex goes to ER
   5. Translation stops temporarily
   a. 12345
   b. 23451
   c. 32541
   d. 41325

6. Once a secretory protein is in the ER, it cannot get out because it
   a. has a linear shape
   b. is bound to the signal recognition particle
   c. is hydrophobic
   d. lacks a signal sequence

7. Which of the following is not a function of smooth ER?
   a. electron transport
   b. fatty acid desaturation
   c. glycosylation of proteins during translation
   d. glycogen breakdown

8. Which is not true of the cytochrome P450 system? It

a. can be induced by xenobiotics
b. desaturates fatty acids
c. is located in the smooth ER membrane
d. transports electrons from NADPH to oxygen

9. During translation of a secreted protein (e.g. collagen), the following events occur except
   a. glycosylation of the protein
   b. hydroxylation of proline and lysine
   c. signal sequence cleavage
   d. transport to the Golgi

10. The major protein of the sacroplasmic reticulum membrane
    a. acts as a Na^+ channel
    b. actively transports Ca^{++}
    c. binds actin and myosin
    d. synthesizes a neurotransmitter

Chapter 7

1. In the secretory pathway in pancreatic exocrine cells, proteins are initially translated on free ribosomes. Then, the proteins go to the
   a. ER surface
   b. ER lumen
   c. cis Golgi
   d. trans Golgi

2. Between which two membranes is the Golgi membrane intermediate in lipid composition?
   a. ER and nuclear
   b. ER and plasma
   c. lysosomal and nuclear
   d. lysosomal and plasma

3. Modifications of proteins as they pass through the Golgi include all of the following except
   a. glycosylation
   b. proteolysis
   c. signal sequence removal
   d. sulfation

4. Cells involved in the synthesis and secretion of proteins are given a short pulse of radioactive amino acids, then chased for various periods in nonradioactive amino acids. Autoradiography

shows a "wave" of labelled proteins passing through various regions of the cell in sequence. What would this sequence be?
1. extracellular medium
2. Golgi condensing vesicles
3. Golgi peripheral vesicles
4. rough ER
5. zymogen storage granules
a. 12345
b. 23451
c. 42531
d. 43251

5. A protein passing through the Golgi and lacking a specific targeting signal will go to the
a. extracellular medium
b. lysosome
c. mitochondrion
d. nucleus

6. Vesicles containing material from the Golgi just prior to secretion are
a. coated with clathrin
b. heterogeneous with regard to their contents
c. immediately secreted
d. less concentrated than Golgi vesicles

7. Plant cell walls contain a proline-rich protein. A plant cell is pulsed with radioactive proline, and then chased in nonradioactive proline. After various periods of chase, cells are fixed and prepared for autoradiography. The labeling of cell compartments is followed:
1. Cell wall
2. ER
3. Golgi
4. Ribosomes

If the sequence of events in the plant cell is the same as that of animal cells, the order of labeling of cell compartments will be
a. 1234
b. 2134
c. 3421
d. 4231

8. The presence of abundant Golgi near the cell plate in the middle of a plant cell at late te-

lophase and cytokinesis of mitosis reflects the role of the Golgi in
a. lysosomal enzyme targeting
b. polysaccharide synthesis
c. protein packaging
d. uptake of membrane lipids

9. A molecule involved in protein glycosylation which is present in ER but not in the Golgi is
a. dolichol phosphate
b. glucosamine
c. mannose
d. serine

10. In the Golgi
a. all sugars are removed from secretory proteins
b. cis elements appear to bud off to form condensing vacuoles
c. information for targeting proteins to lysosomes is added
d. the intracisternal pH is about 3.0

Chapter 8

1. The marker for an enzyme to go from the Golgi to the lysosome is
a. acidic pH
b. glucosamine
c. phosphomannose
d. protease sensitivity

2. A distinctive feature of the lysosome is that it has
a. a lower pH than the cytoplasm
b. a reduced hydrolase activity
c. DNA
d. ribosomes

3. The iron containing protein, transferrin, is taken up by fibroblast cells in culture. In what order will the iron atoms of this molecule follow the heterophagy pathway?
   1. coated vesicle at cell surface
   2. endosome vesicle
   3. primary lysosome
   4. residual body
   5. secondary lysosome
   a. 1254
   b. 1345
   c. 1354
   d. 1253

4. The major type of enzyme present in the lysosome catalyzes
   a. condensation
   b. hydrolysis
   c. polymerization
   d. proton pumping

5. In autophagy,
   a. a single lysosome surrounds a single cell organelle
   b. cellular constituents are recycled
   c. internal membranes can be synthesized
   d. mitochondria are assembled

6. The fact that lysosomal storage diseases (in which a lysosomal hydrolase is inactive) are often fatal indicates that
   a. autophagy and heterophagy are essential cellular processes
   b. lysosomal membranes are leaky to breakdown products
   c. the stored molecules have an important structural role
   d. the diseases must be maternally inherited

7. Functions of the plant cell vacuole include all of the following except
   a. autophagy
   b. heterophagy
   c. proton pumping
   d. solute accumulation

8. The biochemical activity common to all microbodies is the metabolism of

   a. glycolate
   b. diaminobenzidine
   c. hydrogen peroxide
   d. lactate

9. Unlike humans, some plant seeds can convert fat to carbohydrate. The three organelles involved in this conversion include all of the following except
   a. glyoxysome
   b. lipid body
   c. mitochondrion
   d. peroxisome

10. From which two organelles do melanosomes arise?
    a. ER and Golgi
    b. ER and lysosome
    c. Golgi and lysosome
    d. lysosome and peroxisome

Chapter 9

1. Microtubules grow
   a. by adding newly made tubulin, rather than preformed tubulin
   b. from a microtubule organizing center located anywhere in the cell
   c. in both length and thickness
   d. on the surfaces of membrane-bound structures

2. The protein that makes up microtubules
   a. has binding sites for colchicine
   b. is composed of 3 monomers, held by covalent bonds
   c. is produced by the mitochondrial genome
   d. is the only protein in cilia

3. If a growing microtubule is supplied with radioactive GTP, and then chased in nonradioactive GTP, most of
   a. the GTP will be rapidly converted to GDP in the dynamic instability model
   b. the GTP cap will label both ends of the microtubule
   c. the label will all be lost from the growing end of the microtubule

d. the label will move from one end of the microtubule to the other

4. The shear force generated for ciliary motion is primarily due to
a. doublet on filament 1 sliding relative to that of filament 6
b. disassembly of microtubules
c. release of nexin from microtubules
d. subfibers (a and b) in each doublet sliding past each other

5. Microtubule-associated proteins (MAPs) are
a. abundant only in mitochondria
b. aggregated by colchicine
c. involved in tubulin polymerization
d. sensitive to cytochalasin B

6. A difference between actin in nonmuscle cells and muscle actin is that only the latter
a. binds myosin in large amounts in vivo
b. forms polymers
c. has a very different amino acid sequence
d. has binding sites for associated proteins

7. A cellular function which is inhibited by colchicine and cytochalasin together, but not separately, probably involves
a. both microtubules and microfilaments
b. intermediate filaments
c. microtubules only
d. no component of the cytoskeleton

8. A possible role for intermediate filaments is in
a. ciliary movements
b. movements of whole cells
c. slow active transport across the plasma membrane
d. the Z line of muscle

9. An epithelial cell is stained with anti-actin, anti-tubulin and anti-keratin antibodies. Under electron microscopy, the antibodies will be seen to bind to all of the cytoskeleton except
a. intermediate filaments
b. microfilaments
c. microtrabeculae
d. microtubules

10. A person with immotile cilia syndrome, a genetic disease, might lack
a. desmin
b. dynein
c. neurofilaments
d. vinculin

Chapter 10

1. The inner and outer membranes of the nuclear envelope can be differentiated in that the
a. inner lipid bilayer is thicker
b. inner membrane contains pores
c. outer membrane is associated with ribosomes
d. outer membrane has active transport proteins

2. Which of the following is not a property of the nuclear pores?
a. constant number per nucleus in a cell's lifetime
b. diameter from membrane to membrane of 70 nanometers
c. occupy 5%-30% of the nuclear surface area
d. open channel diameter of about 5 nanometers

3. Histone and nonhistone proteins differ in that only histones
a. are basically charged
b. can bind to DNA
c. occur in chromatin
d. occur in many copies of a few types

4. An Ameba nucleus is transplanted into an enucleated frog oocyte which has the following radioactively labelled substances:

    1: enkephalin (mol. wt. = 600)
    2: ferritin (mol. wt. = 400,000)
    3: ovalbumin (mol. wt. = 70,000)
    4: sodium ion
    5: rRNA (mol. wt. = 700,000)

Which substances will most easily enter the Ameba nucleus?
  a. 1 and 4
  b. 2 and 3
  c. 2 and 4
  d. 3 and 5

5. Proteins are targeted to the nucleus
  a. by a hydrophobic signal
  b. by binding via a signal to the nuclear envelope
  c. by the presence of a post-translational modification
  d. following insertion into the ER

6. Nuclear DNA is isolated and sheared into pieces of constant size. The DNA is denatured and then slowly renatured. The first sequences to renature will be
  a. globin genes
  b. localized in the nucleolus
  c. highly repetitive in frequency
  d. unique in frequency

7. The nucleosome
  a. contains only DNA and nonhistone proteins
  b. has a core of histones with DNA wound around it
  c. is fully responsible for DNA packaging into chromosomes
  d. surrounds nuclear pores

8. The three DNA sequences which define a chromosome include all of the following except:
  a. centromere
  b. enhancer
  c. origin of DNA replication
  d. telomere

9. The nuclear matrix is not
  a. composed primarily of proteins
  b. composed of folded nucleosomes
  c. the binding site of certain hormones
  d. the binding site of replicating DNA

10. Cell I with a block of constitutive heterochromatin (containing highly repetitive DNA) is fused with cell II, which does not contain this block. The result will probably be that
  a. cell I's nucleus will become entirely heterochromatic
  b. cell I will synthesize a protein to cause cell II's nucleus to become partially heterochromatic
  c. mRNA will not be made by nucleus I
  d. the nucleus I will retain its heterochromatic block

Chapter 11

1. The granular component of the nucleolus is mostly
  a. DNA
  b. mRNA
  c. incomplete ribosomes
  d. nucleosomes

2. Ribosomes can be disaggregated by detergents into their component proteins and RNAs. This indicates that
  a. precursor ribosomes are larger than the mature form
  b. protein synthesis can occur without ribosomes that are intact
  c. ribosomes are held together non-covalently
  d. the RNA molecules are ionically bound to each other

3. Cytoplasmic 18S and 28S ribosomal RNA
  a. derives from the processing of a large precursor
  b. is always smaller than tRNA
  c. is cotranslationally processed in the ER
  d. surrounds all of the ribosomal proteins

4. An antibody is produced which binds to cucumber ribosomal protein L6. Cucumber ri-

bosomes are incubated with the antibody and then visualized under electron microscopy. Many single ribosomes are seen; there are no aggregates. This experiment shows that protein L6 is
a. bound to rRNA
b. in the interior of the ribosome
c. on the surface of the large subunit
d. part of the tRNA binding site

5. The ribosome appears to have a third tRNA binding site, in addition to the A and P sites already well described. The proposed role of this third, or E site, may be to
a. bind tRNA after it releases its amino acid
b. bind the initiator tRNA only
c. initiate protein synthesis by binding to T factor
d. serve as a peptidyl transferase

6. The genes for 28S and 18S ribosomal RNA
a. are present in one copy per haploid genome
b. are distant from each other on the chromosome
c. are transcribed separately from each other
d. occur in the nucleolar organizer region

7. A human cell is given a short pulse with radioactive uridine. Much of the label immediately becomes associated with the nucleolus. The size of the rRNA at this stage is about
a. 5.8S
b. 18S
c. 28S
d. 41S

8. A frog egg contains about a trillion ribosomes. The large amount of rRNA for these ribosomes is made via
a. conversion of mRNA into rRNA
b. import of DNA from the neighboring nurse cells
c. more rapid transcription of rRNA genes in the nucleolar organizer
d. selective amplification of the rRNA genes

9. A mutation in frogs causes a deletion of all of the nucleolar organizer DNA. A zygote homozygous for this mutation would show

a. increased synthesis of 5S rRNA in the nucleolar organizer region
b. more ribosomes per cell than the wild type
c. no survival beyond early embryology
d. twice as much hybridization of 18S rRNA to DNA as wild type

10. The diameter of a typical nucleolus is about
a. 0.1 nm
b. 50 nm
c. 1 μm
d. 50 μm

Chapter 12

1. During embryonic development, the total cell cycle time for a given cell type can change. These changes are primarily due to variations in
a. G1
b. G2
c. M
d. S

2. A synchronous population of cells is given a pulse of radioactive thymidine. Following the pulse, the cells are chased in nonradioactive thymidine. Cells are sampled for autoradiography at various times of chase, with the following results:

Time of chase (min.)	Labelled mitoses (%)
0	0
30	0
60	0
90	100
120	0

These data show that
a. DNA is synthesized during mitosis
b. G1 is very short
c. G2 is about 1.5 hours long
d. mitosis is more than 2 hours long

3. Cells in G_0
   a. can be stimulated to enter S
   b. have the tetraploid amount of DNA
   c. must accumulate division potential before entering M
   d. occur in rapidly dividing tissues

4. A temperature-sensitive mutant of the hamster fibroblast cell cycle shows a normal phenotype at 35°. But at 45°, it fails to form the nuclear envelope at telophase. If cells are kept at 45° during G1 and S, and then switched to 35° for the rest of the cycle, they show the mutant phenotype. However, if the switch is made before S begins, the phenotype is normal. These observations suggest that
   a. all components needed for nuclear envelope assembly are made during mitosis
   b. a protein needed for nuclear envelope formation acts during S
   c. nuclear envelope assembly requires a protein made during G2
   d. temperature affects the phenotype of normal but not mutant cells

5. Eukaryotic DNA replication
   a. has one origin of replication per chromosome
   b. is unidirectional along a chromosome
   c. occurs during only part of interphase
   d. occurs only during S phase of meiosis

6. The major events of mitotic prophase include all of the following except
   a. chromosome coiling
   b. DNA replication
   c. nuclear envelope breakdown
   d. nucleolar disaggregation

7. A cell in mitosis is irradiated with a laser microbeam so that the spindle is bleached midway between the kinetochores and the pole. During anaphase, the chromosomes move toward the bleached region, but this region itself stays in the same position relative to the pole. This experiment indicates that
   a. chromosomes have actin to power their way to the pole

   b. microtubules do not depolymerize at the pole during anaphase
   c. microtubules must slide past one another during anaphase
   d. the plus ends of microtubules are at the poles

8. Which statement most accurately describes the protein kinase(K)-cyclin(C) model for cell cycle control?
   a. K is most active during G1, when it stimulates C
   b. K promotes the phosphorylation of histones, which allows for transcription
   c. the activity of K is constant throughout the cell cycle, but C varies
   d. when activated by C, K stimulates the transition from G2 to M

9. A difference between the meiotic and mitotic cell cycles in mammals is that in the meiotic cycle
   a. G1 and S phases occur in interphase
   b. cycling occurs only in germ cells
   c. spindle fibers are attached to kinetochores
   d. the spindle forms during prometaphase

10. If cells are cultured in colchicine, they will undergo complete cell cycles, except for the fact that the chromosomes do not migrate to the poles during anaphase. This results in polyploid cells. One can conclude that
   a. colchicine does not affect the microtubules of the mitotic spindle
   b. DNA replication does not necessarily result in chromosome separation
   c. the centromeres represent under-replicated DNA
   d. the spindle is made after interphase

Chapter 13

1. A unique component of certain plant and animal extracellular proteins is
   a. glucose
   b. glycine
   c. hydroxyproline
   d. more than one polypeptide chain

2. The following events occur in the assembly of collagen. What is their proper order?
   1. association of three polypeptide chains
   2. cleavage of termini to form tropocollagen
   3. hydroxylation of lysine residues
   4. secretion to the extracellular medium
   a. 1234
   b. 1324
   c. 2143
   d. 3142

3. If mammalian cells are treated with an antibody to the amino acid sequence, RGD (arginine-glycine-aspartate), they
   a. bind to each other more tightly
   b. detach from their extracellular matrix
   c. invade other tissues
   d. lose their integrin

4. People with Ehlers-Danlos disease, type V, lack lysyl oxidase activity. Therefore, in these people one might expect
   a. brittle bones
   b. blood vessels that are not elastic
   c. collagen without carbohydrate
   d. lack of a basement membrane

5. The ground substance of the extracellular matrix of mammalian fibrous tissues is composed of
   a. crosslinked collagen
   b. elastin and desmosine
   c. polycationic hyaluronic acid
   d. proteins linked to sulfated glucosaminoglycans

6. The incorrect pair of structure-function is:
   a. basal lamina - Bowman's capsule of the kidney form
   b. collagen - tendon integrity
   c. elastin - skeletal muscle stretching
   d. ground substance - filtration of bacteria

7. In the molecular pathway of binding of the proteoglycan to the cell surface are the following components:
   1. cell membrane
   2. fibronectin
   3. integrin
   4. proteoglycan

   The order of these components is:
   a. 1234
   b. 1324
   c. 2134
   d. 4321

8. Primary cell walls of plants differ from the secondary walls in that the primary walls
   a. are present in actively growing cells
   b. cannot be expanded
   c. contain more cellulose compared to matrix molecules
   d. lack glycoproteins

9. The most abundant organic molecule synthesized by living systems is
   a. cellulose
   b. chitin
   c. collagen
   d. Rubisco

10. Organelles required for cellulose synthesis include
    a. ER and Golgi
    b. Golgi and plasma membrane
    c. Golgi only
    d. microtubules and plasma membrane

Chapter 14

1. Desmosomes differ from tight junctions in that desmosomes
   a. allow molecules to pass in the intercellular space
   b. are non-communicating
   c. are present in plants
   d. lack proteins

2. Tight junctions
   a. are essential for metabolic coupling
   b. do not occur in vertebrates
   c. have the closest approach of two plasma membranes of any junction
   d. surround connexons

3. In ionic coupling between cells,
   a. cells must be connected by desmosomes
   b. ions flow through junctions between cells
   c. resistance between cells increases
   d. V2/V1 is > 1.0

4. Which statement is not true of gap junctions? They
   a. allow the passage of proteins between cells
   b. are essential for metabolic cooperation between cells
   c. occur in the developing embryo
   d. synchronize the beating of heart muscle cells

5. The most important difference between gap junctions between animal cells and plasmodesmata in plants is that in plasmodesmata
   a. ionic coupling occurs
   b. metabolic cooperation occurs
   c. pore diameter is 1 mm
   d. two adjacent plasma membranes are fused

6. A sponge is dissociated into single cells at 25°C by passing it through a mesh in seawater which lacks Ca^{++} and Mg^{++}. The sponge will reform by cell aggregation if
   a. aggregation factor and ions are added at any temperature
   b. cells are placed in a shaking water bath at any temperature
   c. the temperature is lowered to 4°C
   d. the two ions are added back at 4°C

7. Which statement about a cell surface recognition factor is incorrect?
   a. Cadherins require calcium for activity
   b. LCAM spans the lipid bilayer
   c. NCAM is involved in the formation of the neuromuscular junction
   d. Sponge aggregation factor has only one receptor per sponge cell

8. Plants with both sexes on the same organism prevent self-fertilization by
   a. forming glycoproteins on pollen and not the stigma
   b. forming pollen and eggs in different flowers
   c. having identical CAM's on the cell surfaces of reproductive tissues
   d. not coding for callose, which normally enhances fertilization

9. The colonization of plant roots by nitrogen-fixing bacteria
   a. allows only the bacteria to survive
   b. has two receptors with a lectin between them
   c. involves the S gene system
   d. occurs in all plant roots

10. All of the following may have roles in the invasion of a metastatic tumor into a host tissue except
    a. fibronectin
    b. laminin
    c. LCAM
    d. proteases

Answers

1.	D	B	A	C	D	B	D	B	D	C
2.	C	A	B	C	B	B	B	C	A	C
3.	B	C	C	C	C	B	D	C	D	C
4.	A	B	C	B	D	C	D	A	B	D
5.	B	B	C	C	D	B	A	D	B	C
6.	A	B	A	B	C	D	C	B	D	B
7.	A	B	C	D	A	B	D	B	A	C
8.	C	A	A	B	B	A	B	C	D	A
9.	B	A	D	A	C	A	A	D	C	B
10.	C	A	D	A	B	C	B	B	B	D
11.	C	C	A	B	A	D	D	D	C	C
12.	A	C	A	B	C	B	B	D	B	B
13.	C	D	B	B	D	C	B	A	A	D
14.	A	C	B	A	D	A	D	C	B	C